Essential Mathematics for A Level

David C Taylor BA

Head of Mathematics, Epsom College

Ivor S Atkinson BSc F.I.M.A.

Headmaster, Kelvin Hall, Senior High School, Kingston upon Hull

Nelson

Thomas Nelson and Sons Ltd
Nelson House Mayfield Road
Walton-on-Thames Surrey KT12 5PL

P.O. Box 18123 Nairobi Kenya

116-D JTC Factory Building
Lorong 3 Geylang Square Singapore 1438

Thomas Nelson Australia Pty Ltd
19–39 Jeffcott Street West Melbourne Victoria 3003

Nelson Canada Ltd
81 Curlew Drive Don Mills Ontario M3A 2R1

Thomas Nelson (Hong Kong) Ltd
Watson Estate Block A 13 Floor
Watson Road Causeway Bay Hong Kong

Thomas Nelson (Nigeria) Ltd
8 Ilupeju Bypass PMB 21303 Ikeja Lagos

First published 1981

ISBN 0-17-431280-6

NCN 200-3213-0

Typeset by Santype International Ltd., Salisbury, Wilts.
Printed and bound in Hong Kong.

To the memory of Valerie

Contents

4 Angles

5 Analytical Geometry

6 Trigonometry

7 Curve Sketching

8 Sums, Areas and Integration

9 Vectors

10 Further Differentiation

11 Complex Numbers

12 Further Integration

13 Approximate Methods

14 Two Special Functions

Preface

This book is intended to cover the basic essentials of all Advanced-level mathematics courses. In recent years, much work has been done by teachers in schools, universities and polytechnics as well as examining boards to see what common ground there is within various Advanced-level syllabuses. The Standing Conference on University Entrance and the Council for National Academic Awards has produced a 'Core Syllabus' which it is felt contains all the ingredients needed by prospective undergraduates. This text covers work on all the proposed topics in the core syllabus.

The book has been written as a teaching text, with understanding as its chief aim, and ample practice offered through the worked examples and exercises. Much of the work has been tried in the classroom, and the approaches offered, whilst not particularly 'new' or 'original', are based on several years' classroom experience by the authors. Within the text, there are no questions taken from examination papers of the GCE Boards, but the appendix contains a selection of questions drawn from recent Advanced-level papers. These are given as a miscellaneous set of exercises, and any student who has worked through the book should be able to tackle them all.

It is intended that this core text shall be supplemented by companion texts on Mechanics and Statistics, so that the application of the various mathematical techniques can be mastered and practised.

Glossary of Mathematical Symbols

Logic and relationship symbols

$\Rightarrow$	implies
$\Leftarrow$	is implied by
$\Leftrightarrow$	implies and is implied by
$=$	equals
$\neq$	does not equal
$\approx$	is approximately equal
$\equiv$	is identically equal
$<$	is less than
$>$	is greater than
$\leqslant$	is less than or is equal to
$\geqslant$	is greater then or is equal to

Sets

$P \equiv \{a, b, c, \ldots\}$	The set P consisting of elements a, b, c, ...
$n(P)$	The number of elements in P
$\in$	is a member of
$\notin$	is not a member of
$\{x: x \in P\}$	The set of elements x such that $x \in P$
$\emptyset$	The null (or empty) set
$\mathscr{E}$	The universal set
$\mathbb{N}$	The set of natural (counting) numbers
$\mathbb{Z}$	The set of integers
$\mathbb{Q}$	The set of rational numbers
$\mathbb{R}$	The set of real numbers
$\mathbb{C}$	The set of complex numbers
$\mathbb{Z}^+(\mathbb{Z}^-)$	The set of positive (negative) integers (zero excluded in both cases)

Functions

$f(x)$	The function of x
$f: x \rightarrow \cdots$	The function f which maps x onto ...
$gf(x)$	The combined function g of f of x
$f'(x)$	The derived function of $f(x)$
$[x]$	The integral part of x
z	The complex number $x + iy$
z^*	The complex conjugate of z, $z^* = x - iy$
$\mathscr{R}e(z)$	The real part of z, $\mathscr{R}e(z) = x$
$\mathscr{I}m(z)$	The imaginary part of z, $\mathscr{I}m(z) = y$
$\lvert z \rvert$	The modulus of the complex number z, $\lvert z \rvert = \sqrt{(x^2 + y^2)}$
$\operatorname{Arg} z$	The argument of z

$\arg z$	The principal value of the argument of z
δx	a small increment in x
$\rightarrow$	tends to
$\rightarrow \infty$	becomes infinitely large
$\lim_{x \to 0}[f(x)]$	limiting value of $f(x)$ as x approaches zero
$\frac{dy}{dx}$	The derivative of y with respect to x
$\frac{d^n y}{dx^n}$	The nth derivative of y with respect to x
$\int f(x)\, dx$	The integral function of $f(x)$ with respect to x
$\int_a^b f(x)\, dx$	The definite integral of $f(x)$ for values of x from a to b

Algebra and vectors

$\sum_{r=1}^{n} u_r$	$u_1 + u_2 + \cdots + u_n$
$n!$	$n(n-1)(n-2) \cdots (2)(1)$
${}^nC_r = \frac{n!}{r!(n-r)!}$	The number of combinations of n objects taken r at a time.
$\overrightarrow{OP}$	The vector from O to P
$\mathbf{r}$	The vector $\mathbf{r}$
$\lvert\mathbf{r}\rvert$	The magnitude of vector $\mathbf{r}$
$\begin{pmatrix} a \\ b \end{pmatrix} = a\mathbf{i} + b\mathbf{j}$	The vector which is equivalent to a units in the x-direction and b units in the y-direction.
$\left\lvert\begin{pmatrix} a \\ b \end{pmatrix}\right\rvert$	The magnitude of vector $\begin{pmatrix} a \\ b \end{pmatrix}$.
$\mathbf{a} \cdot \mathbf{b} = \lvert\mathbf{a}\rvert \cdot \lvert\mathbf{b}\rvert \cos\theta$	The scalar product of the two vectors $\mathbf{a}$ and $\mathbf{b}$.

Acknowledgements

We should like to take this opportunity of thanking those people who have helped in the writing of this book: Dr John Hyslop and Mr Mike Page who, as referees, gave valuable comments on the manuscript; Dr Mary Bradburn for verifying the answers to the set exercises; and above all Mrs Valerie Taylor who so painstakingly typed and prepared the manuscript and who, tragically, did not live to see it published.

Finally we are grateful to the following Examination Authorities for permitting the use of examination questions in the Further Exercises included to amplify the content of the text:

Associated Examining Board
Joint Matriculation Board
University of London University Entrance and School Examinations Council
Oxford and Cambridge Joint Board.

D. C. Taylor
I. S. Atkinson

Algebraic Manipulations

CHAPTER ONE

1.1 Introduction

It is assumed that you have studied the application of the four basic rules of arithmetic to the set of integers

$$\{\ldots -3,\ -2,\ -1, 0, 1, 2, 3 \ldots\}$$

and to the set of rational numbers. Also that you have extended these ideas to algebraic work. The first section of this text is intended to consolidate this earlier study so that subsequent manipulation in algebraic exercises will be performed accurately, and with confidence. We identify firstly four predominant types of algebraic expression and emphasise the difference in nature between them.

a The equation

When we substitute, in turn, the values $x = 3$ and $x = 1$ in the expression

$$(x - 3)^2 = 6 - 2x$$

both the left-hand side (L.H.S.) and right-hand side (R.H.S.) of the expression have the same values.

$$\begin{aligned} x = 3:&\quad \text{L.H.S.} = 0, && \text{R.H.S.} = 0,\\ x = 1:&\quad \text{L.H.S.} = (-2)^2 = 4, && \text{R.H.S.} = 6 - 2 = 4 \end{aligned}$$

Any other value substituted for x would produce different results for the L.H.S. and R.H.S., (try for example $x = 2, 4, -1$) and the equality of the two sides would not be established.

An algebraic form of this nature, in which the expression is true for a **particular set of values** of the variable x, is called **an equation**.

b The identity

If, however, we consider the expression

$$(x + 1)^2 = x^2 + 2x + 1$$

it can be verified that for any value selected and substituted for x, the L.H.S. and R.H.S. of the equation have the same value. (Try some values for yourself.)

Expressions of this type, which are true for **all values** of the variable x, are called **identities** and we make use of a special symbol $\equiv$ to show this. We write

$$(x + 1)^2 \equiv x^2 + 2x + 1$$

for such forms.

c The inequality

The third type of expression contains the inequality symbol(s), that is, either $>$ (greater than), or $<$ (less than).

Thus an expression of the form

$$x - 2 \geqslant 3$$

requires identification of all values of x such that $x - 2$ is either equal to 3 or greater than 3. Obviously when $x = 5$, $x - 2 = 3$ and when $x > 5$, $x - 2 > 3$. The solution of this problem is therefore $x \geqslant 5$, or all those values of x as shown in figure 1.1 on the real number line.

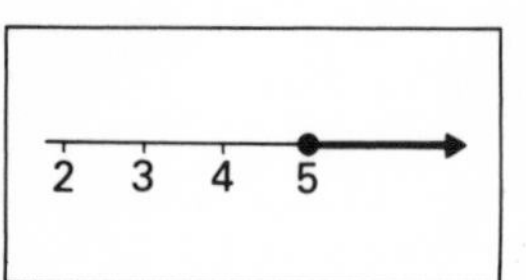

Figure 1.1

Such expressions, satisfied by a **range of values of** x, are called **inequalities.** A fuller definition of the term range can be found in section 3.1, but it should be realised that the range may be **open**, as in this example, or **closed**, as in the example which follows.

Example 1

Find x so that $3 < x < 4$

All values of x, which are less than 4 and at the same time greater than 3, are shown on the real number line in figure 1.2.

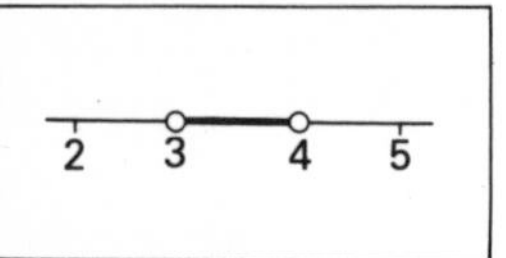

Figure 1.2

Note that in figure 1.1, since $x = 5$ is **included** in the solution set, the diagram shows inclusion of this fact by appropriate shading. In figure 1.2, $x = 3$ and $x = 4$ are **excluded** from the solution and are shown to be so by the absence of shading.

d The polynomial

Many algebraic expressions consist of a collection of terms, all of the form ax^n, where a is some fixed numerical value (called the **coefficient** of the term) and $n \in \mathbb{Z}^+$, i.e. n is a positive integer. Usually these expressions are arranged systematically, either in ascending order of size of powers of x, or in descending order of size of such powers.

For example, we would write

as $6x^5 + 5x^4 + 4x^3 + 3x^2 + 2x + 1$,
or $1 + 2x + 3x^2 + 4x^3 + 5x^4 + 6x^5$.

These expressions are called **polynomials** ('poly' meaning many and 'nomials' meaning numbers or terms).

The highest power of x involved in the polynomial defines the **degree** (or order) of the polynomial. Thus, in the example, the polynomial is of degree 5. We refer to polynomials of degree 2, involving three terms, as **quadratic trinomials**, or briefly, **quadratics**; polynomials of degree 3 as being **cubic polynomials**, or briefly, **cubics**, and so on.

Note that such expressions of the type

$$x + \frac{1}{x}, \quad \sqrt{(x + 1)}, \quad x^{4/3}$$

are not polynomials, since here all the powers of $x \notin \mathbb{Z}^+$.

1.2 Revision of the four rules related to polynomials

Addition and subtraction

The principles governing the addition and subtraction of algebraic expressions are basically the associative and commutative properties. Using these properties, terms of the same degree may be collected together and then combined by the distributive law.

Example 1
Given $P \equiv 2x^3 - x^2 + 4x - 2$ and $Q \equiv x^3 - 3x^2 - x + 4$
find $P + Q$ and $P - Q$.

$$\begin{aligned} P + Q &\equiv (2x^3 - x^2 + 4x - 2) + (x^3 - 3x^2 - x + 4) \\ &\equiv (2x^3 + x^3) - (x^2 + 3x^2) + (4x - x) + (-2 + 4) \\ &\equiv 3x^3 - 4x^2 + 3x + 2 \\ P - Q &\equiv (2x^3 - x^2 + 4x - 2) - (x^3 - 3x^2 - x + 4) \\ &\equiv (2x^3 - x^3) + (-x^2 + 3x^2) + (4x + x) - (2 + 4) \\ &\equiv x^3 + 2x^2 + 5x - 6 \end{aligned}$$

Example 2
Find, in its simplest form,

$$(a^2 - ab + a) + (ab + ac + b) - (ac - c^2 - c)$$

$$\begin{aligned} \text{Expression} &\equiv a^2 - ab + a + ab + ac + b - ac + c^2 + c \\ &\equiv a^2 + c^2 + (-ab + ab) + (ac - ac) + a + b + c \\ &\equiv a^2 + c^2 + a + b + c \end{aligned}$$

Multiplication

The further principle governing this operation is the distributive property of multiplication, and is illustrated as follows:

Example 3
Reduce to its simplest form,

$$P \equiv x^2(x - x - 2) + x(x^3 - x^2 + 1) - 2(x^4 - x^3 - 1)$$

$$\begin{aligned} P &\equiv (x^4 - x^3 - 2x^2) + (x^4 - x^3 + x) - (2x^4 - 2x^3 - 2) \\ &\equiv (x^4 + x^4 - 2x^4) + (-x^3 - x^3 + 2x^3) - 2x^2 + x + 2 \\ &\equiv -2x^2 + x + 2 \end{aligned}$$

Example 4
Multiply $(x^2 - 3x + 2)$ by $(x^2 + x - 3)$

$$\begin{aligned} &(x^2 - 3x + 2) \,.\, (x^2 + x - 3) \\ &\quad\equiv (x^2 - 3x + 2) \,.\, x^2 + (x^2 - 3x + 2) \,.\, x + (x^2 - 3x + 2) \,.\, (-3) \\ &\quad\equiv (x^4 - 3x^3 + 2x^2) + (x^3 - 3x^2 + 2x) + (-3x^2 + 9x - 6) \\ &\quad\equiv x^4 + (-3x^3 + x^3) + (2x^2 - 3x^2 - 3x^2) + (2x + 9x) - 6 \\ &\quad\equiv x^4 - 2x^3 - 4x^2 + 11x - 6 \end{aligned}$$

Example 5

Multiply $a^2 - ab + b^2$ by $a + b$

$$\begin{aligned}(a^2 - ab + b^2) . (a + b) \\ &\equiv (a^2 - ab + b^2) . a + (a^2 - ab + b^2) . b \\ &\equiv (a^3 - a^2b + ab^2) + (a^2b - ab^2 + b^3) \\ &\equiv a^3 + (-a^2b + a^2b) + (ab^2 - ab^2) + b^3 \\ &\equiv a^3 + b^3\end{aligned}$$

The process of multiplication may be performed simply as an arithmetical one, using and detaching the coefficients of the terms only. For example, we could set out the coefficients of the multiplication exercise of Example 4 in the following way, and carry out a 'long multiplication' sum with them.

	x^4	x^3	x^2	x	1	
$x^2 - 3x + 2$			1	-3	2	
$x^2 + x - 3$			1	1	-3	$\times$
			-3	9	-6	i.e. $-3(x^2 - 3x + 2)$
		1	-3	2		$+x(x^2 - 3x + 2)$
	1	-3	2			$+x^2(x^2 - 3x + 2)$
	1	-2	-4	11	-6	

i.e. $x^4 - 2x^3 - 4x^2 + 11x - 6$

Division

Since multiplication and division are reverse processes, we can relate the process of the division of one polynomial by another polynomial to the long division process in arithmetic. If we use the above expressions to illustrate this, we have (using detached coefficients):

	x^2 x 1	x^4	x^3	x^2	x	1	
				1	-3	$+2$	Quotient
Divisor	$1 + 1 - 3$)	1	-2	-4	$+11$	-6	
		1	$+1$	-3			(Divisor multiplied by 1)
			-3	-1	$+11$	-6	(Subtract)
			-3	-3	$+9$		(Divisor multiplied by -3)
				2	$+2$	-6	(Subtract)
				2	$+2$	-6	(Divisor multiplied by $+2$)
				.	.	.	(Subtract)

Thus $(x^4 - 2x^3 - 4x^2 + 11x - 6) \div (x^2 + x - 3) \equiv x^2 - 3x + 2$.

As we would expect from the chosen expressions, the division is exact, the **quotient** being $x^2 - 3x + 2$ and there is no **remainder**.

We notice that each term in the quotient is found by making the first term in the **divisor** divide exactly each time. It is clear that for the division of one polynomial $f(x)$ by a second one $g(x)$ to be possible, the degree of $f(x)$ must be greater than, or equal to, the degree of $g(x)$. Also, that it will be only in a limited number of cases that this division will be exact. Generally, if the division results in the derivation of a quotient $q(x)$ and a remainder $r(x)$, we may write

$$\frac{f(x)}{g(x)} \equiv q(x) + \frac{r(x)}{g(x)}.$$

The division process will terminate as soon as the degree of $r(x)$ is less than the degree of $g(x)$.

Example 6

Determine the quotient and remainder (if any) when $2x^3 - 7x^2 - 9x + 38$ is divided by $x - 3$.

Using detached coefficients, and writing both the divisor and expression to be divided in descending powers of x,

	x^3	x^2	x	1	
		2	−1	−12	Quotient
1 − 3)	2	−7	−9	+38	
	2	−6			
		−1	−9	+38	
		−1	+3		
			−12	+38	
			−12	+36	
			·	+2	

the quotient is $2x^2 - x - 12$ and the remainder is $+2$.

Example 7

Divide $2x^6 - 3x^4 - 27$ by $x^2 - 3$

In this example, terms in descending powers of x in both polynomials are absent. We treat such terms as having zero coefficients and write thus:

	x^6	x^5	x^4	x^3	x^2	x	1	
			2	0	3	0	9	Quotient
1 0 −3)	2	0	−3	0	0	0	−27	
	2	0	−6					
			3	0	0	0	−27	
			3	0	−9			
					9	0	+27	
					9	0	−27	
					·	·	·	

Thus the division is exact, with a quotient of $2x^4 + 3x^2 + 9$. (Care is needed in deriving the second term in the quotient, i.e. $3x^2$, from the division of $3x^4$ by x^2.)

A further method of deriving the quotient in cases where the division of one polynomial by another is known to be exact, is illustrated in this next example.

Example 8

Consider the division of $x^3 - 4x^2 + 5x - 2$ by $x - 2$, and suppose the result to be exact, that is, there is no remainder.

Then $x^3 - 4x^2 + 5x - 2 \equiv q(x)(x - 2)$

and it is clear that since the highest power of x in $q(x)$ must be 2, $q(x)$ must be of the form $ax^2 + bx + c$.

$$\begin{aligned}\text{Thus}\quad x^3 - 4x^2 + 5x - 2 &\equiv (ax^2 + bx + c)(x - 2)\\ &\equiv ax^3 + bx^2 + cx - 2ax^2 - 2bx - 2c\\ &\equiv ax^3 + (b - 2a)x^2 + (c - 2b)x - 2c\end{aligned}$$

Thus comparing the values of the coefficients on both sides of this identity.

$$\begin{aligned} x^3:&\quad 1 = a\\ x^2:&\quad -4 = b - 2a\\ x:&\quad 5 = c - 2b\\ 1:&\quad -2 = -2c\end{aligned}$$

So, $a = 1$, $c = 1$, and $-4 = b - 2 \Rightarrow b = -2$. The quotient is therefore, $x^2 - 2x + 1$.

Example 9

Divide $6x^3 + 5x^2 - 8x - 3$ by $2x + 3$

Assuming that the division is exact, so that there is no remainder, as before, let

$$\begin{aligned}6x^3 + 5x^2 - 8x - 3 &\equiv (ax^2 + bx + c)(2x + 3)\\ &\equiv 2ax^3 + (3a + 2b)x^2 + (3b + 2c)x + 3c\end{aligned}$$

condensing the multiplication on the R.H.S.

Comparing the coefficients of x^3, x^2, and 1, on each side of this identity (it is necessary to do this three times to acquire three equations for the unknowns a, b, c), we see that

$$\begin{aligned} x^3:&\quad 6 = 2a \qquad \Rightarrow a = 3\\ x^2:&\quad 5 = 3a + 2b \Rightarrow 5 = 9 + 2b \quad b = -2\\ 1:&\quad -3 = 3c \qquad \Rightarrow c = -1.\end{aligned}$$

(We may check these values by also comparing the coefficient of x, i.e. $-8 = 3b + 2c$. This is so when $b = -2$, $c = -1$.)

Therefore, the division is exact, and the quotient is $3x^2 - 2x - 1$.

1 Add together
a) $2x^2 - 5x - 1$, $3x^2 + 2x + 4$
b) $x^2 + 4x - 5$, $4x^2 - 7x - 6$, $-2x^2 + x + 10$
c) $7a + 3b - 5c$, $-4a + 2b + c$, $2a - 6b + 3c$

2 Subtract
a) $3x^2 + 7x - 2$ from $4 + 8x - x^2$
b) $4x^3 - 5x^2 + 3x + 7$ from $7 - 2x + 6x^2 - 8x^3$
c) $x^2 + 4xy - y^2 - 2$ from $3y^2 + 1 + 2xy - 4x^2$

3 Simplify
a) $x^2(x^2 - x + 2) + x(x^3 - x^2 + 1) - 2(x^4 - x^3 - 1)$
b) $(x + 1)(x + 1) + (x - 1)(x - 1) - 2(x + 1)(x - 1)$
c) $(4x - 1)(3x + 2) + (x - 2)(x + 3) + 8$

4 Multiply
a) $x^3 - 2x^2 - 2x + 5$ by $x^2 - x + 2$
b) $x^3 + 3x^2 - 2$ by $x^2 + 2x - 4$
c) $x^2 + y^2 + z^2 - xy - yz - zx$ by $x + y + z$

5 Divide
a) $x^3 + x^2 - x + 2$ by $x + 2$
b) $6x^3 + x^2 + 15$ by $2x + 3$
c) $x^3 + 4x^2 - 3x + 1$ by $x - 2$
d) $9x^3 + 6x^2 - 11x + 1$ by $3x^2 + 4x - 1$
e) $x^4 - 12x^2 + 21x - 10$ by $x^2 + 3x - 5$

6 Assuming that the division is exact and by using the method of comparing coefficients, find the quotient when:
a) $3x^3 - 4x^2 - 8x + 8$ is divided by $x - 2$
b) $4x^3 + 4x^2 + 3x + 9$ is divided by $2x + 3$
c) $x^4 - 7x^2 + 1$ is divided by $x^2 - 3x + 1$
d) $x^5 - x^4 - x^3 + 2x^2 - 2x + 1$ is divided by $x^2 - 2x + 1$

1.3 Synthetic division

The method of detached coefficients may also be used in another way, which generally shortens the working.

Suppose that the polynomial

$$f(x) \equiv a_0 x^n + a_1 x^{n-1} + a_2 x^{n-2} \cdots + a_n$$

is to be divided by $x - \lambda$.

Then $$\frac{f(x)}{x - \lambda} \equiv q(x) + \frac{R}{x - \lambda}$$

where $q(x)$ will be a polynomial of degree $n - 1$, and R will be a numerical constant. If we let

$$q(x) = b_0 x^{n-1} + b_1 x^{n-2} + b_3 x^{n-3} \cdots + b_{n-1},$$

then since $f(x) \equiv (x-\lambda)q(x) + R$,

$$\begin{aligned} &a_0x^n + a_1x^{n-1} + a_2x^{n-2} \cdots + a_n \\ &\equiv (x-\lambda)(b_0x^{n-1} + b_1x^{n-2} + \cdots + b_{n-1}) + R \\ &\equiv b_0x^n + (b_1 - \lambda b_0)x^{n-1} + (b_2 - \lambda b_1)x^{n-2} \\ &\quad \cdots + (b_{n-1} - \lambda b_{n-2})x + (R - \lambda b_{n-1}). \end{aligned}$$

Equating coefficients of x^n, $x^{n-1} \cdots x$, 1

$$a_0 = b_0,\ a_1 = b_1 - \lambda b_0,\ a_2 = b_2 - \lambda b_1, \cdots$$

So

$$\begin{aligned} b_0 &= a_0 \\ b_1 &= a_1 + \lambda b_0 \\ b_2 &= a_2 + \lambda b_1 \quad \text{etc.} \end{aligned}$$

The relationship between the a's and b's may therefore be written as

$$\begin{array}{cccccc} a_0 & a_1 & a_2 & a_3 & \cdots & a_n \\ & \lambda b_0 & \lambda b_1 & \lambda b_2 & \cdots & \lambda b_{n-1} \\ \hline b_0 & b_1 & b_2 & b_3 & & R \\ \hline \end{array} \Bigg\} \text{(add)}$$

Thus, for example, if $x^3 + 2x^2 - x + 4$ is to be divided by $x - 1$, we write

$$\begin{array}{cccc} 1 & +2 & -1 & +4 \\ & 1 & 3 & 2 \\ \hline 1 & +3 & +2 & 6 \\ \hline \end{array} \quad (\lambda = 1)$$

and the quotient is $x^2 + 3x + 2$, remainder is 6.

Example 1
Divide $2x^3 - 7x^2 - 9x + 36$ by $x - 3$

We write

$$\begin{array}{cccc} 2 & -7 & -9 & +36 \\ & +6 & -3 & -36 \\ \hline 2 & -1 & -12 & 0 \\ \hline \end{array} \quad (\lambda = 3)$$

So by this method of **synthetic division** the quotient is $2x^2 - x - 12$ and the remainder is 0.

Example 2
Divide $3x^3 - 11x^2 + 2x + 5$ by $3x + 1$.

To compare the divisor with $x - \lambda$ we write

$3x = y$, or $x = \dfrac{y}{3}$. Then

$$\begin{aligned} 3x^3 - 11x^2 + 2x + 5 &\equiv 3\left(\frac{y}{3}\right)^3 - 11\left(\frac{y}{3}\right)^2 + 2\left(\frac{y}{3}\right) + 5 \\ &\equiv \tfrac{1}{9}\{y^3 - 11y + 6y + 45\} \end{aligned}$$

which is to be divided by $y + 1$, so using the detached coefficients

$$\begin{array}{rrrrl} 1 & -11 & +6 & +45 & \\ & -1 & +12 & -18 & (\lambda = -1) \\ \hline 1 & -12 & +18 & 27 & \\ \hline \end{array}$$

The quotient is

$\frac{1}{9}(y^2 - 12y + 18)$ and remainder $\frac{27}{9}$

or writing $y = 3x$,

$x^2 - 4x + 2$ and remainder 3

Example 3
Divide $x^5 + 1$ by $x + 1$

Here

$$\begin{array}{rrrrrrl} 1 & 0 & 0 & 0 & 0 & 1 & \\ & -1 & +1 & -1 & +1 & -1 & (\lambda = -1) \\ \hline 1 & -1 & +1 & -1 & +1 & 0 & \\ \hline \end{array}$$

The division is exact, with quotient

$x^4 - x^3 + x^2 - x + 1.$

Exercise B

Use the method of synthetic division to carry out the following.

1) $2x^3 + 3x^2 - 7x + 10 \div x - 2$
2) $3x^3 + 12x^2 + 4x + 16 \div x + 4$
3) $x^3 - 2x^2 - 5 \div 3x - 2$
4) $6x^4 - 17x^3 + 22x - 9 \div 2x - 3$
5) $x^6 - 2x^3 + 1 \div x - 1$

1.4 The remainder theorem

From the expression

$$\frac{f(x)}{x - \lambda} \equiv q(x) + \frac{R}{x - \lambda}$$

$$f(x) \equiv (x - \lambda)q(x) + R$$

Since an identity is true for all values of x, it will be true for any particular value.

So, if we replace x by λ

$$f(\lambda) = (\lambda - \lambda)q(\lambda) + R$$
$$\Rightarrow f(\lambda) = 0 + R$$

i.e. the remainder $R = f(\lambda)$

Example

Find the remainder when

a) $f(x) \equiv x^3 - 3x^2 + 4x - 2$ is divided by

i) $x - 1$

ii) $x + 2$

b) $f(x) \equiv 2x^3 + 3x^2 - 4x + 1$ is divided by $2x - 1$

a) Comparing with $x - \lambda$, in i) $\lambda = 1$, and in ii) $\lambda = -2$.

i) $R = f(1) \quad = 1^3 - 3(1)^2 + 4(1) - 2 = 0$

ii) $R = f(-2) = (-2)^3 - 3(-2)^2 + 4(-2) - 2$
$= -8 - 12 - 8 - 2 = -30$

b) Since

$$\frac{f(x)}{2x-1} \equiv q(x) + \frac{R}{2x-1}$$

$$f(x) \equiv (2x - 1)q(x) + R$$

When $2x - 1 = 0$, i.e. $x = \frac{1}{2}$, $R = f(\frac{1}{2})$.

$$\begin{aligned} R = f(\tfrac{1}{2}) &= 2(\tfrac{1}{2})^3 + 3(\tfrac{1}{2})^2 - 4(\tfrac{1}{2}) + 1 \\ &= \tfrac{1}{4} + \tfrac{3}{4} - 2 + 1 \\ &= 0 \end{aligned}$$

1.5 The factor theorem

If $x - \lambda$ is a factor of $f(x)$, R will be zero when $f(x) \div (x - \lambda)$. i.e. $R = f(\lambda) = 0$. So if for a given polynomial $f(x)$, **$f(\lambda) = 0$ then $(x - \lambda)$ is a factor of $f(x)$.**

Example 1

Test $x - 1$, $x + 1$, $x + 2$, for factors of $f(x) \equiv 3x^3 - 2x^2 - 5x + 3$.

$x - 1$: $f(1) = 3 - 2 - 5 + 3 \neq 0 \Rightarrow x - 1$ is not a factor
$x + 1$: $f(-1) = -3 - 2 + 5 + 3 \neq 0 \Rightarrow x + 1$ is not a factor
$x + 2$: $f(-2) = 3(-2)^3 - 2(-2)^2 - 5(-2) + 3$
$= -24 - 8 + 10 + 3 \neq 0 \Rightarrow x + 2$ is not a factor

Example 2

Verify that $x + 2$ is a factor of $f(x) \equiv 2x^3 - 3x^2 - 11x + 6$, and factorise $f(x)$ completely.

$$\begin{aligned} f(-2) &= 2(-2)^3 - 3(-2)^2 - 11(-2) + 6 \\ &= -16 - 12 + 22 + 6 - 0 \Rightarrow x + 2 \text{ is a factor} \end{aligned}$$

By synthetic division

$$\begin{array}{rrrrl} 2 & -3 & -11 & +6 & \\ & -4 & +14 & -6 & (\lambda = -2) \\ \hline 2 & -7 & +3 & 0 & \\ \hline \end{array}$$

Quotient is $2x^2 - 7x + 3$

and $2x^3 - 3x^2 - 11x + 6 \equiv (x + 2)(2x^2 - 7x + 3)$

We know that the quadratic trinomial $2x^2 - 7x + 3$ factorises to $(2x - 1)(x - 3)$.
So $2x^3 - 3x^2 - 11x + 6 \equiv (x + 2)(2x - 1)(x - 3)$.
Note Supposing that a polynomial $f(x)$ factorises into the form

$$(x \pm \alpha)(x \pm \beta)(x \pm \gamma) \cdots \equiv f(x) \equiv a_0 x^n + a_1 x^{n-1} + \cdots + a_n$$

multiplying out the L.H.S. the constant term will be of the form $\pm\alpha\beta\gamma \cdots$. This must be equal to a_n,

i.e. $\pm\alpha\beta\gamma \cdots \equiv a_n$

So $\alpha, \beta, \gamma, \ldots$ or $-\alpha, -\beta, -\gamma, \ldots$ must be factors of a_n. This information may be used as follows:

Example 3
Factorise $f(x) \equiv x^3 - 2x^2 - 5x + 6$

Factors of 6 are $\pm 1, \pm 2, \pm 3, \pm 6$.
We try them in turn

$$\begin{aligned} f(1) &= 1 - 2 - 5 + 6 = 0 \Rightarrow x - 1 && \text{is a factor} \\ f(-1) &= -1 - 2 + 5 + 6 \neq 0 \Rightarrow x + 1 && \text{is not a factor} \\ f(2) &= 8 - 8 - 10 + 6 \neq 0 \Rightarrow x - 2 && \text{is not a factor} \\ f(-2) &= -8 - 8 + 10 + 6 = 0 \Rightarrow x + 2 && \text{is a factor} \\ f(3) &= 27 - 18 - 15 + 6 = 0 \Rightarrow x - 3 && \text{is a factor} \end{aligned}$$

$\Rightarrow f(x) \equiv x^3 - 2x^2 - 5x + 6 \equiv (x - 1)(x + 2)(x - 3)$.

Example 4
Factorise $f(x) \equiv x^4 - x^2 + 4x - 4$

Factors of 4 are $\pm 1, \pm 2, \pm 4$.

$$\begin{aligned} f(1) &= 1 - 1 + 4 - 4 = 0 \Rightarrow x - 1 && \text{is a factor} \\ f(-2) &= 16 - 4 - 8 - 4 = 0 \Rightarrow x + 2 && \text{is a factor} \end{aligned}$$

Let $x^4 - x^2 + 4x - 4 \equiv (x - 1)(x + 2)(ax^2 + bx + c)$
$\equiv (x^2 + x - 2)(ax^2 + bx + c)$

Comparing coefficients of x^4, x^3, x^2

x^4: $1 = a$
x^3: $0 = a + b \Rightarrow b = -1$
x^2: $-1 = -2a + b + c \Rightarrow c = 2$

(or by comparing coefficients of the constant term, $-4 = -2c$ giving $c = 2$, directly).
Hence $x^4 - x^2 + 4x - 4 \equiv (x - 1)(x + 2)(x^2 - x + 2)$.

1.6 Factors of some special polynomials

It has been assumed already that you are able to factorise quadratic expressions, and particularly the difference of two squares

$$x^2 - y^2 \equiv (x - y)(x + y).$$

Two further expressions are of interest, namely the sum and difference of two cubes.

If $f(x) \equiv x^3 + y^3$, $f(-y) = -y^3 + y^3 = 0 \Rightarrow x + y$ is a factor

$$x^3 + y^3 \equiv (x+y)(ax^2 + bxy + cy^2)$$

Comparing coefficients of x^3, x^2y, and y^3.

$$\begin{aligned} x^3&: 1 = a \\ x^2y&: 0 = a + b \Rightarrow b = -1 \\ y^3&: 1 = c \end{aligned}$$

So	$x^3 + y^3 \equiv (x+y)(x^2 - xy + y^2)$
Similarly	$x^3 - y^3 \equiv (x-y)(x^2 + xy + y^2)$

Example 1

Factorise a) $8x^3 + 1$ b) $x^6 + y^6$ c) $x^3 - y^3 - x^2y + xy^2$

a) $8x^3 + 1 \equiv (2x)^3 + 1 \equiv (2x+1)(4x^2 - 2x + 1)$

b) $x^6 + y^6 \equiv (x^2)^3 + (y^2)^3 \equiv (x^2 + y^2)(x^4 - x^2y^2 + y^4)$

c) $$\begin{aligned} x^3 - y^3 - x^2y + xy^2 &\equiv (x-y)(x^2 + xy + y^2) - xy(x-y) \\ &\equiv (x-y)(x^2 + xy + y^2 - xy) \\ &\equiv (x-y)(x^2 + y^2) \end{aligned}$$

Example 2

Simplify

a) $E \equiv \dfrac{(x^3 + y^3)(x-y)}{(x^3 - y^3)(x+y)}$ b) $E \equiv \dfrac{x^3 - y^3}{x - y} - \dfrac{x^3 + y^3}{x + y}$

a) $$\begin{aligned} E \equiv \frac{(x^3 + y^3)(x-y)}{(x^3 - y^3)(x+y)} &= \frac{(x+y)(x^2 - xy + y^2)(x-y)}{(x-y)(x^2 + xy + y^2)(x+y)} \\ &= \frac{x^2 - xy + y^2}{x^2 + xy + y^2}, \text{ by cancelling common factors} \end{aligned}$$

b) $$\begin{aligned} E \equiv \frac{x^3 - y^3}{x - y} - \frac{x^3 + y^3}{x + y} &= \frac{(x-y)(x^2 + xy + y^2)}{(x-y)} - \frac{(x+y)(x^2 - xy + y^2)}{(x+y)} \\ &= (x^2 + xy + y^2) - (x^2 - xy + y^2) \\ &= 2xy \end{aligned}$$

Example 3

Simplify

$$E \equiv (x^2 - xy + y^2)\left(\frac{1}{x} + \frac{1}{y}\right) \div \left(\frac{x}{y^2} + \frac{y}{x^2}\right)$$

$$\begin{aligned} E &\equiv \frac{(x^2 - xy + y^2)(y + x)}{xy} \div \frac{(x^3 + y^3)}{x^2y^2} \\ &\equiv \frac{(x^3 + y^3)}{xy} \div \frac{(x^3 + y^3)}{x^2y^2} \\ &\equiv \frac{(x^3 + y^3)}{xy} \times \frac{x^2y^2}{(x^3 + y^3)} \end{aligned}$$

Hence $E \equiv xy$ (by cancelling like factors)

1 Derive the remainders when the expression

$2x^3 + x^2 - 13x + 6$ is divided by

a) $x - 2$ b) $x + 2$ c) $x - 4$ d) $2x - 1$

2 Verify that
a) $x - 3$ is a factor of $4x^3 - 4x^2 - 21x - 9$
b) $x + 3$ is a factor of $2x^3 + 7x^2 + 2x - 3$

3 Factorise completely
a) $3x^3 - x^2 - 8x - 4$ c) $4x^3 - 7x^2 - 21x + 18$
b) $2x^3 + x^2 - 18x - 9$ d) $x^4 + 4x^3 - 8x - 32$

4 Factorise completely
a) $x^3 + 27$ d) $125x^3 + 8y^3$
b) $x^6 - 1$ e) $x^3 + y^3 + x + y$
c) $x^4 - xy^3$ f) $x^3 + y^3 + 3x^2y + 3xy^2$

5 Simplify

a) $\dfrac{x^3 - y^3}{x^2y + y^3} \div \dfrac{x - y}{x^2 + y^2}$ c) $\dfrac{x^3 + y^3}{x^3 - y^3} - \dfrac{x^2}{x^2 + xy + y^2}$

b) $\dfrac{x^2 - y^2}{x + y} \times \dfrac{x^2 + xy + y^2}{x^3 - y^3}$ d) $\dfrac{x^3 - y^3}{x^2 - 2xy + y^2} - (x - y)$

1.7 Quadratic equations

An equation of the form $ax^2 + bx + c = 0$ where $a \neq 0$, is called a quadratic equation. If b and c both exist the equation is a **quadratic trinomial**. If however, b or c are zero, the equation is a **quadratic binomial**. When $b = c = 0$ the equation is simply $x^2 = 0$, $\Rightarrow x = 0$ (since $a \neq 0$).

We can examine methods of solving the quadratic equation.

a Quadratic binomials

$c = 0$

The equation is now $ax^2 + bx = 0$

$$x(ax + b) = 0$$

$$\text{i.e. } x = 0 \text{ or } x = -\frac{b}{a}$$

$b = 0$

The equation is $ax^2 + c = 0$

or $x^2 = -\dfrac{c}{a}$

Provided that $-\dfrac{c}{a} > 0$, i.e. a, c are of opposite signs,

$$x = \pm\sqrt{\left(\frac{-c}{a}\right)}$$

If, however, a and c are of the same sign, x^2 is negative and we would be faced with calculating the square root of a negative number. Such a value is not determinable in terms of numbers existing on the real number line and hence the quadratic does not have real number solutions.

b Quadratic trinomials

Method 1 Solution by factors

Consider the equation $2x^2 - 5x + 2 = 0$
The L.H.S. of the equation factorises to $(2x - 1)(x - 2)$
Hence the equation becomes $(2x - 1)(x - 2) = 0$
Thus, either $(2x - 1) = 0$, or $(x - 2) = 0$

$$(2x - 1) = 0 \Rightarrow 2x = 1 \Rightarrow x = \tfrac{1}{2}$$
$$(x - 2) = 0 \Rightarrow x = 2$$

The solutions are therefore $x = \frac{1}{2}$ or 2
Again, the equation $2x^2 - 5x - 12 = 0$
factorises to form $(2x - 3)(x + 4) = 0$
Hence $2x - 3 = 0 \Rightarrow x = \frac{3}{2} = 1{\cdot}5$
or $x + 4 = 0 \Rightarrow x = -4$
Thus the solutions are $1{\cdot}5$ or -4

Such a method of solution is quite straightforward, provided that the factors exist and are found carefully. You are advised to check your assumed factors by multiplying out the terms within the brackets. If, however, factorisation is not possible, the quadratic may be solved by the following method.

Method 2 Solution by formula

From $ax^2 + bx + c = 0$

$$x^2 + \frac{b}{a}x + \frac{c}{a} = 0, \text{ dividing throughout by } a.$$

Then $$x^2 + \frac{b}{a}x = -\frac{c}{a}.$$

The L.H.S. of this equation becomes a **perfect square** if we add $\left(\frac{b}{2a}\right)^2$ to it.

For $$x^2 + \frac{b}{a}x + \left(\frac{b}{2a}\right)^2 \equiv \left(x + \frac{b}{2a}\right)^2.$$

Thus adding the term $\left(\frac{b}{2a}\right)^2$ to each side of the equation, it will become

$$\left(x + \frac{b}{2a}\right)^2 = \left(\frac{b}{2a}\right)^2 - \frac{c}{a}$$
$$= \frac{b^2}{4a^2} - \frac{c}{a}$$
$$= \frac{b^2 - 4ac}{4a^2}$$

So taking square roots of both sides of this result,

$$x + \frac{b}{2a} = \frac{\pm\sqrt{(b^2 - 4ac)}}{2a}$$

or $$x = \frac{-b \pm \sqrt{(b^2 - 4ac)}}{2a}$$

Generally all quadratic equations may be solved by replacing a, b, c in this formula by their given values, or by carrying out the steps of **completing the square** as shown.

Example 1

Solve the equation $2x^2 + x - 8 = 0$ using the formula, giving your answers correct to two decimal places.

By comparison with $ax^2 + bx + c = 0$, we identify $a = 2$, $b = 1$ and $c = -8$

And $$x = \frac{-b \pm \sqrt{(b^2 - 4ac)}}{2a}$$

Hence, $$x = \frac{-1 \pm \sqrt{\{1^2 - (4)(2)(-8)\}}}{4}$$

$$= \frac{-1 \pm \sqrt{65}}{4}$$

$$= \frac{-1 \pm 8{\cdot}0623}{4}.$$

i.e. $x = 1{\cdot}7656$ or $-2{\cdot}2656$
or $x = 1{\cdot}77$ or $-2{\cdot}27$ to two decimal places

Example 2

Solve $4x^2 + 3x - 5 = 0$ by completing the square, giving your answers correct to two decimal places.

From $4x^2 + 3x - 5 = 0$

$$x^2 + \tfrac{3}{4}x = \tfrac{5}{4}$$

Adding $(\frac{3}{8})^2$ to each side of the equation

$$x^2 + \tfrac{3}{4}x + (\tfrac{3}{8})^2 = \tfrac{5}{4} + \tfrac{9}{64}$$

$$(x + \tfrac{3}{8})^2 = \tfrac{89}{64}$$

Taking square roots

$$x + \frac{3}{8} = \pm\frac{\sqrt{89}}{8} = \pm\frac{9{\cdot}434}{8}$$

$$\Rightarrow x = \frac{-3}{8} \pm \frac{9{\cdot}434}{8}$$

$$x = \frac{6{\cdot}434}{8} \quad \text{i.e.} \quad x = 0{\cdot}804$$

or $x = \dfrac{-12{\cdot}434}{8}$ i.e. $x = -1{\cdot}554$

So $x = 0{\cdot}80$ or $-1{\cdot}55$ to two decimal places

In the solution by the formula, the expression $b^2 - 4ac$ is significant in that it distinguishes between the nature of the solutions (or roots) of the equation $ax^2 + bx + c = 0$. For if $b^2 - 4ac > 0$, i.e. $b^2 > 4ac$ the square root of $b^2 - 4ac$ exists and two distinct values of x will follow by calculation.

If, however, $b^2 - 4ac = 0$, the roots will be

$x = \dfrac{-b \pm 0}{2a}$, i.e. both will be $x = \dfrac{-b}{2a}$ and the roots are **identical**.

But if $b^2 - 4ac < 0$, i.e. $b^2 < 4ac$, we would again be faced with calculating the square root of a negative number. Again such a value is not determinable in terms of numbers existing on the real number line and the quadratic does not possess real number solutions. We shall see in chapter 11 a further consideration of this general problem.

We refer to the expression $b^2 - 4ac$ as the **discriminant** of the quadratic equation, the equation having two **real, identical** or **non-real** solutions, according to whether the discriminant is **positive, zero,** or **negative**.

Example 3

Distinguish between the nature of the roots of the quadratic equations
a) $3x^2 - 5x + 1 = 0$ b) $x^2 - 14x + 49 = 0$ c) $4x^2 - 7x + 5 = 0$

In a) $b^2 - 4ac = (-5)^2 - 4 \times 3 \times 1 = 13$
The roots are therefore both real and distinct.

In b) $b^2 - 4ac = (-14)^2 - 4 \times 49 = 0$
The roots are therefore real and identical.

In c) $b^2 - 4ac = (-7)^2 - 4 \times 4 \times 5 = -31$
The roots are therefore both non-real.

Exercise D

1 Solve the following quadratic binomials
a) $4x^2 - 9 = 0$ c) $4x^2 - 25 = 0$
b) $3x^2 + 4x = 0$ d) $8x^2 - 32 = 0$

2 Solve the following quadratic equations by the method of factorisation.
a) $x^2 - 2x - 15 = 0$ d) $3x^2 + 8x + 4 = 0$
b) $2x^2 + 7x - 15 = 0$ e) $3x^2 - 16x + 5 = 0$
c) $4x^2 + 5x - 6 = 0$

3 Solve the following quadratic equations by completing the square. Give your answers correct to two decimal places.
a) $x^2 - 8x - 18 = 0$ b) $3x^2 - 8x + 2 = 0$

4 Solve the next group of quadratic equations using the formula, again giving answers correct to two decimal places.
a) $2x^2 + 4x - 5 = 0$ c) $3x^2 = 8x - 1$
b) $4x^2 + 4x - 3 = 0$ d) $5x^2 + 12 = 19x$

5 Distinguish between the nature of the roots of the following quadratic equations. It is not necessary to solve the equations.

a) $6x^2 - 8x - 1 = 0$ d) $2x^2 - 5x + 8 = 0$
b) $4x^2 + 7x + 1 = 0$ e) $7x^2 + 3x = -2$
c) $4x^2 - 12x + 9 = 0$

Method 3 Solution by graphs.
We have already considered how to solve the equation $2x^2 + x - 8 = 0$, using the formula, and have seen that, corrected to two decimal places, the roots are 1·77 or −2·27.

If we write $f(x)$ for $2x^2 + x - 8$, a set of values of $f(x)$ may be written down corresponding to any chosen set of values for x.

Thus, for example, using x: −3, −2, −1, 0, 1, 2, the results may be associated in a table of values:

x	−3	−2	−1	0	1	2
$f(x)$	7	−2	−7	−8	−5	2

If we now plot these values of $f(x)$ against the values of x on graph paper, using suitably positioned axes and appropriate scales, a set of points is obtained: this set of points outlines the shape of the given quadratic function. Joining these points by a smooth curve, as shown in figure 1.3, values of $f(x)$ for any value of x in the range of plotted values may be found approximately.

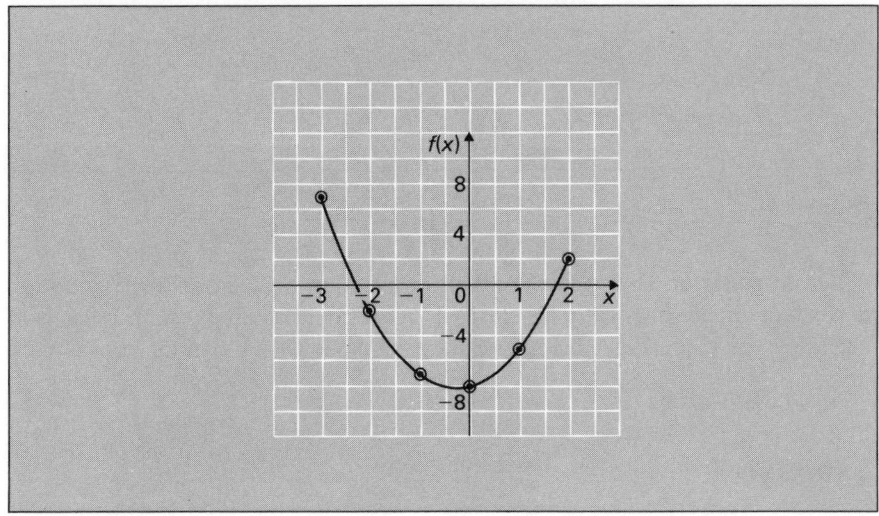

Figure 1.3

Particularly, the values of x for which $f(x) = 0$ are found by reading off the x-values where the curve crosses the x-axis. These are seen to be *about* 1·75 and −2·25 and are in error by about 1 per cent of the more exact values calculated by the formula. Obviously graphical solutions of quadratic equations can only lead to approximate roots and the method must compare unfavourably with *Method 2*. Nevertheless, the idea outlined is useful, and coupled with methods of improving the approximations included in chapter 12, will assist in finding solutions of equations when exact methods do not exist, or are beyond your scope at this stage.

It is interesting to note the shape of the curve of $f(x) \equiv -2x^2 - x + 8$, i.e. the previous expression with the signs reversed. The table of values for the same set of values of x as before is

x	-3	-2	-1	0	1	2
$f(x)$	-7	2	7	8	5	-2

in which the signs of $f(x)$ are reversed. The graph of these values of $f(x)$ against x, as shown in figure 1.4, crosses the x-axis for the same two values $x = 1{\cdot}75$ and $-2{\cdot}25$ and is the previous graph inverted.

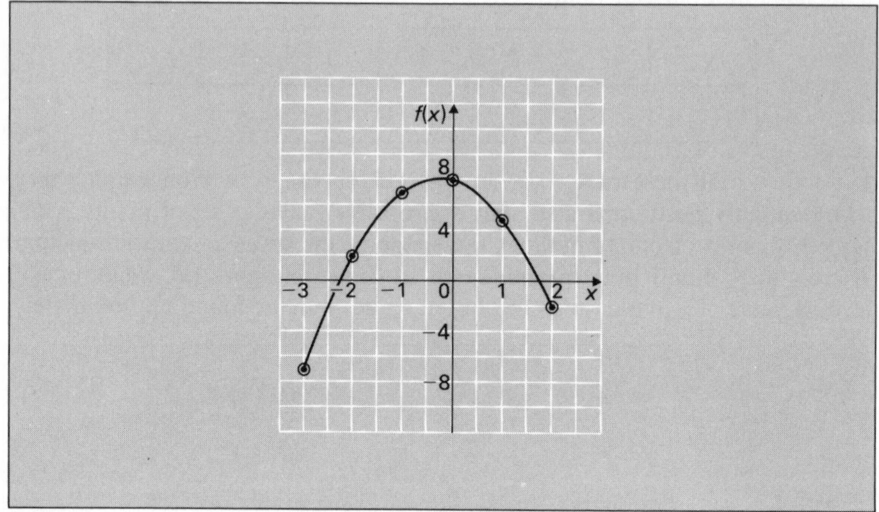

Figure 1.4

Before going on to the final part of this section on the quadratic equation and shape of the quadratic function, it is worthwhile comparing the levels of accuracy you can achieve using graphical methods with the more formal ones.

Exercise E

Solve the following equations a) by routine methods, b) using graph paper, choosing as large a scale as you can to improve the degree of your accuracy.

1 $2x^2 - 3x - 2 = 0$ 2 $x^2 - 3x - 2 = 0$ 3 $3x^2 - x - 7 = 0$

1.8 Quadratic functions

We now look at the graph of the quadratic function $f(x) \equiv ax^2 + bx + c$, and particularly consider the significance of the sign of the coefficient of x^2, related to the alternative forms of the discriminant $b^2 - 4ac$.

Repeating the ideas of completing the square we can rewrite $f(x)$ as follows:

$$\begin{aligned} f(x) &\equiv ax^2 + bx + c \\ &\equiv a\left\{x^2 + \frac{b}{a}x + \frac{c}{a}\right\} \\ &\equiv a\left\{\left(x + \frac{b}{2a}\right)^2 + \frac{4ac - b^2}{4a^2}\right\}, \text{ by completing the square.} \end{aligned}$$

Now if $b^2 < 4ac$, $\dfrac{4ac - b^2}{4a^2}$ is positive, since a^2 is positive.

Also for all values of x excepting $-\dfrac{b}{2a}$, when the value is zero, $\left(x + \dfrac{b}{2a}\right)^2$ being a squared number is positive. Thus $f(x)$ is the constant a multiplied by a positive number. Clearly $f(x)$ is then greater than 0 if $a > 0$, and less than 0 if $a < 0$. The alternative forms of the graph of $f(x)$ against x will then be as shown in figure 1.5.

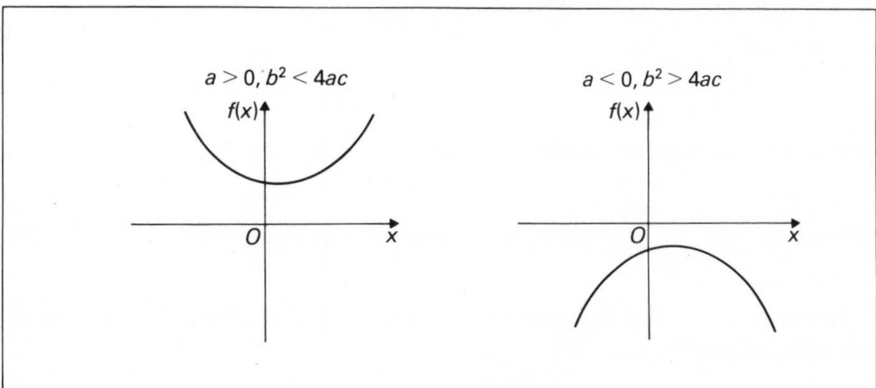

Figure 1.5

But if $b^2 = 4ac$, $\dfrac{4ac - b^2}{4a^2}$ is zero

and $f(x)$ is then $a\left(x + \dfrac{b}{2a}\right)^2$.

When $x = -\dfrac{b}{2a}$, $f(x) = 0$,

otherwise $\left(x + \dfrac{b}{2a}\right)^2$ is positive for all x.

Then $f(x)$ is the constant a multiplied by a positive number as before. So if $a > 0$, $f(x) > 0$, and if $a < 0$, $f(x) < 0$. The alternative forms will then be as shown in figure 1.6.

If, however, $b^2 - 4ac > 0$, $\dfrac{4ac - b^2}{4a^2} < 0$

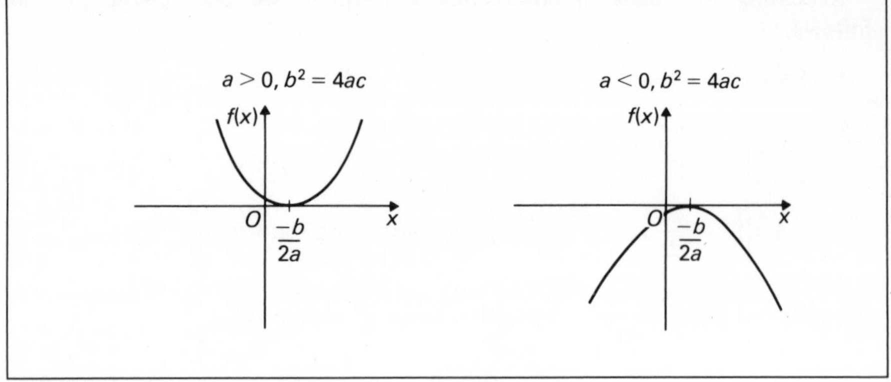

Figure 1.6

and $f(x) \equiv 0$ will have two solutions given by

$$x = \frac{-b \pm \sqrt{(b^2 - 4ac)}}{2a}$$

$f(x)$ is now identically equal to

$$a\left\{\left(x + \frac{b}{2a}\right)^2 - N\right\}$$

where N is the positive number $\dfrac{b^2 - 4ac}{4a^2}$.

When $x = -\dfrac{b}{2a}$ the expression in brackets has its least value and is $(0 - N)$ and $f(x) = -aN$.

Hence if $a > 0$, $f(x) < 0$ and if $a < 0$, $f(x) > 0$. The graphs will then be of the form shown in figure 1.7.

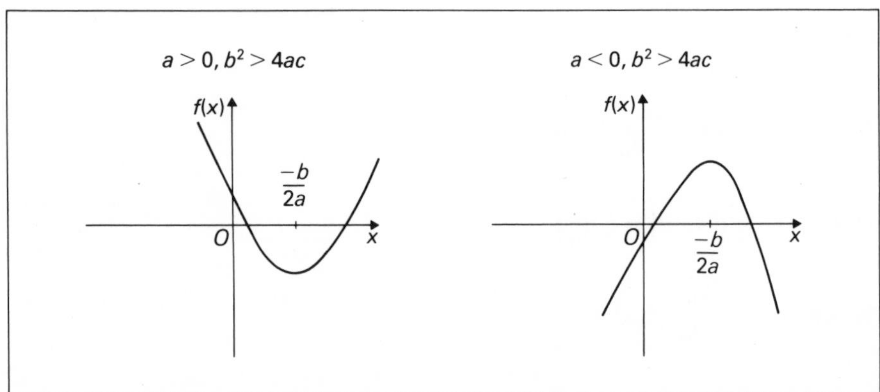

Figure 1.7

Thus in all cases if $\boldsymbol{a > 0}$ the quadratic function for variations in x possesses a **minimum value**, as shown previously, and if $\boldsymbol{a < 0}$ the quadratic function possesses a **maximum value**.

These values occur for $x = -\dfrac{b}{2a}$ in all cases.

1.9 Relationship between the roots of a quadratic equation and the coefficients of the terms of the equation

Suppose that α, β are the roots of $ax^2 + bx + c = 0$. Then $(x - \alpha)(x - \beta) = 0$ and $ax^2 + bx + c = 0$ have the same solutions α, β.

i.e. $x^2 - (\alpha + \beta)x + \alpha\beta = 0$ and $x^2 + \frac{b}{a}x + \frac{c}{a} = 0$

have the same solutions (reducing the coefficient of x^2 in the second equation to unity).

Hence $x^2 - (\alpha + \beta)x + \alpha\beta \equiv x^2 + \frac{b}{a}x + \frac{c}{a}$

Comparing the coefficient of x and 1 on both sides of this identity

$$x:\quad -(\alpha + \beta) = \frac{b}{a} \quad \Rightarrow \quad \alpha + \beta = -\frac{b}{a}$$

$$1:\quad \alpha\beta = \frac{c}{a}$$

Using these results it is possible to write down other symmetrical expressions involving α and β without determining α and β explicitly.

For example:

$$\alpha^2 + \beta^2 \equiv \alpha^2 + 2\alpha\beta + \beta^2 - 2\alpha\beta$$

(adding and subtracting $2\alpha\beta$)

$$\equiv (\alpha + \beta)^2 - 2\alpha\beta$$

$$= \left(-\frac{b}{a}\right)^2 - 2\frac{c}{a} = \frac{b^2 - 2ac}{a^2}$$

$\alpha^3 + \beta^3 \equiv (\alpha + \beta)(\alpha^2 - \alpha\beta + \beta^2)$ as shown in 1.6

$$\equiv -\frac{b}{a}\left(\frac{b^2 - 2ac}{a^2} - \frac{c}{a}\right)$$

$$= \frac{-b(b^2 - 3ac)}{a^3}$$

$$\alpha - \beta \equiv \sqrt{(\alpha - \beta)^2} = \sqrt{(\alpha^2 - 2\alpha\beta + \beta^2)}$$

$$\equiv \sqrt{\left(\frac{b^2 - 2ac}{a^2} - 2\frac{c}{a}\right)}$$

$$\equiv \frac{\sqrt{(b^2 - 4ac)}}{a}$$

$$\frac{1}{\alpha} + \frac{1}{\beta} \equiv \frac{\alpha + \beta}{\alpha\beta} \equiv -\frac{\frac{b}{a}}{\frac{c}{a}}$$

$$= -\frac{b}{c}$$

and so on.

Furthermore, if the roots of the second quadratic are symmetrical functions of α and β the roots of $ax^2 + bx + c = 0$, it is possible to write down this second quadratic equation without finding α and β explicitly.

For if λ, μ are these roots, the new equation will be

$$(x - \lambda)(x - \mu) = 0$$

i.e. $x^2 - (\lambda + \mu)x + \lambda\mu = 0$

or $x^2 - (\text{sum of new roots})x + (\text{product of new roots}) = 0$

These ideas are illustrated in the following examples.

Example 1

If α, β are the roots of $4x^2 - 6x + 1 = 0$, find

a) $\dfrac{1}{\alpha^2} + \dfrac{1}{\beta^2}$ b) $\alpha^3 - \beta^3$ c) $\left(\alpha^2 + \dfrac{2}{\beta^2}\right)\left(\beta^2 + \dfrac{2}{\alpha^2}\right)$

From the given equation $\alpha + \beta = \dfrac{-b}{a} = \dfrac{6}{4} = \dfrac{3}{2}$

and $\alpha\beta = \dfrac{c}{a} = \dfrac{1}{4}$

Now

a) $\dfrac{1}{\alpha^2} + \dfrac{1}{\beta^2} = \dfrac{\beta^2 + \alpha^2}{\alpha^2\beta^2} = \dfrac{(\alpha + \beta)^2 - 2\alpha\beta}{\alpha^2\beta^2}$

$= \{(\frac{3}{2})^2 - \frac{2}{4}\} \div (\frac{1}{4})^2$

$= \frac{7}{4} \div \frac{1}{16} = 28$

b) $\alpha^3 - \beta^3 = (\alpha - \beta)(\alpha^2 + \alpha\beta + \beta^2)$

$= \{\sqrt{(\alpha - \beta)^2}\} \times \{(\alpha + \beta)^2 - \alpha\beta\}$

$= \{\sqrt{(\alpha^2 + \beta^2 - 2\alpha\beta)}\} \times \{(\alpha + \beta)^2 - \alpha\beta\}$

$= \sqrt{\{(\alpha + \beta)^2 - 4\alpha\beta\}} \times \{(\alpha + \beta)^2 - \alpha\beta\}$

$= \sqrt{\{\frac{9}{4} - 1\}} \times \{\frac{9}{4} - \frac{1}{4}\}$

$= \dfrac{\sqrt{5}}{2} \times 2 = \sqrt{5}$

c) $\left(\alpha^2 + \dfrac{2}{\beta^2}\right)\left(\beta^2 + \dfrac{2}{\alpha^2}\right) = \alpha^2\beta^2 + 2 + 2 + \dfrac{4}{\alpha^2\beta^2}$

$= \frac{1}{16} + 4 + \dfrac{4}{\frac{1}{16}}$

$= 68\frac{1}{16}$

Example 2

If α, β are the roots of the equation $5x^2 - x - 7 = 0$, form the quadratic equation whose roots are α^2, β^2.

We have $\alpha + \beta = \frac{1}{5}$ $\alpha\beta = -\frac{7}{5}$ from $5x^2 - x - 7 = 0$

Sum of new roots $= \alpha^2 + \beta^2 = (\alpha + \beta)^2 - 2\alpha\beta$

$= \frac{1}{25} + \frac{14}{5} = \frac{71}{25}$

Product of new roots $= \alpha^2\beta^2 = \frac{49}{25}$

So the new equation is

$$x^2 - x(\text{sum of new roots}) + (\text{product of new roots}) = 0$$

i.e.
$$x^2 - \tfrac{71}{25}x + \tfrac{49}{25} = 0$$
$$25x^2 - 71x + 49 = 0$$

Example 3

Determine the equation whose roots are 5 less than those of $x^2 + 10x + 4 = 0$.

If the roots are α_1, β_1, then $\alpha_1 = \alpha - 5$, $\beta_1 = \beta - 5$ where α, β are the roots of $x^2 + 10x + 4 = 0$.

Hence $\alpha = \alpha_1 + 5$ and $\beta = \beta_1 + 5$.
But α, β satisfy $x^2 + 10x + 4 = 0$

i.e. $\alpha^2 + 10\alpha + 4 = 0$
and $\beta^2 + 10\beta + 4 = 0$

Hence $(\alpha_1 + 5)^2 + 10(\alpha_1 + 5) + 4 = 0$ and similarly for β_1.
So α_1, β_1 satisfy

$$(x + 5)^2 + 10(x + 5) + 4 = 0$$
$$x^2 + 10x + 25 + 10x + 50 + 4 = 0$$
$$x^2 + 20x + 79 = 0$$

Example 4

Derive a relation between a, b and c if one root of the equation $ax^2 + bx + c = 0$ is five times the other.

If one root is α, the other will be 5α.

So $\alpha + 5\alpha = 6\alpha = \dfrac{-b}{a}$

and $\alpha \,.\, 5\alpha = 5\alpha^2 = \dfrac{c}{a}$

So substituting $\alpha = -\dfrac{b}{6a}$ from the first equation in the second,

$$5\left(\frac{-b}{6a}\right)^2 = \frac{c}{a}$$

from which $5b^2 = 36ac$

Exercise F

1 If α, β are the roots of $3x^2 + 4x - 2 = 0$, find the values of

a) $\dfrac{1}{\alpha} + \dfrac{1}{\beta}$

b) $(\alpha - 3\beta)(\beta - 3\alpha)$

c) $\left(2\alpha + \dfrac{1}{\beta}\right)\left(2\beta + \dfrac{1}{\alpha}\right)$

d) $\alpha^3 - \beta^3$

e) $\alpha^2 - \beta^2$

2 If α, β are the roots of $2x^2 - 7x + 2 = 0$, form the equation whose roots are

a) $\dfrac{2}{\alpha}, \dfrac{2}{\beta}$ b) $\dfrac{\alpha}{\beta}, \dfrac{\beta}{\alpha}$ c) $\alpha + \dfrac{1}{\beta}, \beta + \dfrac{1}{\alpha}$ d) $\alpha - 2\beta$, $\beta - 2\alpha$

3 Form the equation
a) whose roots are 3 times the roots of $3x^2 + 8x + 3 = 0$
b) whose roots are the reciprocals of the roots of $5x^2 - 6x + 2 = 0$
Write down the sum of the squares of the roots of this new equation.

4 Given that α, β are the roots of $x^2 + ax + b = 0$, form the equation whose roots are:

a) $\alpha\beta + \alpha, \alpha\beta + \beta$ b) $\alpha + \frac{1}{\alpha}, \beta + \frac{1}{\beta}$ c) $\frac{1}{\alpha^2}, \frac{1}{\beta^2}$ d) α^3, β^3

5 If one root of $x^2 + ax + 16 = 0$ is four times the other root, find the possible values of a.

6 Given that α, β are the roots of $x^2 - ax + b = 0$, derive the equation whose roots are

$$\alpha^2 + \frac{1}{\beta^2}, \beta^2 + \frac{1}{\alpha^2}$$

Show that, provided that a and b are real numbers, this second equation will have equal roots if $a = 0$ or $a^2 = 4b$.

1.10 Relationship between the roots of a cubic equation and the coefficients of the terms of the equation

Extending the ideas used with the quadratic equation we suppose that α, β, γ are the roots of

$$ax^3 + bx^2 + cx + d = 0.$$

Then $(x - \alpha)(x - \beta)(x - \gamma) = 0$

and $x^3 + \frac{b}{a}x^2 + \frac{c}{a}x + \frac{d}{a} = 0$ are identical.

Now
$$\begin{aligned}(x - \alpha)(x - \beta)(x - \gamma) &\equiv [x^2 - (\alpha + \beta)x + \alpha\beta](x - \gamma)\\ &\equiv x^3 - (\alpha + \beta + \gamma)x^2 + (\alpha\beta + \beta\gamma + \gamma\alpha)x - \alpha\beta\gamma\end{aligned}$$

So comparing coefficients of x^2, x and 1

x^2: $-(\alpha + \beta + \gamma) = \frac{b}{a} \Rightarrow$

x :

1 : $-\alpha\beta\gamma = \frac{d}{a} \Rightarrow$

$$\alpha + \beta + \gamma = -\frac{b}{a}$$

$$\alpha\beta + \beta\gamma + \gamma\alpha = \frac{c}{a}$$

$$\alpha\beta\gamma = -\frac{d}{a}$$

These results give

i) the sum of the roots of the cubic $= -\frac{b}{a}$

ii) the sum of the products of the roots taken in pairs $= \frac{c}{a}$

iii) the product of the roots $= -\frac{d}{a}$

Clearly the method may be extended to establish similar results for a polynomial of any degree, but the counterparts of these results become extensive and complicated when in use. We limit the illustration of the cubic results to the following cases.

Example 1

Given that α, β, γ are the roots of $x^3 - x^2 - 7 = 0$, determine the values of

a) $\dfrac{1}{\alpha} + \dfrac{1}{\beta} + \dfrac{1}{\gamma}$ b) $\alpha^2 + \beta^2 + \gamma^2$ c) $\dfrac{1}{\alpha\beta} + \dfrac{1}{\beta\gamma} + \dfrac{1}{\gamma\alpha}$

Using the basic results

$$\alpha + \beta + \gamma = -(-1) = 1, \ \alpha\beta + \beta\gamma + \gamma\alpha = 0,$$
$$\alpha\beta\gamma = -(-7) = 7$$

a) $\dfrac{1}{\alpha} + \dfrac{1}{\beta} + \dfrac{1}{\gamma} = \dfrac{\beta\gamma + \alpha\gamma + \alpha\beta}{\alpha\beta\gamma} = \dfrac{0}{7} = 0$

b) $\alpha^2 + \beta^2 + \gamma^2 = (\alpha + \beta + \gamma)^2 - 2(\alpha\beta + \beta\gamma + \gamma\alpha)$

(by multiplying out the R.H.S.)

$= 1 - 2 \times 0 = 1$

c) $\dfrac{1}{\alpha\beta} + \dfrac{1}{\beta\gamma} + \dfrac{1}{\gamma\alpha} = \dfrac{\gamma + \alpha + \beta}{\alpha\beta\gamma} = \dfrac{1}{7}$

Example 2

Determine the cubic equation whose roots are the reciprocals

$\dfrac{1}{\alpha}, \dfrac{1}{\beta}, \dfrac{1}{\gamma}$ of the roots α, β, γ of the equation $x^3 - 4x^2 + 5x - 2 = 0$

Hence determine $\dfrac{1}{\alpha^2} + \dfrac{1}{\beta^2} + \dfrac{1}{\gamma^2}$

If the roots of the new equation are α_1, β_1, γ_1,

$$\alpha_1 = \frac{1}{\alpha}, \ \beta_1 = \frac{1}{\beta}, \ \gamma_1 = \frac{1}{\gamma}$$

i.e. $\alpha = \dfrac{1}{\alpha_1}, \ \beta = \dfrac{1}{\beta_1}, \ \gamma = \dfrac{1}{\gamma_1}$

But α, β, γ satisfy $x^3 - 4x^2 + 5x - 2 = 0$

Hence α_1, β_1, γ_1 satisfy $\left(\dfrac{1}{x}\right)^3 - 4\left(\dfrac{1}{x}\right)^2 + 5\left(\dfrac{1}{x}\right) - 2 = 0$

i.e. $1 - 4x + 5x^2 - 2x^3 = 0$
or $2x^3 - 5x^2 + 4x - 1 = 0$

Now $\dfrac{1}{\alpha^2} + \dfrac{1}{\beta^2} + \dfrac{1}{\gamma^2} = \alpha_1^2 + \beta_1^2 + \gamma_1^2$

$$= (\alpha_1 + \beta_1 + \gamma_1)^2 - 2(\alpha_1\beta_1 + \beta_1\gamma_1 + \gamma_1\alpha_1)$$

But $\alpha_1 + \beta_1 + \gamma_1 = \frac{5}{2}\left(\text{i.e. } -\frac{b}{a}\right)$,

$$\alpha_1\beta_1 + \beta_1\gamma_1 + \gamma_1\alpha_1 = \frac{4}{2} = 2\left(\text{i.e. } \frac{c}{a}\right)$$

Hence $\frac{1}{\alpha^2} + \frac{1}{\beta^2} + \frac{1}{\gamma^2} = \left(\frac{5}{2}\right)^2 - 2(2) = \frac{9}{4}$

Example 3

The roots of the equation $x^3 - 3ax^2 + 23x - 15 = 0$ differ by 2. Determine them, and the value of a.

Since the roots differ by 2 let them be $\alpha - 2, \alpha, \alpha + 2$

Then $\alpha - 2 + \alpha + \alpha + 2 = 3\alpha = 3a \Rightarrow \alpha = a$.

$$(\alpha - 2)\alpha + (\alpha - 2)(\alpha + 2) + \alpha(\alpha + 2) = 3\alpha^2 - 4 = 23 \Rightarrow \alpha = \pm 3$$

and $(\alpha - 2)\alpha(\alpha + 2) = 15$

If $\alpha = 3$, L.H.S. $= 1 \,.\, 3 \,.\, 5 = 15 =$ R.H.S.

$\alpha = -3$, L.H.S. $= (-5)(-3)(-1) = -15 \neq$ R.H.S.

Hence $\alpha = 3$ and roots are 1, 3, 5, and $a = 3$

Example 4

If α, β, γ are the roots of the equation $x^3 - 2x^2 + 3x - 4 = 0$, find the equation whose roots are $\beta + \gamma, \gamma + \alpha, \alpha + \beta$

From the given equation

$$\alpha + \beta + \gamma = 2, \alpha\beta + \beta\gamma + \gamma\alpha = 3, \alpha\beta\gamma = 4$$

The new equation is

$$x^3 - x^2(\text{sum of new roots}) + x(\text{product, two at a time, of new roots}) - (\text{triple product of roots}) = 0$$

Now

$$\begin{aligned}
\text{sum of new roots} &= (\beta + \gamma) + (\gamma + \alpha) + (\alpha + \beta) \\
&= 2(\alpha + \beta + \gamma) = 4 \\
\text{products, two at a time} &= (\beta + \gamma)(\gamma + \alpha) + (\gamma + \alpha)(\alpha + \beta) + (\alpha + \beta)(\beta + \gamma) \\
&= (2 - \alpha)(2 - \beta) + (2 - \beta)(2 - \gamma) + (2 - \gamma)(2 - \alpha) \\
&\qquad (\text{using } \alpha + \beta + \gamma = 2) \\
&= 4 - 2(\alpha + \beta) + \alpha\beta + 4 - 2(\beta + \gamma) \\
&\qquad + \beta\gamma + 4 - 2(\gamma + \alpha) + \alpha\gamma \\
&= 12 - 4(\alpha + \beta + \gamma) + \alpha\beta + \beta\gamma + \gamma\alpha \\
&= 12 - 4 \times 2 + 3 = 7
\end{aligned}$$

Likewise, the triple product

$= (2 - \alpha)(2 - \beta)(2 - \gamma) = 8 - 4(\alpha + \beta + \gamma) + 2(\alpha\beta + \beta\gamma + \gamma\alpha) - \alpha\beta\gamma$

$= 8 - 4 \times 2 + 2 \times 3 - 4 = 2$

Hence the equation is

$$x^2 - 4x^2 + 7x - 2 = 0$$

An alternative method of dealing with this example is as follows:

Since $\alpha + \beta + \gamma = 2$ we can write in turn

$$\beta + \gamma = 2 - \alpha$$
$$\alpha + \gamma = 2 - \beta$$

and $\alpha + \beta = 2 - \gamma$

Then since α, β, γ are roots of $x^3 - 2x^2 + 3x - 4 = 0$ and $\beta + \gamma, \alpha + \gamma, \alpha + \beta$ are roots of the new equation we can write more generally

$$y = 2 - x \text{ or } x = 2 - y$$

So the new equation is

$$(2 - y)^3 - 2(2 - y)^2 + 3(2 - y) - 4 = 0$$

Now $(2 - y)^3 = (2 - y)^2(2 - y) = (4 - 4y + y^2)(2 - y)$
$= 8 - 12y + 6y^2 - y^3$

Therefore the new equation is

$$8 - 12y + 6y^2 - y^3 - 2(4 - 4y + y^2) + 6 - 3y - 4 = 0$$

i.e. $8 - 12y + 6y^2 - y^3 - 8 + 8y - 2y^2 + 6 - 3y - 4 = 0$

or $-y^3 + 4y^2 - 7y + 2 = 0$

which we can write as $y^3 - 4y^2 + 7y - 2 = 0$

or $x^3 - 4x^2 + 7x - 2 = 0$

since the symbol used for the variable in the equation is of no consequence. This approach to the solution of this problem is, of course, an extension of the idea used in Example 3 of the earlier section 1.9.

Exercise G

1 Show that if α, β, γ are the roots of $ax^3 + bx^2 + cx + d = 0$,

a) $\dfrac{1}{\alpha\beta} + \dfrac{1}{\gamma\alpha} + \dfrac{1}{\gamma\beta} = \dfrac{b}{d}$ c) $\alpha^3 + \beta^3 + \gamma^3 = (3abc - b^3 - 3a^2d)/a^3$

b) $\alpha^2 + \beta^2 + \gamma^2 = \dfrac{b^2 - 2ac}{a^2}$

2 Given α, β, γ are the roots of $x^3 - 3x^2 - x + 2 = 0$, determine
a) $\alpha^2 + \beta^2 + \gamma^2$ b) $\alpha^2\beta^2 + \beta^2\gamma^2 + \gamma^2\alpha^2$ c) $\alpha^2\beta^2\gamma^2$
and write down the equation whose roots are $\alpha^2, \beta^2, \gamma^2$

3 Write down the equation whose roots are a) greater by 1, b) less by 2 than those of the equation $x^3 + 3x^2 + 2x + 4 = 0$.

1.11 Partial fractions

We have seen in the examples in section 1.6 the processes involved in combining a set of algebraic fractions and reducing them to a single fraction.

Thus, for example

$$f(x) \equiv \frac{2}{x - 1} - \frac{3}{x + 3} + \frac{4}{x - 5}$$

$$\equiv \frac{2(x + 3)(x - 5) - 3(x - 1)(x - 5) + 4(x - 1)(x + 3)}{(x - 1)(x + 3)(x - 5)}$$

$$\equiv \frac{2(x^2 - 2x - 15) - 3(x^2 - 6x + 5) + 4(x^2 + 2x - 3)}{(x - 1)(x + 3)(x - 5)}$$

$$\equiv \frac{3x^2 + 22x - 57}{(x - 1)(x + 3)(x - 5)}$$

We now consider the reverse process, for it is often important as a step in some algebraic exercise to break down complicated fractions such as

$$\frac{3x^2 + 22x - 57}{(x-1)(x+3)(x-5)}$$

and express them as the sum (or difference) of simpler proper algebraic fractions. This reverse process is called expressing $f(x)$ in terms of its **partial fractions.** We limit the exercise in this chapter to expressions $f(x)$ which have linear factors only in their denominators, such expressions $f(x)$ being of proper algebraic form. An additional step in the process, should $f(x)$ be improper, would be to determine the quotient and remainder (say by division) and then apply the process to the remainder.

The process is illustrated in the following examples.

When $f(x)$ has linear unrepeated factors in the denominator

Example 1

Express $\dfrac{x+1}{(x-1)(x+2)}$ in terms of its partial fractions.

It is clear that separate fractions with denominators $(x - 1)$ and $(x + 2)$ when combined would give a single fraction with $(x - 1)(x + 2)$ as a common denominator. Moreover, following our earlier definition, for a fraction with $(x - 1)$ to be proper, the numerator of the fraction can only be a numerical constant.

We write
$$\frac{x+1}{(x-1)(x+2)} \equiv \frac{A}{(x-1)} + \frac{B}{(x+2)}$$

where A, B are numerical values to be found. Bringing the R.H.S. to the same common denominator as the L.H.S., the expression will be an identity if the numerators are the same, namely

$$x + 1 \equiv A(x+2) + B(x-1)$$

We can derive A, B by comparing coefficients of the two expressions on each side of this identity (as shown in the last example demonstrating the division of polynomials), or by substitution of specific values.

By comparing coefficients

of x: $\quad 1 = A + B$

of 1: $\quad 1 = 2A - B$

$\Rightarrow A = \frac{2}{3}$ and $B = \frac{1}{3}$

(Or *by substituting specific values of* x; suitable values are those giving zeros of the individual factors. From $x + 2 = 0$, $x = -2$, and from $x - 1 = 0$, $x = 1$

When $x = -2$, $\quad -2 + 1 = A \,.\, 0 \quad + B(-2-1) \Rightarrow B = \frac{1}{3}$

$x = \;\;1$, $\quad 1 + 1 = A(1+2) + B(1-1) \quad \Rightarrow A = \frac{2}{3}$)

Thus we would write

$$\frac{x+1}{(x-1)(x+2)} \equiv \frac{2}{3(x-1)} + \frac{1}{3(x+2)}$$

Example 2

Express $\dfrac{2x^2 - x - 6}{(x-1)(x^2-9)}$ in terms of its partial fractions.

Since $x^2 - 9 \equiv (x-3)(x+3)$

$$\frac{2x^2 - x - 6}{(x-1)(x^2-9)} \equiv \frac{A}{(x-1)} + \frac{B}{(x-3)} + \frac{C}{(x+3)}$$

Thus $2x^2 - x - 6 \equiv A(x-3)(x+3) + B(x-1)(x+3) + C(x-1)(x-3)$ where A, B and C are numerical coefficients.

Should we decide to find A, B and C by comparing coefficients of x^2, x and 1 we would extract three simultaneous equations in the three unknowns. Elimination of one of these unknowns would reduce this problem to the usual one of solving two such equations in two unknowns. It is clear, however that this process is more 'long-winded' than it need be, for the substitution of the values $x = 1$ (from $x - 1 = 0$), $x = 3$ (from $x - 3 = 0$) and $x = -3$ (from $x + 3 = 0$) yields the values A, B and C directly

$$x = 1: \quad 2 - 1 - 6 = A(1-3)(1+3) + B\,.\,0(1+3) + C\,.\,0(1-3)$$

$$\Rightarrow A = \tfrac{5}{8}$$

And $x = 3$: $\quad 9 = A\,.\,0 + B\,.\,12 + C\,.\,0 \Rightarrow B = \tfrac{3}{4}$
$x = -3$: $\quad 15 = A\,.\,0 + B\,.\,0 + 24C \Rightarrow C = \tfrac{5}{8}$

Hence $$\frac{2x^2 - x - 6}{(x-1)(x^2-9)} \equiv \frac{5}{8(x-1)} + \frac{3}{4(x-3)} + \frac{5}{8(x+3)}.$$

When $f(x)$ has linear repeated factors

Example 1

Express $\dfrac{x}{(x+1)^2(x-2)}$ in terms of its partial fractions.

We would expect the denominators to yield as possible partial fractions

$$\frac{A}{(x-2)} + \frac{B}{(x+1)} + \frac{Cx + D}{(x+1)^2}$$

the first two fractions $\dfrac{A}{(x-2)}$ and $\dfrac{B}{(x+1)}$ maintaining the previous idea whilst $\dfrac{Cx+D}{(x+1)^2}$ would be a possible form of the fraction with a quadratic denominator.

Now we can write

$$\frac{B}{(x+1)} + \frac{Cx+D}{(x+1)^2} \quad \text{in the form} \quad \frac{B}{(x+1)} + \frac{Cx + C + D - C}{(x+1)^2}$$

or $$\frac{B}{(x+1)} + \frac{C(x+1)}{(x+1)^2} + \frac{D-C}{(x+1)^2}$$

or $$\frac{B+C}{(x+1)} + \frac{D-C}{(x+1)^2}$$

Now since B, C and D are numerical values to be found, $B + C$ and $D - C$ are only simple combinations of them and the fractions which would ultimately appear would be of the form

$$\frac{\text{a number}}{(x+1)} + \frac{\text{a second number}}{(x+1)^2}$$

It is therefore sufficient to express the fractions in such a form at the beginning of the exercise. We let therefore,

$$\frac{x}{(x+1)^2(x-2)} \equiv \frac{a}{(x-2)} + \frac{b}{(x+1)} + \frac{c}{(x+1)^2}$$

Hence $\quad x \equiv a(x+1)^2 + b(x-2)(x+1) + c(x-2)$

Neither of the methods used for determining a, b, c result in the immediate calculation of all their values.

Substituting $x = -1 \quad -1 = -3c \Rightarrow c = \frac{1}{3}$

$\qquad\qquad\quad x = 2 \quad2 = 9a \Rightarrow a = \frac{2}{9}$

gives two of the three values.

We can now choose any other value for x, or resort to comparison of coefficients. In the outcome the results are the same, for choosing

$$\begin{aligned} x = 0:\quad 0 &= a - 2b - 2c \\ 2b &= a - 2c \\ &= \tfrac{2}{9} - \tfrac{2}{3} \\ \Rightarrow b &= -\tfrac{2}{9} \end{aligned} \qquad\qquad \begin{aligned} \text{comparing } x^2:\quad 0 &= a + b \\ b &= -a \\ &= -\tfrac{2}{9} \end{aligned}$$

Thus $$\frac{x}{(x+1)^2(x-2)} \equiv \frac{2}{9(x-2)} - \frac{2}{9(x+1)} + \frac{1}{3(x+1)^2}$$

Example 2

Express $\dfrac{4x-3}{x^3(x+1)}$ in terms of its partial fractions.

By an extension of the argument used in Example 1

let $$\frac{4x-3}{x^3(x+1)} \equiv \frac{A}{x} + \frac{B}{x^2} + \frac{C}{x^3} + \frac{D}{(x+1)}$$

Then $\quad 4x - 3 \equiv Ax^2(x+1) + Bx(x+1) + C(x+1) + Dx^3$

When $x = 0, \quad -3 = C \Rightarrow C = -3$

$\qquad\quad x = -1, \quad -7 = -D \Rightarrow D = 7$

Comparing coefficients of

x^3: $\quad 0 = A + D \Rightarrow A = -7$

x^2: $\quad 0 = A + B \Rightarrow B = 7$

Thus $$\frac{4x-3}{x^3(x+1)} \equiv \frac{-7}{x} + \frac{7}{x^2} - \frac{3}{x^3} + \frac{7}{(x+1)}$$

Exercise H

Express $f(x)$ in terms of its partial fractions given that $f(x)$ is

1 $\dfrac{x}{(x-1)(x+2)}$

2 $\dfrac{7(x+1)}{(x+3)(2x-1)}$

3 $\dfrac{2x-1}{(x+1)(x+3)(x-2)}$

4 $\dfrac{x-1}{x^2-4}$

5 $\dfrac{x^2-2x+4}{x(x^2-2x-3)}$

6 $\dfrac{x+1}{x^2(x-1)}$

7 $\dfrac{9}{(x-1)(x+2)^2}$

8 $\dfrac{3x^2+x-2}{(2x-1)(x-2)^2}$

9 $\dfrac{x^2-1}{x^2(2x+1)}$

10 $\dfrac{1}{x^2(x^2-1)^2}$

CHAPTER TWO

Logarithms and Indices

2.1 Rules of indices

One of the simplest and most commonly used of all the identities we have met so far is the one illustrated by

$$x \times x \times x \equiv x^3$$

This is followed immediately by, for example,

$$x^4 \times x^3 \equiv \{x \times x \times x \times x\} \times \{x \times x \times x\} \equiv x^7$$

$$x^5 \div x^3 \equiv \frac{x \times x \times x \times x \times x}{x \times x \times x} \equiv x^2$$

$$(x^3)^2 \equiv (x^3) \times (x^3) \equiv x^6$$

These three results can of course be generalised immediately to give the first three of the basic rules of indices, namely

Rule 1

$x^m \times x^n \equiv x^{m+n}$ i.e. to multiply powers of the same base **add indices.**

Rule 2

$x^m \div x^n \equiv x^{m-n}$ i.e. to divide, **subtract indices.**

Rule 3

$(x^m)^n \equiv x^{mn}$ i.e. to raise a power of a base to a second index **multiply the indices.**

It is possible that you will also have studied the application of these basic rules for values of $m, n \in \mathbb{Z}$ and also for $m, n \in \mathbb{Q}$.

Associated with the main set of rules of indices is a subsequent set: familiarity with these is expected. They are particularly:

Rule 4

$x^{-m} \equiv \dfrac{1}{x^m} \quad (m > 0)$ i.e. the interpretation of a **negative index.**

Rule 5

$x^0 \equiv 1$ i.e. the interpretation of the special **index zero.**

Rule 6

$x^{1/m} \equiv \sqrt[m]{x} \quad (m > 0)$ i.e. the interpretation of a **simple fractional index.**

Rule 7

$x^{m/n} \equiv \sqrt[n]{x^m} \equiv \{\sqrt[n]{x}\}^m \quad (n > 0)$ i.e. the interpretation of **rational indices.**

Example 1

Evaluate the following numbers expressing the result in the simplest possible form

a) 4^{-2} e) $\left(\frac{16}{81}\right)^{3/4}$

b) $16^{1/2}$ f) $\left(\frac{25}{36}\right)^{-1/2}$

c) 10^0 g) $9^{-3/2} \times 27^{2/3}$

d) $8^{2/3}$ h) $64^{-1/3} + \left(\frac{1}{4}\right)^{3/2} - 8^{-1}$

Answers

a) $4^{-2} = \frac{1}{4^2} = \frac{1}{16}$

b) $16^{1/2} = \sqrt{16} = \pm 4$

c) $10^0 = 1$ (anything to the index zero = 1)

d) $8^{2/3} = (8^{1/3})^2 = 2^2 = 4$

(or $= (8^2)^{1/3} = (64)^{1/3} = 4$)

e) $\left(\frac{16}{81}\right)^{3/4} = \frac{16^{3/4}}{81^{3/4}} = \frac{(16^{1/4})^3}{(81^{1/4})^3} = \frac{2^3}{3^3} = \frac{8}{27}$

f) $\left(\frac{25}{36}\right)^{-1/2} = \left(\frac{36}{25}\right)^{1/2} = \pm\frac{6}{5}$ $\left(\text{using } \left(\frac{a}{b}\right)^{-1} = \frac{b}{a}\right)$

g) $9^{-3/2} \times 27^{2/3} = \frac{1}{(9^{1/2})^3} \times (27^{1/3})^2 = \frac{9}{27} = \frac{1}{3}$

h) $64^{-1/3} + \left(\frac{1}{4}\right)^{3/2} - 8^{-1} = \frac{1}{64^{1/3}} + \frac{1}{(4^{1/2})^3} - \frac{1}{8}$

$= \frac{1}{4} + \frac{1}{8} - \frac{1}{8} \qquad = \frac{1}{4}$

Example 2

Simplify the following algebraic expressions

a) $x^2 \times x^{2/3}$ d) $(2x^2)^{-1} \div (-2x)^{-2}$

b) $x^{1/6} \times x^{-2/3} \div x^{1/4}$ e) $x^{1/2}(x^{1/2} - x^{-1/2})$

c) $8x^4 \times 2x^2 \times (2x)^{-2}$ f) $(x^{1/3} - x^{-1/3})(x^{2/3} + 1 + x^{-2/3})$

Answers

a) $x^2 \times x^{2/3} = x^{8/3}$ (applying Rule 1)

b) $x^{1/6} \times x^{-2/3} \div x^{1/4} = x^{-1/2} \div x^{1/4}$ (applying Rule 1)

$= x^{-3/4}$ (applying Rule 2)

c) $8x^4 \times 2x^2 \times (2x)^{-2} = 16x^6 \times \frac{1}{(2x)^2}$ (applying Rules 1 and 4)

$= 16x^6 \times \frac{1}{4x^2} = 4x^4$

d) $(2x^2)^{-1} \div (-2x)^{-2} = \dfrac{1}{2x^2} \times (-2x)^2$ (applying Rule 4)

$= \dfrac{1}{2x^2} \times 4x^2 = 2$

e) $x^{1/2}(x^{1/2} - x^{-1/2}) = x^{1/2+1/2} - x^{1/2-1/2}$ (removing brackets)

$= x^1 - x^0$

$= x - 1$ (applying Rule 5)

f) $(x^{1/3} - x^{-1/3})(x^{2/3} + 1 + x^{-2/3}) = x - x^{1/3} + x^{1/3} - x^{-1/3} + x^{-1/3} - x^{-1}$
(multiplying out and applying Rule 1)

$= x - x^{-1}$

Exercise A

1 Find the value of

a) $4^{3/2}$

b) $9^{-1/2}$

c) $27^{5/3}$

d) $125^{-4/3}$

e) $\left(\dfrac{1}{2}\right)^{-2}$

f) $\left(\dfrac{27}{8}\right)^{-1/3}$

g) $4^{-3} \times 2^0$

h) $8^{4/3} \times 4^{-2}$

i) $\dfrac{8^{1/3} \times 81^{1/4}}{36^{1/2}}$

j) $16^{-1/2} \times 64^{1/3} \div 32^{1/5}$

2 Reduce the following expressions to their simplest forms.

a) $(x^2)^5 \times (x^3)^2$

b) $(x^7)^2 \div (x^3)^4$

c) $(3x^3)^3 \times \dfrac{1}{3x^3}$

d) $4x^{-3/2} \times 2x^{1/2}$

e) $5x^{-3/2} \div x^{-1/2}$

f) $5x^{-4} \times 2x^2 \div (4x^{-3})$

g) $6x^{-4} - 3x^{-2}$

h) $x^{-1} + 2x^{-2} - 3x^{-3}$

i) $\dfrac{2x^{-2/3}y^{3/2}}{3x^{1/3}y^{-1/2}}$

2.2 Surds

Amongst the set numbers defined by $\sqrt{x}$, for $x \in \mathbb{Z}^+$, certain members of the set have exact values. These members are $\sqrt{1}$, $\sqrt{4}$, $\sqrt{9}$, $\sqrt{16}$, $\sqrt{25}$, ... All other members of the set are not rational and it is frequently more convenient to leave them in their basic $\sqrt{x}$ form. As such these numbers are called **surds**, and expressions involving them, **surd expressions**. Again it is desirable to condense such expressions to their simplest forms, and the following examples indicate some of the procedures and techniques used. These are primarily dependent upon two rules which are counterparts of the 1st and 2nd rules of indices.

Rule 1

$\sqrt{(x)} \times \sqrt{(y)} = \sqrt{(xy)}$ (from $(xy)^{1/2} = x^{1/2}y^{1/2}$)

Rule 2

$$\sqrt{(x)} \div \sqrt{(y)} = \sqrt{\left(\frac{x}{y}\right)} \quad \left(\text{from } \left(\frac{x}{y}\right)^{1/2} = \frac{x^{1/2}}{y^{1/2}}\right)$$

Example 1

Reduce the following surd numbers and expressions to their simplest forms

a) $\sqrt{80}$
b) $\frac{1}{2}\sqrt{12}$
c) $2\sqrt{18} \times 3\sqrt{2}$
d) $2\sqrt{3}(\sqrt{27} + \sqrt{3})$
e) $(2\sqrt{3} + 1)(3\sqrt{3} - 2)$
f) $\sqrt{50} + \sqrt{98} - \sqrt{32} - \sqrt{72} + \sqrt{8}$

Answers

a) $\sqrt{80} = \sqrt{(5 \times 16)} = \sqrt{5} \times \sqrt{16} = 4\sqrt{5}$

b) $\frac{1}{2}\sqrt{12} = \frac{1}{\sqrt{4}}\sqrt{12} = \sqrt{\frac{12}{4}} = \sqrt{3}$

c) $2\sqrt{18} \times 3\sqrt{2} = 2\sqrt{(9 \times 2)} \times 3\sqrt{2} = 6\sqrt{2} \times 3\sqrt{2}$
$= 18\sqrt{(2 \times 2)} = 18 \times 2 = 36$

d) $2\sqrt{3}(\sqrt{27} + \sqrt{3}) = 2\sqrt{3}\sqrt{27} + 2\sqrt{3}\sqrt{3}$
$= 2\sqrt{81} + 2\sqrt{9} = 2 \times 9 + 2 \times 3 = 24$

e) $(2\sqrt{3} + 1)(3\sqrt{3} - 2) = 6\sqrt{3}\sqrt{3} - 4\sqrt{3} + 3\sqrt{3} - 2$
$= 18 - \sqrt{3} - 2 = 16 - \sqrt{3}$

f) $\sqrt{50} + \sqrt{98} - \sqrt{32} - \sqrt{72} + \sqrt{8}$
$= \sqrt{(2 \times 25)} + \sqrt{(2 \times 49)} - \sqrt{(2 \times 16)} - \sqrt{(2 \times 36)} + \sqrt{(2 \times 4)}$
$= 5\sqrt{2} + 7\sqrt{2} - 4\sqrt{2} - 6\sqrt{2} + 2\sqrt{2}$
$= 4\sqrt{2}$

Example 2

Express the following surd expressions in a form in which the denominators are rational numbers. (This process is called rationalisation of the expression.)

a) $\dfrac{2}{\sqrt{12}}$

b) $\dfrac{3}{\sqrt{75}}$

c) $\dfrac{1}{\sqrt{3} - 1}$

d) $\dfrac{\sqrt{3}}{3\sqrt{3} + 2\sqrt{2}}$

e) $\dfrac{\sqrt{5}}{\sqrt{5} - 1} + \dfrac{\sqrt{5}}{\sqrt{5} + 1}$

Answers

a) Multiplying the expression in both numerator and denominator by the same value leaves the fraction unaltered.

Thus $\dfrac{2}{\sqrt{12}} = \dfrac{2}{\sqrt{12}} \times \dfrac{\sqrt{12}}{\sqrt{12}} = \dfrac{2\sqrt{12}}{12} = \dfrac{\sqrt{12}}{6} = \dfrac{\sqrt{(4 \times 3)}}{6} = \dfrac{2\sqrt{3}}{6} = \dfrac{\sqrt{3}}{3}$

b) $\dfrac{3}{\sqrt{75}} = \dfrac{3}{\sqrt{(25 \times 3)}} = \dfrac{3}{5\sqrt{3}} = \dfrac{3\sqrt{3}}{5\sqrt{3}\sqrt{3}} = \dfrac{3\sqrt{3}}{5 \times 3} = \dfrac{\sqrt{3}}{5}$

c) To rationalise expressions of this type we make use of the fact that $(x-y)(x+y)=x^2-y^2$

$$\frac{1}{\sqrt{3}-1}=\frac{1}{(\sqrt{3}-1)}\times\frac{(\sqrt{3}+1)}{(\sqrt{3}+1)}$$ (multiplying both numerator and denominator by $(\sqrt{3}+1)$)

$$=\frac{\sqrt{3}+1}{(\sqrt{3})^2-1}=\frac{\sqrt{3}+1}{3-1}=\frac{\sqrt{3}+1}{2}$$

d) $$\frac{\sqrt{3}}{3\sqrt{3}+2\sqrt{2}}=\frac{\sqrt{3}(3\sqrt{3}-2\sqrt{2})}{(3\sqrt{3}+2\sqrt{2})(3\sqrt{3}-2\sqrt{2})}$$

(multiplying both numerator and denominator by $(3\sqrt{3}-2\sqrt{2})$)

$$=\frac{\sqrt{3}(3\sqrt{3}-2\sqrt{2})}{(3\sqrt{3})^2-(2\sqrt{2})^2}$$

$$=\frac{\sqrt{3}(3\sqrt{3}-2\sqrt{2})}{27-8}$$

$$=\frac{3\sqrt{9}-2\sqrt{6}}{19}=\frac{9-2\sqrt{6}}{19}$$

e) Rationalising each fraction as a separate item

$$\frac{\sqrt{5}}{\sqrt{5}-1}+\frac{\sqrt{5}}{\sqrt{5}+1}=\frac{\sqrt{5}(\sqrt{5}+1)}{(\sqrt{5}-1)(\sqrt{5}+1)}+\frac{\sqrt{5}(\sqrt{5}-1)}{(\sqrt{5}+1)(\sqrt{5}-1)}$$

$$=\frac{5+\sqrt{5}}{4}+\frac{5-\sqrt{5}}{4}$$

$$=\frac{5+\sqrt{5}+5-\sqrt{5}}{4}=\frac{10}{4}=2\tfrac{1}{2}$$

Example 3

Simplify

a) $(\sqrt{x}+1)(\sqrt{x}-1)$ b) $\sqrt{(1-x)}+\dfrac{x}{\sqrt{(1-x)}}$

Rationalise

c) $\dfrac{2}{\sqrt{x}-1}+\dfrac{2}{\sqrt{x}+1}$

Answers

a) $(\sqrt{x}+1)(\sqrt{x}-1)=(\sqrt{x})^2-1^2=x-1$

b) Bringing to a common denominator

$$\sqrt{(1-x)}+\frac{x}{\sqrt{(1-x)}}=\frac{\{\sqrt{(1-x)}\}^2+x}{\sqrt{(1-x)}}$$

$$=\frac{1-x+x}{\sqrt{(1-x)}}$$

$$=\frac{1}{\sqrt{(1-x)}}\left(\text{or }\frac{\sqrt{(1-x)}}{1-x}\text{ in rational form}\right)$$

c) Rationalising each fraction separately

$$\frac{2}{\sqrt{x}-1}+\frac{2}{\sqrt{x}+1}=\frac{2\{\sqrt{x}+1\}}{(\sqrt{x}-1)(\sqrt{x}+1)}+\frac{2\{\sqrt{x}-1\}}{(\sqrt{x}+1)(\sqrt{x}-1)}$$

$$=\frac{2\sqrt{x}+2}{x-1}+\frac{2\sqrt{x}-2}{x-1}$$

$$=\frac{4\sqrt{x}}{x-1}$$

Exercise B

1 Simplify the following surds, and surd expressions

a) $\sqrt{28}$
b) $\sqrt{75}$
c) $\sqrt{48}$
d) $\sqrt{6}\times\sqrt{8}$
e) $\sqrt{3}\times\sqrt{24}$
f) $\sqrt{112}\div\sqrt{28}$
g) $\sqrt{120}\div\sqrt{24}$
h) $5\sqrt{3}+\sqrt{48}$
i) $\sqrt{32}-2\sqrt{8}$
j) $\sqrt{72}-\sqrt{8}-\sqrt{98}+\sqrt{50}$
k) $\sqrt{3}(3\sqrt{3}-\sqrt{12})$
l) $(\sqrt{2}-1)^2$
m) $(3\sqrt{2}+1)(2\sqrt{2}-1)$

2 Express the following surd numbers with rational denominators.

a) $\frac{3}{\sqrt{2}}$
b) $\frac{8}{\sqrt{12}}$
c) $\frac{9}{\sqrt{3}}$
d) $\frac{\sqrt{40}}{2\sqrt{5}}$
e) $\frac{3}{\sqrt{2}-1}$
f) $\frac{1}{2+\sqrt{3}}$
g) $\frac{2}{3-\sqrt{7}}$
h) $\frac{2+\sqrt{3}}{2-\sqrt{3}}$
i) $\frac{\sqrt{6}-\sqrt{2}}{\sqrt{6}+\sqrt{2}}$
j) $\frac{2}{\sqrt{3}+\sqrt{2}}+\frac{2}{\sqrt{3}-\sqrt{2}}$

3 Simplify the following algebraic expressions, leaving denominators in rational form.

a) $\frac{\sqrt{x}}{\sqrt{x}+1}$
b) $\frac{(\sqrt{x}+2)(\sqrt{x}-1)}{\sqrt{x}}$
c) $\frac{1}{x+\sqrt{(x^2+1)}}$
d) $\frac{x-y}{\sqrt{x}+\sqrt{y}}$
e) $\left(\sqrt{x}+\frac{1}{\sqrt{x}}\right)\left(\sqrt{x}-1-\frac{1}{\sqrt{x}}\right)$
f) $\frac{1}{1+\sqrt{x}}+\frac{1}{1-\sqrt{x}}+\frac{1}{1-x}$

2.3 Equations involving square roots (surd equations)

The general method of solving such equations is to overcome the difficulty of the surd by isolating it on one side of the equation and then squaring throughout. Unfortunately it is likely that by this approach extra solutions are introduced from a second equation involving a negative sign i.e. $(\sqrt{a})^2$ and $(-\sqrt{a})^2$ both give the same result a. Solutions of surd equations should, therefore, be rechecked in the original equation before being offered as final answers.

Example 1

Solve $\sqrt{(x+13)} - x = 1$, checking that the solution(s) satisfy the equation.

The equation may be rearranged as

$$\sqrt{(x+13)} = (1+x)$$

So squaring: $\quad x + 13 = (1+x)^2$

(**Note** that $-\sqrt{(x+13)} = (1+x)$ when squared would result in the same form.)

$$\begin{aligned}
\text{So} \quad x + 13 &= x^2 + 2x + 1 \\
\Rightarrow \quad x^2 + x - 12 &= 0 \\
\Rightarrow \quad (x+4)(x-3) &= 0 \\
\Rightarrow \quad x &= -4 \text{ or } 3
\end{aligned}$$

When $x = -4$ L.H.S. $= \sqrt{9} + 4 = 7 \neq 1$ (extraneous root)
$x = 3$ L.H.S. $= \sqrt{16} - 3 = 4 - 3 = 1$ (true)
Hence $x = 3$ is the required root.

Example 2

Solve $\sqrt{(x+6)} + \sqrt{(x-1)} = 7$ checking the solution.
Squaring throughout $\{\sqrt{(x+6)} + \sqrt{(x-1)}\}^2 = 49$

$$\begin{aligned}
\text{i.e.} \quad x + 6 + 2\sqrt{(x+6)}\sqrt{(x-1)} + x - 1 &= 49 \\
\text{i.e.} \quad 2\sqrt{\{(x+6)(x-1)\}} + 2x + 5 &= 49 \\
\Rightarrow \quad 2\sqrt{\{(x+6)(x-1)\}} &= 49 - 2x - 5 \\
\Rightarrow \quad 2\sqrt{\{(x+6)(x-1)\}} &= 44 - 2x \\
\Rightarrow \quad \sqrt{\{(x+6)(x-1)\}} &= 22 - x \text{ (dividing throughout by 2)}
\end{aligned}$$

$$\begin{aligned}
\text{Squaring} \quad (x+6)(x-1) &= (22-x)^2 \\
x^2 + 5x - 6 &= 484 - 44x + x^2 \\
\Rightarrow \quad 49x &= 490 \\
\Rightarrow \quad x &= 10
\end{aligned}$$

$$\begin{aligned}
\text{(Checking} \quad \text{L.H.S.} &= \sqrt{(10+6)} + \sqrt{(10-1)} \\
&= \sqrt{16} + \sqrt{9} \\
&= 4 + 3 = 7, \text{ true.)}
\end{aligned}$$

Exercise C

Solve the following equations checking your solutions in each case.

1 $\sqrt{(x-1)} = 3$
2 $\sqrt{(2x+5)} - 1 = x$
3 $\sqrt{(3x-5)} + 3 = x$
4 $\sqrt{(2x+7)} - \sqrt{x} = 2$
5 $\sqrt{(x+2)} - \sqrt{(x-3)} = 1$
6 $\sqrt{(2x+1)} - \sqrt{x} = \sqrt{(x-3)}$

2.4 Logarithms

Logarithms, were invented by John Napier (*c* 1614) and developed by Professor Henry Briggs (who devised the decimal form, or common logarithms, used to this day as a simplifying aid in computational exercises). The term 'logarithm' is an alternative word for an **index** or **power of a given positive number base**.

For example, since $2^4 = 16$, we define the index 4 to be the logarithm of 16 to the base 2 and write

$$4 = \log_2 16.$$

Likewise, since $5^2 = 25$, we may write $2 = \log_5 25$. Furthermore, using the rules of negative indices as in the previous section,

$(\frac{1}{2})^{-2} = 4$ and we may write $\log_{1/2} 4 = -2$

As a general rule, given that a **base a raised to the index x yields a result y**, we define x to be $\mathbf{\log_a y}$.
Thus $a^x = y \Leftrightarrow x = \log_a y$

Example
Evaluate
a) $\log_4 256$ b) $\log_{0\cdot25} 8$ c) $\log_{\sqrt{3}} \{27\sqrt{3}\}$.

a) Let $x = \log_4 256.$
Then $4^x = 256 = 4^4$
Comparing indices, $x = 4$

b) Let $x = \log_{0\cdot25} 8$

Then $(0\cdot25)^x = 8$

i.e. $\left(\frac{1}{4}\right)^x = 8$

$(2^{-2})^x = 2^3$

$2^{-2x} = 2^3$

So comparing indices

$-2x = 3 \Rightarrow x = -1\cdot5$

c) Let $x = \log_{\sqrt{3}} \{27\sqrt{3}\}$
$(\sqrt{3})^x = 27\sqrt{3}$
$(3^{1/2})^x = 3^3 \,.\, 3^{1/2}$
$3^{x/2} = 3^{3\frac{1}{2}}$

$\Rightarrow \quad \frac{x}{2} = 3\frac{1}{2} \Rightarrow x = 7$

Exercise D

1 Express in logarithmic form

a) $3^3 = 27$ f) $5^{-1} = 0\cdot2$

b) $2^5 = 32$ g) $2^{-2} = \frac{1}{4}$

c) $6^3 = 216$ h) $\left(\frac{2}{3}\right)^{-2} = 2\cdot25$

d) $8^{2/3} = 4$ i) $4^0 = 1$

e) $\left(\frac{1}{3}\right)^2 = \frac{1}{9}$ j) $4^{-1/2} = 0\cdot5$

2 Evaluate

a) $\log_5 125$ f) $\log_{0\cdot 1} 100$

b) $\log_4 4$ g) $\log_{1/2} 4$

c) $\log_{10} 0{\cdot}001$ h) $\log_{2\sqrt{3}} 1728$

d) $\log_2 \sqrt{2}$ i) $\log_6 \dfrac{1}{216}$

e) $\log_{121} 11$ j) $\log_{27} \dfrac{1}{81}$

2.5 Rules of logarithms

1 Addition of logarithms

Let $\log_a x \equiv m, \quad \log_a y \equiv n$

Then $x \equiv a^m$ and $y \equiv a^n$

So $xy \equiv a^m \times a^n \equiv a^{m+n}$

Thus $\log_a (xy) \equiv m + n$ from the basic definition. So:

$$\log_a (xy) \equiv \log_a x + \log_a y$$

2 Subtraction of logarithms

Similarly

$$\frac{x}{y} \equiv \frac{a^m}{a^n} \equiv a^{m-n}$$

Hence $\log_a \left(\dfrac{x}{y}\right) \equiv m - n$ or

$$\log_a \left(\frac{x}{y}\right) \equiv \log_a x - \log_a y$$

3 Logarithms of powers of numbers

From $\log_a x = m \Leftrightarrow x = a^m$

$x^p \equiv (a^m)^p = a^{mp}$

Hence $\log_a (x^p) \equiv mp$ or

$$\log_a (x^p) \equiv p \log_a x$$

4 Change of base of a logarithm

Again from $\log_a x = m \Leftrightarrow x = a^m$, consider the effect of taking logarithms to a different base b of the form $x = a^m$

i.e. $\log_b x \equiv \log_b (a^m)$
$\equiv m \log_b a$ from Rule 3

$$\Rightarrow \log_b x \equiv \log_a x \times \log_b a$$

i.e. to change the base of the logarithm of x from a to b, we multiply by $\log_b a$.
In this result if we replace x by a,

$$\log_b a \equiv \log_a a \times \log_b a$$

or dividing by $\log_b a$, given that this is non-zero,

$$1 \equiv \log_a a$$

And if $x = b$,

$$\log_b b \equiv 1 \equiv \log_a b \times \log_b a$$

We see that by using these rules, we can simplify logarithmic expressions in a similar way to the simplification of indices by the use of index rules, as in Example 2 of 2.1, page 33.

Example 1

Simplify

1 a) $2 \log_a 3 + 3 \log_a 2$
b) $\log_{10} \frac{75}{16} - 2 \log_{10} \frac{5}{9} + \log_{10} \frac{32}{243}$
c) $2 \log_d a + \log_d b - 3 \log_d c$

Answers

a) $2 \log_a 3 + 3 \log_a 2 = \log_a 3^2 + \log_a 2^3$ (applying Rule 3)
$= \log_a 9 + \log_a 8$
$= \log_a 72$ (applying Rule 1)

b) $\log_{10} \dfrac{75}{16} - 2 \log_{10} \dfrac{5}{9} + \log_{10} \dfrac{32}{243}$

$= \log_{10} \dfrac{75}{16} - \log_{10} \dfrac{5^2}{9^2} + \log_{10} \dfrac{32}{243}$ (applying Rule 1)

$= \log_{10} \left(\dfrac{75}{16} \times \dfrac{9^2}{5^2} \times \dfrac{32}{243}\right)$ (applying Rules 1 and 2)

$= \log_{10} 2$ (cancelling numerical values)

c) $2 \log_d a + \log_d b - 3 \log_d c = \log_d \left(\dfrac{a^2 b}{c^3}\right)$ (applying Rule 3, then rules 1 and 2 in one step).

Example 2

Using the rules of logarithms only, write down the value of
$N = (\log_5 49) \times (\log_7 125)$

$N = (\log_5 7^2) \times (\log_7 5^3) = 2(\log_5 7) \times 3(\log_7 5)$
$= 6(\log_5 7)(\log_7 5) = 6 \log_7 7$ by Rule 4
$= 6$ since $\log_7 7 = 1$

Exercise E

1 Using the rules of logarithms, simplify

a) $\log_{10} 7 + 2 \log_{10} 2 - \log_{10} 280$

b) $2 \log_x 5 - 2 \log_x 15 + 3 \log_x 3 - \log_x 6$

c) $\dfrac{3 \log_{10} 2 - \log_{10} 4}{\log_{10} 4 - \log_{10} 2}$

d) $2 \log_{10} (\frac{5}{3}) - \log_{10} (\frac{5}{18})$

2 Evaluate

a) $(\log_x y^3) \times (\log_y x^4)$

b) $(\log_2 x) \div (\log_8 x)$

c) uvw, given that $u = \log_v w$, $v = \log_w u$, $w = \log_u v$

2.6 Common logarithms

It is extremely likely that you will have used already the system of common logarithms as an aid to easing decimal calculations. The system uses of course a base 10, so that, as $10^2 = 100$, we can write $\log_{10} 100 = 2$ in the common logarithm notation.

Furthermore as
$$\begin{aligned}
10^{-2} &= 0{\cdot}01 &&\Rightarrow \log_{10} 0{\cdot}01 &&= -2\\
10^{-1} &= 0{\cdot}1 &&\Rightarrow \log_{10} 0{\cdot}1 &&= -1\\
10^{0} &= 1 &&\Rightarrow \log_{10} 1 &&= 0\\
10^{1} &= 10 &&\Rightarrow \log_{10} 10 &&= 1\\
10^{3} &= 1000 &&\Rightarrow \log_{10} 1000 &&= 3\\
10^{4} &= 10000 &&\Rightarrow \log_{10} 10000 &&= 4
\end{aligned}$$

and so on.

Now as any denary number can be expressed in standard (index) form (i.e. $a \times 10^n$ where $1 \leqslant a < 10$), the common logarithm of such a number is

$$\begin{aligned}
&\log_{10} (a \times 10^n)\\
&= \log_{10} a + n \log_{10} 10\\
&= n + \log_{10} a \quad (\text{since } \log_{10} 10 = 1)
\end{aligned}$$

i.e. the logarithm consists of the sum of the relevant power of 10, together with a (decimal) value representing the logarithm of some number between 1 and 10. The part n of this result is called the **characteristic**, and the part $\log_{10} a$ is called the **mantissa** of the logarithm of the given number.

The mantissa is obtained from tables and the characteristic must be evaluated separately. (**Note** that electronic calculators give both parts together in the display.)

Thus for example:

$$343{\cdot}5 = 3{\cdot}435 \times 10^2 \text{ and } \log_{10} 343{\cdot}5 = 0{\cdot}5359 + 2 = 2{\cdot}5359$$
$$\text{and } 0{\cdot}003435 = 3{\cdot}435 \times 10^{-3} \Rightarrow \log_{10} 0{\cdot}003435 = 0{\cdot}5359 - 3 = -2{\cdot}4641$$

In this second illustration it is generally more convenient to write, and use, the result in the form $\bar{3}{\cdot}5359$ where the $\bar{3}$, read as **bar three**, means -3.

Conversely when we have the common logarithm of a given number, to find the number to which this corresponds we would use the reverse process, generally employing anti-logarithm tables to convert the $\log_{10} a$ back to the part a of the standard form. (On a calculator, we use either the '10^x' or the 'INV LOG' button.)

For example, if
$$\begin{aligned} \log_{10} x &= 3{\cdot}2678 \\ &= 3 + 0{\cdot}2678 \\ x &= (\text{antilog } 0{\cdot}2678) \times 10^3 \\ &= 1{\cdot}852 \times 10^3 \\ &= 1852 \end{aligned}$$

And when
$$\begin{aligned} \log_{10} x &= \bar{2}{\cdot}8762 \\ &= (-2) + 0{\cdot}8762 \\ &= (\text{antilog } 0{\cdot}8762) \times 10^{-2} \\ &= 7{\cdot}519 \times 10^{-2} \\ &= 0{\cdot}07519 \end{aligned}$$

Once the principles of converting numbers to their common logarithms and converting common logarithms back to their corresponding forms have been mastered, we use the first three rules of logarithms in the form

Task		Operation
$\times$	$\Rightarrow$	$+$
$\div$	$\Rightarrow$	$-$
$(\ )^n$	$\Rightarrow$	$\times$ by n
$\sqrt[n]{(\ )}$	$\Rightarrow$	$\div$ by n

and the work is greatly eased. Developing the sequence of operations in the form of a chain of events also systematises the overall task.

It is common practice that the notation log x, or simply lg x, is used for common logarithms, so from now on the base will not be included in such cases. However, whenever logarithms of alternative bases are being used we shall abide by the basic convention.

Example 1

Evaluate $\left(\dfrac{15{\cdot}15 \times 212{\cdot}0}{1039}\right)^3$

The steps of the chain in the sequence of operations is as shown by the tinted areas.

log 15·15 [+] log 212·0 [−] log 1039 = [ ] [× by 3] = Answer (Antilog →)

and would be suitably written

No	Log	
15·15	1·1804	
212·0	2·3263	+
	3·5067	
1039	3·0165	−
	0·4902	
	3	×
Ans = 29·54 ←	1·4706	

Example 2

Evaluate $\dfrac{278{\cdot}4 \times (0{\cdot}0825)^2}{\sqrt[4]{0{\cdot}3644}}$

The chain is log 278·4 [+] [2] log 0·0825 [−] [$\frac{1}{4}$] log 0·3644 $\xrightarrow[\text{Antilog}]{}$ = Answer

No	Log	
0·0825	$\bar{2}{\cdot}9165$	
	2	×
$(0{\cdot}0825)^2$	$\bar{3}{\cdot}8330$	
278·4	2·4446	+
	0·2776	
$\sqrt[4]{0{\cdot}3644}$	$\bar{1}{\cdot}8904$	−
Ans = 2·439 ←	0·3872	

No	Log
0·3644	$\div 4)\ \bar{1}{\cdot}5616 = 4)\ -0{\cdot}4384$
←	$\bar{1}{\cdot}8904 \qquad -0{\cdot}1096$

(An alternative method for $\sqrt[4]{0{\cdot}3644}$ is to write

$$4)\ \bar{1}{\cdot}5616 = 4)\ \bar{4} + 3{\cdot}5616 = \bar{1}{\cdot}8904)$$

Example 3

Evaluate $\log_4 5$

If $\quad x = \log_4 5$

$\quad 4^x = 5$

Taking logs to base 10

$$\log_{10} 4^x = \log_{10} 5$$

$$\Rightarrow x \log_{10} 4 = \log_{10} 5$$

or $\qquad x = \dfrac{\log 5}{\log 4}$ (dropping bases)

$$= \frac{0{\cdot}6990}{0{\cdot}6021}$$

No	Log
0·6990	$\bar{1}{\cdot}8445$
0·602(1)	$-\bar{1}{\cdot}7797$
1·161 ←	0·0648

Hence $\log_4 5 = 1{\cdot}161$

Exercise F

1 Work out the values of

a) $\dfrac{309{\cdot}2 \times 86{\cdot}82}{2635}$

b) $\left(\dfrac{292{\cdot}4}{83{\cdot}44}\right)^2$

c) $\sqrt[3]{\{202{\cdot}6 \times 18{\cdot}42\}}$

d) $16{\cdot}42 \times 2{\cdot}58 \times \sqrt{18{\cdot}66}$

e) $\sqrt[3]{\left\{\dfrac{9{\cdot}68 \times 14{\cdot}66}{65{\cdot}62}\right\}}$

f) $\dfrac{67{\cdot}82 \times 318{\cdot}4}{(54{\cdot}92)^2}$

g) $0{\cdot}3845 \times (0{\cdot}8921)^2$

h) $\sqrt{(3{\cdot}462 \times 0{\cdot}0581 \times 0{\cdot}9212)}$

i) $\dfrac{1{\cdot}592^2 + 1}{1{\cdot}592^2 - 1}$

j) $\sqrt{\left\{(2{\cdot}921)^2 + \dfrac{38{\cdot}96}{29{\cdot}92}\right\}}$

2 Evaluate

a) $\log_2 7$
b) $\log_5 10$
c) $\log_4 17$
d) $\log_{0\cdot3} 8$
e) $\log_{1/4} 13$

2.7 Solution of exponential equations in a single unknown

These equations are typified by the fact that the unknown in the equation occurs as an index.

Example 1

Solve the equation $3^x = 7$

Taking logs to base 10 (**Note** any base could be used, but 10 is convenient.)

$$\log (3^x) = \log 7$$
$$\Rightarrow x \log 3 = \log 7$$
$$\Rightarrow \quad x = \frac{\lg 7}{\lg 3} = \frac{0{\cdot}8451}{0{\cdot}4771}$$

Hence $x = 1{\cdot}771$

Example 2

Solve the equation $3^{x+1} = 4^{2x-1}$

Again by taking logs to base 10

$$\log (3^{x+1}) = \log (4^{2x-1})$$
$$\Rightarrow (x + 1) \log 3 = (2x - 1) \log 4$$

From which

$$\log 3 + \log 4 = x (2 \log 4 - \log 3)$$

i.e. $\log 12 = x \log \frac{16}{3}$ (applying the basic rules)

$$\Rightarrow x = \frac{\lg 12}{\lg 5{\cdot}33} = \frac{1{\cdot}0792}{0{\cdot}7270}$$

And $x = 1{\cdot}484$

Other equations of this type are of a polynomial form and ultimately the solution depends upon the methods of the last two examples.

Example 3

Solve $2^{2x} - 5(2^x) + 6 = 0$

If we write y for 2^x in this equation, since $2^{2x} = (2^x)^2 = y^2$ the equation becomes

$$y^2 - 5y + 6 = 0$$

i.e. $(y - 2)(y - 3) = 0$

So $y = 2^x = 2$ or $2^x = 3$.

Clearly $2^x = 2^1$ and comparing indices, $x = 1$

Taking logs to base 10 of $2^x = 3$

$$x \log 2 = \log 3$$

and $x = \dfrac{\lg 3}{\lg 2} = \dfrac{0{\cdot}4771}{0{\cdot}3010}$

giving $x = 1{\cdot}585$

and the equation then has the two solutions 1 and 1·585. Clearly, dependent upon the nature of the coefficients of such 'quadratic indicial' equations we can expect two, one (i.e. two coincident), or no real solutions in any range of examples.

Example 4

Solve $10^{3x} - 5(10^{2x}) + 8(10^x) - 4 = 0$

Writing $y = 10^x$, $y^2 = 10^{2x}$ and $y^3 = 10^{3x}$, the equation becomes

$$f(y) \equiv y^3 - 5y^2 + 8y - 4 = 0$$

Since $f(1) = 1 - 5 + 8 - 4 = 0$,

$y - 1$ is a factor of this cubic polynomial.

Hence $y^3 - 5y^2 + 8y - 4 \equiv (y - 1)(Ay^2 + By + C)$

and comparing coefficients:

of y^3, $1 = A$

of y^2, $-5 = -A + B \Rightarrow B = -4$

of y, $8 = -B + C \Rightarrow C = 4$.

So $y^3 - 5y^2 + 8y - 4 \equiv (y - 1)(y^2 - 4y + 4)$

$\equiv (y - 1)(y - 2)^2$.

Hence, $y = 10^x = 1$ or $10^x = 2$ (twice)

Now we can write $10^0 = 1$

$\Rightarrow 10^x = 10^0 \Rightarrow x = 0$

From $10^x = 2$ taking logs to base 10

$$x = \frac{\lg 2}{\lg 10} = \lg 2 \text{ since } \log_{10} 10 = 1$$

$$x = 0{\cdot}301$$

The solutions of the equation are therefore 0 and 0·301 (twice).

Exercise G

Solve the following exponential (indicial) equations for the unknown value x.

1 $4^x = 6$
2 $5^{2x+1} = 3$
3 $2(7^x) = 95$
4 $2^{2x} - 4(2^x) + 3 = 0$
5 $3^{2x} - 2(3^{x+1}) + 5 = 0$
6 $2(10^{2x}) - 5(10^x) + 2 = 0$
7 $3^{3x} - 6(3^{2x}) + 11(3^x) - 6 = 0$

2.8 Logarithmic equations involving one or more unknown values

Generally, appropriate use of one or more of the fundamental logarithmic rules will convert such equations into a standard algebraic form, from which the solution(s) may be derived.

Example 1

Solve the equation $\log_2 x + \log_x 2 = 2$

The difficulty in this example is the use of different bases in the terms on the L.H.S. of the equation.

Using Rule 4 for the change of base of a logarithm

$$\log_2 2 \equiv \log_x 2 \times \log_2 x$$

And since

$$\log_2 2 = 1$$

$$\log_x 2 = \frac{1}{\log_2 x}$$

the equation becomes

$$\log_2 x + \frac{1}{\log_2 x} = 2$$

i.e. $(\log_2 x)^2 - 2\log_2 x + 1 = 0$
or $(\log_2 x - 1)^2 = 0$

Then $\log_2 x = 1$ (twice)
and $x = 2^1 = 2$ using the basic definition.

Example 2

Determine the positive number x which satisfies the equation

$$\log_3 x = \log_9 (x + 6)$$

Again this is mainly a problem of mixed bases.

Using Rule 4

$$\begin{aligned}\log_3 x &\equiv \log_9 x \times \log_3 9\\ &\equiv \log_9 x \times \log_3 (3^2)\\ &\equiv 2\log_9 x \times \log_3 3\\ &\equiv 2\log_9 x\\ &\equiv \log_9 (x^2)\end{aligned}$$

The equation then becomes

$$\log_9 (x^2) = \log_9 (x + 6)$$

or simply $x^2 = x + 6$

Hence $\quad x^2 - x - 6 = 0$

$\Rightarrow (x - 3)(x + 2) = 0$

And $\quad x = 3 \quad$ (x cannot $= -2$. Why?)

Example 3

Solve the simultaneous equations for positive values of x and y,

$$\left.\begin{aligned} xy &= 16 \\ \log_x y &= 3 \end{aligned}\right\}$$

Since $\log_x y = 3 \Rightarrow y = x^3$

So replacing y by this value in the first equation

$$x \times x^3 = 16$$
$$x^4 = 16$$

and $\quad x = 2$, as x is positive.

Hence $\quad y = x^3 = 8$

and the solution is $x = 2$, $y = 8$

Exercise H

1 Solve the equations for the unknown value x.
a) $4 \log_x 2 - \log_2 x - 3 = 0$
b) $\log_2 x = \log_4 (8x - 16)$
c) $\log_2 2x - 2 \log_8 x = 4$
d) $\log_4 (3x + 4) = \log_2 x$ and hence find the value of $\log_4 x$.

2 Solve the simultaneous equations for positive values of the unknowns x and y.

a) $\left.\begin{aligned} xy &= 4 \\ \log x + 2 \log y &= 1 \end{aligned}\right\}$ b) $\left.\begin{aligned} x - y &= 21 \\ \log x + \log y &= 2 \end{aligned}\right\}$ c) $\left.\begin{aligned} x + y &= 10 \\ \log_2 x &= \log_8 y \end{aligned}\right\}$

2.9 Binomial expressions. Pascal's triangle

In Chapter 1 we considered two special binomial forms of the general quadratic equation. We are now concerned with the process of multiplying out (or expanding) powers of a general binomial expression $(x + y)^n$.

We see that starting with $(x + y)^0$ a succession of results can be built up by multiplying the terms of each previous result in turn by $x + y$.

Thus

$$(x + y)^0 \equiv 1$$
$$(x + y)^1 \equiv 1(x + y) \equiv x + y$$
$$(x + y)^2 \equiv (x + y)(x + y) \equiv x^2 + 2xy + y^2$$
$$(x + y)^3 \equiv (x + y)(x^2 + 2xy + y^2) \equiv x^3 + 3x^2y + 3xy^2 + y^3$$
$$(x + y)^4 \equiv (x + y)(x^2 + 3x^2y + 3xy^2 + y^3)$$
$$\equiv x^4 + 4x^3y + 6x^2y^2 + 4xy^3 + y^4$$

and so on.

A process of developing these expressions without the tedium of extensive multiplication was devised by Blaise Pascal (1623 – 1662) and is referred to as Pascal's triangle. In this process we detach the coefficients from the terms of the expansion and write them in a triangular pattern.

Thus for $n = 0, 1, 2, 3, 4, 5$ the pattern is

$n = 0$: 1
$n = 1$: 1 1
$n = 2$: 1 2 1
$n = 3$: 1 3 3 1
$n = 4$: 1 4 6 4 1
$n = 5$: 1 5 10 10 5
. . . .

The relationship between the coefficients on any row and one on a subsequent row can be seen to be one of simple addition and can be used to develop the expansion as far as we wish to go with it. Thus, adding terms of one row, we obtain the next row, as shown:

$n = 2$, 1 2 1

$n = 3$, 1 3 3 1

$n = 4$, 1 4 6 4 1

From the row $n = 5$, 1 5 10 10 5 1 we would then get

$n = 6$: 1 6 15 20 15 6 1 and
$n = 7$: 1 7 21 35 35 21 7 1
$n = 8$: 1 8 28 56 70 56 28 8 1
$n = 9$: 1 9 36 84 126 126 84 36 9 1

etc.

The general form of this result was established by Sir Isaac Newton (1642 – 1727) and is left for consideration until chapter 15 of this book.

Example 1

Expand the following binomial expressions

a) $(1 + 2x)^3$ b) $(2x - y)^4$ c) $\left(2x + \frac{1}{x}\right)^5$

Answers

a) Since the coefficients of the terms in the expansion are 1 3 3 1,

$$(1 + 2x)^3 \equiv 1 + 3(2x) + 3(2x)^2 + (2x)^3$$

i.e. $$(1 + 2x)^3 \equiv 1 + 6x + 12x^2 + 8x^3$$

b) The coefficients in this expansion are 1 4 6 4 1

$$(2x - y)^4 \equiv 1(2x)^4 + 4(2x)^3(-y) + 6(2x)^2(-y)^2 + 4(2x)(-y)^3 + 1(-y)^4$$
$$\equiv 16x^4 - 32x^3y + 24x^2y^2 - 8xy^3 + y^4$$

c) The coefficients are 1 5 10 10 5 1

Hence $\left(2x+\frac{1}{x}\right)^5 \equiv 1(2x)^5 + 5(2x)^4\left(\frac{1}{x}\right) + 10(2x)^3\left(\frac{1}{x}\right)^2$

$$+ 10(2x)^2\left(\frac{1}{x}\right)^3 + 5(2x)\left(\frac{1}{x}\right)^4 + 1\left(\frac{1}{x}\right)^5$$

$$\equiv 32x^5 + 80x^3 + 80x + \frac{40}{x} + \frac{10}{x^3} + \frac{1}{x^5}$$

Example 2

Obtain the coefficients of x, x^2 and x^3 in the expansion of $(3-2x)(1-5x)^8$ in ascending powers of x.

To derive these coefficients we must first establish the pattern of the coefficients in $(x+y)^8$.

Starting with $(x+y)^4$ the form of Pascal's triangle is

$n=4$,	1	4	6	4	1	
$n=5$,	1	5	10	10	5	1
$n=6$,	1	6	15	20	15	...
$n=7$,	1	7	21	35	35	....
$n=8$,	1	8	28	56	70	.....

Hence $(1-5x)^8 = 1 + 8(-5x) + 28(-5x)^2 + 56(-5x)^3 + \cdots$
$= 1 - 40x + 700x^2 - 7000x^3 + \cdots$

So $(3-2x)(1-5x)^8 = (3-2x)(1 - 40x + 700x^2 - 7000x^3 + \cdots)$
$= 3 - 120x + 2100x^2 - 21\,000x^3 + \cdots$
$- 2x + 80x^2 - 1400x^3 + \cdots$

by multiplying the two expressions.

Hence $(3-2x)(1-5x)^8 = 3 - 122x + 2180x^2 - 22\,400x^3 + \cdots$

and the required coefficients are -122, 2180 and $-22\,400$.

Example 3

Find the coefficient of x in the expansion of

$$E \equiv \left(1+\frac{1}{x}\right)^2(2-3x)^6$$

Expanding both brackets

$$E \equiv \left(1+\frac{2}{x}+\frac{1}{x^2}\right)$$
$$\times\ (2^6 + 6\,.\,2^5(-3x) + 15\,.\,2^4(-3x)^2 + 20\,.\,2^3(-3x)^3 + \cdots)$$

We now 'pick out' the required coefficient by realising that the 'x' term is found from

1 (of the first bracket) $\times$ coefficient of x in the second

$+\frac{2}{x}$ (of the first bracket) $\times$ coefficient of x^2 in the second

$+\frac{1}{x^2}$ (of the first bracket) $\times$ coefficient of x^3 in the second

Thus required coefficient

$= 6 \,.\, 2^5(-3) + 2 \,.\, 15 \,.\, 2^4(-3)^2 + 20 \,.\, 2^3(-3)^3$
$= -576 + 4320 - 4320$
$= -576$

Example 4

Evaluate $(2\sqrt{2} + 1)^4 + (2\sqrt{2} - 1)^4$ using the binomial expansion.

$$(2\sqrt{2} + 1)^4 = (2\sqrt{2})^4 + 4(2\sqrt{2})^3 + 6(2\sqrt{2})^2 + 4(2\sqrt{2}) + 1$$
$$= 64 + 64\sqrt{2} + 48 + 8\sqrt{2} + 1$$

and $(2\sqrt{2} - 1)^4 = 64 - 64\sqrt{2} + 48 - 8\sqrt{2} + 1$
(since $(x - y)^4 = x^4 - 4x^3y + 4x^2y^2 - \ldots$)

Hence, adding

$$(2\sqrt{2} + 1)^4 + (2\sqrt{2} - 1)^4 = 128 + 96 + 2 = 226$$

Exercise I

1 Write down the expansions of

a) $(1 + 2x)^5$ d) $\left(2 - \dfrac{x}{2}\right)^4$

b) $(2x - 3)^4$ e) $\left(x + \dfrac{2}{x}\right)^4 + \left(2x - \dfrac{1}{x}\right)^4$

c) $\left(x + \dfrac{1}{x}\right)^6$

2 Calculate the value of the term not involving x in the expansion of

$$\left(\frac{2x}{3} - \frac{3}{2x}\right)^6$$

3 Find the coefficient of x^4 in the expansion of
a) $(1 + 2x)(1 + x)^5$ b) $(1 - x + 2x^2)(3 - 2x)^4$

4 Determine the values of a and b if

$$(2 + \sqrt{3})^5 \equiv a + b\sqrt{3}$$

5 By writing $(1 + x + x^2)$ in the form $\{1 + x(1 + x)\}$ determine the expansion of $(1 + x + x^2)^4$ in ascending powers of x.

6 Calculate, using the binomial expansion

$$(2\sqrt{3} - 1)^5 + 12(\sqrt{3} - 2)^4 - 6(3 - \sqrt{3})^3$$

CHAPTER THREE

Differentiation

3.1 Introduction

The concept of a function in mathematics is one which seems to vary according to the book being read. The variety of notations, too, seems to make the topic more confused. You may have met the idea first as a mapping, in which you would have used the notation

$$f : x \to 2x$$

This notation will be associated with the mapping diagram on two parallel number lines (see figure 3.1).

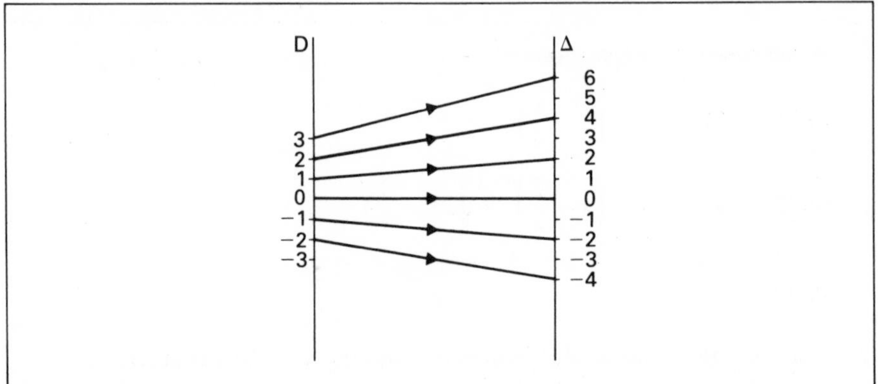

Figure 3.1

In this type of diagram (or graph), the left hand number line is called the domain (denoted by D) and the right hand number line is called the Codomain* (denoted by Δ).

On the other hand, you may have been introduced to the idea of a function using the notation

$$f(x) = 2x$$

This will probably have been associated with a diagram, or graph, using a pair of perpendicular lines as axes (Figure 3.2).

Here we have a representation of all the points which satisfy the functional relation $f(x) = 2x$. By comparison with the mapping diagram (Figure 3.1), we see that the x-axis corresponds to the domain and the $f(x)$-axis corresponds to the codomain.

You may be more familiar with the second diagram when the line is defined by the relation $y = 2x$, where the $f(x)$-axis is called, more simply, the y-axis.

* Some texts use Range for this set. Strictly, D maps **into** Δ and **onto** the Range i.e. $R \subseteq \Delta$

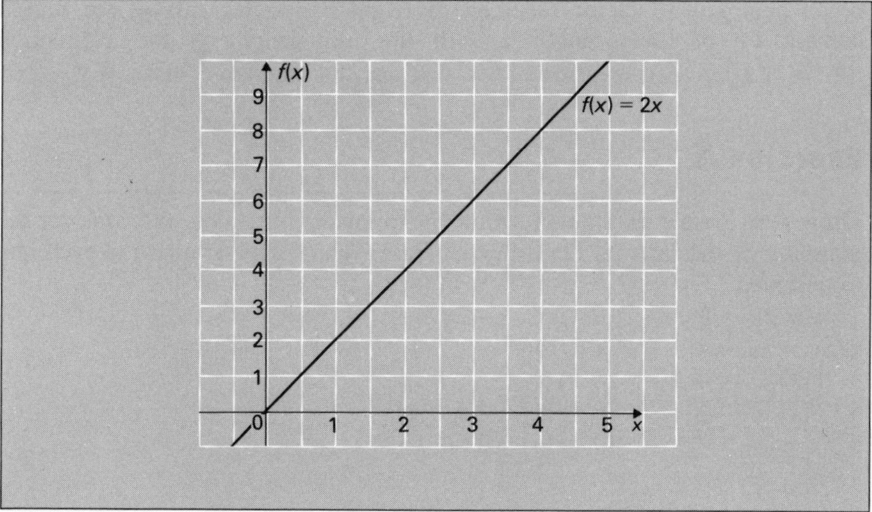

Figure 3.2

Comparative rates of growth

Whichever of the above approaches we may have met, we are drawing a diagram which shows how one variable varies with another. This may be the daily change of the F.T. index, or a hospital patient's daily temperature change, or a comparison of the amount of fuel used in a blast furnace related to the temperature achieved in the furnace. In each of these, we are comparing mathematically the way in which one variable changes in relation to another. Our graphs show us how they vary one with the other, but it is often necessary to know the rate at which one changes in relation to the other.

For example, in measuring speed, we compare the rate at which displacement changes in relation to time.

In the two diagrams of figure 3.3 we see that the interval [2, 3] of the domain or x-axis is mapped into the interval [4, 6] of the codomain or y-axis, i.e. an interval of length 1 unit of the domain is mapped into an interval of length 2 units of the codomain. Check for yourself that any interval maps into an interval of twice the length. On the mapping diagram, we could speak

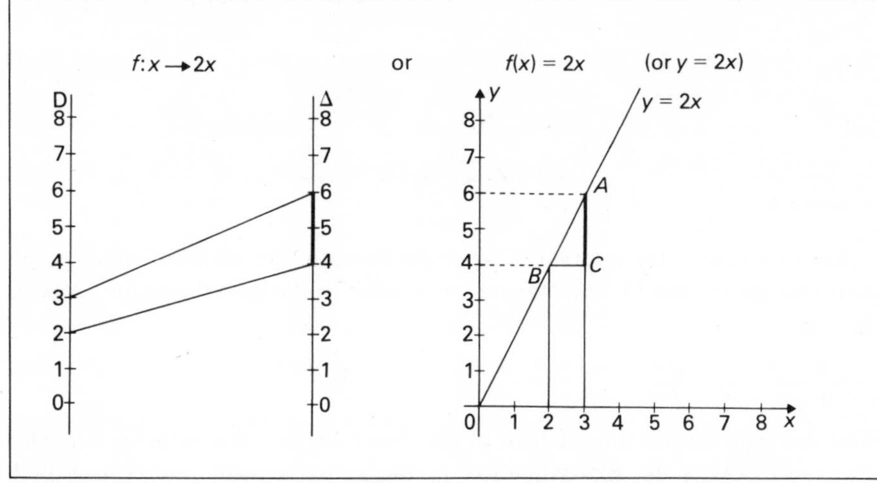

Figure 3.3

of an enlargement factor of 2; on the right hand diagram, we see that a comparison of the y-length 2 with the unit length in the x-direction (i.e. AC/CD) gives a measure of tan $A\hat{B}C$, or the gradient of the line $y = 2x$.

Exercise A

Draw appropriate diagrams to show the following functions, and consider the mapping of the interval [2, 3] in each case. What is the gradient of each line represented?

1 $y = 2x - 1$
2 $y = 3x - 4$
3 $f(x) = \frac{1}{2}x + 2$
4 $f(x) = 4$
5 $y = 2$
6 $f: x \to -2x + 1$
7 $f: x \to -x$
8 $y = x$
9 $f(x) = 3(x - 1)$
10 $f(x) = 3x + 1$

3.2 Gradients of curves

In each of these cases, we have compared the growth rate of y related to the comparable change in x values. For straight line graphs this has presented no difficulties. Consider now the graphs in figure 3.4.

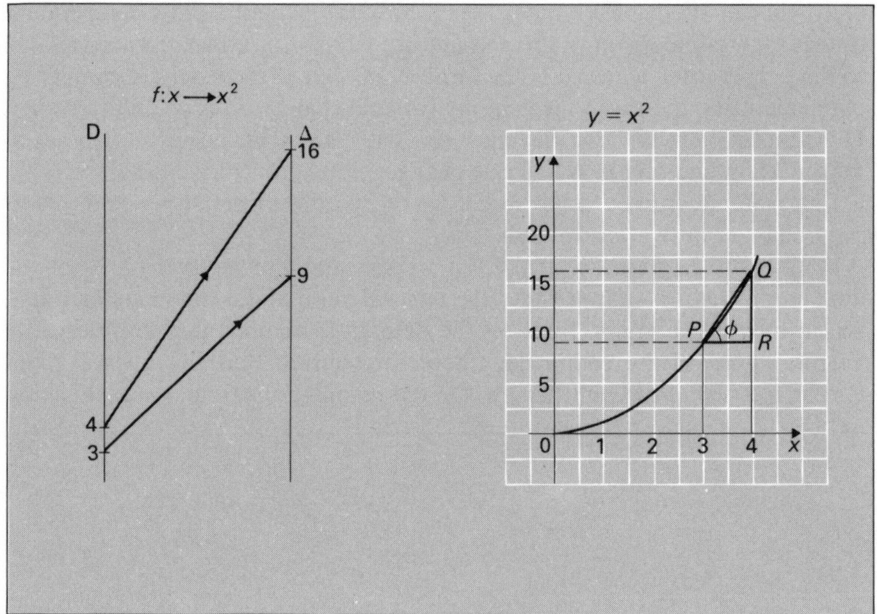

Figure 3.4

Let us consider the interval [3, 4] of the domain. We see from either graph that this maps onto [9, 16], and we could refer to the growth rate of y related to x as

$$\frac{16 - 9}{4 - 3} \quad \text{or} \quad \frac{QR}{PR}$$

and this tells us that the gradient of the chord PQ is 7, that is $\tan \phi = 7$. This however, is only an approximation to the actual growth rate. The growth occurs along the curve, but here we have measured the growth along the chord. We can improve the approximation by taking the point Q nearer to the point P. Here are the growth rates we obtain for smaller and smaller intervals of the domain.

$[3, 4] \rightarrow \quad [9, 16] \quad$ growth rate $\dfrac{16 - 9}{4 - 3} = 7$

$[3, 3{\cdot}5] \rightarrow \quad [9, 12{\cdot}25] \quad$ growth rate $\dfrac{12{\cdot}25 - 9}{3{\cdot}5 - 3} = 6{\cdot}5$

$[3, 3{\cdot}4] \rightarrow \quad [9, 11{\cdot}56] \quad$ growth rate $\dfrac{11{\cdot}56 - 9}{3{\cdot}4 - 3} = 6{\cdot}4$

$[3, 3{\cdot}1] \rightarrow \quad [9, 9{\cdot}61] \quad$ growth rate $\dfrac{9{\cdot}61 - 9}{3{\cdot}1 - 3} = 6{\cdot}1$

$[3, 3{\cdot}01] \rightarrow [9, 9{\cdot}0601]$ growth rate $\dfrac{9{\cdot}0601 - 9}{3{\cdot}01 - 3} = 6{\cdot}01$

It seems that the nearer we move the point Q to P, the nearer our growth rate comes to 6. Geometrically, as Q approaches P, the chord PQ approaches the tangent at P. (Use a ruler on your own graph, keeping it fixed at P but moving it as Q moves down the curve.)

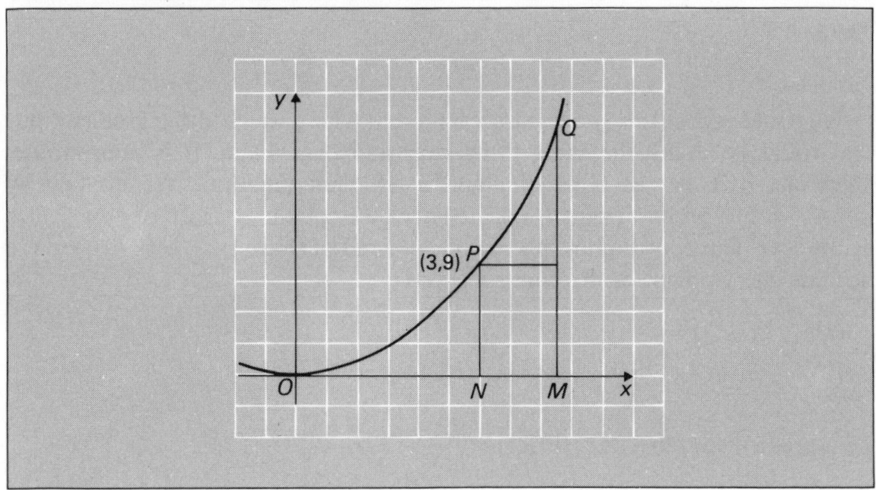

Figure 3.5

Let us look again at the growth rate algebraically to establish the result. Consider again the curve $y = x^2$ shown in figure 3.5. Suppose Q is just a little way along the curve from P, so that its x coordinate is a bit more than 3, say $(3 + h)$. Then the y coordinate of Q will be $(3 + h)^2$, because $y = x^2$. Show for yourself that QR is of length $(3 + h)^2 - 9$ and that PR is of length h. Then the gradient of chord PQ is

$$\frac{(3 + h)^2 - 9}{h} = \frac{9 + 6h + h^2 - 9}{h}$$

$$= \frac{6h + h^2}{h}$$

$$= 6 + h$$

When Q moves down the curve to approach P, then h decreases and approaches zero. We say that the gradient of the tangent at P is the limiting value of $(6 + h)$ as $h \to 0$, and we write this as $\lim_{h\to 0}(6 + h)$. Clearly this approaches 6, as we expected.

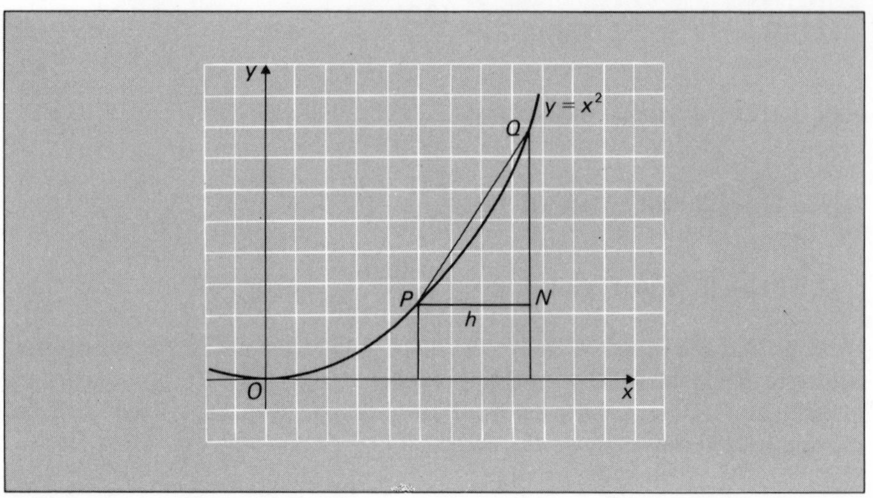

Figure 3.6

We could repeat this process for each point where we need the gradient, but we would be doing the same thing over and over again. It is appropriate, therefore, that we see if we can find a 'gradient function'. We do this by considering a general point P on the curve $y = x^2$ and a nearby point Q (see figure 3.6). The coordinates of P are (x, x^2) and of Q are $(x + h, (x + h)^2)$. We see that the gradient of the chord PQ is

$$\frac{QN}{PN} = \frac{(x + h)^2 - x^2}{h}$$

i.e. gradient of tangent at P is $\lim_{h\to 0}\left\{\frac{(x + h)^2 - x^2}{h}\right\}$

$$= \lim_{h\to 0}\left\{\frac{x^2 + 2xh + h^2 - x^2}{h}\right\}$$

$$= \lim_{h\to 0}\left\{\frac{2xh + h^2}{h}\right\}$$

$$= \lim_{h\to 0}\{2x + h\}$$

$$= 2x$$

so for a **general** point (x, x^2) on the curve $y = x^2$, the gradient is $2x$. We say that $2x$ is the **gradient function** for the function $f(x) = x^2$. We denote the gradient function by $f'(x)$.

So, if $f(x) = x^2$ then $f'(x) = 2x$

We can carry out a similar process for $f(x) = x^3$, and see that

$$f'(x) = \lim_{h \to 0} \left\{ \frac{(x+h)^3 - x^3}{h} \right\}$$

$$= \lim_{h \to 0} \left\{ \frac{x^3 + 3x^2h + 3xh^2 + h^3 - x^3}{h} \right\}$$

$$= \lim_{h \to 0} \left\{ \frac{3x^2h + 3xh^2 + h^3}{h} \right\}$$

$$= \lim_{h \to 0} \{3x^2 + 3xh + h^2\}$$

$$= 3x^2$$

That is, if $f(x) = x^3$ then $f'(x) = 3x^2$

In general, for a function $y = f(x)$, the gradient function is defined in a similar way:

$$f'(x) = \lim_{h \to 0} \left\{ \frac{f(x+h) - f(x)}{h} \right\}$$

Exercise B

Use this definition to find the gradient function for the following.

1 $f(x) = x$
2 $f(x) = -(x^2)$
3 $f(x) = 2x + 3$
4 $f(x) = x + x^2$
5 $f(x) = x^4$
6 $f(x) = 3x^2$
7 $f(x) = 3x^2 + 2x$
8 $f(x) = 2$
9 $f(x) = x^3 - x^2$
10 $f(x) = 3(x^3 - x^2)$

3.3 The general result

From this exercise, you will have noticed a pattern about the answers. It seems that if $f(x) = x^n$, then the gradient function

$$f'(x) = nx^{n-1}$$

We shall now prove this result for cases when $n \in \mathbb{Z}^+$.

In chapter 2, Pascal's triangle was used to give the binomial coefficients for expansions of the type $(x + h)^n$, where $n \in \mathbb{Z}^+$. In every case, the expansion begins as $x^n + nx^{n-1}h + \ldots$, and these are the only terms that need concern us when finding the gradient function of $f(x) = x^n$. Consider

$$f'(x) = \lim_{h \to 0} \left\{ \frac{f(x+h) - f(x)}{h} \right\}, \text{ where } f(x) = x^n$$

$$f'(x) = \lim_{h \to 0} \left\{ \frac{(x+h)^n - x^n}{h} \right\}$$

$$f'(x) = \lim_{h \to 0} \left\{ \frac{(x^n + nx^{n-1}h + \cdots) - x^n}{h} \right\}$$

where the terms represented by the dots all contain h^2, at least.

$$\text{Hence}\quad f'(x) = \lim_{h \to 0} \left\{ \frac{nx^{n-1}h + \cdots}{h} \right\}$$

$$f'(x) = \lim_{h \to 0} \{nx^{n-1} + \cdots\}$$

where the missing terms here all contain h, at least.

Taking the limit, we have

$f'(x) = nx^{n-1}$ (all other terms vanish, as they all contain a factor of h)

and our general result is proved for integral values of the index n. The result is true for *any* value of the index n, but it is beyond the scope of our work here to prove that result.

Your answers to 4, 7, 9 and 10 in Exercise B will have shown you that the gradient function for the sum of two functions is the sum of the two separate gradient functions.

That is, if $f(x) = g(x) + h(x)$
then $f'(x) = g'(x) + h'(x)$

Gradient functions are also called **derived functions or derivatives**. We see that the following table of derived functions is obtained.

$f(x)$	$f'(x)$
x	1
x^2	$2x$
x^3	$3x^2$
x^n	nx^{n-1}
ax^n	anx^{n-1}
$\frac{1}{x} = x^{-1}$	$-x^{-2} = -\frac{1}{x^2}$
$x = x^{1/2}$	$\frac{1}{2}x^{-1/2} = \frac{1}{2\sqrt{x}}$

Note These last two results assume that the result $f(x) = x^n, f'(x) = nx^{n-1}$ is valid for all $n \in \mathbb{R}$.

You may have thought it strange to let the point Q move towards P in order to establish the gradient functions. There are other methods and approaches which yield the same answers. Let us consider an alternative approach for the function $y = x^2$ (figure 3.7).

We wish to calculate the gradient at P whose coordinates are (x, x^2). Consider the chord RQ,

where R is $\{x - h, (x - h)^2\}$
and Q is $\{x + h, (x + h)^2\}$

then the gradient of chord RQ is

$$\left\{ \frac{(x + h)^2 - (x - h)^2}{2h} \right\}$$

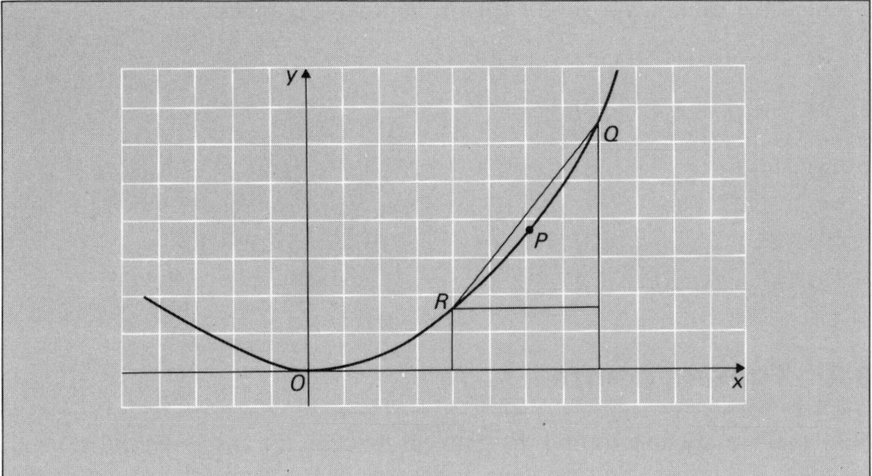

Figure 3.7

and the chord $RQ \to$ tangent at P as $h \to 0$
so

$$
\begin{aligned}
f'(x) &= \lim_{h\to 0}\left\{\frac{(x+h)^2-(x-h)^2}{2h}\right\} \\
&= \lim_{h\to 0}\left\{\frac{x^2+2xh+h^2-x^2+2xh-h^2}{2h}\right\} \\
&= \lim_{h\to 0}\left(\frac{4xh}{2h}\right) \\
&= \lim_{h\to 0}(2x) \\
&= 2x,
\end{aligned}
$$

and we have the expected result.

There are occasions when we may find this alternative approach more manageable than our first method. (See, for example, the section on derived functions of the circular functions in chapter 4.)

Exercise C

1 Find derived functions for the following.

a) $x^2 - 2x + 4$ f) $x^{1/2} + x^{-1/2}$

b) $x^3 - 27x$ g) $16x^2 + \dfrac{16}{x^2}$

c) $x^2 - 9x + 2$ h) $3x^{-3} - 2x^{-2}$

d) $x + \dfrac{1}{x}$ i) $(2x-1)(x-2)$

e) x j) $(x-1)(x+1)$

2 Find the value of $f(2)$ and $f'(2)$ for the following functions.

a) $x^2 - 4x + 4$
b) $x^3 - 12x + 8$
c) $x^2 + \frac{1}{x}$
d) x
e) $\frac{1}{2}x^2 - 2x + 3$
f) $7x - 14$
g) $14x - 7$
h) $2x^4 - 4x^3$
i) $x^2 - 2x$
j) $x^2 - 4$

3.4 Turning points

Now that we are able to find the gradient function for any given polynomial function, this means that we have a measure of the changing pattern of the graph of the function. In particular, we can determine when a curve has a **maximum** or a **minimum** point.

We see from figure 3.8 that at these maximum or minimum points, the tangent to the curve is parallel to the x-axis. That is, the gradient function, or derived function is zero, because $\tan 0° = 0$. We note that these maximum points are only **local** maxima or minima. That is, they are maximum or minimum only relative to nearby points; they are not necessarily overall maxima or minima, as we can see from figure 3.8.

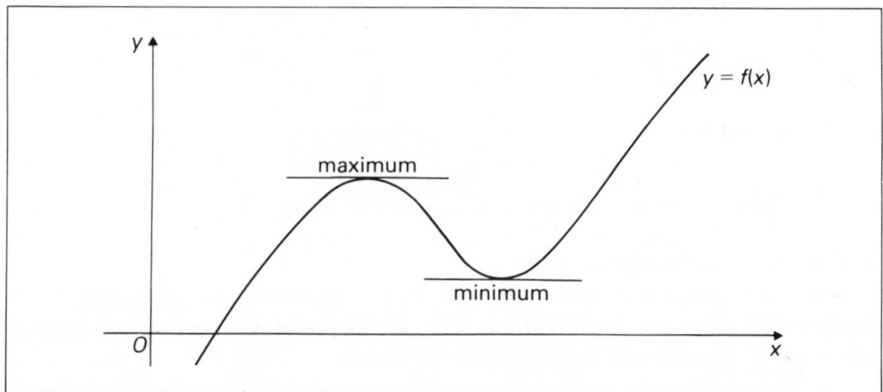

Figure 3.8

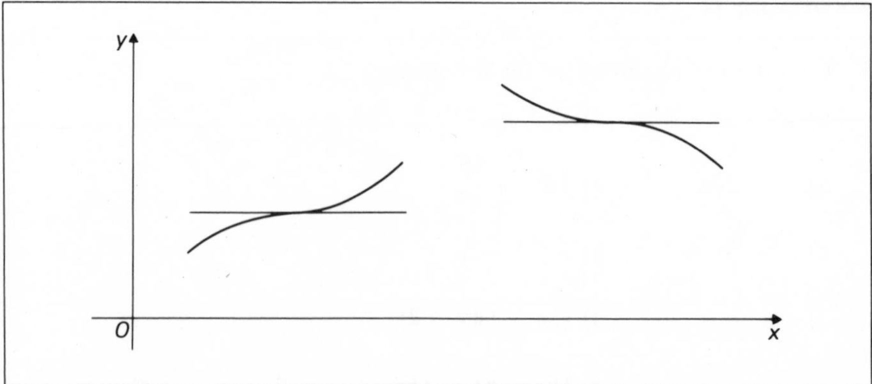

Figure 3.9

There are other points on graphs where the tangent is parallel to the x-axis, as we can see from figure 3.9. These points, where the tangent is parallel to the x-axis *and* it cuts the curve, are called **points of inflexion**.

Note Points of inflexion are any points where the tangent to a curve cuts the curve, whatever the gradient. We are concerned here only with those points of inflexion where the tangent is also parallel to the x-axis. All points at which the tangent is parallel to the x-axis are called **turning points**.

Example

Find the coordinates of the turning points for the function

$$f(x) = x^4 - 8x^3 + 22x^2 - 24x + 7$$

Our first step is to find the gradient function or derivative

$$f'(x) = 4x^3 - 24x^2 + 44x - 24$$

Now we seek the values of x which make $f'(x)$ zero. That is, we need to solve the equation

$$4x^3 - 24x^2 + 44x - 24 = 0$$

This factorises into

$$4(x-1)(x-2)(x-3) = 0$$

So the roots are $x = 1$, $x = 2$, $x = 3$, and we know that there are turning points for these values of x. The y coordinates are found by evaluating $f(1)$, $f(2)$, $f(3)$ i.e.

$$f(1) = 1^4 - 8(1)^3 + 22(1)^2 - 24(1) + 7 = -2$$
$$f(2) = 2^4 - 8(2)^3 + 22(2)^2 - 24(3) + 7 = -1$$
$$f(3) = 3^4 - 8(3)^3 + 22(3)^2 - 24(3) + 7 = -2$$

Hence the coordinates of the turning points are $(1, -2)$, $(2, -1)$, $(3, -2)$, and a sketch of the curve shows its general shape (figure 3.10).

We need, however, to find an analytical method of determining whether turning points are maxima or minima. To do this, we consider how the gradient changes as the curve passes through the turning point.

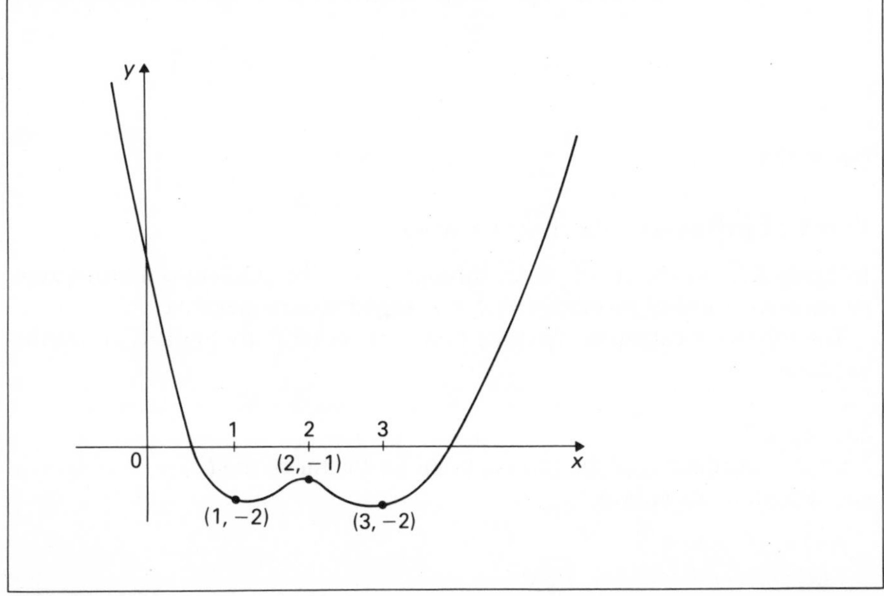

Figure 3.10

Maximum point

In figure 3.11, if there is a maximum at $x = a$, then $f'(a) = 0$. Just to the left of this point, we see that the tangents to the curve make an **acute** angle with the positive x-axis, so that the gradient is **positive**; whilst just to the right of $x = a$, the angle is obtuse, and the gradient is **negative**. We recall that for $90° < \theta° < 180°$, $\tan \theta° < 0$. Hence, as we move from left to right through the maximum point, the gradient changes from **positive** to **negative**.

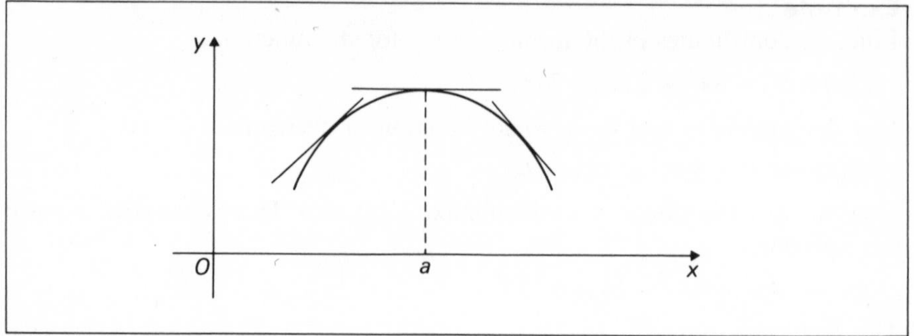

Figure 3.11

Minimum point

Similarly, we see that at the minimum point shown at $x = b$ in figure 3.12, the gradient is zero; just to the left of $x = b$ it is **negative**, and just to the right it is **positive**. The condition for a **minimum** is, then, that the gradient changes from **negative** to **positive**, as we move from left to right.

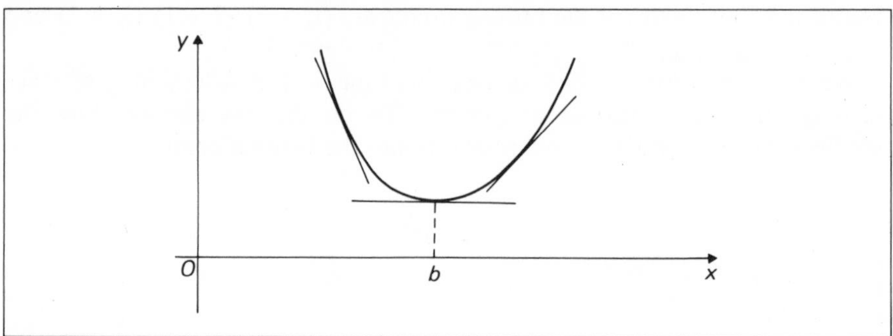

Figure 3.12

Point of inflexion (Parallel to x-axis)

In figure 3.13, as the curve passes through $x = c$, the gradient is **positive, zero, positive**. As it passes through $x = d$, it is **negative, zero, negative**.

The following examples illustrate how these criteria are applied in individual cases.

Example 1

Find the coordinates of the turning point for the function $f(x) = x^2 - 4x + 3$, and determine its nature.

$$f(x) = x^2 - 4x + 3$$
$$f'(x) = 2x - 4$$

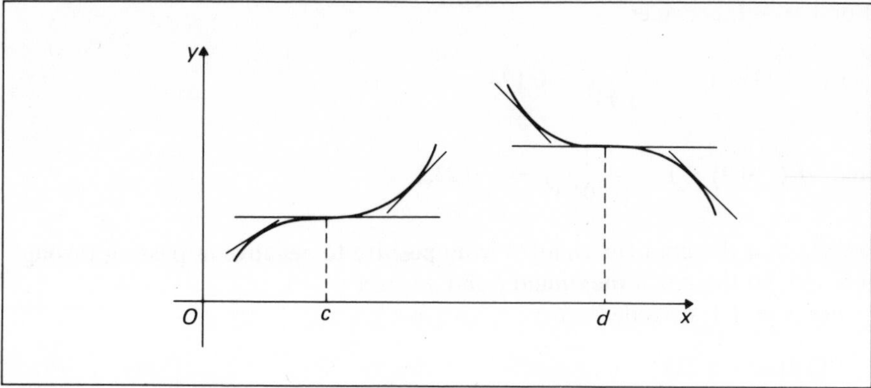

Figure 3.13

Hence the derived function is zero when $2x - 4 = 0$, that is, when $x = 2$.

We consider a point *just* to the left of $x = 2$ and a point *just* to the right. For example, we take $x = 1{\cdot}8$ and $x = 2{\cdot}1$

$$f'(1{\cdot}8) = 2(1{\cdot}8) - 4 = -0{\cdot}4$$
$$f'(2{\cdot}1) = 2(2{\cdot}1) - 4 = 0{\cdot}2$$

and we see that the gradient changes from **negative** to **positive** in passing through the point where $x = 2$. There is a **minimum** point when $x = 2$ and $f(2) = 2^2 - 4(2) + 3 = -1$. The curve can be sketched as shown in figure 3.14.

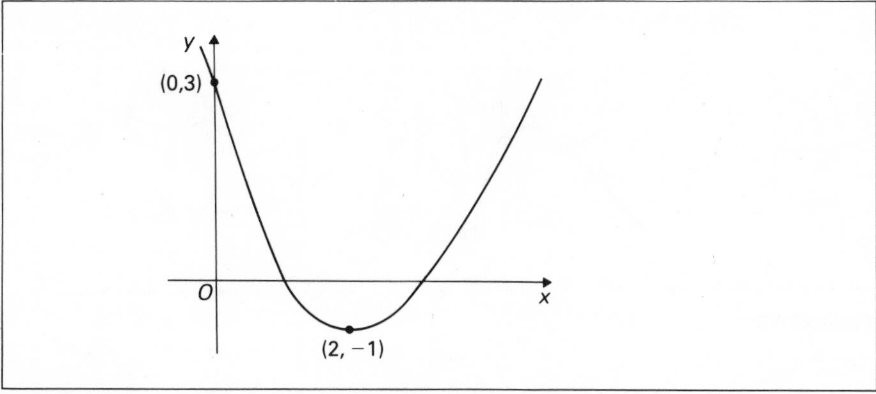

Figure 3.14

Example 2

Investigate the turning points for the function $f(x) = x + \dfrac{1}{x}$

$$f'(x) = 1 - \frac{1}{x^2}$$

$$= 0 \text{ when } x = 1 \text{ or } -1.$$

For $x = -1$, consider

$$f'(-1{\cdot}1) = 1 - \frac{1}{(-1{\cdot}1)^2} = 0{\cdot}174,$$

and $$f'(-0{\cdot}9) = 1 - \frac{1}{(-0{\cdot}9)^2} = -0{\cdot}235,$$

we see that the gradient changes from **positive** to **negative** in passing through $x = -1$, so there is a **maximum** point when $x = -1$.

For $x = +1$, consider

$$f'(0{\cdot}9) = -0{\cdot}235$$
$$f'(1{\cdot}1) = 0{\cdot}174$$

and this gives a **minimum** point.

Check for yourself that the coordinates are **minimum** at (1, 2) and **maximum** at (−1, −2).

The curve has the shape shown in figure 3.15. (See chapter 7 for details on sketching curves of this type.)

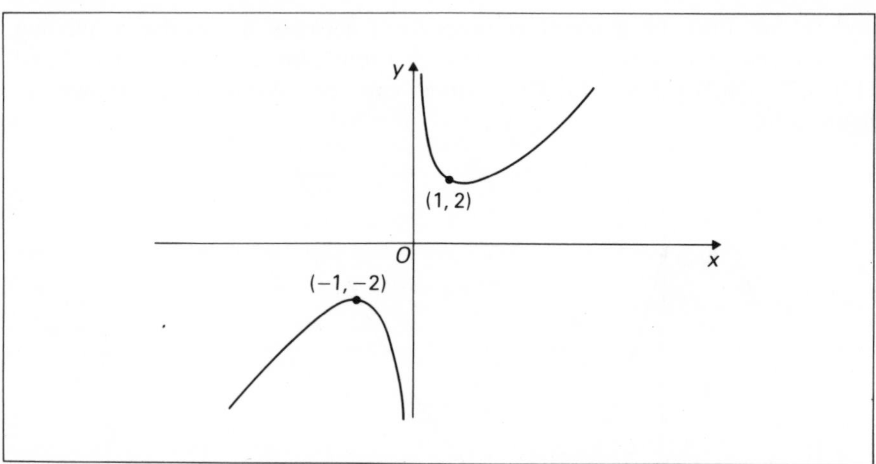

Figure 3.15

Example 3

Find what kind of turning point there is for the function $f(x) = x^3$

$$f'(x) = 3x^2$$

This is zero when $x = 0$

Consider $f'(-0{\cdot}1) = 3(-0{\cdot}1)^2 = 0{\cdot}03$

and $f'(0{\cdot}1) = 3(0{\cdot}1)^2 = 0{\cdot}03$

Here the gradient is positive both to the left *and* to the right of the point in question, and there is, therefore, a **point of inflexion** at (0, 0).

Exercise D

Investigate the nature of the turning points for the following functions. Give their coordinates and sketch the curves.

1 $f(x) = 2 + 6x - x^2$

2 $f(x) = x^3 - 7{\cdot}5x^2 + 18x - 7$

3 $f(x) = \frac{1}{x} + 4x$

4 $f(x) = x^4 - 2x^2$

5 $f(x) = 2x - 2$

6 $f(x) = 4x^2 - 3$

7 $f(x) = x^3 - x^2 + 9$

8 $f(x) = x^4 - x^2$

3.5 Applications

We can apply the same ideas to practical problems. This is illustrated in the following example.

From each corner of a square sheet of metal measuring 40 cm by 40 cm, squares are cut. The sides are then bent up to form an open box. Find the size of each square to be cut out to obtain the maximum volume of the box.

Figure 3.16 is a representation of the situation. When the sides are bent up, we have a box as shown in figure 3.17.

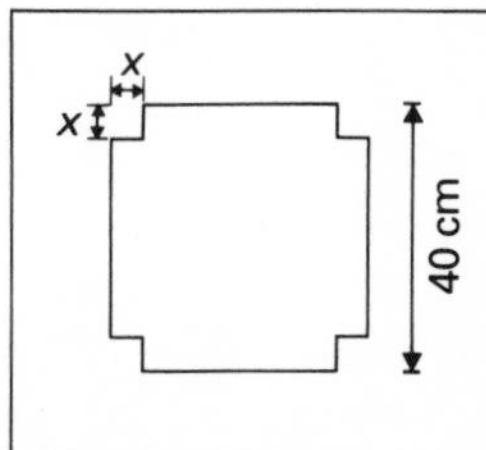

Figure 3.16

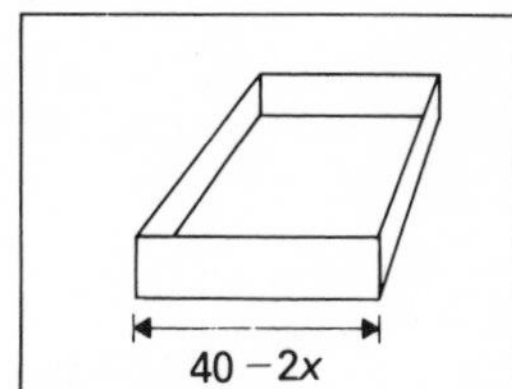

Figure 3.17

$$\text{The volume} = x(40 - 2x)(40 - 2x)$$
$$\text{or volume} = f(x) = x(40 - 2x)(40 - 2x)$$
$$f(x) = 1600x - 160x^2 + 4x^3$$
$$f'(x) = 1600 - 320x + 12x^2$$
$$f'(x) = 0, \quad 12x^2 - 320x + 1600 = 0$$
$$\text{or } 3x^2 - 80x + 400 = 0$$
$$(3x - 20)(x - 20) = 0$$

i.e. $x = \frac{20}{3} = 6\frac{2}{3}$ or $x = 20$
For $x = 6\frac{2}{3}$, consider

$$f'(6) = 1600 - 320(6) + 12(6^2) = 112$$
$$f'(7) = 1600 - 320(7) + 12(7^2) = -52$$

so the gradient of $f(x)$ changes from **positive** to **negative**, and there is a maximum when $x = 6\frac{2}{3}$.

Clearly from the physical situation, $x = 20$ gives a **minimum**, so here, we do not need to investigate further.

Exercise E

1 A piece of wire 40 cm long is bent to form a rectangle. Let x cm = the length of the rectangle, and determine the expression in terms of x for the area A of the rectangle. Prove that A is a maximum when $x = 10$ cm, that is when the rectangle is a square.

2 A cylinder is to be made to contain a volume of 250 cm^3. Let x cm = base radius and h cm = height. Find h in terms of x. Hence write down an expression in terms of x for the total surface area (both ends included) of the cylinder and find the minimum value of this area.

3 A water tank with a square base and vertical sides has an open top and its volume is 13·5 m^3. It has to be lined internally to avoid corrosion and pollution and the lining costs £1·50 per m^2. Find what dimensions for the tank will keep this cost as low as possible.

3.6 An alternative notation for derived functions

In order that we may deal more readily with applications of this work we need a notation which shows clearly the two variables being considered. We introduce this different notation as follows.

Our original scheme was to have two nearby points P and Q on the curve $y = f(x)$. We now take the coordinates of P to be (x, y) and as we increase x by 'a little bit of x' we reach Q. We use the symbol δx to stand for 'a little bit of x' and similarly δy means 'a little bit of y'.

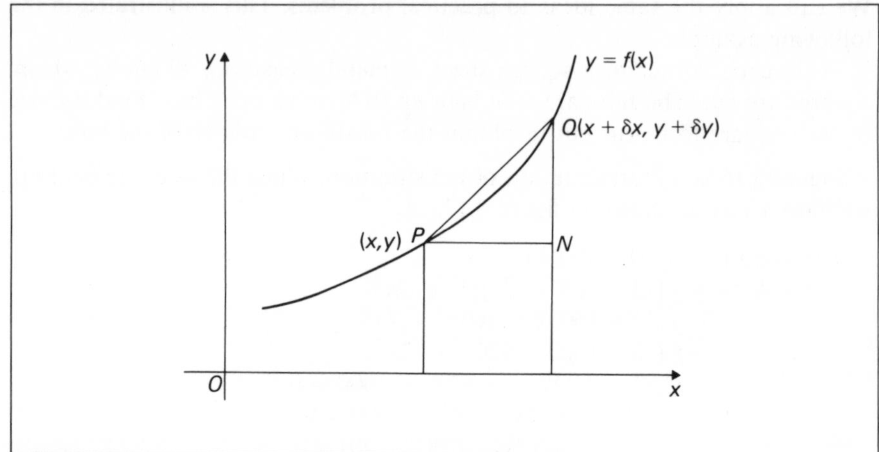

Figure 3.18

In figure 3.18, the gradient of PQ is

$$\frac{NQ}{PN} = \frac{\delta y}{\delta x}$$

and the gradient of the tangent at P is

$$\lim_{\delta x \to 0} \left(\frac{\delta y}{\delta x}\right)$$

that is, our derivative of y (in relation to x) is

$$\lim_{\delta x \to 0} \left(\frac{\delta y}{\delta x}\right)$$

but, as this expression is cumbersome, we use the shorthand expression $\frac{dy}{dx}$ in its place.

Note that although $\frac{dy}{dx}$ looks like a fraction, it is not. It is merely a symbol which describes the derivatives of y in relation to x. It is now clear that by writing the derivative in this form we are comparing the rate of change of y values with the change in x values. So that $\frac{dy}{dx}$ means 'the rate of change of y with respect to x', and in that form, many applications come to mind.

1 In economics, 'the rate of change of inflation with respect to time' is a vital consideration – to politicians, at least!
2 In biology, the rate of growth of a bacteriological culture with respect to time (or with respect to various surroundings) is an important study.
3 In geography, the rate of change of gradient of land in relation to surrounding land is important to the consideration of land use.
4 In physics, the rate of change of displacement with respect to time (called velocity) is an all-important and vital part of the physicists' tool kit.

There are many advantages in using the $\frac{dy}{dx}$ notation, as it labels the two variables being considered. For example, if we consider the volume of a cylinder, we have the relation

$$V = \pi r^2 h,$$

and V is expressed in terms of the *two* variables r and h. We cannot find $\frac{dV}{dh}$ or $\frac{dV}{dr}$ until we know how the 'other letter' (h or r respectively) behaves. If we are told that h remains fixed (or constant), i.e. it is a number, then

$$V = \pi h(r^2) \quad \text{and} \quad \frac{dV}{dr} = \pi h(2r) = 2\pi rh$$

This gives a measure of the growth of the volume in relation to the radius of the base.

A more realistic situation is that of blowing up a balloon (assumed perfectly spherical!)

$$V = \frac{4}{3}\pi r^3$$

then $\frac{dV}{dr} = 4\pi r^2$

and the rate of increase of volume is proportional to the square of the radius – quite a thought, when you are puffing to increase the size!

Exercise F

1 Find $\frac{dy}{dx}$ in the following.

a) $y = x^2 + 3x + 7$ d) $y = \frac{18}{x} - x$

b) $y = 3x^{1/2} - x$ e) $y = \frac{3}{x^2} - 2x$

c) $2y = 7x^2$ f) $y = \frac{2}{x}$

2 If $x = t^2$ and $y = 2t^4$, find

a) $\frac{dx}{dt}$ b) $\frac{dy}{dt}$ c) $\frac{dy}{dx}$

Comment on anything you notice about your answers.

(Hint: for c) find y as a function of x, and then find $\frac{dy}{dx}$.)

3 A particle moves so that its displacement x metres from the starting point t sec later is

$$x = \frac{t}{24}(36 + 12t - t^2)$$

a) Find the speed and acceleration (i.e. the derivative of speed with respect to t)
b) Calculate the speed when $t = 1$, and the acceleration when $t = 2$.

3.7 Higher derivatives

There are occasions when we have to differentiate a function several times (as in question 3 in exercise F). Our notation is

$\frac{dy}{dx}$ is the 1st derivative of y, with respect to x,

$\frac{d}{dx}\left(\frac{dy}{dx}\right)$ is the 2nd derivative of y, with respect to x.

The expression $\frac{d}{dx}\left(\frac{dy}{dx}\right)$ is written as $\frac{d^2y}{dx^2}$.

This again is only a *symbol* and stands for the second derivative of y with respect to x.

In the same way, the third derivative is written

$\frac{d^3y}{dx^3}$ and so on.

Example

If $y = 7x^3 + 4x^2 - \frac{11}{x}$

find $\frac{dy}{dx}, \frac{d^2y}{dx^2}, \frac{d^3y}{dx^3}$.

$$\frac{dy}{dx} = 21x^2 + 8x + \frac{11}{x^2}$$

$$\frac{d^2y}{dx^2} = 42x + 8 - \frac{22}{x^3}$$

$$\frac{d^3y}{dx^3} = 42 + \frac{66}{x^4}$$

Exercise G

1 If $y = 3t^2 + 8t^3$ find $\frac{d^3y}{dt^3}$.

2 If $x = t^2$, $y = 2t^4$ (see question 2 of exercise F)

find a) $\dfrac{d^2y}{dt^2}$ b) $\dfrac{d^2x}{dt^2}$ c) $\dfrac{d^2y}{dx^2}$

Are there any similarities between your answers in this case?

3 When $y = x^3$

find a) $\dfrac{dy}{dx}$ b) $\dfrac{d^2y}{dx^2}$ c) $\dfrac{d^3y}{dx^3}$

What is the value of these derivatives when $x = 0$? Sketch the graph of $y = x^3$ and see whether you can deduce any connection between the values of these derivatives and the behaviour of the function at (0, 0).

4 Repeat question 3 for the curve $y = x^4$. How does this differ from $y = x^3$?

5 Given that $f(x) = x^5 - 10x^4 + 40x^3 - 80x^2 + 80x - 32$, differentiate $f(x)$ until the nth derivative becomes zero for all values of x.

There is one value of x for which the function and the first four of its derivatives are zero. Find this value.

3.8 Another method of determining the nature of stationary points

We have seen that at a maximum or minimum point, $\dfrac{dy}{dx} = 0$, and as we travel along a curve through a **maximum** point (from left to right), we see that the gradient is at first positive, becomes zero and then becomes negative (see Figure 3.11). That is, as we pass through a maximum point, the gradient decreases, i.e. the rate of change of the gradient is negative or

$$\frac{d}{dx}\left(\frac{dy}{dx}\right) < 0.$$

We have, then at a maximum point,

$$\frac{dy}{dx} = 0, \quad \frac{d^2y}{dx^2} < 0$$

Similarly, for a minimum point (see Figure 3.12). As we move from left to right, the gradient changes from negative through zero and becomes positive. So the rate of change of the gradient is positive, or

$$\frac{d}{dx}\left(\frac{dy}{dx}\right) > 0.$$

So for a minimum point,

$$\frac{dy}{dx} = 0 \quad \text{and} \quad \frac{d^2y}{dx^2} > 0$$

It is beyond the scope of this book to consider the complex problem of behaviour of higher derivatives at points of inflexion. Suffice it to say that at a point of inflexion (where the tangent is parallel to the x-axis),

$$\frac{dy}{dx} = 0; \quad \frac{d^2y}{dx^2} = 0 \text{ provided that } \frac{d^3y}{dx^3} \neq 0.$$

Consider again the curves given in Exercise D (page 65) using this alternative method for determining the nature of the turning points.

Angles

CHAPTER FOUR

4.1 Introduction

We know what an angle is, but may find it very difficult to give a useful or helpful definition. Many of us will have met the idea of 'an amount of turning' required to move from one thing to another. We may have talked of 'fractions of a complete turn', before the idea of a method of angle measurement was introduced.

The common measurement used is the degree, where 360° make a complete turn. The idea of 360° was related to the thought that there were 360 days in a year, and in a year, the earth made a complete revolution round the sun. An alternative measure is used by some, where there are 100 grades in a right angle. This was an early attempt to metricate angle measure.

In more advanced work, both of these measures are cumbersome. If we were to continue with them, we would find that certain constants would occur frequently in our work. We shall, therefore, define a new unit of angle measure.

A **radian** is defined as that sectorial angle which cuts off an arc equal in length to the radius (figure 4.1), i.e. in the circle, if arc $AB = r$, then θ is called 1 radian. Clearly, if we double this angle, we double the arc length. So an angle of 2 radians cuts off an arc of $2r$. In general, an angle of ϕ cuts off an arc of length ϕr.

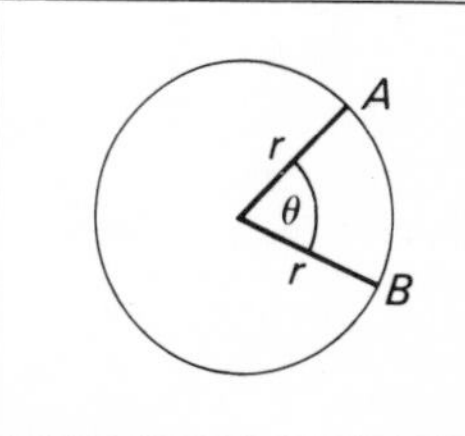

Figure 4.1

For a complete turn, the arc length will equal the circumference, i.e. $2\pi r$, so the angle for a complete turn is 2π. We say that 2π radians is equivalent to 360°; or π radians is equivalent to 180°. There is a symbol for angles measured in radians. They should be written as 1^c, 2^c, π^c etc. but in practice, this symbol is omitted, and the angles written simply as 1, 2, π etc. (The 'c' stands for 'circular measure', since the radian is based on the circle for its definition.)

Exercise A

1 Find the radian measures of the following angles.
a) 30° b) 270° c) 45° d) 245°

2 Find the degree equivalents of the following radians.

a) 1 b) $2\frac{1}{4}$ c) $\frac{\pi}{2}$ d) $\frac{\pi}{3}$

3 In a circle of radius 3 cm, what arc length is cut off by a sectorial angle of

a) 1 b) π c) $\frac{\pi}{3}$ d) 3?

4 In a sector of a circle the angle is $\pi/6$ and the arc length is 5 cm. Find the radius.

5 In a circle of radius 4 cm, a sector has an arc length of 6 cm. Find the angle of the sector a) in radians b) in degrees

4.2 Sectorial area

The calculation of sectorial areas becomes more straightforward with angles measured in radians rather than degrees. In figure 4.2 we wish to calculate the shaded area, and if θ is measured in degrees, this would be $\frac{\theta^\circ}{360^\circ} \times \pi r^2$. If θ is measured in radians, however, the area is given by the fraction $\frac{\theta}{2\pi}$ of the area of the circle, i.e. $\frac{\theta}{2\pi} \times \pi r^2$. This simplifies to $\frac{1}{2}\theta r^2$.

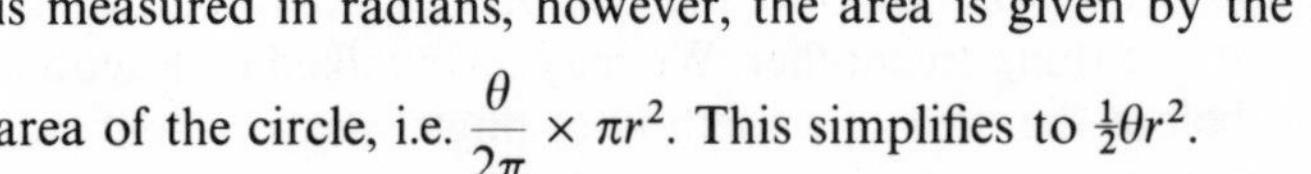

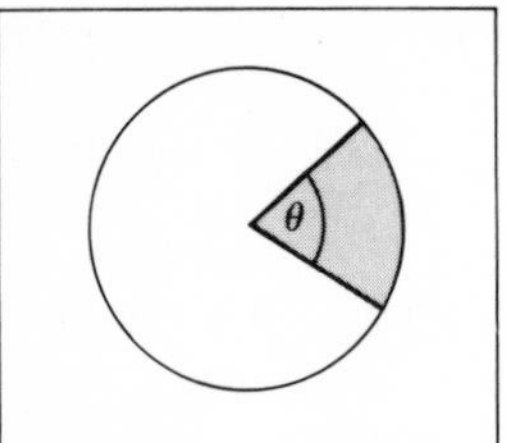

Figure 4.2

This is an easy formula for sectorial area; both to remember and to use. It is here that the beauty of this type of angle measure can be appreciated, because all the π's cancel, and there are no clumsy 360's occurring in fractions.

Exercise B

1 Calculate the sectorial area when

a) $r = 5, \theta = 2$ b) $r \doteq 6, \theta = 1\frac{1}{2}$ c) $r = 1{\cdot}5, \theta = \frac{\pi}{2}$

2 Calculate the sectorial angle when
a) $A = 14, r = 4$ b) $A = 10, r = 5$ c) $A = 6, r = 6$

3 Calculate the radius when
a) $A = 7, \theta = 1$ b) $A = 17, \theta = 2$ c) $A = 3{\cdot}14, \theta = 1{\cdot}57$

4.3 The circular functions

There is a variety of ways of defining the sine, cosine and tangent of an acute angle at an elementary level. It is important for us to have a definition which is appropriate for angles of any size.

The Sine function

Consider a rod OA of length 1 unit fixed at one end O, where O is a fixed point on the fixed line OX (figure 4.3). As OA rotates, so the angle θ changes (θ is always measured anticlockwise from the line OX). We define the sine of

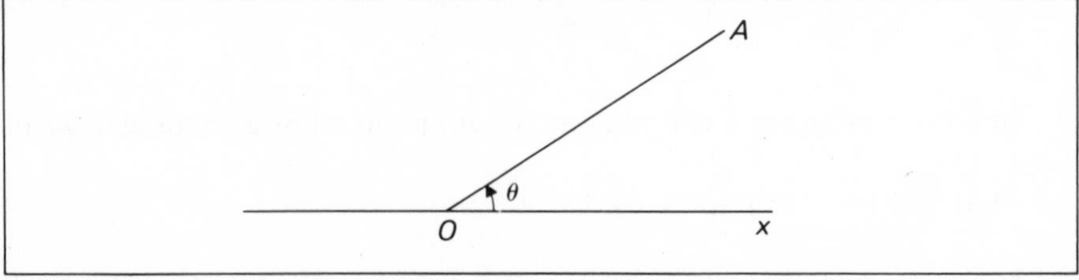

Figure 4.3

the angle θ as the vertical height of A above OX. (If A is below OX, then the vertical height becomes negative). We could construct our own rather crude table of sines from such a device. We see that there will be a certain amount of symmetry in the values obtained, and if we draw the graph of $f(x) = \sin x$, we can again see the symmetry (figure 4.4).

We see, too, that the curve repeats itself every 360° or 2π radians. We would expect this if we look again at figure 4.3, as the rotating line OA repeats previous positions after rotating through 360°. We say that the sine function has a period of 2π.

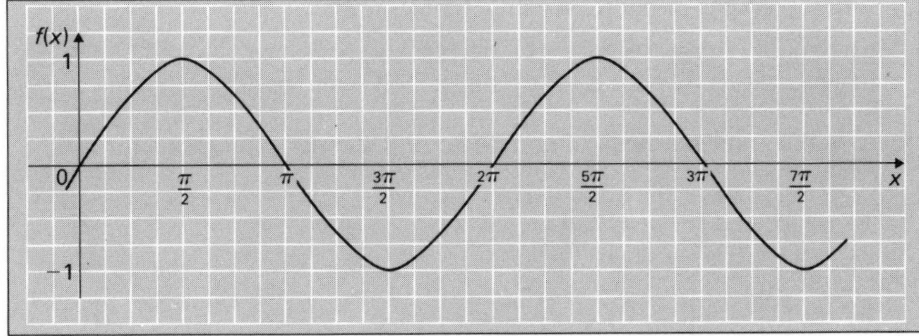

Figure 4.4

The Cosine function

We define the cosine function also by considering the rotating line OA in figure 4.3. The cosine of θ is defined as the projection of OA onto the fixed line OX. This will be positive when the projection is to the right of O and negative when the projection is to the left.

Again, we could use this device to construct our own cosine tables and to draw the graph of $f(x) = \cos x$ (figure 4.5). We see the symmetry of the function and also the similarity of this function with the sine function. The cosine function also repeats after 360° (2π radians) and we say that the cosine function is periodic with period 2π.

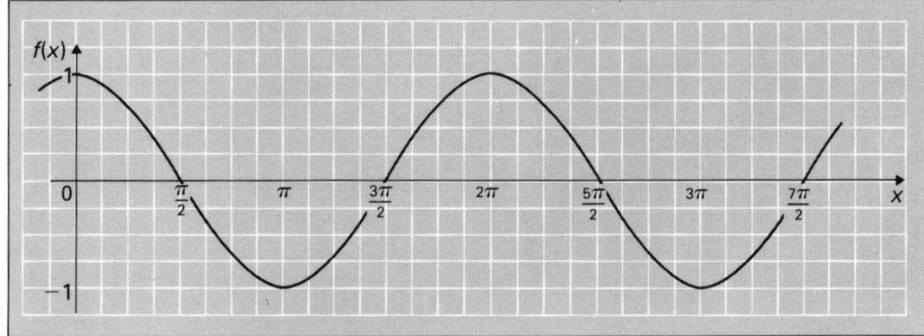

Figure 4.5

Exercise C

1 Use tables of sine and cosine to find

a) $\sin 210°$
b) $\sin 289°$
c) $\sin 108°$
d) $\sin \frac{3\pi}{4}$
e) $\sin \frac{5\pi}{6}$
f) $\cos 210°$
g) $\cos 634°$
h) $\cos 724°$
i) $\cos \frac{3\pi}{4}$
j) $\cos \frac{5\pi}{6}$

2 Use the graphs in figures 4.4 and 4.5 to show that

a) $\cos(2\pi - \alpha) = \cos \alpha$
b) $\cos(\pi - \alpha) = -\cos \alpha$
c) $\sin(2\pi - \alpha) = -\sin \alpha$
d) $\sin(\pi - \alpha) = \sin \alpha$

3. Use the graphs in figs. 4.4 and 4.5 to find simpler expressions for

a) $\sin\left(\frac{\pi}{2} + \alpha\right)$
b) $\cos\left(\frac{3\pi}{2} + \alpha\right)$
c) $\sin\left(\frac{3\pi}{2} - \alpha\right)$
d) $\cos\left(\frac{7\pi}{2} - \alpha\right)$.

4.4 Other circular functions

There are four other circular functions which can be defined from our definitions of the sine and cosine functions.

The Tangent function

$$\tan x = \frac{\sin x}{\cos x}$$

and the graph of $f(x) = \tan x$ is shown in figure 4.6. Here we see some symmetry again, and notice that the tangent function has a period of π.

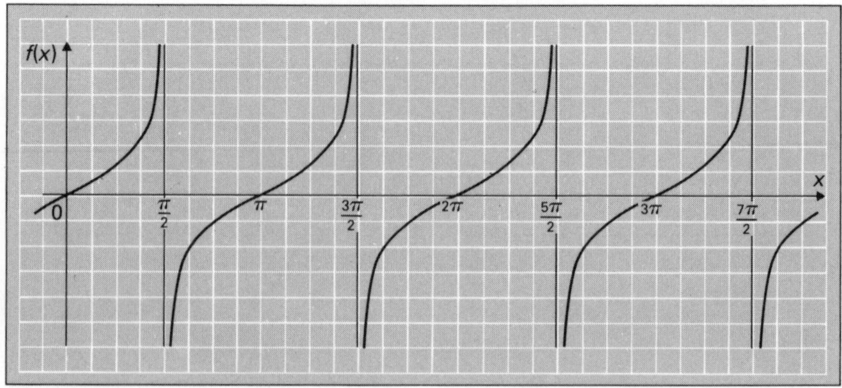

Figure 4.6

The Cotangent function

$$\cot x = \frac{\cos x}{\sin x}$$

and its graph is shown in figure 4.7. The period of the cotangent function is again π.

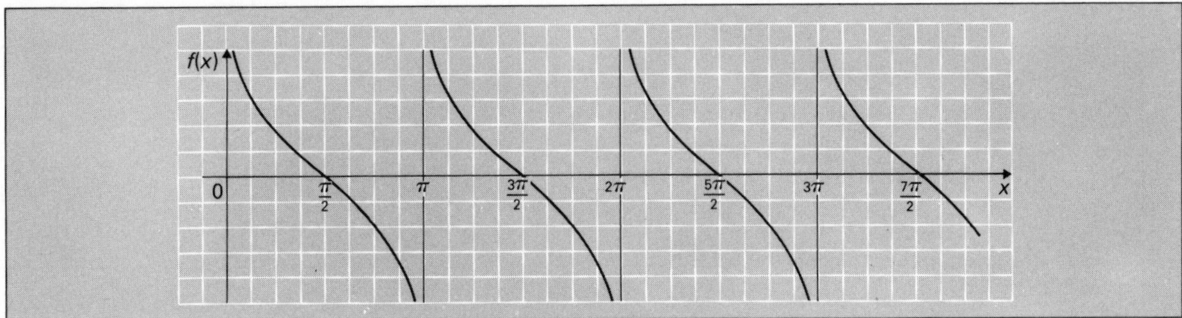

Figure 4.7

The Secant function

$$\sec x = \frac{1}{\cos x}$$

and its graph is shown in figure 4.8.

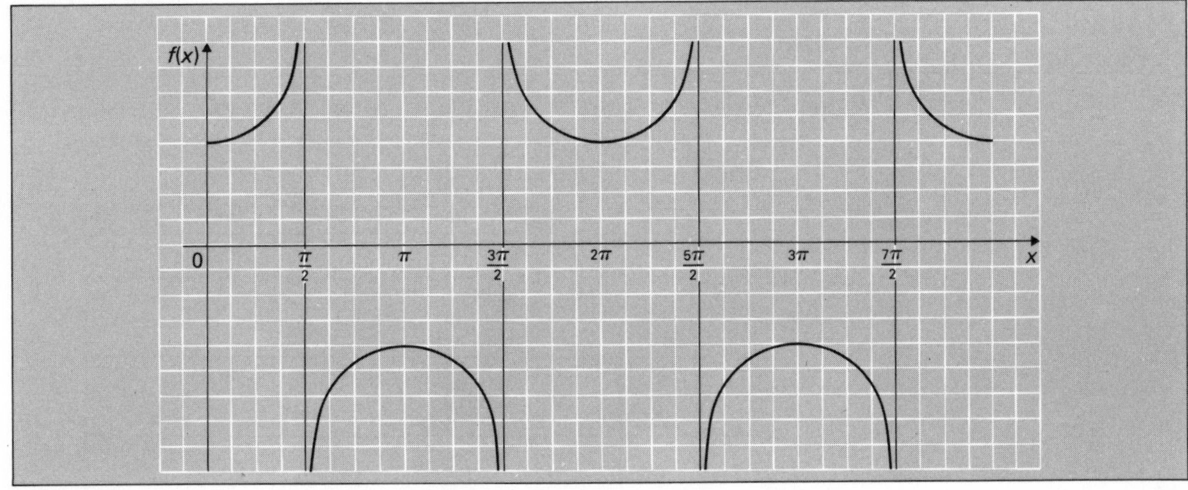

Figure 4.8

The Cosecant function

$$\text{cosec } x = \frac{1}{\sin x}$$

(**Note** cosec x is more usually written csc x.)
The cosecant graph is shown in figure 4.9.

It is very helpful to learn the values of the trigonometric functions of certain angles, as they occur so frequently. These can be obtained without the use of tables.

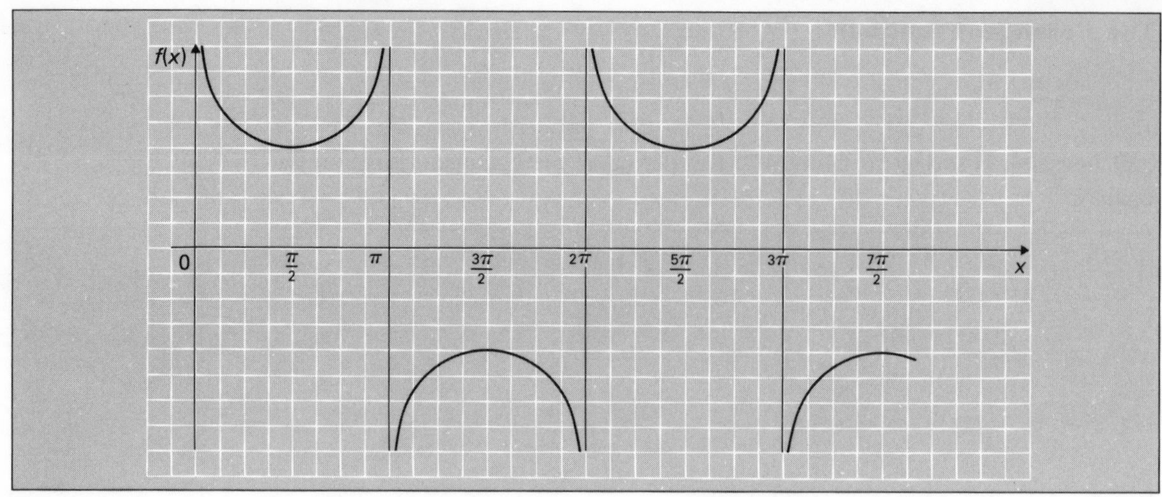

Figure 4.9

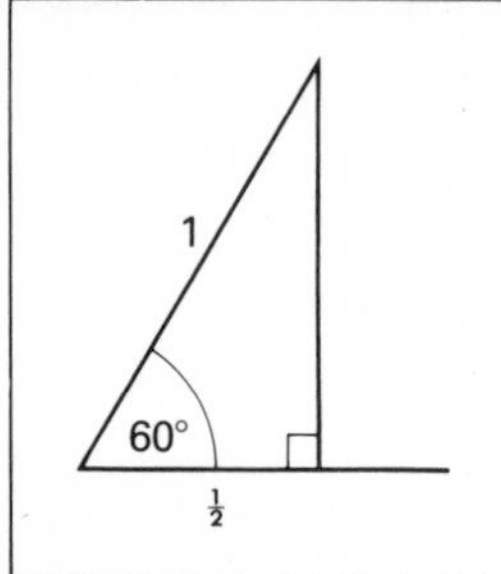

Figure 4.10

Either by referring to figure 4.10 or by constructing an equilateral triangle, can you show that $\cos 60° = \frac{1}{2}$? In figure 4.10, we see that when $\theta = 60°$, and $\cos 60° = \frac{1}{2}$, we can use Pythagoras' theorem to show that the other side of the triangle is $\frac{1}{2}\sqrt{3}$ so that $\sin 60° = \frac{1}{2}\sqrt{3}$.

Can you construct an appropriate triangle to obtain the trigonometric ratios for 45°?

Exercise D

Without recourse to tables or a calculator, complete the following table.

Angle x deg	rad	$\sin x$	$\cos x$	$\tan x$	$\cot x$	$\sec x$	$\csc x$
0°	0	0	1				
30°	$\frac{\pi}{6}$	$\frac{1}{2}$	$\frac{\sqrt{3}}{2}$				
45°	$\frac{\pi}{4}$		$\frac{1}{\sqrt{2}}$	1			
60°	$\frac{\pi}{3}$						
90°	$\frac{\pi}{2}$	1	0	∞			
120°	$\frac{2\pi}{3}$						
135°	$\frac{3\pi}{4}$						
150°	$\frac{5\pi}{6}$						
180°	π	0	-1				

4.5 The compound angle formulae

Many situations demand that we manipulate formulae containing a mixture of trigonometric functions of single and multiple angles. We must seek, therefore, some results which we can use. First, we need to find expressions for $\sin(A+B)$ and $\cos(A+B)$ in terms of $\sin A$, $\cos A$, $\sin B$ and $\cos B$.

There are various methods of establishing these results. We shall use the ideas of rotations established in O-level work. We recall that the matrix which represents a rotation through an angle θ about 0 is

$$\begin{pmatrix} \cos\theta & -\sin\theta \\ \sin\theta & \cos\theta \end{pmatrix}$$

Now, a rotation through angle A followed by a rotation through angle B (both about 0) is clearly equivalent to a single rotation through angle $(A+B)$ about 0. In matrix form, this is

$$\begin{pmatrix} \cos B & -\sin B \\ \sin B & \cos B \end{pmatrix}\begin{pmatrix} \cos A & -\sin A \\ \sin A & \cos A \end{pmatrix} = \begin{pmatrix} \cos(A+B) & -\sin(A+B) \\ \sin(A+B) & \cos(A+B) \end{pmatrix}$$

When we multiply the two matrices on the left, we obtain:

$$\begin{pmatrix} \cos B\cos A - \sin B\sin A & -\cos B\sin A - \sin B\cos A \\ \sin B\cos A + \cos B\sin A & -\sin B\sin A + \cos B\cos A \end{pmatrix}$$

$$= \begin{pmatrix} \cos(A+B) & -\sin(A+B) \\ \sin(A+B) & \cos(A+B) \end{pmatrix}$$

Two matrices are equal only if their corresponding elements are equal, hence

$$\cos(A+B) = \cos B\cos A - \sin B\sin A$$
$$\sin(A+B) = \sin B\cos A + \cos B\sin A$$

The remaining entries merely give a repetition of these results.

It is more usual to write the right hand sides of these expressions in the more logical order:

$$\sin(A+B) = \sin A\cos B + \cos A\sin B$$
$$\cos(A+B) = \cos A\cos B - \sin A\sin B$$

Similar formulae for $\sin(A-B)$ and $\cos(A-B)$ can be obtained either by replacing B by $-B$ in the last two results, or by considering a rotation through angle A followed by a rotation of angle B **clockwise** and using the appropriate matrices. That is we consider a pair of rotations through A and $-B$ giving the matrix equation

$$\begin{pmatrix} \cos(A-B) & -\sin(A-B) \\ \sin(A-B) & \cos(A-B) \end{pmatrix}$$

$$= \begin{pmatrix} \cos A & -\sin A \\ \sin A & \cos A \end{pmatrix}\begin{pmatrix} \cos(-B) & -\sin(-B) \\ \sin(-B) & \cos(-B) \end{pmatrix}$$

and the last matrix is equivalent to

$$\begin{pmatrix} \cos B & \sin B \\ -\sin B & \cos B \end{pmatrix}$$

since from the definition in **4.3**, $\sin(-B) = -\sin B$ and $\cos(-B) = \cos B$.

Work out the matrix product

$$\begin{pmatrix} \cos A & -\sin A \\ \sin A & \cos A \end{pmatrix}\begin{pmatrix} \cos B & \sin B \\ -\sin B & \cos B \end{pmatrix}$$

and establish the required results

$$\cos (A - B) = \cos A \cos B + \sin A \sin B$$

$$\sin (A - B) = \sin A \cos B - \cos A \sin B$$

Exercise E

1 Use the formulae just established to check that
a) $\sin 110° = \sin 70° \cos 40° + \cos 70° \sin 40°$
b) $\cos 60° = \cos 90° \cos 30° + \sin 90° \sin 30°$
c) $\sin (0{\cdot}7) = \sin 1 \cos (0{\cdot}3) - \cos 1 \sin (0{\cdot}3)$
d) $\cos (0{\cdot}5) = \cos (0{\cdot}2) \cos (0{\cdot}3) - \sin (0{\cdot}2) \sin (0{\cdot}3)$

2 Use these formulae to check earlier results that
a) $\sin (\pi - \theta) = \sin \theta$
b) $\cos (2\pi - \theta) = \cos \theta$
c) $\cos (\pi - \theta) = -\cos \theta$
d) $\sin (2\pi - \theta) = -\sin \theta$

3 Using $\sin 30° = \frac{1}{2}$, $\cos 30° = \frac{1}{2}\sqrt{3}$, prove with the relevant formulae that $\cos 60° = \frac{1}{2}$ and $\sin 60° = \frac{1}{2}\sqrt{3}$

Hence check that $\sin 90° = 1$ and $\cos 90° = 0$.

A number of important results follow directly from these formulae. If we write $A = B = \theta$, where θ is *any* angle, in the formula

$$\cos (A - B) = \cos A \cos B + \sin A \sin B$$

we have $\cos (\theta - \theta) = \cos \theta \cos \theta + \sin \theta \sin \theta$

$$\Rightarrow \cos 0 = \cos^2\theta + \sin^2 \theta$$

$$\Rightarrow 1 = \cos^2 \theta + \sin^2 \theta$$

This statement is true for **any value of θ whatever**. If you have learnt your elementary definitions of sines and cosines of acute angles in a right-angled triangle, you will see that the above result is a statement of the Theorem of Pythagoras.

Pythagoras gives $c^2 = a^2 + b^2$ (figure 4.11)

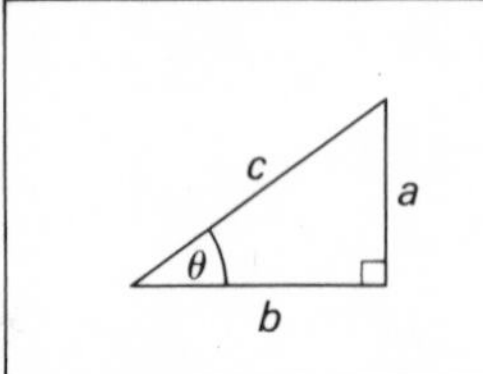

Figure 4.11

since $c \neq 0$, $1 = \left(\frac{a}{c}\right)^2 + \left(\frac{b}{c}\right)^2$

$$1 = \sin^2 \theta + \cos^2 \theta$$

This demonstrates the result for an acute angle θ only. Our earlier method shows it to be true for any θ.

Again writing $A = B = \theta$ in the other formulae, we have:

$$\sin(\theta + \theta) = \sin\theta\cos\theta + \cos\theta\sin\theta$$

$$\sin 2\theta = 2\sin\theta\cos\theta.$$

Also,

$$\cos(\theta + \theta) = \cos\theta\cos\theta - \sin\theta\sin\theta$$

$$\cos 2\theta = \cos^2\theta - \sin^2\theta$$

(These results should confirm your answers to question 3 of the previous exercise.)

Doing the same in the remaining formula demonstrates only that $\sin 0 = 0$ for

$$\sin(\theta - \theta) = \sin\theta\cos\theta - \cos\theta\sin\theta$$

$$\sin 0 = 0$$

All these results are needed frequently in mathematics, and it is a great help to have them committed to memory. They are summarised here as a reminder.

$$\sin(A + B) = \sin A\cos B + \cos A\sin B$$
$$\cos(A + B) = \cos A\cos B - \sin A\sin B$$
$$\sin(A - B) = \sin A\cos B - \cos A\sin B$$
$$\cos(A - B) = \cos A\cos B + \sin A\sin B$$

$$\sin^2\theta + \cos^2\theta = 1 \quad (\text{all } \theta)$$

$$\sin 2\theta = 2\sin\theta\cos\theta \quad (\text{all } \theta)$$
$$\cos 2\theta = \cos^2\theta - \sin^2\theta \quad (\text{all } \theta)$$

Exercise F

1 Find a formula for $\sin 3\theta$ in terms of $\sin\theta$ (write $\sin 3\theta = \sin(2\theta + \theta)$).

2 Find a formula for $\cos 3\theta$ in terms of $\cos\theta$.

3 Prove that
a) $\cos 2\theta = 2\cos^2\theta - 1$ b) $\cos 2\theta = 1 - 2\sin^2\theta$

(These are important alternative results for $\cos 2\theta$.)

4 If $\theta \neq \pi/2$, prove $\sec^2\theta = 1 + \tan^2\theta$.
(The restriction is necessary here for the deduction of this result from $\sin^2\theta + \cos^2\theta = 1$.)

5 Prove that $(\sin\theta + \cos\theta)^2 = 1 + \sin 2\theta$.

6 Without finding the value of θ, find $\sin 4\theta$ and $\cos 4\theta$, given that $\sin\theta = \frac{3}{5}$ and $0° < \theta < 90°$. (Hint: You should first find $\sin 2\theta$ and $\cos 2\theta$.)

7 Use your answers to question 6 to evaluate
a) $\sin 3\theta$ b) $\cos 3\theta$ c) $\sin 5\theta$ d) $\cos 5\theta$

8 Solve the equation

$\frac{1}{2}\cos\theta + \frac{1}{2}\sqrt{3}\sin\theta = 1$ $(0° < \theta < 180°)$

by rewriting the equation as

$\sin 30° \cos\theta + \cos 30° \sin\theta = 1$, hence $\sin(30° + \theta) = 1$.

(For more work of this nature see Chapter 6)

4.7 Sum and difference formulae

There are occasions when we need to convert the sum (or difference) of the sines of two angles into a product of sines and cosines. We establish the results by an extension of the formulae we have just derived.
Consider

$$\sin(A + B) = \sin A \cos B + \cos A \sin B \qquad \mathbf{i}$$
$$\sin(A - B) = \sin A \cos B - \cos A \sin B \qquad \mathbf{ii}$$

Adding **i** and **ii**, we have

$$\sin(A + B) + \sin(A - B) = 2\sin A \cos B$$

Subtracting **ii** from **i**, we have

$$\sin(A + B) - \sin(A - B) = 2\cos A \sin B$$

rewriting these results with $A + B = P$; $A - B = Q$ we obtain

$$\sin P + \sin Q = 2\sin\left(\frac{P+Q}{2}\right)\cos\left(\frac{P-Q}{2}\right)$$

and

$$\sin P - \sin Q = 2\cos\left(\frac{P+Q}{2}\right)\sin\left(\frac{P-Q}{2}\right)$$

Applying a similar procedure to the two corresponding cosine formulae, we obtain

$$\cos P + \cos Q = 2\cos\frac{P+Q}{2}\cos\frac{P-Q}{2}$$

and

$$\cos P - \cos Q = -2\sin\frac{P+Q}{2}\sin\frac{P-Q}{2}$$

(**Note** the negative sign in the last of these results)

1 Prove that $\dfrac{\sin 7\theta - \sin 5\theta}{\cos 7\theta + \cos 5\theta} = \dfrac{\sin \theta}{\cos \theta}$.

2 Prove that $\dfrac{(\sin 7\theta - \sin \theta) \cos 5\theta}{\cos 4\theta} = \sin 8\theta - \sin 2\theta$.

3 Prove that $\dfrac{\cos \theta - \cos 2\theta}{\sin \theta + \sin 2\theta} = \tan (\tfrac{1}{2}\theta)$.

4 Prove that

$$\sin \theta + \sin 2\theta + \sin 3\theta + \sin 4\theta = 4 \cos \frac{\theta}{2} \cos \theta \sin \frac{5\theta}{2}.$$

5 Simplify a) $\dfrac{\sin A + \sin B}{\cos A + \cos B}$

b) $\dfrac{\sin A - \sin B}{\cos A - \cos B}$

6 Find the value of $\sin 10° + \sin 50° - \sin 70°$, without using a calculator or tables.

7 Solve the equation $\sin \theta + \sin 3\theta + \sin 5\theta = 0$, giving all the solutions in the range $0° \leqslant \theta \leqslant 360°$.

8 If A, B, C are the angles of a triangle, prove that

$$\sin A + \sin B + \sin C = 4 \cos \frac{A}{2} \cos \frac{B}{2} \cos \frac{C}{2}.$$

4.8 Tangent formulae

We can also establish formulae for tangents of angles by recalling that we define the tangent function as the quotient

$$\frac{\text{sine}}{\text{cosine}}.$$

We may write $\tan (A + B) = \dfrac{\sin (A + B)}{\cos (A + B)}$

$$\Rightarrow \tan (A + B) = \frac{\sin A \cos B + \cos A \sin B}{\cos A \cos B - \sin A \sin B}$$

Dividing both numerator and denominator of the fraction on the right hand side by $\cos A \cos B$, we obtain

$$\tan (A + B) = \frac{\dfrac{\sin A}{\cos A} + \dfrac{\sin B}{\cos B}}{1 - \dfrac{\sin A \sin B}{\cos A \cos B}}$$

$$\tan (A + B) = \frac{\tan A + \tan B}{1 - \tan A \tan B}$$

Similarly,

$$\tan(A - B) = \frac{\tan A - \tan B}{1 + \tan A \tan B}$$

The double angle formula for tangents is obtained by writing $A = B = \theta$ in the formula

$$\tan(A + B) = \frac{\tan A + \tan B}{1 - \tan A \tan B}$$

$$\text{i.e. } \tan 2\theta = \frac{\tan \theta + \tan \theta}{1 - \tan \theta \tan \theta}$$

$$\Rightarrow \tan 2\theta = \frac{2 \tan \theta}{1 - \tan^2 \theta}$$

This is a particularly exciting result because of the mathematical beauty of the applications. Mathematicians often like to juggle with results to see whether there is any connection with other branches of the subject. We have obtained a formula for $\tan 2\theta$ in terms of $\tan \theta$. We could equally well have obtained a result for $\tan \phi$ in terms of $\tan \frac{\phi}{2}$. We do this by writing $2\theta = \phi$ (or $\theta = \frac{\phi}{2}$), and this gives

$$\tan \phi = \frac{2 \tan \frac{\phi}{2}}{1 - \tan^2 \frac{\phi}{2}}$$

For brevity, we write $t = \tan \frac{\phi}{2}$ so that

$$\tan \phi = \frac{2t}{1 - t^2}$$

From Exercise E question 4, it was established that

$$\sec^2 \phi = 1 + \tan^2 \phi,$$

Hence

$$\sec^2 \phi = 1 + \left(\frac{2t}{1 - t^2}\right)^2$$

$$\sec^2 \phi = \frac{(1 - t^2)^2 + 4t^2}{(1 - t^2)^2}$$

$$\sec^2 \phi = \frac{1 + 2t^2 + t^4}{(1 - t^2)^2}$$

$$\sec^2 \phi = \frac{(1 + t^2)^2}{(1 - t^2)^2}$$

$$\sec \phi = \frac{1 + t^2}{1 - t^2}$$

$$\cos \phi = \frac{1 - t^2}{1 + t^2}$$

We use the relationship that $\sin^2 \phi + \cos^2 \phi = 1$ to obtain $\sin \phi$ in terms of t. You should check for yourself that

$$\sin \phi = \frac{2t}{1 + t^2}$$

These three expressions for $\sin \phi$, $\cos \phi$ and $\tan \phi$ are particularly useful in parts of the calculus (see Chapter 12). It is again helpful in later work if these can be memorised:

$$\sin \phi = \frac{2t}{1 + t^2}, \quad \cos \phi = \frac{1 - t^2}{1 + t^2}, \quad \tan \phi = \frac{2t}{1 - t^2}$$

4.9 Derivatives of sin *x* and cos *x*

In Chapter 3 we learnt how to differentiate elementary algebraic functions. We now need to find the derivatives of the trigonometric functions. We shall find, using our definition of a derivative,

$$f'(x) = \lim_{h \to 0} \left\{ \frac{f(x + h) - f(x - h)}{2h} \right\}$$

that we will need to evaluate

$$\lim_{\theta \to 0} \left\{ \frac{\sin \theta}{\theta} \right\}, \text{ where } \theta \text{ is measured in radians.}$$

We give here an illustration of how the expression $\dfrac{\sin \theta}{\theta}$ behaves as $\theta \to 0$. Note that this is *not* a rigorous proof of the result. Consider figure 4.12 where

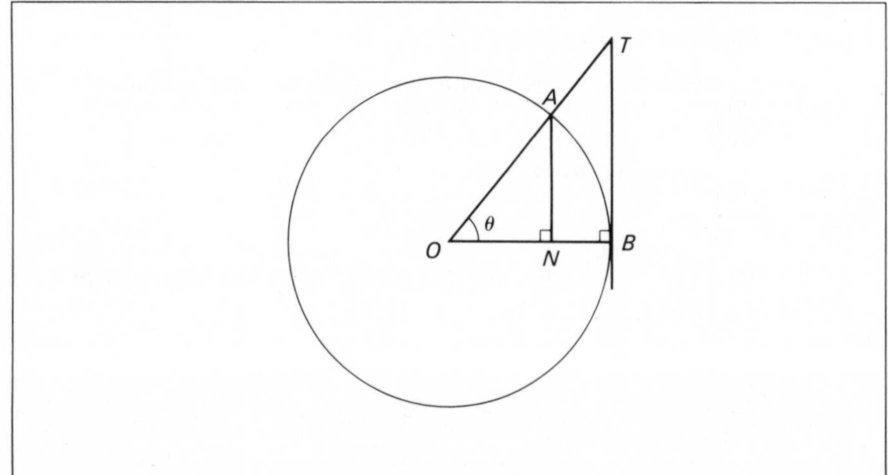

Figure 4.12

the circle is of radius r. It is clear that

$$\triangle\text{OAN} < \text{sectorial area OAB} < \triangle\text{OTB}$$
$$\tfrac{1}{2}r^2 \sin\theta\cos\theta < \tfrac{1}{2}r^2\theta < \tfrac{1}{2}r^2\tan\theta$$
$$\sin\theta\cos\theta < \theta < \tan\theta$$

Divide through by $\sin\theta$

$$\cos\theta < \frac{\theta}{\sin\theta} < \frac{1}{\cos\theta} \quad \left(\text{valid since } \sin\theta \geqslant 0 \text{ for } 0 \leqslant \theta < \frac{\pi}{2}\right)$$

Invert

$$\frac{1}{\cos\theta} > \frac{\sin\theta}{\theta} > \cos\theta$$

$$\lim_{\theta\to 0}\left(\frac{1}{\cos\theta}\right) \geqslant \lim_{\theta\to 0}\left(\frac{\sin\theta}{\theta}\right) \geqslant \lim_{\theta\to 0}(\cos\theta)$$

now

$$\cos\theta \to 1 \text{ as } \theta \to 0, \text{ so}$$

$$1 \geqslant \lim_{\theta\to 0}\left(\frac{\sin\theta}{\theta}\right) \geqslant 1$$

Hence

$$\lim_{\theta\to 0}\left(\frac{\sin\theta}{\theta}\right) = 1$$

Note: an alternative approach is to consider the inequalities derived from $AN < arc\ AB < BT$.

Having established this limiting value for the quotient $\frac{\sin\theta}{\theta}$, we are now able to find the derivative of $\sin x$, as follows:

$$f(x) = \sin x, \quad f(x+h) = \sin(x+h), \; f(x-h) = \sin(x-h)$$

$$\Rightarrow f'(x) = \lim_{h\to 0}\left\{\frac{\sin(x+h) - \sin(x-h)}{2h}\right\}^*$$

Using the formulae of this chapter

$$f'(x) = \lim_{h\to 0}\left\{\frac{\sin x\cos h + \cos x\sin h - \sin x\cos h + \cos x\sin h}{2h}\right\}$$

$$f'(x) = \lim_{h\to 0}\left\{\frac{2\cos x\sin h}{2h}\right\}$$

$$f'(x) = \cos x \lim_{h\to 0}\left\{\frac{\sin h}{h}\right\}, \text{ since } \cos x \text{ is unaffected by the behaviour of } h.$$

Hence

$$f'(x) = \cos x$$

* **Note** This symmetrical definition of $f'(x)$ is preferable to the more conventional definition because of the simplicity of the resulting trigonometric manipulation.

We treat the derivative of cos x in a similar way.

$$f(x) = \cos x, \quad f(x+h) = \cos(x+h), \quad f(x-h) = \cos(x-h)$$

$$\Rightarrow f'(x) = \lim_{h\to 0}\left\{\frac{\cos(x+h) - \cos(x-h)}{2h}\right\}$$

$$f'(x) = \lim_{h\to 0}\left\{\frac{(\cos x \cos h - \sin x \sin h) - (\cos x \cos h + \sin x \sin h)}{2h}\right\}$$

$$f'(x) = \lim_{h\to 0}\left\{\frac{-2 \sin x \sin h}{2h}\right\}$$

$$f'(x) = -\sin x \lim_{h\to 0}\left\{\frac{\sin h}{h}\right\}$$

$$f'(x) = -\sin x$$

We have shown that the derivative of sin x is cos x and the derivative of cos x is $-\sin x$. It is important that these are learnt carefully, paying special attention to the negative sign which occurs in the derivative of cos x.

Exercise H

1 Find for $0 \leqslant x \leqslant 2\pi$, the maximum and the minimum values of sin x, and state the values of x for which they occur.

2 Repeat question 1 for cos x, where $\pi \leqslant x \leqslant 3\pi$.

3 Differentiate
a) $3x + \sin x$ b) $4 \cos x + 2$

and find in b) the values of x for which the derivative is zero.

4 Find the turning point (for $0 \leqslant x \leqslant \pi$) of the function $\sin x + \cos x$. State whether it is maximum or minimum. For what value of x does the next turning point occur? Is it a maximum or minimum? How are the positions of these turning points related to the period of the function

$f: x \to \sin x + \cos x$?

CHAPTER FIVE

Analytical Geometry

5.1 Introduction

When dealing with the concept of a function in section 3.1 (figure 3.2), we made use of a system of perpendicular axes, scaled appropriately, to show in a diagrammatic way the relation $y = 2x$. Figure 3.1 of section 3.1 displayed the association between values of x in the domain D, and the values of y in the codomain Δ derived from them.

This method of representing graphically the relationship between connected variables was invented by the French mathematician and philosopher René Descartes (1596–1650) and opened the way for the creation of a 'new' form of geometry which has since had an immense influence on the development of pure mathematics.

We now look at some of the more important ideas connected with Descartes' system, referred to universally as a **Cartesian system of coordinates.**

The system

Once a point O (called the **origin**) and a pair of perpendicular lines Ox, Oy, uniformly scaled, (called the **coordinate axes**) have been marked in a plane, every other point may be defined by means of the ordered pairs of numbers (x, y). The value of these **coordinates** are, in effect, the lengths of the intervals appropriately directed, starting from the origin O of the domain for x, and the codomain for y.

Figure 5.1 shows a number of points A, B, C, D, E, F located by this system.

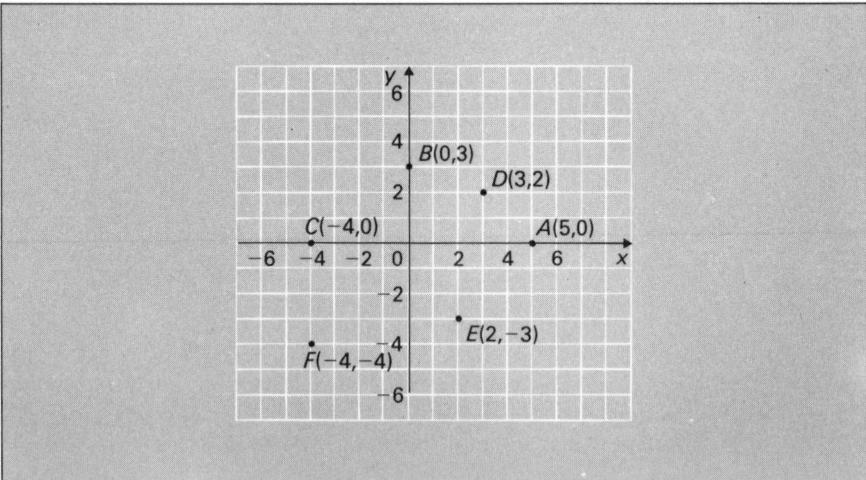

Figure 5.1

a Section

We are here concerned with finding the coordinates of the point K (x, y), as shown in figure 5.2, dividing the line joining $A(x_1, y_1)$ to $B(x_2, y_2)$ into two sections AK, KB such that $AK : KB = p : q$.

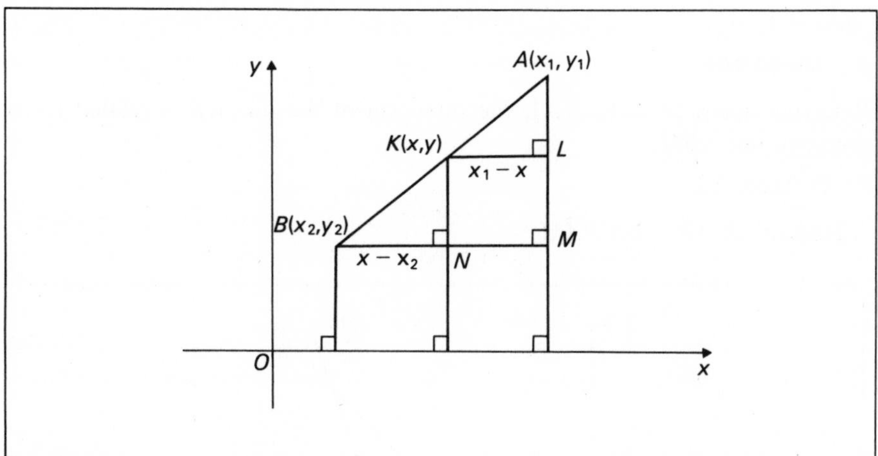

Figure 5.2

From figure 5.2,

$$\frac{p}{q} = \frac{AK}{KB} = \frac{KL}{BN} = \frac{x_1 - x}{x - x_2}$$

$$\Rightarrow p(x - x_2) = q(x_1 - x)$$

$$px + qx = qx_1 + px_2$$

$$x = \frac{qx_1 + px_2}{p + q}.$$

Likewise, using $AK : KB = AL : KN$

$$y = \frac{qy_1 + py_2}{p + q}$$

and the coordinates of K are

$$\left(\frac{qx_1 + px_2}{p + q}, \frac{qy_1 + py_2}{p + q}\right)$$

In the special case where K is the mid-point of AB, $p : q = 1 : 1$ and the mid-point has coordinates

$$\left(\frac{x_1 + x_2}{2}, \frac{y_1 + y_2}{2}\right)$$

b Distance

Using figure 5.2 again, the distance AB between two points with given co-ordinates may be determined as follows:

$$AB^2 = AM^2 + BM^2 \text{ by Pythagoras' Theorm in } \triangle ABM$$
$$\Rightarrow AB^2 = (x_1 - x_2)^2 + (y_1 - y_2)^2$$
$$\Rightarrow AB = \sqrt{\{(x_1 - x_2)^2 + (y_1 - y_2)^2\}}$$

In particular, we see that the distance of a point (X, Y) from the origin $(0, 0)$ is $\sqrt{(X^2 + Y^2)}$.

c Direction

Referring again to section 3.1, the direction of the line AB is related to its gradient, $\tan A\hat{B}M$.

From figure 5.2,

$$\text{gradient of } AB = \tan B\hat{A}M = \frac{y_1 - y_2}{x_1 - x_2}.$$

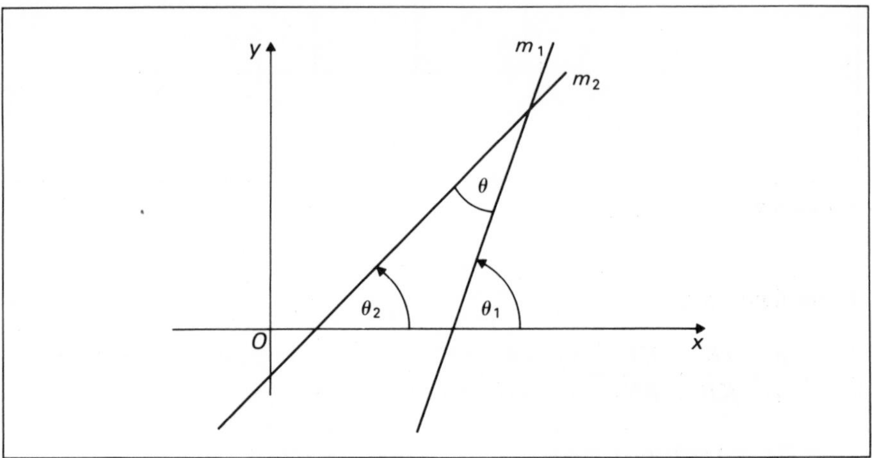

Figure 5.3

Furthermore, if two lines with gradients $m_1 = \tan \theta_1$, and $m_2 = \tan \theta_2$ are given (see figure 5.3) then

i) the lines are parallel if $\theta_1 = \theta_2$

that is $\tan \theta_1 = \tan \theta_2$

$$\Rightarrow m_1 = m_2$$

i.e. **gradients are equal**

ii) the lines are perpendicular if the angle between them, θ, is 90°.

$$\theta = 90°$$
$$\Rightarrow \theta_1 = 90 + \theta_2$$
$$\tan \theta_1 = \tan(90 + \theta_2)$$
$$= -\cot \theta_2$$
$$\Rightarrow m_1 = -\frac{1}{m_2}$$

$$m_1 m_2 = -1.$$

i.e. **the product of the gradients of the lines = −1.**

iii) generally when the lines are oblique as in figure 5.3,

$$\theta = \theta_1 - \theta_2$$

$$\Rightarrow \tan\theta = \tan(\theta_1 - \theta_2)$$

$$= \frac{\tan\theta_1 - \tan\theta_2}{1 + \tan\theta_1 \tan\theta_2} \quad \text{(Using the results obtained in chapter 4)}$$

giving

$$\theta = \tan^{-1}\left\{\frac{m_1 - m_2}{1 + m_1 m_2}\right\}$$

an expression from which the angle between the lines may be calculated whenever m_1 and m_2 are known.

Example 1

Using the coordinates of the points defined in figure 5.1,

i) the mid-point of CE is

$$\left(\frac{-4+2}{2}, \frac{0-3}{2}\right) \quad \text{i.e. } (-1, -1{\cdot}5)$$

ii) the coordinates of a point G dividing AF such that $AG : GF = 1 : 2$ are

$$\left(\frac{1 \times (-4) + 2 \times 5}{1+2}, \frac{1 \times (-4) + 2 \times 0}{1+2}\right) \quad \text{i.e.} \quad (2, -1\tfrac{1}{3})$$

iii) $CD = \sqrt{\{(-4-3)^2 + (0-2)^2\}} = \sqrt{53}$

iv) gradient of $AF = \dfrac{0+4}{5+4} = \dfrac{4}{9}$

v) gradient of OD, $m_1 = \dfrac{2-0}{3-0} = \dfrac{2}{3}$

gradient of OE, $m_2 = \dfrac{0+3}{0-2} = \dfrac{-3}{2}$

and $m_1 m_2 = \dfrac{2}{3} \times \dfrac{-3}{2} = -1$

$\Rightarrow OD$ and OE are perpendicular

vi) gradient of $OD = \dfrac{0-2}{-4-3} = \dfrac{2}{7}$

gradient of $ED = \dfrac{-3-2}{2-3} = 5$

$$\Rightarrow \tan E\hat{D}C = \frac{5 - \frac{2}{7}}{1 + 5 \times \frac{2}{7}} = \frac{33}{7} \times \frac{7}{17}$$

$$= \frac{33}{17}$$

$$\Rightarrow E\hat{D}C = 62{\cdot}74°.$$

Example 2

For the $\triangle ABC$ (figure 5.4) where A, B, C are the points (x_1, y_1), (x_2, y_2), (x_3, y_3) determine the coordinates of the geometric centre G.

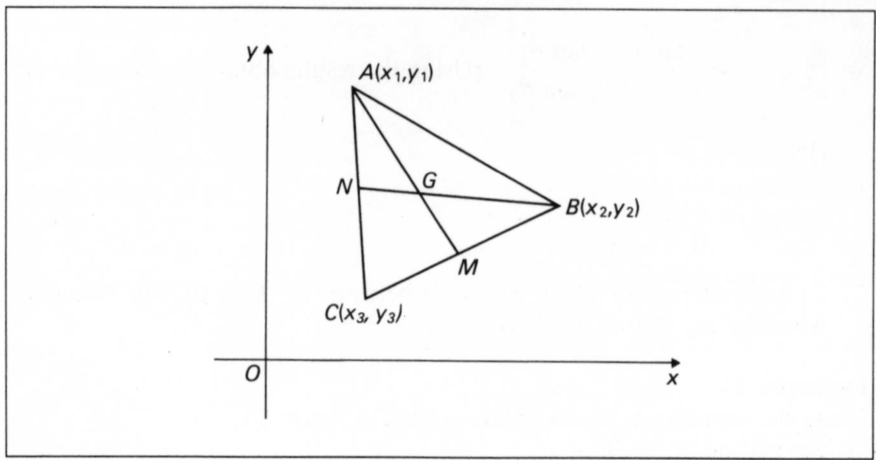

Figure 5.4

The geometric centre is the point of intersection of the medians AM and BN of the $\triangle ABC$, and since

$$CM : MB = CN : NA, \quad NM \| AB \text{ and } NM = \tfrac{1}{2}\, AB$$

It follows that $AG : GM = 2 : 1$ from the similar triangles MNG and ABG.

Using the mid-point result, the coordinates of M are

$$\{\tfrac{1}{2}(x_3 + x_2), \tfrac{1}{2}(y_3 + y_2)\}$$

Hence the coordinates of G are

$$\left\{\frac{1}{2+1}[2 \times \tfrac{1}{2}(x_3 + x_2) + 1 \times x_1], \frac{1}{2+1}[2 \times \tfrac{1}{2}(y_3 + y_2) + 1 \times y_1]\right\}$$

i.e. $\{\tfrac{1}{3}(x_1 + x_2 + x_3), \tfrac{1}{3}(y_1 + y_2 + y_3)\}$

Example 3

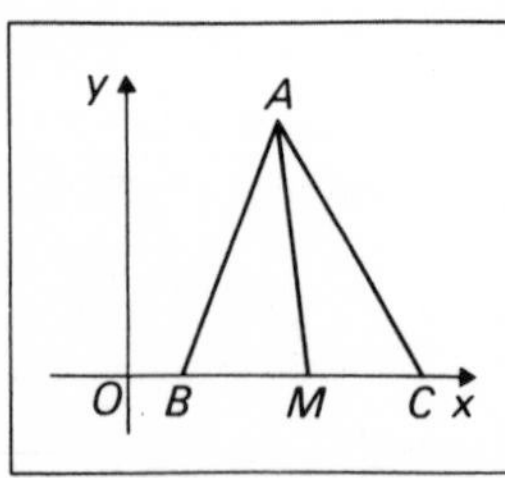

Figure 5.5

For the $\triangle ABC$ shown in figure 5.5, where A is any point (x, y), B is $(b, 0)$, C is $(c, 0)$ and M is the mid-point of BC, verify that

$$AB^2 + AC^2 = 2(AM^2 + BM^2) \quad \text{(Apollonius' Theorem).}$$

Using the mid-point result, M is the point $\{\tfrac{1}{2}(b + c), 0\}$ and using the distance results

$$AB^2 = (x - b)^2 + (y - 0)^2 = x^2 + y^2 - 2bx + b^2$$
$$AC^2 = (x - c)^2 + (y - 0)^2 = x^2 + y^2 - 2cx + c^2$$
$$\Rightarrow AB^2 + AC^2 = 2x^2 + 2y^2 - 2bx - 2cx + b^2 + c^2$$

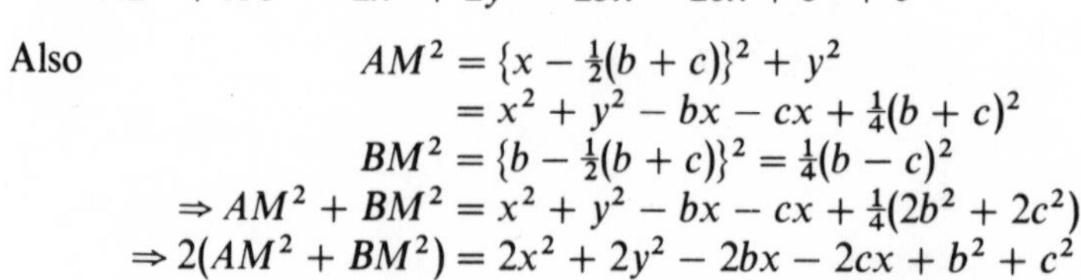

Also

$$AM^2 = \{x - \tfrac{1}{2}(b + c)\}^2 + y^2$$
$$= x^2 + y^2 - bx - cx + \tfrac{1}{4}(b + c)^2$$
$$BM^2 = \{b - \tfrac{1}{2}(b + c)\}^2 = \tfrac{1}{4}(b - c)^2$$
$$\Rightarrow AM^2 + BM^2 = x^2 + y^2 - bx - cx + \tfrac{1}{4}(2b^2 + 2c^2)$$
$$\Rightarrow 2(AM^2 + BM^2) = 2x^2 + 2y^2 - 2bx - 2cx + b^2 + c^2$$

Thus

$$AB^2 + AC^2 = 2(AM^2 + BM^2).$$

Example 4

Prove that the four points $A(-2, 3)$, $B(0, -1)$, $C(8, 3)$ and $D(6, 7)$ are the four vertices of a rectangle.

$$\text{Gradient of } AB = \frac{3+1}{-2-0} = -2 \quad \text{Gradient of } BC = \frac{-1-3}{0-8} = \frac{1}{2}$$

$$\text{Gradient of } CD = \frac{3-7}{8-6} = -2 \quad \text{Gradient of } DA = \frac{7-3}{6+2} = \frac{1}{2}$$

These results imply
i) AB is parallel to CD (both gradients equal to -2)
ii) BC is parallel to DA (both gradients equal to $\frac{1}{2}$)

$\Rightarrow ABCD$ is a parallelogram.

But (gradient of AB) $\times$ (gradient of BC) $= -2 \times \frac{1}{2} = -1$

$\Rightarrow AB$ is perpendicular to BC
$\Rightarrow ABCD$ is a parallelogram whose angles are 90° i.e. a rectangle.

Exercise A

1 Calculate the lengths of the lines joining the points
a) $(3, 4)$ and $(7, 8)$ b) $(-7, 6)$ and $(2, 3)$ c) $(-1, -7)$ and $(6, -2)$

2 Prove that the points $A(1, -1)$, $B(-1, 1)$ and $C(-\sqrt{3}, -\sqrt{3})$ are the vertices of an equilateral triangle.

3 L is a point on the line joining $A(0, 2)$ to $B(4, 8)$ such that $AL : LB = 2 : 3$. Calculate the coordinates of L.

4 Given that A, B, C, D are the points $(-2, 2)$, $(3, 1)$, $(5, 4)$ and $(-4, -1)$ show that AB and CD bisect each other.

5 Find, in its simplest form, the distance between the points

$$P(ap^2, 2ap) \text{ and } Q\left(\frac{a}{p^2}, \frac{-2a}{p}\right).$$

6 Find the gradients of the lines joining the following pairs of points
a) $(6, 6)$ $(1, 3)$ c) $(-2, 0)$, $(4, 3)$
b) $(-2, 1)$, $(3, 4)$ d) $(ap^2, 2ap)$, $(aq^2, 2aq)$

7 Prove that AB is parallel to DC and perpendicular to BC, when A, B, C, D are the four points $(-1, 3)$, $(1, 0)$, $(4, 2)$ and $(0, 8)$.

8 Show, using the idea of gradients, that $(5, 7)$, $(-3, 1)$ and $(-7, -2)$ lie on the same straight line.

9 Prove that $A(1, 4)$, $B(-4, -1)$ and $C(2, 1)$ are the vertices of a right-angled triangle. Calculate the length of the hypotenuse of this triangle.

10 Define completely the four-sided figure whose vertices are $(4, 1)$, $(1, 6)$, $(-4, 3)$ and $(-1, -2)$. Justify your result.

11 a) Find the angle between the lines AB and BC of gradients 2 and -3 respectively.
b) Prove that the acute angle between the lines with gradients $-\frac{1}{2}$ and $\frac{1}{3}$ is 45°.

12 Given that A, B, C are the points $(-3, 2)$, $(1, -5)$ and $(5, 3)$, prove that $\tan A\hat{B}C = 1{\cdot}5$.

5.3 Idea of a locus

If Q, R are two fixed points in a plane, let us now consider the form of the path traced out by a variable point P, such that $PQ = PR$. (We may remember the geometrical answer to this problem, since the idea forms the basis of drawing the mediator of the line QR, using ruler and compasses.)

Let Q be (a_1, b_1) and R be (a_2, b_2) and suppose for any position of P the coordinates of P are (x, y), x and y being variable.

$$PQ = PR$$
squaring: $PQ^2 = PR^2$

and using the distance formula,

$$(x - a_1)^2 + (y - b_1)^2 = (x - a_2)^2 + (y - b_2)^2$$
$$x^2 - 2a_1x + a_1^2 + y^2 - 2b_1y + b_1^2 = x^2 - 2a_2x + a_2^2 + y^2 - 2b_2y + b_2^2$$
i.e. $$2(a_1 - a_2)x + 2(b_1 - b_2)y = (a_1^2 - a_2^2) + (b_1^2 - b_2^2)$$
or $$Ax + By = C$$

where A, B, C represent the numerical values

$$2(a_1 - a_2),\ 2(b_1 - b_2) \text{ and } (a_1^2 - a_2^2) + (b_1^2 - b_2^2).$$

Thus the locus of P (the mediator of QR) is given by a linear relation between the coordinates of any point on its path.

Again let us consider the form of the path traced out by a variable point P such that PO is of constant length, O being the origin of coordinates. (We should realise again that this is the 'centre-radius' definition of a circle, whose centre is O and radius is the given constant length.)

If we choose P to be (x, y) and r to be the constant length,

Then $PO = r$
Squaring $PO^2 = r^2$

and using the distance formula,

$$(x - 0)^2 + (y - 0)^2 = r^2$$
i.e. $$x^2 + y^2 = r^2.$$

The inference of these two investigations is that whenever $P(x, y)$ is a variable point in a Cartesian system, any equation connecting x and y will be representative of some locus in the plane of coordinates. This locus may be a straight line, as in the first illustration, or some more complicated shape in the form of a curve. In the next section of this chapter we examine closely the nature of some of these loci.

1 Determine the equation of the locus of the point $P(x, y)$ which moves so that $PA = PB$ where A is the point (2, 1), and B the point (0, 3).

2 C is the fixed point (4, 3) and $P(x, y)$ a variable point which moves so that $PC = 5$. Determine the equation of the locus of P and show that this locus passes through O.

3 $P(x, y)$ is equidistant from the y-axis and from the point (2, 0). Determine the equation of the locus of P.

4 A and B are the fixed points (1, 0), (5, 6). The point P moves so that $A\hat{P}B = 90°$ for all positions of P. Determine the equation of the locus of P.

5 P is the variable point (x, y) and $\triangle PAB$, $\triangle PCD$ are equal in area where A, B, C, D are the points (1, 0), (5, 0), (0, 2), (0, 4) respectively. Determine the equation of the locus of P.

5.4 The straight line

a) Of given gradient and located to pass through a fixed point (or fixed points)

Example

Determine the equation of a line which passes through the fixed point A (2, 1) and with gradient $\frac{1}{3}$.

If any point on the line is $P(x, y)$, in figure 5.6 the gradient of the line is $\tan P\hat{A}N$, which is given equal to $\frac{1}{3}$.

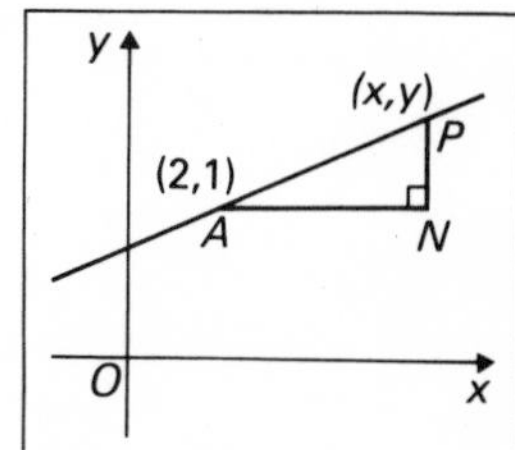

Figure 5.6

$$\tan P\hat{A}N = \frac{y-1}{x-2} = \frac{1}{3}$$

$$3y - 3 = x - 2$$

$$3y = x + 1$$

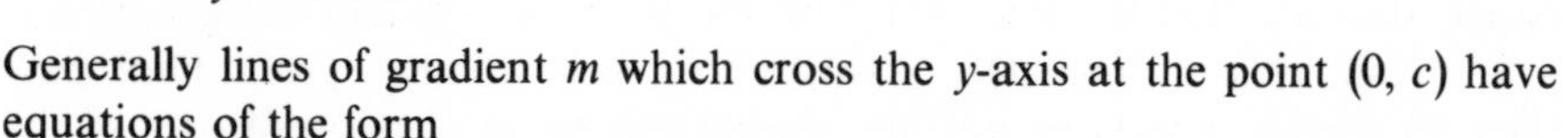

Generally lines of gradient m which cross the y-axis at the point $(0, c)$ have equations of the form

$$\frac{y-c}{x} = m \quad \text{i.e.}$$

$$y = mx + c$$

Those which pass through the fixed point (x_1, y_1) of gradient m have equations of the form

$$\frac{y-y_1}{x-x_1} = m \quad \text{i.e.}$$

$$y - y_1 = m(x - x_1)$$

and those whose directions are fixed by two points (x_1, y_1), (x_2, y_2) through which the lines pass have equations of the form

$$\frac{y - y_1}{x - x_1} = \frac{y_2 - y_1}{x_1 - x_1} \quad \text{i.e. } y - y_1 = \frac{y_2 - y_1}{x_2 - x_1}(x - x_1)$$

b) Cutting off specific intercepts *a* and *b* on the *x*- and *y*-axes

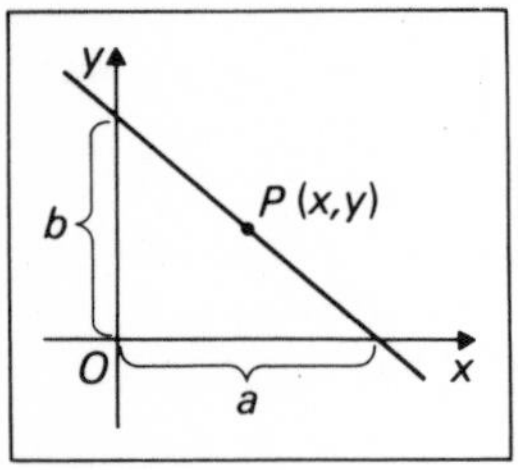

Figure 5.7

Such lines (figure 5.7) pass through the points $(a, 0)$ and $(0, b)$. Therefore, using the previous result the equation of the line is

$$\frac{y - b}{x - 0} = \frac{0 - b}{a - 0}$$

$$\Rightarrow -a(y - b) = bx$$

$$bx + ay = ab$$

or, dividing by ab,

$$\frac{x}{a} + \frac{y}{b} = 1 \quad \text{i.e.}$$

$$\frac{x}{\text{intercept on } x\text{-axis}} + \frac{y}{\text{intercept on } y\text{-axis}} = 1$$

These four general linear forms enable us to write down in such defined circumstances the equations of straight line loci. Moreover, given a linear locus, it is a simple task to locate the position and direction of the line by comparison, in particular, with the form $y = mx + c$. For example, the line $4x + 3y + 2 = 0$ may be rearranged in the form

$$y = -\frac{4}{3}x - \frac{2}{3}$$

and we can immediately sketch the line (as shown in figure 5.8) on graph paper, showing that the intercept c on the y-axis is $-\frac{2}{3}$, and the gradient is $-\frac{4}{3}$. Again, it is immediately possible to check whether given lines pass through specific points or not, for should this be so, the coordinates of the point must satisfy the equation of the locus.

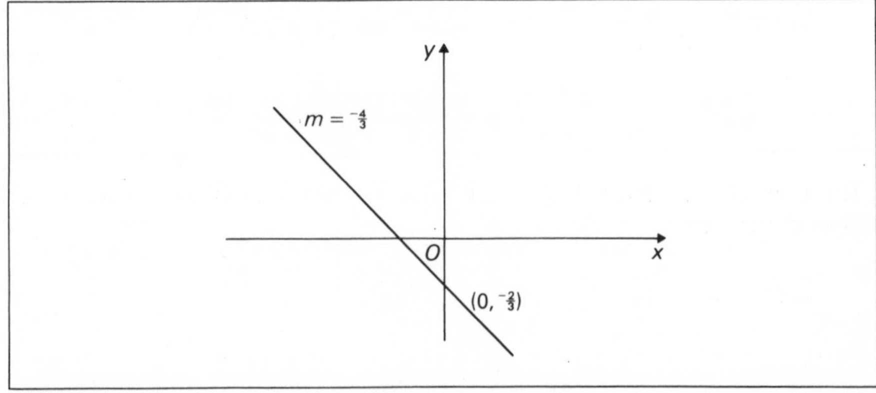

Figure 5.8

Thus $(1, -2)$ lies on $4x + 3y + 2 = 0$, since $4(1) + 3(-2) + 2$ does equal 0, whereas $(-1, 2)$ does not lie on the same line since $4(-1) + 3(2) + 2$ does not equal 0.

Since many geometric problems are concerned with figures basically consisting of points and lines, the strength of Descartes' system now becomes apparent. Some illustrations of these processes follow in the next examples. You should derive equations to the line locus using the specific definitions.

Exercise C

In questions 1 to 5 find the equations of the lines.

1 a) Passing through $(1, 1)$ and having gradient 3.
b) Passing through $(4, 5)$ and having gradient $\frac{1}{2}$.
c) Passing through $(4, -3)$ and having gradient $-\frac{3}{4}$.

2 a) Passing through the points $(-3, 2)$, $(4, -5)$.
b) Passing through the points $(4, 3)$, $(1, -1)$.
c) Passing through the points $(-2, 1)$, $(2, -3)$.

3 a) x-intercept $= 3$, y-intercept $= -5$.
b) x-intercept $= -\frac{1}{2}$, y-intercept $= -\frac{1}{3}$.
c) x-intercept $= \dfrac{-c}{t}$, y-intercept $= c$.

4 a) Passing through $(1, 2)$ and inclined to the x-axis at 45°.
b) Passing through $(0, 4)$ and inclined to the x-axis at 135°.

5 a) Passing through the point $(2, 5)$ and making an angle with the x-axis whose tangent is $\frac{1}{2}$.
b) Passing through the point $(3, 1)$ and making an angle with the x-axis whose tangent is -3.

6 Verify that
a) $(3, 4)$ lies on the line $3x - 2y - 1 = 0$.
b) $(-2, -4)$ lies on the line $3y = 5x - 2$.

7 Determine the equation of the straight line through the points $(2, 3)$, $(4, 1)$ and verify that the line passes through the point $(-1, 6)$.

Example 1

Determine the equation of the line passing through the points $A(1, 2)$, $B(5, 6)$ and the equation of the line through M, the mid-point of AB, perpendicular to AB. If these two lines cut the y-axis at U, V respectively, calculate the area of $\triangle UMV$.

Equation of AB is $\dfrac{y-2}{x-1} = \dfrac{6-2}{5-1} = 1$

$$\Rightarrow y = x + 1$$

Mid-point of AB is $\{\frac{1}{2}(1+5), \frac{1}{2}(2+6)\}$ i.e. $M(3, 4)$ and gradient of $AB = 1$. Thus the gradient of the line through M perpendicular to AB is -1 (using $m_1 m_2 = -1$).

Hence the equation of this perpendicular is $y - 4 = -(x - 3)$

i.e. $y = -x + 7$

AB and the perpendicular to AB through M, meet the y-axis at $U(0, 1)$ (from $y = x + 1$) and $V(0, 7)$ (from $y = -x + 7$).

$$\text{Area of } \triangle UMV = \tfrac{1}{2}UV \times (x\text{-coordinate of } M)$$
$$= \tfrac{1}{2}6 \times 3 = 9 \text{ square units}$$

Example 2

The equations of two lines LM, PQ are $2x + y = 4$ and $4x + 2y = 2$. Prove that the lines are parallel, and find the distance between them.

Since LM can be written in the form $y = -2x + 4$, its gradient is -2.

And PQ is $2y = -4x + 2$

i.e. $y = -2x + 1$ in which the gradient is also -2

$\Rightarrow LM$ and PQ are parallel.

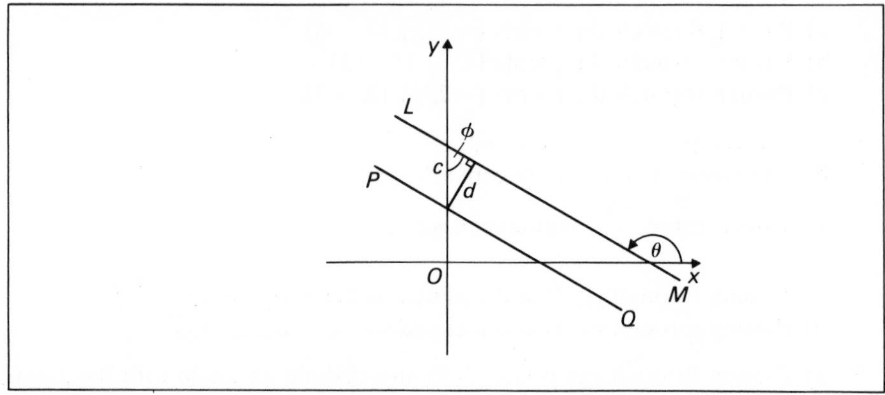

Figure 5.9

LM cuts the y-axis at $(0, 4)$, and PQ cuts the y-axis at $(0, 1)$. Hence the length cut off on the y-axis between the lines is $c = 4 - 1 = 3$, and using the details in figure 5.9, the distance between the lines is d

where $d = c \sin\phi = 3 \sin\phi$.

But $\phi = \theta - 90°$ where $\tan\theta = -2$

and $\tan\phi = -\cot\theta = \frac{1}{2}$

$$\sin\phi = \frac{1}{\sqrt{5}}$$

$$d = \frac{3}{\sqrt{5}} = \frac{3}{5}\sqrt{5}$$

Example 3

$PQRS$ is a quadrilateral in which P is the point $(1, 0)$, Q the point $(2, -4)$, and S the point $(-1, -2)$. PR is perpendicular to QS and QR is parallel to PS. Determine the coordinates of R.

$$\text{Gradient of } QS = \frac{-4+2}{2+1} = -\tfrac{2}{3}$$

Hence gradient of $PR = \frac{3}{2}$

$\Rightarrow$ equation to PR is $y = \frac{3}{2}(x - 1)$

Gradient of $PS = \dfrac{0 + 2}{1 + 1} = 1$

Hence gradient of QR is 1

$\Rightarrow$ equation to QR is $y + 4 = x - 2, \quad \Rightarrow y = x - 6$

$\Rightarrow QR$ and PR meet at the point of intersection of $y = \frac{3}{2}(x - 1)$ and $y = x - 6$

i.e. $3(x - 1) = 2(x - 6) \Rightarrow x = -9$
and $y = -15$

$\Rightarrow R$ is the point $(-9, -15)$

Example 4

Determine the area of the triangle (figure 5.10) with sides

$L_1 \equiv 7x + 8y - 5 = 0$
$L_2 \equiv 11x + y + 50 = 0$
$L_3 \equiv 4x - 7y - 26 = 0$

Solving the pairs of lines simultaneously, we obtain the coordinates of the vertices of $\triangle ABC$

L_1 and L_2 meet at $A(-5, 5)$
L_2 and L_3 meet at $B(-4, -6)$
L_3 and L_1 meet at $C(3, -2)$

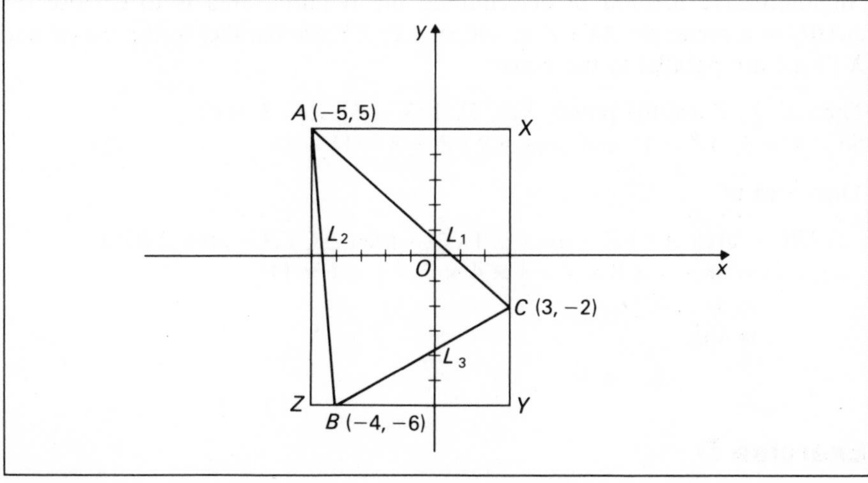

Figure 5.10

There are many ways of approaching this problem. Suppose we decide upon '$\frac{1}{2}$ base × height' with BC as base and 'height' the perpendicular from A to BC.

$BC = \sqrt{\{(-4 - 3)^2 + (-6 + 2)^2\}} = \sqrt{65}$

and BC may be written in the form

$y = \frac{4}{7}x - \frac{26}{7} \Rightarrow$ gradient $\frac{4}{7}$

Hence the gradient of the perpendicular to BC is $-\frac{7}{4}$ and the equation of this perpendicular through A is

$$y - 5 = \frac{-7}{4}(x + 5)$$

i.e. $\quad 4y = -7x - 15$

This line meets BC at the point $N(-\frac{1}{65}, -\frac{242}{65})$

$AN = \sqrt{\{(-5 + \frac{1}{65})^2 + (5 + \frac{242}{65})^2\}} = \frac{81}{65}\sqrt{65}$

Hence area $\triangle ABC = \frac{1}{2} \times \sqrt{65} \times \frac{81}{65}\sqrt{65} = 40{\cdot}5$ square units.

It is worth noting that the length of the perpendicular drawn from a point (x_1, y_1) to the line $ax + by + c = 0$ is given by the positive value of the expression

$$\pm\frac{ax_1 + by_1 + c}{\sqrt{(a^2 + b^2)}}$$

(Prove this result using the method shown for AN or otherwise.) So we could have written down directly for the point $A(-5, 5)$ and the line $L_3 = 4x - 7y - 26 = 0$.

$$AN = \text{positive value of } \pm\frac{4(-5) - 7(5) - 26}{\sqrt{(4^2 + 7^2)}}$$

$$= \frac{81}{\sqrt{65}}$$

An alternative method of determining the required area is to enclose the $\triangle ABC$ in a rectangle $AXYZ$ in which AX, ZY are parallel to the x-axis and XY, AZ are parallel to the y-axis.

Then X, Y, Z are the points $X(3, 5)$, $Y(3, -6)$, $Z(-5, -6)$.
So $AX = 8$, $AZ = 11$ and area $AXYZ = 8 \times 11 = 88$

Then area of

$$\begin{aligned}\triangle ABC &= \text{area } AXYZ - \text{area } \triangle AXC - \text{area } \triangle CYB - \text{area } \triangle BZA\\ &= 88 - \tfrac{1}{2} \times 8 \times 7 - \tfrac{1}{2} \times 4 \times 7 - \tfrac{1}{2} \times 1 \times 11\\ &= 88 - 47\tfrac{1}{2}\\ &= 40\tfrac{1}{2}\end{aligned}$$

Exercise D

1 Find the equation of the line which
a) is parallel to $3x + 4y = 12$,
b) cuts off an intercept of 5 units on the x-axis.
Determine also the equation of a line passing through the point (4, 5) and perpendicular to $3x + 4y = 12$.

2 Determine the equation of the line through the point of intersection of the lines $y + x = 3$ and $y = 2x$, which is also parallel to $x + 4y = 1$. If this line passes through $(x_1, 3)$, determine the value of x_1.

3 $ABCD$ is a rectangle in which the vertices are labelled counterclockwise. Given that A is the point $(1, 2)$, and C the point $(5, 5)$, and that both B and D lie on the line $x = 3$, find

a) the coordinates of B and D,
b) the equation of the sides AB and BC of the rectangle,
c) the lengths of the sides of the rectangle.

4 The equations of the sides PQ and QR of a parallelogram $PQRS$ are $3x - y - 7 = 0$ and $x - 3y + 5 = 0$. The point $K(-8, 3)$ lies on the side PS and the point $H(5, 2)$ lies on RS. Calculate the coordinates of the vertex S.

5 P, Q, R, S are the mid-points of the sides AB, BC, CD, DA respectively of the quadrilateral $ABCD$. Given that the coordinates of the vertices of this quadrilateral are $A(2, 2)$, $B(5, 0)$, $C(8, 10)$ and $D(3, 6)$, determine the equations of the diagonals PR and QS of the figure $PQRS$, and the point of intersection of these diagonals. If this point is X, and AC and BD intersect at Y, calculate also the length of XY.

6 $\triangle ABC$ has vertices $A(1, 0)$, $B(2, -4)$ and $C(-5, -2)$. The altitudes from A and B perpendicular to the sides facing these vertices are AH and BK, which meet at W. Derive

a) the equations of AH and BK,
b) the coordinates of W,
c) the area of $\triangle ABC$.

7 In the parallelogram $OABC$ where O is the origin of coordinates, B is the point $(2, 3)$, OA is the line $y = \frac{1}{4}x$ and the gradient of the side OC is -1. Determine the equation of the side BC and calculate the coordinates of the point of intersection of the diagonals of the parallelogram.

Find also the equation of the side CO, and the coordinates of C.

8 Is the figure $ABCD$, where A, B, C, D are the points $A(1, 5)$, $B(6, 2)$, $C(3, -3)$, $D(-2, 0)$ a square? Justify your answer.

9 In the $\triangle ABC$, $A\hat{B}C = 90°$, $AB = 2BC$ and A, B are the points $A(6, 7)$ and $B(0, -1)$. Determine the equation of AB and also its length.

Find also the equation of BC and the coordinates of the two possible positions which may be found for C.

10 The vertices of a $\triangle PQR$ are $P(3, 1)$, $Q(-2, 4)$ and $R(5, 0)$. Determine the equation of the side PR, the length of the perpendicular from Q to PR, and the area of $\triangle PQR$.

11 A line passes through the points $(2, 0)$ and has a given gradient $-\frac{3}{4}$. Write down the equation of the line. Calculate also the coordinates of the two points which lie on the line but are situated at a distance of 10 units from $(2, 0)$.

12 Determine the equation of the line which passes through the point $(-2, 5)$ and is perpendicular to the line $2y = x - 4$. Calculate the coordinates of the point of intersection of the two lines.

Prove also that the point $(-6, 1)$ is equidistant from the two lines. Hence find the equation of the bisector of the angle between the lines.

13 The perpendicular drawn from the origin to the lines $2x + y - 3 = 0$ and $x - 2y + 5 = 0$ meet these lines at L and N respectively. Determine the equation of the line LN.

14 $L_1 \equiv 4x + 5y - 18 = 0$ and $L_2 \equiv 7x + 2y = 0$ are two adjacent sides of a parallelogram. The equation of a diagonal of the parallelogram is, therefore, $11x + 7y - 9 = 0$. Calculate the coordinates of the vertices of the parallelogram.

15 The equations of the sides of a $\triangle ABC$ are $7x - 3y = 0$, $2x - 5y = 0$ and $5x + 2y = 29$. Determine the coordinates of A, B and C and the equations of the three medians of the triangle. Verify that the medians meet at a point (the geometric centre of the triangle), and calculate the coordinates of this point.

5.5 Linear inequalities

Before we leave the discussion on basic ideas associated with the straight line equation, it is worthwhile considering the geometric interpretation of the solution of allied inequalities. For example, $x + y = 3$, representing a line cutting off intercepts of 3 units on both x- and y-axes, is satisfied by an infinite number of points expressible in the form of an ordered pair (x, y). We must therefore examine the two further relations $x + y > 3$ and $x + y < 3$, which immediately arise for such ordered pairs.

It is easy to write down such points as (0, 4), (1, 3), (2, 2) and (−1, 5) all of which satisfy $x + y > 3$. Furthermore, such values as (0, 4·1), (0, 4·2), (0, 4·3), ..., give further emphasis to the simplicity of this task. From figure 5.11(a) it

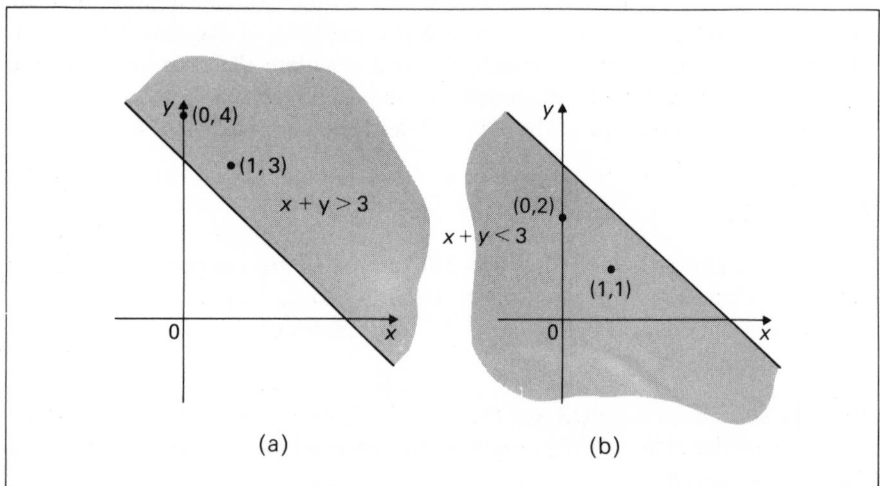

Figure 5.11

is clear that all of these points lie in, and hence define, a region of the x-y plane 'above', or to the right of the line $x + y = 3$. This region is shown shaded in figure 5.11(a) and excludes the line $x + y = 3$ itself, which serves as a boundary.

(**Note** that the inequality $x + y \geqslant 3$ is satisfied by all points in the shaded region, together with those lying on the line itself.)

Furthermore, the inequality $x + y < 3$, satisfied by such pairs of values as (0, 2), (1, 1), (2, 0) and (1, 1·9), (1, 1·8), (1, 1·7), ..., has an infinite number of solutions, all located in the region of the coordinate plane 'below' or to the left of, the line $x + y = 3$. (And $x + y \leqslant 3$ includes points on the line as before.) Such a region is shown shaded in figure 5.11(b).

Problems on compound inequalities are solved by the methods illustrated in the following example.

Example
Determine the region of the x-y plane for which the inequalities $x + y < 1$, $x - y > -1$ and $3y > x - 1$ are satisfied simultaneously.
Taking the 'boundary' lines in pairs,

$x + y = 1$ and $x - y = -1$ intersect at $(0, 1)$
$x - y = -1$ and $3y = x - 1$ intersect at $(-2, -1)$
and $3y = x - 1$ and $x + y = 1$ intersect at $(1, 0)$

For $x + y < 1$, points (x, y) lie to the left of the line $x + y = 1$,
For $x - y > -1$, points (x, y) lie to the right of the line $x - y = -1$,
and for $3y > x - 1$ such points all lie 'above' i.e. to the left of, the line $3y = x - 1$. Combining together, the region as shaded in figure 5.12 is defined by those values of (x, y) which satisfy the three inequalities simultaneously. This region excludes all three boundary lines.

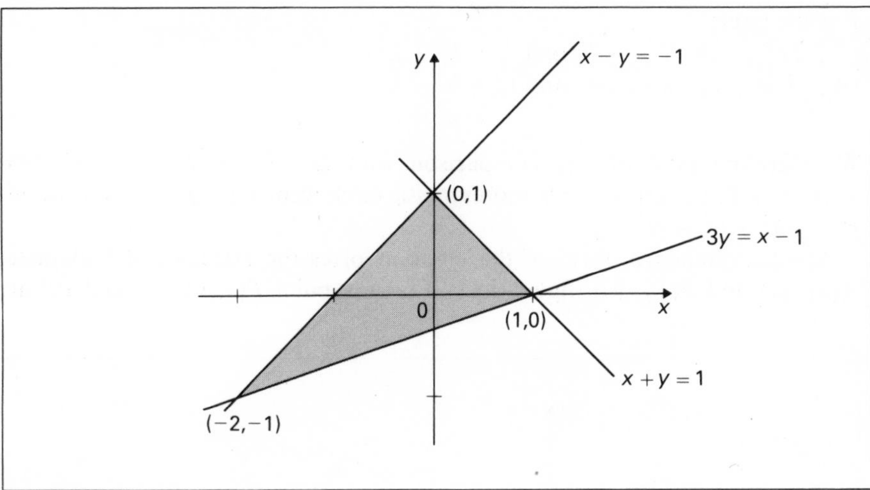

Figure 5.12

5.6 The circle

The second of the examples in section 5.3 illustrated the form of the equation of a circle, centre at O and radius r. Should the centre now be at the point $C(h, k)$, and radius be r (figure 5.13), for any point $P(x, y)$, $PC = r$.
Squaring $PC^2 = r^2$

and using the distance formula

the equation of the circle is $(x - h)^2 + (y - k)^2 = r^2$

This equation may be written as $x^2 - 2hx + h^2 + y^2 - 2ky + k^2 - r^2 = 0$

or $$x^2 + y^2 - 2hx - 2ky + (h^2 + k^2 - r^2) = 0$$

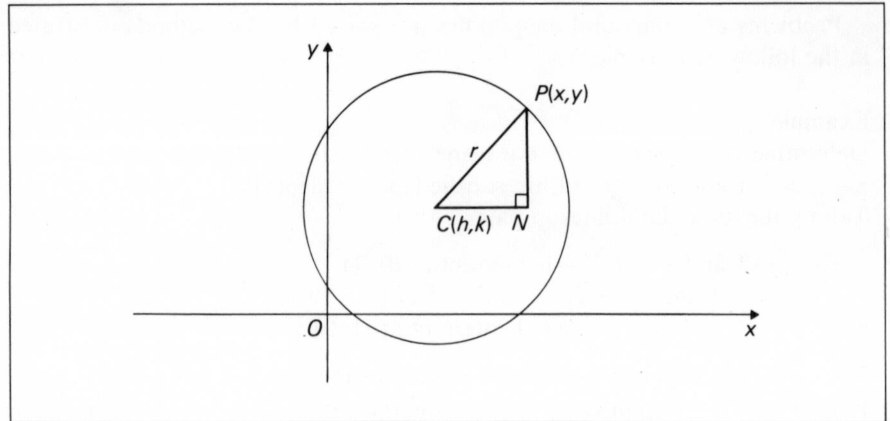

Figure 5.13

This is clearly of the form $x^2 + y^2 + 2gx + 2fy + c = 0$, g, f and c being numerical values.

For the circle

$$2g = -2h,\ 2f = -2k \text{ and } c = h^2 + k^2 - r^2$$

i.e. $\quad h = -g,\quad k = -f$ and $r^2 = h^2 + k^2 - c$

$$= g^2 + f^2 - c$$

We therefore establish, by comparison with $(x - h)^2 + (y - k)^2 = r^2$, that $x^2 + y^2 + 2gx + 2fy + c = 0$ represents a circle centre $(-g, -f)$ and radius $r = \sqrt{(g^2 + f^2 - c)}$.

Another 'standard' form of the circle involves the extremes of a diameter $A(x_1, y_1)$ and $B(x_2, y_2)$ (figure 5.14). For any point $P(x, y)$, AP and BP are

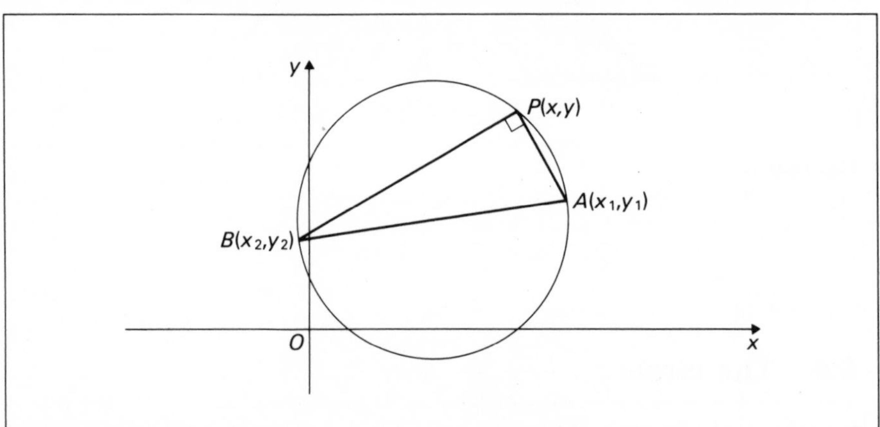

Figure 5.14

perpendicular chords (angle in a semicircle is equal to 90°). Therefore the product of the gradients of AP and $BP = -1$, i.e.

$$\left(\frac{y - y_1}{x - x_1}\right) \times \left(\frac{y - y_2}{x - x_2}\right) = -1$$

$$\Rightarrow (y - y_1)(y - y_2) = -(x - x_1)(x - x_2)$$

or $\quad (x - x_1)(x - x_2) + (y - y_1)(y - y_2) = 0$

Example 1

Write down the equation of a circle centre $(-2, 3)$ and radius 4.

The circle is $(x+2)^2 + (y-3)^2 = 4^2$

i.e. $x^2 + 4x + 4 + y^2 - 6y + 9 = 16$

$x^2 + y^2 + 4x - 6y - 3 = 0$

Example 2

A circle is drawn such that $(0, -1)$ and $(2, 3)$ are the ends of a diameter. Determine the equation of the circle and the length of the chord of the circle which is a part of the line $y = -x$.

The equation of the circle is

$$(x-0)(x-2) + (y+1)(y-3) = 0$$

i.e. $x^2 - 2x + y^2 - 2y - 3 = 0$

$x^2 + y^2 - 2x - 2y - 3 = 0$

The circle is cut by the line $y = -x$ at the points (x, y) where

$$x^2 + x^2 - 2x + 2x - 3 = 0$$
$$2x^2 - 3 = 0 \Rightarrow x = \pm\sqrt{\tfrac{3}{2}}$$
$$\Rightarrow y = \mp\sqrt{\tfrac{3}{2}}$$
$$\Rightarrow \text{length of chord} = \sqrt{\{(\sqrt{\tfrac{3}{2}} + \sqrt{\tfrac{3}{2}})^2 + (-\sqrt{\tfrac{3}{2}} - \sqrt{\tfrac{3}{2}})^2\}}$$
$$= 2\sqrt{3} \text{ units}$$

Example 3

Determine the equation of the circumcircle of the $\triangle ABC$ where A, B, C are the points $(5, 0)$, $(6, 0)$ and $(8, 6)$. Write down the centre and radius of this circle. Suppose that the circle has the equation

$$x^2 + y^2 + 2gx + 2fy + c = 0$$

Since A, B, C lie on the circle, the coordinates of the three points satisfy the equation

$A(5, 0) \Rightarrow 25 + 10g + c = 0$

$B(6, 0) \Rightarrow 36 + 12g + c = 0$

from which $g = -5{\cdot}5$, $c = 30$

$C(8, 6) \Rightarrow 64 + 36 + 16g + 12f + c = 0$

thus $100 - 88 + 12f + 30 = 0$

and $f = -3{\cdot}5$

The equation of the circle is therefore

$$x^2 + y^2 - 11x - 7y + 30 = 0$$

The centre is '$(-g, -f)$' i.e. $(5{\cdot}5, 3{\cdot}5)$ and the radius

$$= \text{‘}\sqrt{(g^2 + f^2 - c)}\text{’}$$
$$= \sqrt{(\tfrac{121}{4} + \tfrac{49}{4} - 30)}$$
$$= \sqrt{\tfrac{50}{4}}$$
$$= \tfrac{5}{2}\sqrt{2} \text{ units}$$

Exercise E

1 Write down the equations of the circles defined as follows
  a) Centre at O, radius 5 units.
  b) Centre $(2, -3)$, radius 3 units.
  c) Centre at O, passing through $(-3, 4)$.
  d) Centre at $(-1, 2)$, passing through $(2, 6)$.
  e) With the line joining $(4, 3)$ and $(2, -1)$ as diameter.

2 Write down the coordinates of the centre of the following circles, and their radii.
  a) $(x+1)^2 + (y-2)^2 = 5$
  b) $(x+1)(x-3) + (y-2)(y-4) = 0$
  c) $x^2 + y^2 + 4x - 6y - 3 = 0$

3 Determine the equations of each of the following circles, and their centres and radii, given that the circles pass through the points
  a) $(-1, 3), (2, 2), (1, 4)$
  b) $(2, 1), (0, 5), (-1, 2)$
  c) $(-6, 5), (-3, -4), (2, 1)$

Geometrical problems involving straight lines and circles will frequently contain a determination of their coordinates of intersection (if any). Since these coordinates satisfy both equations

i.e. $Ax + By + C = 0$
and $x^2 + y^2 + 2gx + 2fy + c = 0$

in general forms, the elimination of x (or y) will yield a quadratic equation in y (or x). This quadratic may have
a) two real distinct roots, showing two points of intersection of the line and circle, or
b) two real coincident roots, showing that the two points of intersection coincide, and hence the line is a tangent, or
c) two imaginary roots, showing that the line does not cross the circle at all.

The following examples illustrate these and other geometric properties of the circle.

Example 1

Find the length of the chord whose equation is $4x - 3y - 5 = 0$ of the circle $x^2 + y^2 + 3x - y - 10 = 0$.

From $4x - 3y - 5 = 0$, or $3y = 4x - 5$
and $9x^2 + 9y^2 + 27x - 9y - 90 = 0$
$\Rightarrow 9x^2 + (4x-5)^2 + 27x - 3(4x-5) - 90 = 0$
i.e. $25x^2 - 25x - 50 = 0$
$x^2 - x - 2 = 0$
$(x-2)(x+1) = 0$
$\Rightarrow x = 2, -1$
$\Rightarrow$ Points of intersection are $(2, 1), (-1, -3)$
$\Rightarrow$ The length of the chord joining them is
$\sqrt{\{(2+1)^2 + (1+3)^2\}}$
$= 5$

Example 2

Prove that the line $5x + 12y = 169$ is a tangent to the circle $x^2 + y^2 = 169$. Find the coordinates of the point of contact of the tangent.

The line cuts the circle at the points whose y values are given by

$$\tfrac{1}{25}(169 - 12y)^2 + y^2 = 169$$

i.e. $169^2 - 2 \times 12 \times 169y + 144y^2 + 25y^2 = 25 \times 169$

i.e. $y^2 - 24y + 144 = 0$ on dividing by 169

so $(y - 12)^2 = 0$

Since this is a perfect square, both values of y of the points of intersection are the same. Therefore the line is a tangent to the circle.

When $y = 12$, $5x + 144 = 169$

$$x = 5$$

And the point of contact of the tangent is $(5, 12)$.

Example 3

Calculate the length of the tangent from the point $A(8, 7)$ to the circle $x^2 + y^2 - 6x - 2y + 1 = 0$.

The centre of the given circle is $C(3, 1)$, and the radius of the circle r, is

$$\sqrt{\{(-3)^2 + (-1)^2 - 1\}} = 3$$

By Pythagoras, $AC^2 = r^2 + t^2$, where t is length of the tangent from A to the circle.

$$AC^2 = \{(8 - 3)^2 + (7 - 1)^2\} = 61$$
$$61 = 9 + t^2$$
$$t = 2\sqrt{13}$$

Example 4

Determine the equations of the two tangents to the circle $x^2 + y^2 = 25$ which are parallel to the line $4x - 3y - 2 = 0$.

The given line may be written in the form

$y = \frac{4}{3}x - \frac{2}{3}$, i.e. it has a gradient of $\frac{4}{3}$

Now, let the equation of a line parallel to this line be $y = \frac{4}{3}x + c$. This line meets the circle in points (x, y), where

$$x^2 + (\tfrac{4}{3}x + c)^2 = 25$$
$$\tfrac{25}{9}x^2 + \tfrac{8}{3}cx + c^2 - 25 = 0$$
$$25x^2 + 24cx + 9(c^2 - 25) = 0$$

For $y = \frac{4}{3}x + c$ to be a tangent, this quadratic must be a perfect square

i.e. $(24c)^2 = 4 \times 25 \times 9(c^2 - 25)$

$$16c^2 = 25(c^2 - 25)$$
$$9c^2 = 25^2$$
$$\Rightarrow c = \pm\tfrac{25}{3}$$

the tangents are $y = \frac{4}{3}x \pm \frac{25}{3}$

i.e. $3y = 4x + 25$ and $3y = 4x - 25$

Exercise F

1 Find the coordinates of the points at which the line $y = -x + 4$ cuts the circle $x^2 + y^2 + 4x - 2y - 20 = 0$.

2 The chord $y = 3x + 5$ of the circle $x^2 + y^2 = 5$ intersects the circle at A, B. Find the length of AB.

3 The line $3y = x + 8$ intersects the circle $9x^2 + 9y^2 - 18x + 6y - 170 = 0$ at A and B. Given that C is the centre of the circle, prove that $A\hat{C}B = 90°$.

4 Calculate the length of the tangent from the point (8, 4) drawn to the circle

$$4x^2 + 4y^2 - 16x - 12y - 63 = 0.$$

5 The line $y = mx + c$ is a tangent to the circle $x^2 + y^2 = 25$, touching the circle at the point $(-3, 4)$. Calculate the values of m and c.

6 The circle

$$x^2 + y^2 - 4x - 6y + \lambda = 0$$

touches the x-axis at the point X. Determine the value of λ and the coordinates of X.

5.7 Parametric coordinates

Much of the work of this chapter so far has been concerned with the locus of a point moving according to given conditions, and the algebraic equation of the locus fulfilled by all points (x, y) obeying these given conditions. We now introduce a further idea and method of defining such loci.

Consider, for example, the equations

$$x = 3t - 2, \ y = 3 - 2t$$

expressing the coordinates (x, y) of a point in terms of another variable t. Giving t a set of numerical values, corresponding values of (x, y) can be calculated.

A few of these values are displayed in the following table.

t	1	2	3	0	-1
x	1	4	7	-2	-5
y	1	-1	-3	3	5

If these points (x, y) are plotted on graph paper (figure 5.15), it is clear that they all lie on a straight line whose equation can be derived as

$$y = -\tfrac{2}{3}x + \tfrac{5}{3} \quad \text{or} \quad 2x + 3y = 5$$

For every value of t, one point (x, y) can be defined this way, and the locus is the infinite set of points corresponding to the infinite set of values of t. We give the name **'parameter'** to this variable t and say that the equations expressing

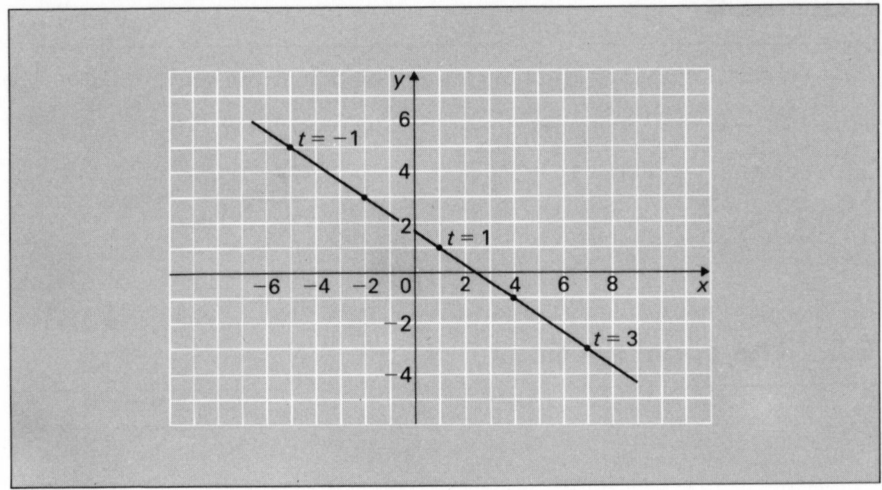

Figure 5.15

x and y in terms of t are the **parametric equations** of the locus. Moreover, from $x = 3t - 2$ and $y = 3 - 2t$ we can write $\frac{1}{3}(x + 2) = t$ and $t = \frac{1}{2}(3 - y)$. Thus, eliminating t,

$$\begin{aligned} &\tfrac{1}{3}(x + 2) = \tfrac{1}{2}(3 - y) \\ \Rightarrow\ &2(x + 2) = 3(3 - y) \\ \Rightarrow\ &2x + 3y = 5 \end{aligned}$$

i.e. the cartesian equation of the locus as already achieved.

It is reasonable to assume, therefore, that it will be possible to express a locus either in the form of its cartesian equation, or in parametric form. The second alternative becomes more important in the final section of this chapter.

A further illustration of this idea is to consider the equations

$$x = 5 \cos \theta, \quad y = 5 \sin \theta$$

which may be considered to define a set of points (x, y) for a set of values of the variable θ.

The nature of the locus of these points is readily determined if we realise that

$$\cos \theta = \frac{x}{5}, \quad \sin \theta = \frac{y}{5},$$

and $\cos^2 \theta + \sin^2 \theta = 1$ for all θ.

Eliminating θ,

$$\frac{x^2}{25} + \frac{y^2}{25} = 1$$

i.e. $x^2 + y^2 = 25$

and the locus is a circle, centre at O, and radius 5 units. The equations $x = 5 \cos \theta$, $y = 5 \sin \theta$ are therefore the parametric equations of this circle.

Exercise G

Find the cartesian equations of the loci whose parametric equations are

1 $x = 2t - 1, \quad y = 5t + 3$

2 $x = 3t + 1, \quad y = 4t - 2$

3 $x = 2t, \quad y = \dfrac{1}{t}$

4 $x = 2t^2, \quad y = 2t$

5 $x = \cos 2\theta, \quad y = \cos^2 \theta$

5.8 The conic sections

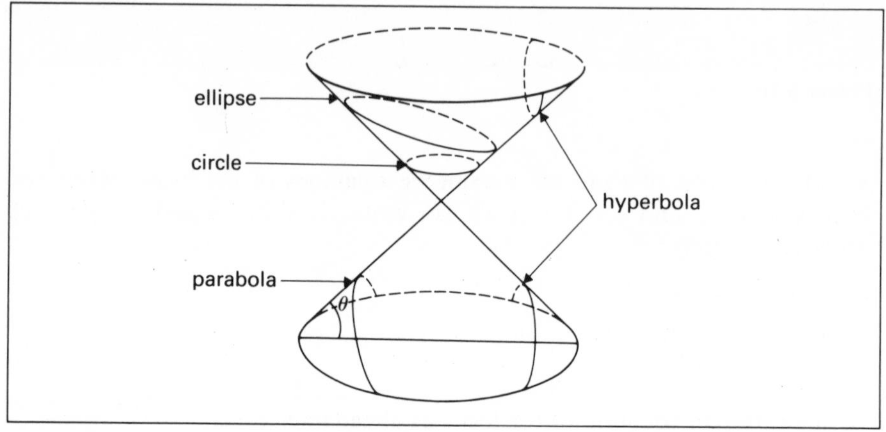

Figure 5.16

When a double cone is cut by a plane at an acute angle α to the base of the cone, the set of cross-sections so obtained is called the set of **conic sections**. The sections are:

a) circle $(\alpha = 0)$
b) ellipse $(\alpha < \theta)$
c) parabola $(\alpha = \theta)$
d) hyperbola $(\alpha > \theta)$
e) pair of straight lines ($\alpha > \theta$, and plane through vertex)
f) single straight line ($\alpha = \theta$, and plane through vertex)
g) single point ($\alpha < \theta$, and plane through vertex)

These sections may also be defined as loci, as follows: 'the locus of a point P such that its distance from a fixed point S, the **focus**, is a constant multiple, e, of its distance from a fixed line UV, called the **directrix**'. The multiple e is called the **eccentricity**. Of the sections above, e), f), g) are degenerate cases and are not considered in this chapter. For the others, we have:

b) an ellipse (for $0 < e < 1$)
c) a parabola (for $e = 1$)
d) a hyperbola (for $e > 1$)

We limit our study here to a consideration of the standard equations of the parabola and ellipse and an examination of some of the more important properties of these two loci.

Given the position of the focus S, and the directrix NA, let SA be the perpendicular drawn from the focus to the directrix to meet it at A. Let $SA = 2a$ and let O be the mid-point of SA. Then, since $OS = OA$, O is a point on the parabola. Now choose OS to define the positive direction of the x-axis, and Oy parallel to the directrix to be the y-axis (figure 5.17). Then S is the

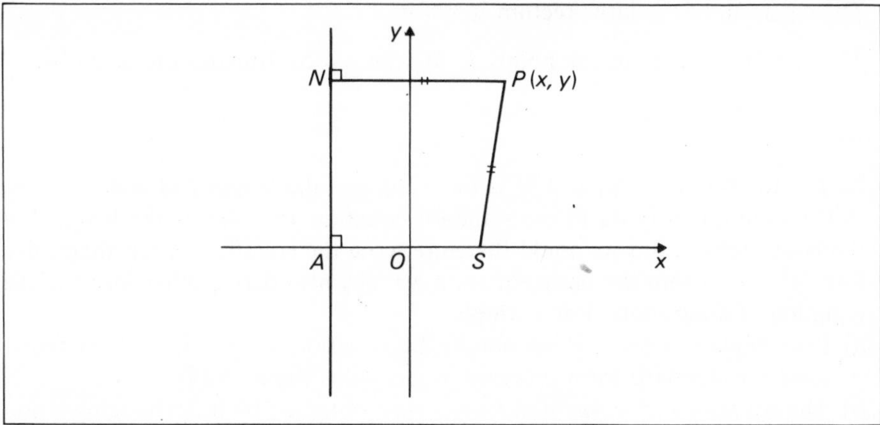

Figure 5.17

point $(a, 0)$. Then any point P on the locus will have coordinates (x, y) referred to this cartesian system, and the locus of P will be defined by $PS = PN$ where N is the foot of the perpendicular drawn from P to the directrix, and has coordinates $(-a, y)$.

Thus $PS = PN$ gives

$$\sqrt{\{(x-a)^2 + y^2\}} = x + a$$

Squaring $\quad (x-a)^2 + y^2 = (x+a)^2$

$$y^2 = (x+a)^2 - (x-a)^2$$
$$= \{(x+a) + (x-a)\}\{(x+a) - (x-a)\}$$
$$= (2x)(2a)$$
$$y^2 = 4ax$$

This is the standard equation of the parabola and we note that

a) it passes through the origin (since $OS = OA$),

b) it is symmetrical about the x-axis.

For all values of x, $y \propto \pm\sqrt{x}$, so as $x \to \infty$, y also $\to \infty$

The shape of the parabola is as shown in figure 5.18.

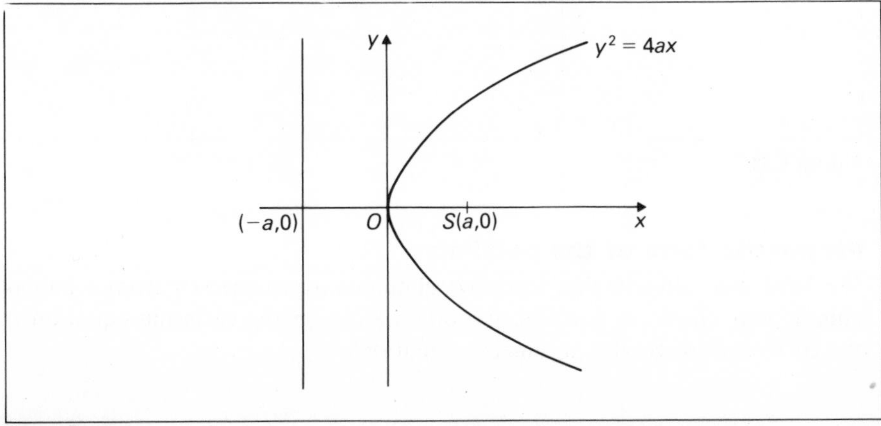

Figure 5.18

The special point O on the parabola is called the **vertex of the parabola**, and the axis Ox about which the curve is symmetrical is called the **axis of the parabola**.

Any line joining two points on the parabola is defined to be a **chord** of the parabola. If a chord passes through the focus of the parabola, it is called a **focal-chord**. Moreover, the particular focal-chord which is perpendicular to the axis of the parabola is the **latus rectum**.

The equation of the latus rectum is $x = a$

This meets $y^2 = 4ax$ at the points L, M whose y coordinates are given by

$$y^2 = 4a^2$$

i.e. $\quad y = \pm 2a$

So L is the point $(a, 2a)$ and M is $(a, -2a)$, and the length $LM = 4a$.

Thus the quantity $4a$ in the standard equation $y^2 = 4ax$ is the length l of the latus rectum, and we could therefore write the equation in the alternative form $y^2 = lx$. From the basic equation we may also derive other forms of the equation of a parabola. For example,

a) If we replace x by $-x$, we obtain the equation $y^2 = -4ax$ which represents the standard form reflected in the y-axis (figure 5.19).

b) The equations $x^2 = 4ay$ and $x^2 = -4ay$, obtained by interchanging x and y, and reflecting in the x-axis, as shown in figure 5.20.

A further form of the equation of the parabola with which we should also be familiar is the equation $(y - \beta)^2 = 4a(x - \alpha)$. In this case the vertex of the parabola has been translated to the point (α, β), the axis of the parabola remaining parallel to Ox, and the latus rectum remaining unchanged (figure 5.21).

(**Note** The equation of any parabola whose axis is oblique to the x-axis is beyond our need at this stage.)

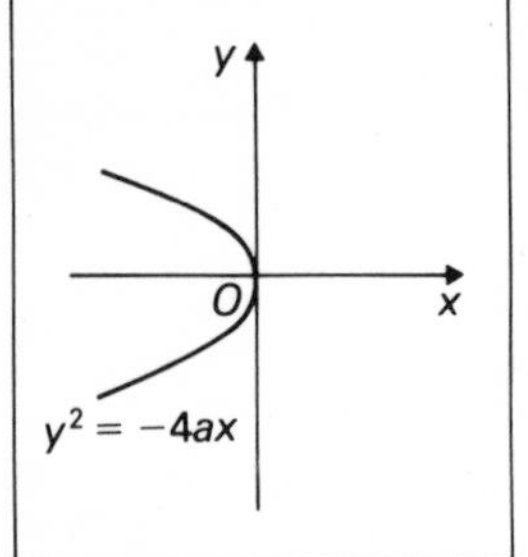

Figure 5.19

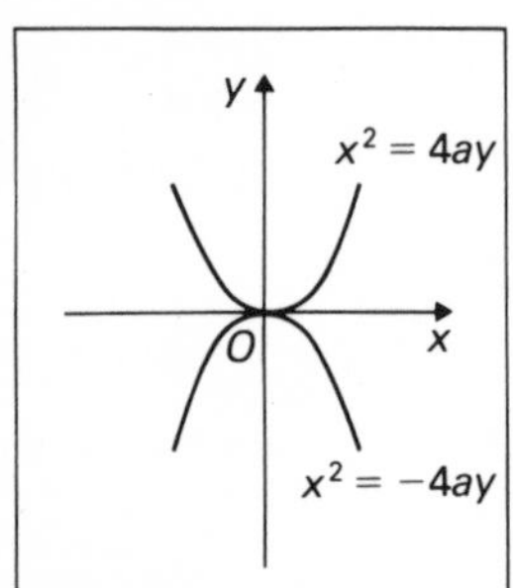

Figure 5.20

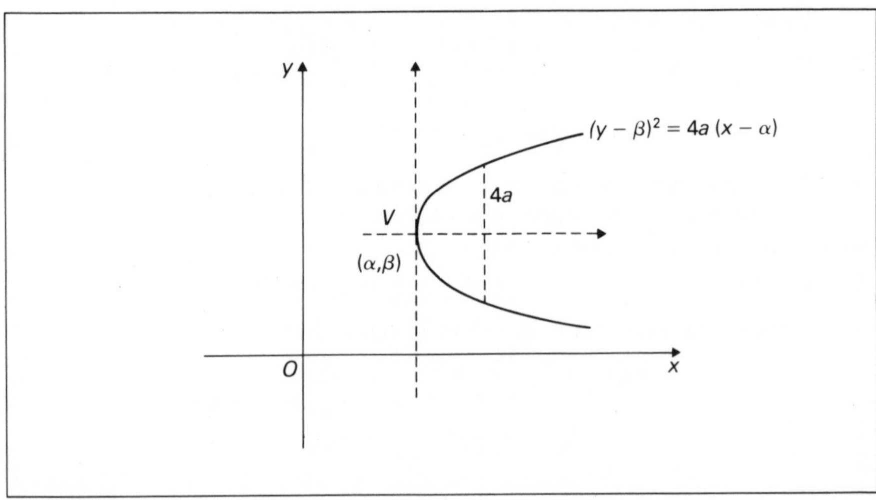

Figure 5.21

Parametric form of the parabola

We have seen already that when we eliminate a parameter t from a pair of equations of the form $x = f(t)$, $y = g(t)$, we obtain the cartesian equation of the curve defined by the parametric equations.

Now, if a curve is defined by the parametric equations $x = at^2$, $y = 2at$, where a is a given constant, we can write

$$t = \frac{y}{2a}.$$

Then $$x = a\left(\frac{y}{2a}\right)^2$$

$$= \frac{y^2}{4a}$$

Hence $y^2 = 4ax$ and the given parametric equations define the parabola as written in its standard form. Hence, we may choose the pair of equations $x = at^2$, $y = 2at$ as an alternative definition of the parabola. Any point on the curve corresponds to a particular value of the parameter t.

Exercise H

1 Write down the equation of the parabolas defined as follows
a) focus (2, 0), directrix $x = -2$
b) focus (0, -1), directrix $y = 1$
c) focus (1, 2), directrix $x + 1 = 0$

2 Sketch the following parabolas showing both the position of the focus and directrix.
a) $y^2 = 25x$ b) $x^2 = 9y$ c) $(y + 1)^2 = 4x$

3 Express the equation $y^2 + 4y = 8x - 12$ in the form $(y + \beta)^2 = 4a(x + \alpha)$. Hence determine the coordinates of the vertex and focus of the parabola $y^2 + 4y = 8x - 12$.

4 Determine the length of the latus rectum of
a) $y^2 = 16x$ b) $x^2 = -8y$ c) $(y - 2)^2 = 2(x - 1)$

5 The point (5, 20) lies on the parabola $y^2 = 4ax$. Determine the value of a and the length of the latus rectum.

6 Determine the equation of a parabola whose focus is the point (2, 1) and directrix is the line $2x + y + 1 = 0$.

7 Determine the cartesian equations of the curves whose parametric equations are
a) $x = t^2$, $y = t$ b) $x = -10t^2$, $y = 5t$ c) $x + 1 = 9t^2$, $y = 3t$

8 Determine the coordinates of the focus of the parabola $y^2 = 12x$. Prove that the point $P(3, 6)$ lies on the parabola and derive the equation of the focal chord which passes through P.

9 The extremes of a focal chord of the parabola $x = at^2$, $y = 2at$ are the points $P(ap^2, 2ap)$ and $Q(aq^2, 2aq)$. By writing down the gradients of PS and QS, where S is the focus, prove that $pq = -1$.

10 Determine a parametric representation of the parabola $x^2 = -4y$.

5.10 The ellipse

Given that the focus-directrix definition of an ellipse is $PS = ePN$, where $e < 1$, S being the focus and N the foot of the perpendicular drawn from the variable point P to the directrix, let the perpendicular from S to the directrix meet it at M. Then the point A, lying on SM such that $AS = eAM$, will be a point on the locus. There is also a second point A' lying on MS produced such that $A'S = eA'M$, which will also lie on the locus. We now choose O, the mid-point of AA', to be the origin of the coordinate system and the line OM to be the positive x-axis. Then the line Oy, parallel to MN, becomes the y-axis of the system (figure 5.22).

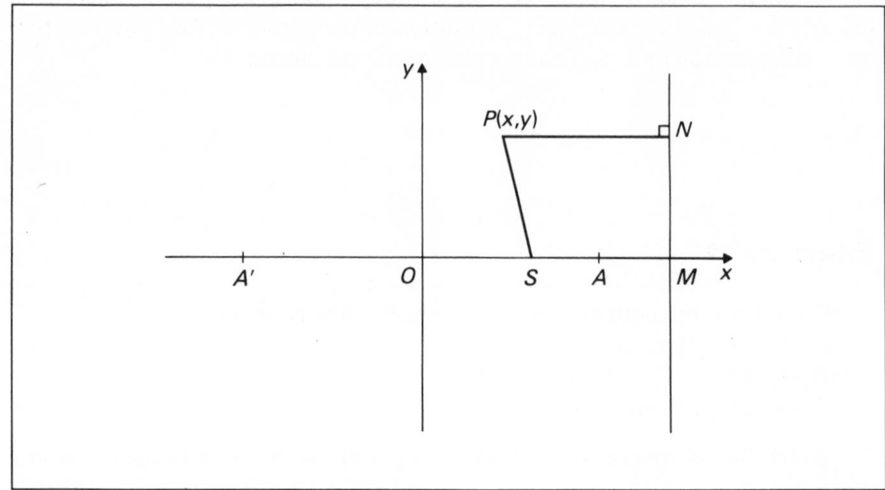

Figure 5.22

The variable point P will now be the point (x, y), and if we let $AA' = 2a$, then A will be $(a, 0)$ and A' will be $(-a, 0)$.

Now let S be the point $(s, 0)$ and M the point $(m, 0)$.

Since $\quad AS = eAM, \quad AO - SO = e(OM - OA)$ or $a - s = e(m - a)$
and since $\quad A'S = eA'M, \quad A'O + OS = e(A'O + OM)$ or $a + s = e(m + a)$
Thus $\quad 2a = 2em$
and $\quad 2s = 2ea$

i.e. $\quad m = \dfrac{a}{e}$ and $s = ae$

So S is the point $(ae, 0)$, M the point $\left(\dfrac{a}{e}, 0\right)$ and N the point $\left(\dfrac{a}{e}, y\right)$.

Now the locus of P is defined by

$$PS = ePN$$

or $\quad PS^2 = e^2PN^2$

Thus $\quad (x - ae)^2 + (y - 0)^2 = e^2\left\{\left(x - \dfrac{a}{e}\right)^2 + (y - y)^2\right\}$

$$(x - ae)^2 + y^2 = (ex - a)^2$$

$$x^2 - 2aex + a^2e^2 + y^2 = e^2x^2 - 2aex + a^2$$

or $\quad (1 - e^2)x^2 + y^2 = a^2(1 - e^2)$

Therefore

$$\frac{x^2}{a^2} + \frac{y^2}{a^2(1-e^2)} = 1 \quad \text{since } e \neq 1$$

We now write $b^2 = a^2(1 - e^2)$ so that the standard equation of the ellipse may be written in the form

$$\frac{x^2}{a^2} + \frac{y^2}{b^2} = 1$$

Now since we may write

$$y = \pm b\sqrt{\left(1 - \frac{x^2}{a^2}\right)},$$

for any value of x such that $\frac{x^2}{a^2} \leqslant 1$,

i.e. $|x| \leqslant a$, y will have two values equal in size but opposite in sign, and the curve

a) is limited to the region $-a \leqslant x \leqslant a$,
b) is symmetrical about the x-axis,
c) crosses the y-axis at $B(0, b)$ and $B'(0, -b)$.

Furthermore, since we may also write

$$x = \pm a\sqrt{\left(1 - \frac{y^2}{b^2}\right)}$$

the curve is limited to the region $-b \leqslant y \leqslant b$, is symmetrical about the y-axis and crosses the x-axis at $A(a, 0)$ and $A'(-a, 0)$.

Because of this symmetry, since $S(ae, 0)$ is a focus of the ellipse, $S'(-ae, 0)$ is also a focus. And $x = \frac{a}{e}$ is a directrix, so $x = -\frac{a}{e}$ is also a directrix. The shape of the locus is as shown in figure 5.23.

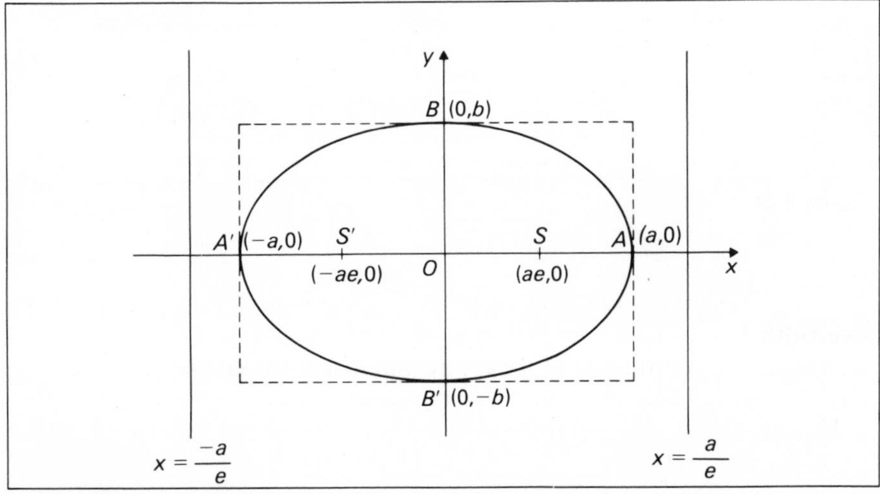

Figure 5.23

It should be noted that the origin O is called **the centre of the ellipse** and that the ellipse is contained entirely within the rectangle whose length is $2a$ and breadth $2b$.
($a > b$ for, since $b^2 = a^2(1 - e^2)$ and $e < 1$, $1 - e^2 < 1$. Thus $a^2 > b^2$ and $a > b$)
The length AA' is called the **major axis** of the ellipse and BB' is the **minor axis.**
If $\alpha^2 = \beta^2(1 - e^2)$, the equation

$$\frac{x^2}{\alpha^2} + \frac{y^2}{\beta^2} = 1$$

represents an ellipse whose major axis is of length 2β, minor axis 2α, foci at the points $(0, \beta e)$, $(0, -\beta e)$ and directrices

$$y = \frac{\beta}{e}, \quad y = -\frac{\beta}{e}.$$

The equation $\dfrac{(x-h)^2}{a^2} + \dfrac{(y-k)^2}{b^2} = 1$

similarly represents an ellipse whose centre is at the point $C(h, k)$, and major and minor axes of length $2a$, $2b$ as shown in figure 5.24.

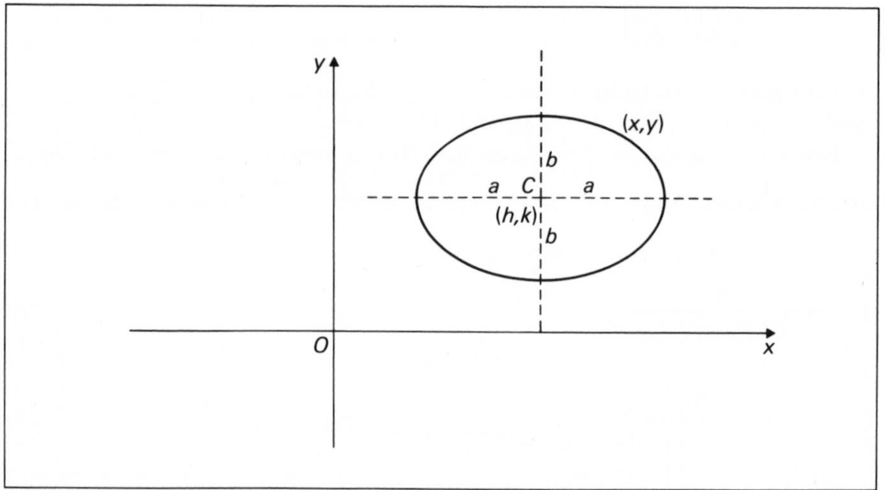

Figure 5.24

Example 1

Determine the coordinates of the centre and foci of the ellipse

$$\frac{(x-1)^2}{9} + \frac{(y-2)^2}{5} = 1$$

Also write down the equation of its directrices.

For this ellipse $a^2 = 9$, $b^2 = 5$

$$b^2 = a^2(1 - e^2) \text{ gives } 5 = 9(1 - e^2)$$
$$9e^2 = 9 - 5 = 4$$
$$e = \sqrt{\frac{4}{9}} = \frac{2}{3}$$

So $\quad ae = 3 \times \frac{2}{3} = 2$ and $\frac{a}{e} = \frac{3}{\frac{2}{3}} = 4{\cdot}5$

Now the centre of the ellipse is the point (1, 2)
Hence the coordinates of the foci are

$(1 + ae, 2)$ and $(1 - ae, 2)$
i.e. $(3, 2)$ and $(-1, 2)$

The directrices are

$$x = 1 + \frac{a}{e} \quad \text{and} \quad x = 1 - \frac{a}{e} \quad \Rightarrow x = 5{\cdot}5 \quad \text{and} \quad x = -3{\cdot}5$$

Example 2
Find the length of the latus rectum of the ellipse

$$\frac{x^2}{16} + \frac{y^2}{7} = 1$$

From $\quad b^2 = a^2(1 - e^2)$
$$7 = 16(1 - e^2) \quad \Rightarrow \quad e = \tfrac{3}{4}$$
So the coordinates of a focus are $(ae, 0)$, i.e. $(3, 0)$ since $a = 4$. When $x = 3$ on the ellipse

$$\frac{9}{16} + \frac{y^2}{7} = 1$$

i.e. $\quad y^2 = 7(1 - \frac{9}{16})$
$\quad\quad = \frac{49}{16}$
hence $\quad y = \pm\frac{7}{4}$

The latus rectum joins the points $(3, \frac{7}{4})$ and $(3, -\frac{7}{4})$ and is of length $\frac{7}{4} - (-\frac{7}{4}) = 3{\cdot}5$.

Parametric equations for the ellipse

The parametric equations $x = a \cos \theta$, $y = b \sin \theta$ will define a curve for a set of values of the parameter θ.

Now $\quad \cos \theta = \frac{x}{a}$ and $\sin \theta = \frac{y}{b}$

Since $\cos^2 \theta + \sin^2 \theta = 1$ for all values of θ

$$\frac{x^2}{a^2} + \frac{y^2}{b^2} = 1$$

Hence the equations $x = a \cos \theta$, $y = b \sin \theta$ define the ellipse parametrically.

We note that we could obtain this equation of the ellipse by applying a 'squashing' transformation defined by

$$\mathbf{M} = \begin{pmatrix} 1 & 0 \\ 0 & \frac{b}{a} \end{pmatrix}$$

on the equation of a circle $x^2 + y^2 = a^2$.

Using this, we see that

$$\begin{pmatrix} x' \\ y' \end{pmatrix} = \begin{pmatrix} 1 & 0 \\ 0 & \frac{b}{a} \end{pmatrix} \begin{pmatrix} x \\ y \end{pmatrix} \Rightarrow x' = x \text{ and } y' = \frac{b}{a} y,$$

i.e. $x = x'$ and $y = \dfrac{ay'}{b}$,

so that the circle $x^2 + y^2 = a^2$ transforms into

$$(x')^2 + \left(\frac{ay'}{b}\right)^2 = a^2$$

$$\Rightarrow \left(\frac{x'}{a}\right)^2 + \left(\frac{y'}{b}\right)^2 = 1$$

which is, again, the standard equation of the ellipse.

The circle $x^2 + y^2 = a^2$ is called the **auxiliary circle**. Corresponding to any point $P(a \cos \theta, b \sin \theta)$ on the ellipse, there is a point $Q(a \cos \theta, a \sin \theta)$ on the auxiliary circle. The angle θ which thus determines the position of Q is called the **eccentric angle** of P.

Note That when $\theta = 0$ $\quad x = a$ and $y = 0$
when $\theta = 90°$ $\quad x = 0$ and $y = b$
when $\theta = 180°$ $\quad x = -a$ and $y = 0$
when $\theta = 270°$ $\quad x = 0$ and $y = -b$
when $\theta = 360°$ $\quad x = a$ and $y = 0$

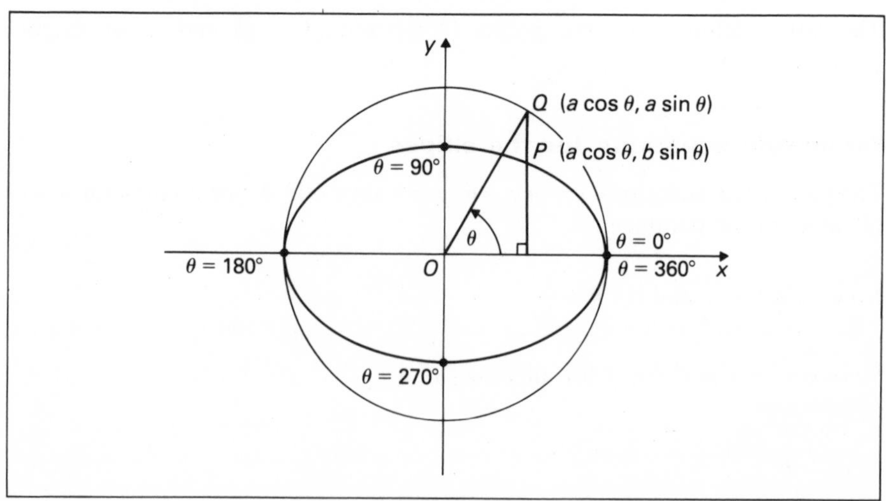

Figure 5.25

The parametric equations $x = a \cos \theta$, $y = b \sin \theta$ therefore define the ellipse traced out anticlockwise, starting from the extreme position on the major axis. Moreover the complete locus is defined for $0° \leqslant \theta° \leqslant 360°$, as shown in figure 5.25.

Exercise I

1 Sketch the curves given by the following equations, showing the lengths of the major and minor axes.

a) $\dfrac{x^2}{25} + \dfrac{y^2}{9} = 1$ b) $\dfrac{x^2}{16} + \dfrac{y^2}{4} = 1$ c) $4x^2 + 9y^2 = 36$

2 Determine the coordinates of the foci and equations of the directrices of the given curves in question 1.

3 Write down the equation of an ellipse, given
a) the major axis is of 12 units length, the minor axis is of 10 units length, and the centre is at the origin.
b) the major axis is 10 units length, the centre is the origin and the eccentricity is $\frac{3}{5}$.
c) the foci are the points $(\pm 2, 0)$, the centre is the origin and the eccentricity is $\frac{4}{5}$.

4 Determine the coordinates of the centres, foci and equations of the directrices of the ellipses whose equations are

a) $\dfrac{(x-2)^2}{9} + (y-1)^2 = 1$ b) $4(x+1)^2 + 9(y-2)^2 = 36$

5 Write down the parametric equations of the following curves.

a) $\dfrac{x^2}{4} + y^2 = 1$ b) $4x^2 + 9y^2 = 9$ c) $\dfrac{(x-1)^2}{9} + \dfrac{(y-2)^2}{4} = 1$

6 Derive the cartesian equations of the loci given by the following parametric equations.

a) $x = 4 \cos \theta$, $y = 3 \sin \theta$
b) $x = 12 \cos \theta$, $y = 5 \sin \theta$
c) $x = 3 \cos \theta$, $y = 1 + 2 \sin \theta$
d) $x = 1 - 2 \sin \theta$, $y = 1 + \cos \theta$

5.11 Tangents and normals to the parabola and ellipse

Parabola

If the line PT is a tangent to $y^2 = 4ax$ touching the curve at P, the line PN perpendicular to the tangent at P is called the **normal** to the curve at P (figure 5.26).

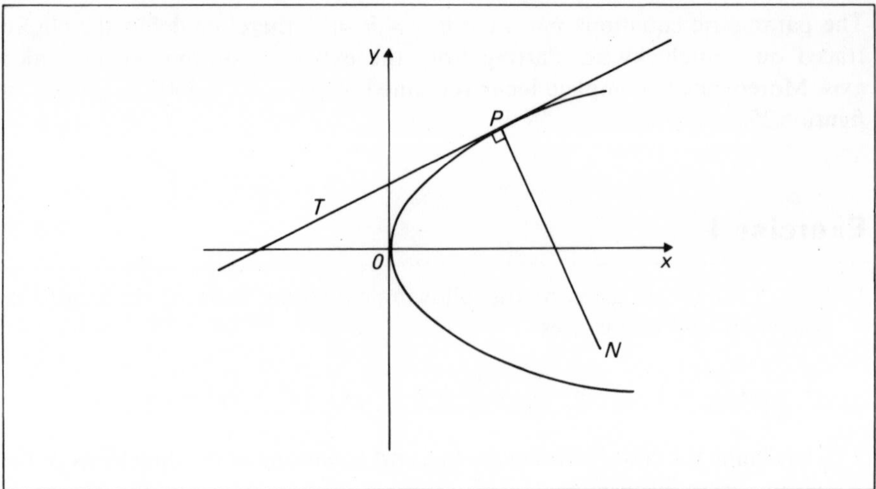

Figure 5.26

We derive the equations of the tangent and normal when

a) P is the point (x_1, y_1) and the curve is defined by its cartesian equation $y^2 = 4ax$.

b) P is the point $(ap^2, 2ap)$ and the curve is defined by its parametric equations $x = at^2$, $y = 2at$.

a) The equation of the tangent will be

$$\frac{y - y_1}{x - x_1} = m \quad \text{where } m \text{ is the gradient of the tangent at } (x_1, y_1)$$

Now, using the ideas of differential calculus

$$m = \left(\frac{dy}{dx}\right)_{(x_1,\, y_1)} \quad \text{where } \left(\frac{dy}{dx}\right)_{(x_1,\, y_1)} \text{ is the value of } \frac{dy}{dx}$$

worked out at the point (x_1, y_1).

We can write $y = \pm 2a^{1/2}x^{1/2}$ so that

$$\frac{dy}{dx} = \pm 2a^{1/2} \cdot \frac{1}{2} x^{-1/2}$$

$$= \pm a^{1/2}x^{-1/2} = \pm \frac{a^{1/2}}{x^{1/2}}$$

$$= 2\frac{a}{y} \text{ since } \pm\frac{1}{x^{1/2}} = 2\frac{a^{1/2}}{y}$$

Hence $\left(\frac{dy}{dx}\right)_{(x_1,\, y_1)} = \frac{2a}{y_1}$

So the tangent is $\frac{y - y_1}{x - x_1} = \frac{2a}{y_1}$

i.e. $yy_1 - y_1^2 = 2a(x_1 - x_1)$

But $y_1^2 = 4ax_1$, since (x_1, y_1) lies on the parabola

$$\Rightarrow yy_1 - 4ax_1 = 2a(x - x_1)$$

$$\Rightarrow yy_1 = 2a(x + x_1)$$

For the normal, the gradient is $-\frac{y_1}{2a}$, since for the normal and tangent (being perpendicular) the product of the gradients is -1.
Hence the equation to the normal is

$$\frac{y-y_1}{x-x_1}=\frac{-y_1}{2a}$$

$$2a(y-y_1)=-y_1(x-x_1)$$

$$xy_1+2ay=y_1(x_1+2a)$$

b) If the curve is defined parametrically, p is the value of the parameter at the point P, and the gradient of the tangent is

$$\frac{2a}{y_1} \quad \text{or} \quad \frac{2a}{2ap}=\frac{1}{p}$$

The tangent is then

$$\frac{y-2ap}{x-ap^2}=\frac{1}{p}$$

or

$$py-2ap^2=x-ap^2 \qquad \text{i.e.}$$

$$py=x+ap^2$$

The gradient of the normal at P is $-p$ and the equation of the normal is

$$y-2ap=-p(x-ap^2) \qquad \text{i.e.}$$

$$px+y=2ap+ap^3$$

Ellipse

The equations to the tangent and normal to the ellipse (figure 5.27) will be of the form

$$\frac{y-y_1}{x-x_1}=m \text{ and } \frac{y-y_1}{x-x_1}=-\frac{1}{m}$$

where m is the gradient of the tangent at the point P on the curve.

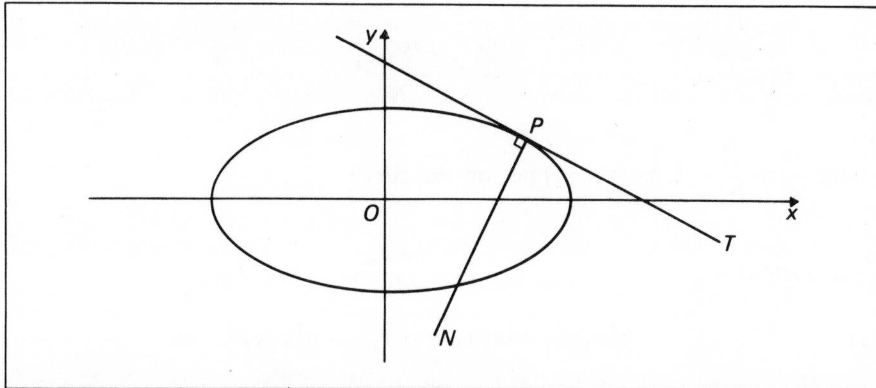

Figure 5.27

As an alternative method to the use of the calculus, we derive the value of m by first finding the gradient of a chord joining P to a point Q on the curve, and then letting the chord become a tangent by making Q approach P on the curve. If P is (x_1, y_1) and $Q(x_2, y_2)$ then the gradient of PQ is

$$\frac{y_1 - y_2}{x_1 - x_2}$$

Now $\quad \dfrac{x_1^2}{a^2} + \dfrac{y_1^2}{b^2} = 1$ and $\dfrac{x_2^2}{a^2} + \dfrac{y_2^2}{b^2} = 1$

So $\quad \dfrac{(x_1^2 - x_2^2)}{a^2} = \dfrac{-(y_1^2 - y_2^2)}{b^2}$

$$\Rightarrow \quad \frac{y_1^2 - y_2^2}{x_1^2 - x_2^2} = -\frac{b^2}{a^2}$$

And $\quad \dfrac{(y_1 - y_2)(y_1 + y_2)}{(x_1 - x_2)(x_1 + x_2)} = -\dfrac{b^2}{a^2}$

$$\frac{y_1 - y_2}{x_1 - x_2} = -\frac{b^2(x_1 + x_2)}{a^2(y_1 + y_2)}$$

Now, when Q approaches P on the curve i.e. x_2 tends to x_1 and y_2 tends to y_1

$$\frac{y_1 - y_2}{x_1 - x_2} \text{ tends to } m$$

Hence $m = -\dfrac{b^2(2x_1)}{a^2(2y_1)} = -\dfrac{b^2 x_1}{a^2 y_1}$

So the equation of the tangent is

$$y - y_1 = -\left(\frac{b^2 x_1}{a^2 y_1}\right)(x - x_1)$$

$$a^2 y y_1 - a^2 y_1^2 = -b^2 x x_1 + b^2 x_1^2$$

$$b^2 x x_1 + a^2 y y_1 = b^2 x_1^2 + a^2 y_1^2$$

or $\quad \dfrac{x x_1}{a^2} + \dfrac{y y_1}{b^2} = \dfrac{x_1^2}{a^2} + \dfrac{y_1^2}{b^2} \qquad$ i.e.

$$\frac{x x_1}{a^2} + \frac{y y_1}{b^2} = 1$$

since $\dfrac{x_1^2}{a^2} + \dfrac{y_1^2}{b^2} = 1$, as (x_1, y_1) lies on the curve.

The normal is $\qquad y - y_1 = \dfrac{a^2 y_1}{b^2 x_1}(x - x_1)$

or $\qquad b^2 x_1 y - b^2 x_1 y_1 = a^2 y_1 x - a^2 x_1 y_1 \qquad$ i.e.

$$a^2 y_1 x - b^2 x_1 y = (a^2 - b^2) x_1 y_1$$

If the curve is defined parametrically by $x = a\cos\theta$, $y = b\sin\theta$ the gradient of the tangent at the point $\theta = \phi$ is

$$-\frac{b^2 a\cos\phi}{a^2 b\sin\phi}$$

$$= \frac{-b\cos\phi}{a\sin\phi}$$

and the tangent is $y - b\sin\phi = \left(-\frac{b\cos\phi}{a\sin\phi}\right)(x - a\cos\phi)$

i.e. $$ay\sin\phi - ab\sin^2\phi = -b\cos\phi + ab\cos^2\phi$$

$$bx\cos\phi + ay\sin\phi = ab(\sin^2\phi + \cos^2\phi)$$

$$= ab$$

i.e.

$$\frac{\cos\phi}{a}x + \frac{\sin\phi}{b}y = 1$$

Similarly the normal is $a^2bx\sin\phi - b^2ay\cos\phi = (a^2 - b^2)ab\sin\phi\cos\phi$

using the form involving (x_1, y_1) as already established. This standard form is more generally written as

$$\frac{ax}{\cos\phi} - \frac{by}{\sin\phi} = a^2 - b^2$$

(dividing throughout by $ab\sin\phi\cos\phi$).

Exercise J

1 Determine the equations of the tangents to the following parabolas at the given points.

a) $y^2 = 4ax$, $(a, -2a)$ c) $y^2 = 5(x + 2)$, $(3, 5)$
b) $y^2 = 9x$, $(4, 6)$

2 Determine the equations of the normals to the parabolas given in question 1 at the points given.

3 Prove that the lines given by the following equations are tangents to the parabolas as defined. Determine the points of contact of the given tangents.

a) $y = x + 1$, $y^2 = 4x$ c) $6y - x - 9 = 0$, $y^2 = x$
b) $3y = 9x + 2$, $y^2 = 8x$ d) $2y = x - 1$, $y^2 = 2x - 6$

4 Determine the equations of the normals to the following parabolas at the points given.

a) $y^2 = x$, $(1, 1)$ b) $y^2 = 4x$, $(4, -4)$ c) $y^2 = 4(x - 1)$, $(5, 4)$

5 Derive the equation of the tangent and normal to the parabolas as given in parametric form, at the point defined by the special value of the parameter.

a) $x = 4t^2$, $y = 8t$, $t = 2$ c) $x = \frac{1}{2}t^2$, $y = t$, $t = -1$

b) $x = 5t^2$, $y = 10t$, $t = \frac{1}{2}$

6 Determine the equations of the tangents and normals at the points given on the following ellipses.

a) $\frac{x^2}{15} + \frac{y^2}{10} = 1$, $(3, -2)$ c) $\frac{x^2}{16} + \frac{y^2}{12} = 1$, $(2, 3)$

b) $\frac{x^2}{8} + \frac{y^2}{2} = 1$, $(-2, -1)$

7 Derive the equations of the tangents and normals at the point with given eccentric angle on the following ellipses

a) $x = 5 \cos \theta$, $y = 3 \sin \theta$, $\theta = \frac{\pi}{3}$

b) $x = 4 \cos \theta$, $y = 3 \sin \theta$, $\tan \theta = \frac{3}{4}$

c) $x = 2 \cos \theta$, $y = \sin \theta$, $\theta = \frac{\pi}{4}$

Trigonometry

CHAPTER SIX

6.1 The sine and cosine rules

If we are given the lengths of the three sides a, b, c of a $\triangle ABC$, the triangle may be drawn uniquely using ruler and compass methods as shown in figure 6.1. The determination of the angles at A, B, C would therefore follow immediately. As the lengths of the sides vary, so will the size of the angles. It

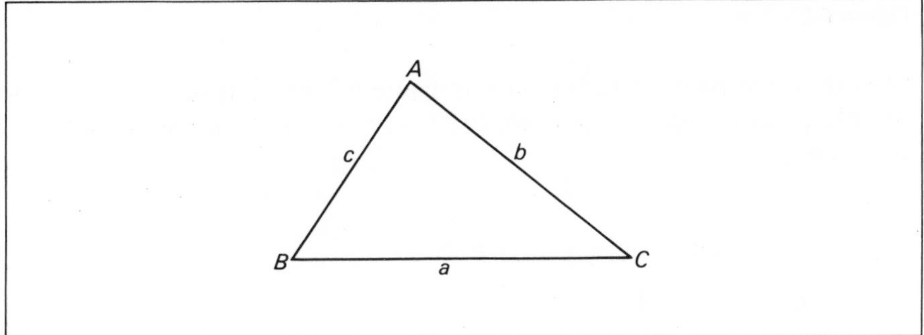

Figure 6.1

may be expected, therefore, that there will exist relationships connecting these sides and angles according to these variations. In fact this is so and a number of these connections can be found quite simply. The two most important of these results are referred to as **the sine rule** and **the cosine rule.** The sine rule is

$$\frac{a}{\sin A} = \frac{b}{\sin B} = \frac{c}{\sin C} = 2R$$

where R is the radius of the circumcircle of $\triangle ABC$. There are three alternative forms of the cosine rule.

$$a^2 = b^2 + c^2 - 2bc \cos A$$
$$b^2 = c^2 + a^2 - 2ca \cos B$$
$$c^2 = a^2 + b^2 - 2ab \cos C$$

and these results may be suitably rearranged to express the cosine of an angle in terms of the sides, and become

$$\cos A = \frac{b^2 + c^2 - a^2}{2bc}$$

$$\cos B = \frac{c^2 + a^2 - b^2}{2ca}$$

$$\cos C = \frac{a^2 + b^2 - c^2}{2ab}$$

To establish the truth of these rules it will be sufficient to consider the $\triangle ABC$ positioned so that AB is placed along the x-axis of a coordinate system, with A at the origin.

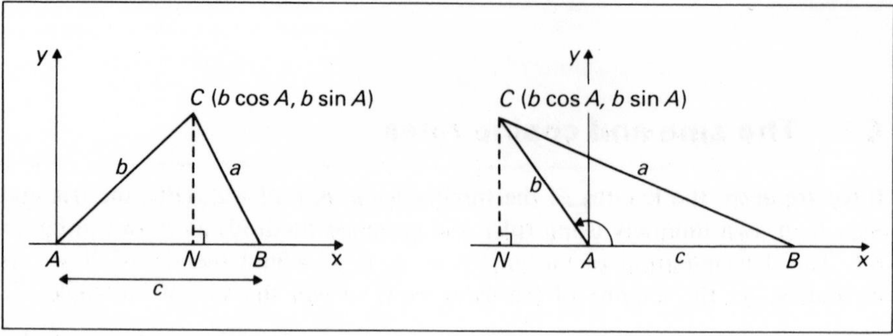

Figure 6.2

In both of the two alternative cases in figure 6.2 where $B\hat{A}C$ is either acute or obtuse, the coordinates of C are (b cos A, b sin A). So, if the ordinate CN of C is drawn

$$CN = b \sin A$$

and from $\triangle CNB$, $\quad CN = a \sin B$

$$\Rightarrow b \sin A = a \sin B$$

$$\Rightarrow \quad \frac{a}{\sin A} = \frac{b}{\sin B}$$

It is clear that by transposing the triangle into the position with B at the origin and BC along the x-axis

$$\frac{b}{\sin B} = \frac{c}{\sin C} \quad \text{also,}$$

and combining together,

$$\frac{a}{\sin A} = \frac{b}{\sin B} = \frac{c}{\sin C}$$

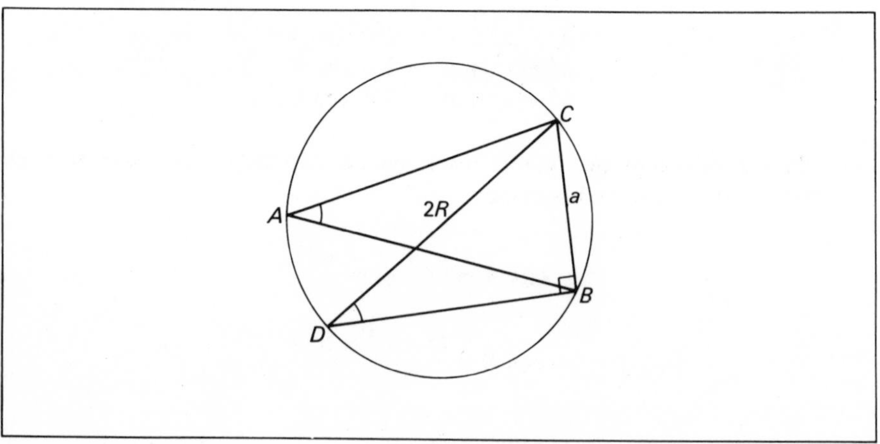

Figure 6.3

Furthermore referring to figure 6.3 in which CD is the diameter of the circumcircle of $\triangle ABC$, $C\hat{A}B = C\hat{D}B$ (both subtended by the arc CB in the same segment of the circle)

and $\quad C\hat{B}D = 90°$

(being subtended by the diameter CD).

So $\quad a = 2R \sin A$ from $\triangle CDB$

Thus $\quad \dfrac{a}{\sin A} = \dfrac{b}{\sin B} = \dfrac{c}{\sin C} = 2R$

Using the same two diagrams with $\triangle ABC$ positioned in the coordinate plane, B is the point $(c, 0)$ and C $(b \cos A, b \sin A)$. Therefore, by using the distance formulae

$$BC^2 = a^2 = (0 - b \sin A)^2 + (c - b \cos A)^2$$

$$\Rightarrow a^2 = b^2 \sin^2 A + b^2 \cos^2 A - 2bc \cos A + c^2$$

$$\Rightarrow a^2 = b^2 (\sin^2 A + \cos^2 A) + c^2 - 2bc \cos A$$

$$\Rightarrow a^2 = b^2 + c^2 - 2bc \cos A$$

The proofs of the alternative forms follow by like methods.

All of the results of the sine and cosine rules considered in their entirety permits the calculation of unknown elements in given triangles for which other information has been offered.

a) For the sine rule to be relevant, a side and facing angle must be known before the calculation of the value of any other element is possible.
b) the cosine rule will be relevant either if two sides and the angle between them are given, or if three sides are given.

It is also important to remember that since $\sin \theta° = \sin (180° - \theta°)$ i.e. the sine of an angle and its supplement are the same, any application of the sine rule leading to the calculation of an angle through its sine, offers two possible alternatives. Such a possibility should not be neglected in practical cases.

Example 1

In $\triangle ABC$ $b = 7$ cm, $\hat{A} = 50°$, $\hat{B} = 70°$. Find the remaining sides and angles.

$$\hat{C} = 180° - (50° + 70°) = 60°$$

$$\frac{a}{\sin 50°} = \frac{7}{\sin 70°} = \frac{c}{\sin 60°}$$

Thus $\quad a = \dfrac{7 \sin 50°}{\sin 70°}$

$$c = \frac{7 \sin 60°}{\sin 70°}$$

Using tables of logarithms or a calculator,

$a = 5{\cdot}706$ cm $\qquad c = 6{\cdot}451$ cm

Example 2

Given that $a = 12{\cdot}2$ cm, $\hat{A} = 45{\cdot}6°$, $b = 9{\cdot}6$ cm, calculate the other angles and side of the $\triangle ABC$.

Here, by the sine rule,

$$\frac{12{\cdot}2}{\sin 45{\cdot}6°} = \frac{9{\cdot}6}{\sin B} = \frac{c}{\sin C}$$

$$\sin B = \frac{9{\cdot}6 \sin 45{\cdot}6°}{12{\cdot}2}$$

$$= 0{\cdot}5622$$

$$\hat{B} = 34{\cdot}2° \quad \text{or} \quad 145{\cdot}8°$$

There are now two possible alternatives, namely

$$\hat{C} = 180° - (45{\cdot}6° + 34{\cdot}2°) = 100{\cdot}2°$$

or $\quad 180° - (45{\cdot}6° + 145{\cdot}8°) = -11{\cdot}4°$ (not admissible)

$$C = \frac{12{\cdot}2}{\sin 45{\cdot}6°} \times \sin 100{\cdot}2° = 16{\cdot}81 \text{ cm}$$

Example 3

Calculate the third side and other angles in a triangle in which sides of length 11 cm and 13 cm enclose an angle of 55°.

Using the cosine rule in the form $a^2 = b^2 + c^2 - 2bc \cos A$

and letting $\quad b = 11$ cm, $c = 13$ cm, $\hat{A} = 55°$

$$\begin{aligned} a^2 &= 11^2 + 13^2 - 2 \times 11 \times 13 \cos 55° \\ &= 121 + 169 - 164{\cdot}05 \\ &= 290 - 164{\cdot}05 \\ &= 125{\cdot}95 \\ a &\approx \sqrt{126} = 11{\cdot}225 \text{ cm} \end{aligned}$$

We may now continue to use the cosine rule to find $\hat{B}$ and/or $\hat{C}$ or choose the sine rule. If we choose the latter method, we will be faced with two alternative sets of values of the angles. The use of the cosine rule in the form

$$\cos B = \frac{a^2 + c^2 - b^2}{2ac} \quad \text{yields}$$

$$\cos B = \frac{126 + 169 - 121}{2 \times 11{\cdot}23 \times 13} = 0{\cdot}5962$$

$$\Rightarrow \hat{B} = 53{\cdot}41°$$

So finally $\quad \hat{C} = 180° - 55° - 53{\cdot}41° = 71{\cdot}59°$

Exercise A

1 Solve completely the following triangles.

a) $b = 5$ cm, $\quad c = 6$ cm, $\quad \hat{A} = 70°$

b) $a = 3$ cm, $\quad b = 10$ cm, $\quad \hat{C} = 45°$

c) $a = 7$ cm, $\quad b = 6$ cm, $\quad c = 9$ cm

d) $a = 3{\cdot}51$ cm, $\hat{B} = 61{\cdot}5°$, $\hat{C} = 75{\cdot}4°$
e) $c = 8{\cdot}53$ cm, $\hat{A} = 117{\cdot}33°$, $\hat{B} = 24{\cdot}17°$
f) $b = 21{\cdot}1$ cm, $\hat{A} = 25{\cdot}17°$, $\hat{C} = 68{\cdot}35°$
g) $a = 4{\cdot}62$ cm, $b = 7{\cdot}10$ cm, $c = 5{\cdot}82$ cm
h) $a = 4{\cdot}78$ cm, $b = 4{\cdot}13$ cm, $c = 1{\cdot}36$ cm
i) $a = 1{\cdot}082$ cm, $b = 2{\cdot}171$ cm, $\hat{B} = 66{\cdot}48°$
j) $b = 3{\cdot}0$ cm, $c = 4{\cdot}4$ cm, $\hat{B} = 38°$

Find the difference between the two values of a in j).

2 In a $\triangle ABC$ values of a, b and $\hat{A}$ are given. Prove that if c_1, c_2 are the two positive values of c which may be calculated using the sine rule

$$c_1 - c_2 = 2\{a^2 - b^2 \sin^2 A\}^{1/2}.$$

3 In the parallelogram $ABCD$, $AB = 7{\cdot}21$ cm, $BC = 8{\cdot}32$ cm. The diagonal $AC = 8{\cdot}56$ cm. Calculate
a) the length of the diagonal BD,
b) the angles of $ABCD$,
c) the acute angle between the diagonals.

4 Calculate the radius of the circumcircle of the $\triangle PRQ$ in which $PQ = 2{\cdot}168$ cm, $PR = 2{\cdot}613$ cm and $Q\hat{P}R = 62{\cdot}6°$.

5 Use the sine rule to show that for any $\triangle$, $R = abc/4\triangle$ where $\triangle$ denotes the area of the triangle.

6.2 Solution of trigonometric equations

Earlier, when discussing the use of the sine rule, we were faced with the problem of finding an angle in a triangle from an equation of the form $\sin x = p$, where $|p| \leqslant 1$. We saw that there were two possible solutions which would fit the triangle namely the acute angle $\theta°$, and the supplementary obtuse angle $180° - \theta°$.

Now let us discuss the solution of the equation $\sin x = p$, in further detail. Broadly, the solutions may be located by reading off the points of intersection of the graphs of $y = \sin x$ and $y = p$. From chapter 4 we know the shape of the infinite sine wave and $y = p$ is a line drawn parallel to the x-axis making an intercept p on the y-axis. Therefore, for a positive value of p, the solutions may be obtained by reading off the x values of the points of intersection of the graphs, as shown in figure 6.4.

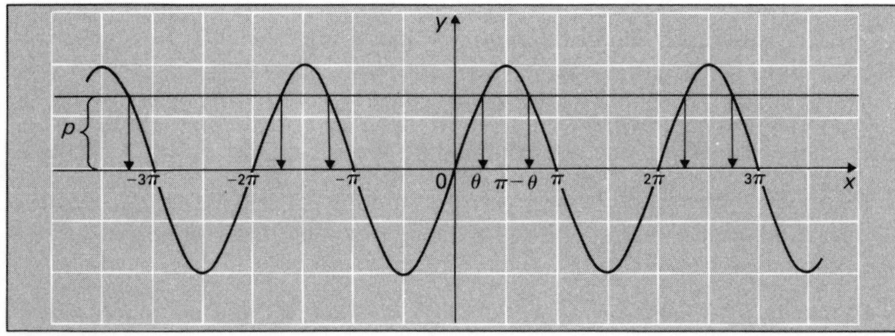

Figure 6.4

If p is negative, but is still such that $|p| \leqslant 1$, the solution set will follow by an identical method. If however $|p| > 1$, the line $y = p$ will not intersect the graph of $y = \sin x$ and the equation $\sin x = p$ would obviously then have no solutions.

Given that the first identifiable answer which can be read off from the graph is $x = \theta$, for values of $x > 0$ the solution set is

$$\{\theta, \pi - \theta, 2\pi + \theta, 3\pi - \theta, 4\pi + \theta, \ldots\}$$

and for $x < 0$ is $\{\ldots, -3\pi - \theta, -2\pi + \theta, -\pi - \theta, \ldots\}$. It is clear that each element of the complete set, for all x, consists of two parts
a) a multiple of π,
b) a quantity which is either $+\theta$ or $-\theta$.
In a) the quantity involving π may be written in the form $n\pi$, where $n = \ldots -3, -2, -1, 0, 1, 2, 3, 4 \ldots$. It is now seen in b) that whenever n is even, the sign of θ is $+$, whereas when n is odd the sign is $-$.

The 'alternating symbol' $(-1)^n$ fulfils this obligation for such values of n.

Thus the general formula for which all solutions of the equation $\sin x = p$ may be written down is

$$x = n\pi + (-1)^n\theta \quad \text{where}$$

n is an integer, positive, negative or zero, and θ any special value which satisfies the original equation $\sin x = p$. If we wish to work in degree measure rather than radians we would use

$$x° = n\ 180° + (-1)^n\theta°$$

to write down such solutions.

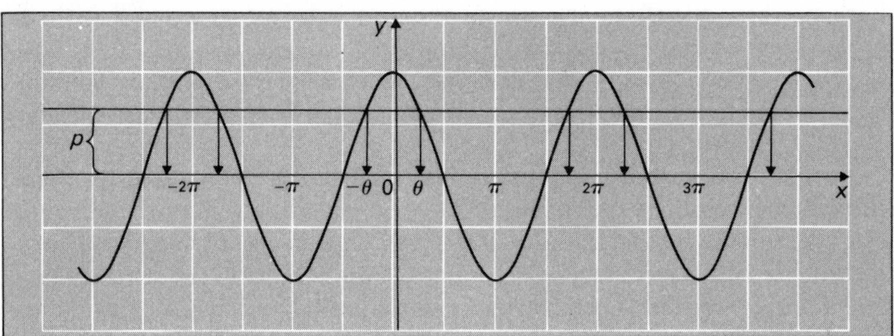

Figure 6.5

We may now turn to consideration of like results which express generally the solutions of the equations,

$\cos x = p$ where $|p| \leqslant 1$
and $\tan x = p$ where p is not bounded by specific limits.

For $\cos x = p$, the graph from which the solution set may be written down is shown in figure 6.5, and supposing that the first identifiable solution is $x = \theta$ the infinite solution set is

$$\{\ldots, -4\pi - \theta, -4\pi + \theta, -2\pi - \theta, -2\pi + \theta, -\theta, \theta, 2\pi - \theta, 2\pi + \theta, \ldots\}$$

Here the part of the general formula associated with π is always even and fits the expression $2n\pi$ for all n. But for every value of n, there are two corresponding values of θ; for example, when $n = 0$ there are the two values $+\theta$ and $-\theta$, when $n = 1$ the two values $+\theta$ and $-\theta$, and so on. This part of the general formula will therefore be written $\pm\theta$ and the whole becomes

$$x = 2n\pi \pm \theta \quad \text{for all } n \in \mathbb{Z}. \text{ (Or } x = 360° \times n \pm \theta° \text{ in degree measure.)}$$

For the solution of $\tan x = p$, we consider the intersecting graphs as shown in figure 6.6. The solution set is

$$\{\ldots, -3\pi + \theta, -2\pi + \theta, -\pi + \theta, \theta, \pi + \theta, 2\pi + \theta, 3\pi + \theta, \ldots\}$$

and this of course is generalised by the formula

$$x = n\pi + \theta \text{ for all } n \in \mathbb{Z}. \text{ (Or } x = 180° \times n + \theta°.)$$

No limitations need be imposed upon the value of p when seeking to solve such tangent equations.

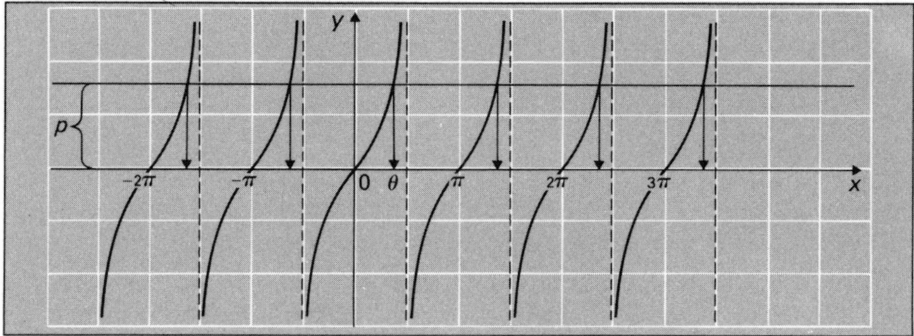

Figure 6.6

It is not necessary to establish further general results for the three reciprocal equations,

$$\csc x = p \quad \sec x = p \text{ and } \operatorname{cotan} x = p$$

since by inversion we would obtain

$$\sin x = \frac{1}{p} \qquad \cos x = \frac{1}{p} \qquad \tan x = \frac{1}{p}$$

However it should be noted that for the first two equations to have solutions, $|p| \geqslant 1$; whereas in the third p is unbounded.

Example 1

Find the general solution of $\tan x = 1$ in radian form and write down the solution set for which $-3\pi \leqslant x \leqslant 2\pi$.

We know that a solution of $\tan x = 1$ is $x = 45°$, i.e. $\frac{\pi}{4}$.

The general solution for x is

$$x = n\pi + \frac{\pi}{4} = \frac{(4n + 1)\pi}{4} \quad \text{for all } n \in \mathbb{Z}.$$

When $n = -4 \quad x = -3\frac{3}{4}\pi$ (outside specified range)
$n = -3 \quad x = -2\frac{3}{4}\pi$
$n = -2 \quad x = -1\frac{3}{4}\pi$
$n = -1 \quad x = -\frac{3}{4}\pi$
$n = 0 \quad x = \frac{1}{4}\pi$
$n = 1 \quad x = 1\frac{1}{4}\pi$
$n = 2 \quad x = 2\frac{1}{4}\pi$ (outside specified range)

$\Rightarrow$ for $-3\pi \leqslant x \leqslant 2\pi$ the solution set is

$\{-2\frac{3}{4}\pi, -1\frac{3}{4}\pi, -\frac{3}{4}\pi, \frac{1}{4}\pi, 1\frac{1}{4}\pi, 2\frac{1}{4}\pi\}$

Example 2

Find the solution set of $\sin x° = 0{\cdot}5$ for $-330° \leqslant x \leqslant 540°$.
A solution of $\sin x° = 0{\cdot}5$ is $x = 30$
The general solution of the equation is
$x = n \times 180 + (-1)^n\, 30$

When $n = -3$, $x = -540 + (-1)^{-3} \times 30 = -540 - 30 = -570$ (outside specified range)
$n = -2$, $x = -360 + (-1)^{-2} \times 30 = -360 + 30 = -330$
$n = -1$, $x = -180 + (-1)^{-1} \times 30 = -180 - 30 = -210$
$n = 0$, $x = 0 + (-1)^{0} \times 30 = 30$
$n = 1$, $x = 180 + (-1)^{1} \times 30 = 180 - 30 = 150$
$n = 2$, $x = 360 + (-1)^{2} \times 30 = 360 + 30 = 390$
$n = 3$, $x = 540 + (-1)^{3} \times 30 = 540 - 30 = 510$
$n = 4$, $x = 720 + (-1)^{4} \times 30 = 720 + 30 = 750$ (outside specified range)

So $\quad x = \{-330, -210, 30, 150, 390, 510\}$

In general practice, once the extremes which fit the range have been established, the intermediate work may be simplified.

The preceding results will enable us to find generally as many solutions as we require of equations which reduce to the standard sine, cosine or tangent form. In practice many equations do this and the following structure typifies the broad categories into which particular examples fall and the methods employed to solve them.

a Equations which are, or may be transformed to, algebraic forms and are then solved by the appropriate algebraic method

Example 1 Using factorisation

Find the values of x which lie between $-90°$ and $+90°$ (inclusive) and which satisfy the equation $2 \sin 2x = \sin x$.

Using the trigonometric identity $\sin 2x = 2 \sin x \cos x$, the equation becomes

$4 \sin x \cos x - \sin x = 0$
$\Rightarrow \sin x\,(4 \cos x - 1) = 0$
$\Rightarrow \sin x = 0 \quad$ or $\cos x = \frac{1}{4}$
From $\sin x = 0 \quad x = n\, 180° + (-1)^n\, 0°$
when $x = 0 \quad x = 0°$

For all other values of n, x lies outside the given range.

For $\cos x = \frac{1}{4}$ $x = n360° \pm 75{\cdot}52°$
when $n = 0$ $x = \pm 75{\cdot}52°$

For all other values of n, x lies outside the given range.

Hence, for $-90° \leqslant x° \leqslant 90°$ the equation has the three solutions $-75{\cdot}52°$, $0°$, and $+75{\cdot}52°$.

Example 2 Reducing to a quadratic trinomial and using factorisation

Solve the equation $\sec^2 \theta = 3 \tan \theta - 1$ giving all values of θ which lie between $0°$ and $360°$ inclusive.

Since $\sec^2 \theta = 1 + \tan^2 \theta$ and using $t = \tan \theta$, the equation may be written as

$$1 + t^2 = 3t - 1$$
$$t^2 - 3t + 2 = 0$$
$$(t-1)(t-2) = 0$$
$$t = 1 \quad \text{or} \quad t = 2$$

From the equation $\tan \theta = 1$, $\theta = 180°n + 45°$. For $n = 0, 1$ the two values $45°$ and $225°$ lie inside the range.

From the equation $\tan \theta = 2$, $\theta = 180°n + 63{\cdot}43°$. Also for $n = 0, 1$ in the formula $63{\cdot}43°$ and $243{\cdot}43°$ lie in the range.

Thus the given equation has four solutions, viz. $45°$, $63{\cdot}43°$, $225°$ and $243{\cdot}43°$ in the range $0° \leqslant \theta° \leqslant 360°$.

Example 3 Reducing to a quadratic binomial

Solve $\tan \alpha \tan 2\alpha = 1$ such that $0 < \alpha < \pi$.

From the 'double angle' formula $\tan 2\alpha = \dfrac{2 \tan \alpha}{1 - \tan^2 \alpha}$

Substituting in the given equation $\dfrac{2 \tan^2 \alpha}{1 - \tan^2 \alpha} = 1$

$$\Rightarrow \quad 2 \tan^2 \alpha = 1 - \tan^2 \alpha$$

$$\Rightarrow \quad 3 \tan^2 \alpha = 1$$

$$\tan \alpha = \pm\frac{1}{\sqrt{3}}$$

From $\tan \alpha = \dfrac{1}{\sqrt{3}}$, $\alpha = n\pi + \dfrac{\pi}{6}$ and $\alpha = \dfrac{\pi}{6}$ for $0 < \alpha < \pi$
$(n = 0)$

From $\tan \alpha = -\dfrac{1}{\sqrt{3}}$, $\alpha = n\pi + \left(\dfrac{-\pi}{6}\right)$ and $\alpha = \dfrac{5\pi}{6}$ for $0 < \alpha < \pi$
$(n = 1)$

The equation has the two solutions $\dfrac{\pi}{6}$ and $\dfrac{5\pi}{6}$, inside the given range.

Example 4 Reducing to non-factorising quadratic trinomials
Solve, for values of x such that $0° \leqslant x° \leqslant 360°$, the equation $\cos 2x + \cos x = 1$.

Using the trigonometrical identity $\cos 2x = 2\cos^2 x - 1$, the equation may be written in the form

$$2\cos^2 x - 1 + \cos x - 1 = 0$$
$$2\cos^2 x + \cos x - 2 = 0$$

Letting $\cos x = c$ this becomes

$$2c^2 + c - 2 = 0$$

$$c = \frac{-1 \pm \sqrt{(1+16)}}{4} = \frac{-1 \pm 4{\cdot}123}{4}$$

$$c = -1{\cdot}2807(5) \quad \text{(inadmissible since } |c| < 1)$$

or $c = 0{\cdot}7807(5)$

A solution of this equation is $38{\cdot}67°$. The general solution of the given equation is

$$x° = 2n \times 180° \pm 38{\cdot}67°$$

When $n = 0$ $\quad x° = 38{\cdot}67°$ or $-38{\cdot}67°$
$n = 1$ $\quad x° = 360° - 38{\cdot}67° = 321{\cdot}33°$
or $\quad x° = 360° + 38{\cdot}67° = 398{\cdot}67°$

For $0° \leqslant x° \leqslant 360°$

$$x = 38{\cdot}67 \text{ or } 321{\cdot}33$$

Example 5 Equations of a polynomial form higher than the quadratic
Find all the angles between 0° and 360° which satisfy the equation $3\tan^3\theta - 3\tan^2\theta - \tan\theta + 1 = 0$.

The equation is a cubic polynomial in $t = \tan\theta$, of the form

$$3t^3 - 3t^2 - t + 1 = 0$$
$$3t^2(t-1) - (t-1) = 0$$
$$(t-1)(3t^2-1) = 0$$
$$3(t-1)\left(t - \frac{1}{\sqrt{3}}\right)\left(t + \frac{1}{\sqrt{3}}\right) = 0$$

For the factor $t - 1 = 0$, $\theta = 45°$ and $225°$ from example 2 above. For the other factors, $\theta = 30°$, $150°$, $210°$ and $330°$, extending the ideas of the general solution of example 3 above. So the equation has the six solutions: 30°, 45°, 150°, 210°, 225° and 330° in the given range.

Exercise B

Solve the following equations for the inclusive ranges as specified.

1 $\sin 2\theta = 2\cos\theta$ $\quad$ (0° to 180°)

2 $\sin\theta + \cos\theta = 0$ $\quad$ (0 to 2π)

3 $3\sin\alpha + \cos^2\alpha = 3$ $\quad$ (0° to 180°)

4	$3\cos^2 x + 5\sin x = 1$	$(0° \text{ to } 360°)$
5	$2\sin x = \sqrt{3}\tan x$	$(-90° \text{ to } 90°)$
6	$8\cos^2\phi - 6\sin\phi = 3$	$\left(-\frac{\pi}{2} \text{ to } \frac{\pi}{2}\right)$
7	$2\sin^2\theta = 2 - \cos\theta$	$(0 \text{ to } 2\pi)$
8	$\cos 2\theta - \cos\theta - 2 = 0$	$(0° \text{ to } 360°)$
9	$\tan x \sec x = 2$	$\left(\frac{\pi}{4} \text{ to } \frac{5\pi}{4}\right)$
10	$2\cos 3\theta = 3\sin 2\theta$	$(0° \text{ to } 360°)$
11	$9\cos 2\alpha + 40\sin\alpha = 30$	$(0° \text{ to } 360°)$
12	$\tan x + \cot x = 4$	$(\pi \text{ to } 2\pi)$
13	$\cos 2\theta + \sin\theta = 1$	$(0 \text{ to } 2\pi)$
14	$\tan 2\theta = \tan\theta$	$\left(-\frac{\pi}{4} \text{ to } \frac{\pi}{4}\right)$
15	$3\sec^2\theta - 5\tan\theta = 1$	$(0° \text{ to } 360°)$

b Equations of a form which may be reduced to a factorisable form, using the identities connecting the sum (or difference) of two sines (or cosines).

It is worth noting that such examples usually have a number of solutions in a fairly narrow range.

Example 1

Give the solution set for $0° \leqslant x° \leqslant 360°$ for which

$$\cos x - \cos 7x = \sin 4x$$

Using the identity for the difference of two cosines, namely

$$\cos\theta - \cos\phi = 2\sin\tfrac{1}{2}(\theta + \phi)\,.\,\sin\tfrac{1}{2}(\phi - \theta)$$

$$\cos x - \cos 7x = 2\sin 4x \sin 3x$$

The equation then becomes

$$2\sin 4x \sin 3x - \sin 4x = 0$$
$$\sin 4x\,(2\sin 3x - 1) = 0$$
$$\Rightarrow \sin 4x = 0 \text{ or } \sin 3x = \tfrac{1}{2}$$

From $\sin 4x = 0$, $4x = n180° + 0° \Rightarrow x = \frac{n}{4} \times 180° = n \times 45°$

For $0° \leqslant x° \leqslant 360°$ solutions from this formula are

0°, 45°, 90°, 135°, 180°, 225°, 270°, 315°, 360°

From $\sin 3x = \frac{1}{2}$, $3x = n180° + (-1)^n 30°$ or

$$x = n60° + (-1)^n 10°$$

for $0° \leqslant x° \leqslant 360°$ the solutions from this formula are

10°, 50°, 130°, 170°, 250°, 290°

Combining the two sets, for $0° \leqslant x° \leqslant 360°$ the equation has the fifteen solutions

$$0°, 10°, 45°, 50°, 90°, 130°, 135°, 170°,$$
$$180°, 225°, 250°, 270°, 290°, 315°, 360°$$

Exercise C

Solve the following equations, for the inclusive ranges as specified.

1	$\cos\theta - 2\cos 2\theta + \cos 3\theta = 0$	$0° \leqslant \theta° \leqslant 180°$
2	$1 + \cos 2\theta + \cos 4\theta + \cos 6\theta = 0$	$-90° \leqslant \theta° \leqslant 90°$
3	$\cos\theta + \cos 5\theta = \cos 2\theta$	$0° \leqslant \theta° \leqslant 360°$
4	$\sin x + \sin 3x = \sin 2x + \sin 4x$	$0° \leqslant x° \leqslant 180°$
5	$\sin 9x + \sin 3x = \sin 6x$	$0° \leqslant x° \leqslant 90°$
6	$\sin 3\theta + \sin\theta = \sin 2\theta$	$-180° \leqslant \theta° \leqslant 180°$
7	$\sin\theta + \sin 7\theta = \sin 4\theta$	$0° \leqslant \theta° \leqslant 180°$
8	$\sin y + \sin 3y + \sin 5y = 0$	$-180° \leqslant y° \leqslant 180°$
9	$\cos 2y + \cos 4y + \cos 6y = 0$	$-90° \leqslant y° \leqslant 90°$
10	$\cos 5y - \cos y = \sin 3y$	$0° \leqslant y° \leqslant 180°$

6.3 The equation $a \cos\theta + b \sin\theta = c$

In this section we consider the solution of equations of the above form, where the numerical values a, b, c are appropriately chosen.

We may write $a\cos\theta + b\sin\theta$ in the form

$$\sqrt{(a^2+b^2)}\left\{\frac{a}{\sqrt{(a^2+b^2)}}\cos\theta + \frac{b}{\sqrt{(a^2+b^2)}}\sin\theta\right\}$$

by simply multiplying and dividing the given expression by

$$\sqrt{(a^2+b^2)}$$

If $\dfrac{a}{\sqrt{(a^2+b^2)}}$ is now defined to be $\cos\alpha$, α being an auxiliary angle introduced through its cosine form for convenience, consideration of the right-angled triangle, as shown in figure 6.7, shows that $\dfrac{b}{\sqrt{(a^2+b^2)}} = \sin\alpha$, PR being equal to b by Pythagoras.

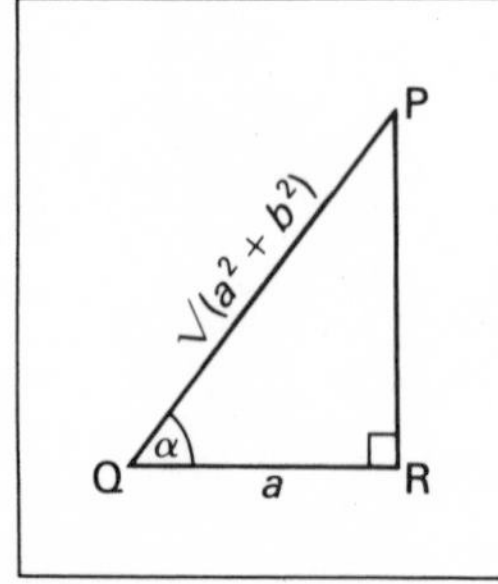

Figure 6.7

Thus $a\cos\theta + b\sin\theta$ may be written as

$$\sqrt{(a^2+b^2)}\{\cos\theta\cos\alpha + \sin\theta\sin\alpha\}$$
or $$\sqrt{(a^2+b^2)}\cos(\theta - \alpha)$$

As an alternative form, the introduction of an auxiliary angle β fulfilling the condition

$\sin\beta = \dfrac{a}{\sqrt{(a^2+b^2)}}$ would likewise define $\cos\beta$ to be $\dfrac{b}{\sqrt{(a^2+b^2)}}$

and the given expression $a \cos \theta + b \sin \theta$ would be written as $\sqrt{(a^2 + b^2)} \sin (\theta + \beta)$.

Using such ideas, the solution of the equation

$$a \cos \theta + b \sin \theta = c$$

would be compressed to solving either

$$\cos (\theta - \alpha) = \frac{c}{\sqrt{(a^2 + b^2)}} \quad \text{or} \quad \sin (\theta + \beta) = \frac{c}{\sqrt{(a^2 + b^2)}}$$

α and/or β being known through the given values defining them. Since the numerical values of $\cos (\theta - \alpha)$ or $\sin (\theta + \beta)$ must lie between ± 1 for these equations to possess solutions it follows that

$$\left|\frac{c}{\sqrt{(a^2 + b^2)}}\right|^2 \leqslant 1$$

i.e. $c^2 \leqslant a^2 + b^2$

Should this condition not be fulfilled, the original equation cannot be solved.

For the graph of the function $a \cos \theta + b \sin \theta$, for varying θ, i.e. a shape which basically consists of the addition of multiples of ordinates of the curves $y = \cos \theta$, $y = \sin \theta$, for varying θ, turning points may be derived by considering the form of $y = \sqrt{(a^2 + b^2)} \cos (\theta - \alpha)$. Because the maximum and minimum values of $\cos (\theta - \alpha)$ are $+1$ and -1 it follows that the maximum value of $a \cos \theta + b \sin \theta$ is $\sqrt{(a^2 + b^2)}$ and the minimum value is $-\sqrt{(a^2 + b^2)}$.

The following examples now illustrate the methods of solving equations of this form.

Example 1

Solve the equation $\sqrt{3} \cos \theta° + \sin \theta° = 1$, giving all solutions for θ between 0° and 360°.

First it is verified that since $(\sqrt{3})^2 + 1^2 \geqslant 1^2$ (i.e. $a^2 + b^2 \geqslant c^2$) the equation is solvable. Now the L.H.S. may be written in the form

$$2\left(\frac{\sqrt{3}}{2} \cos \theta + \frac{1}{2} \sin \theta\right)$$

by multiplying and dividing by $\sqrt{\{(\sqrt{3})^2 + 1^2\}}$.

So, introducing the auxiliary angle α, using $\cos \alpha = \frac{1}{2}\sqrt{3}$, it follows that $\sin \alpha = \frac{1}{2}$ and the L.H.S. of the equation is

$$2\{\cos \theta \cos \alpha + \sin \theta \sin \alpha\}$$

i.e. $2 \cos (\theta - \alpha)$

The given equation then reduces to

$$2 \cos (\theta - \alpha) = 1$$
$$\cos (\theta - \alpha) = \tfrac{1}{2}$$
$$\theta - \alpha = 2n \times 180° \pm 60°$$

But a value of α from $\cos \alpha = \frac{1}{2}\sqrt{3}$ is $\alpha = 30°$.

Hence $\theta = 30° + 2n \times 180° \pm 60°$.

$n = 0$ gives $\theta = 30° + 60° = 90°$ within the defined limits, and $n = 1$ gives $\theta = 30° + 360° - 60° = 330°$ also, all other values of n giving values of θ outside the range. Hence $\theta = 90°$ or $330°$.

Example 2
Solve the equation $3\cos\theta - 4\sin\theta = 2{\cdot}5$, giving all solutions for θ between $-180°$ and $+180°$.
Since $3^2 + (-4)^2 \geqslant 2{\cdot}5^2$ we may proceed. Also $3^2 + (-4)^2 = 5^2$ and therefore dividing throughout by 5

$$\frac{3}{5}\cos\theta - \frac{4}{5}\sin\theta = \frac{2{\cdot}5}{5} = \frac{1}{2}$$

Choose α so that $\sin\alpha = \frac{3}{5}\ \alpha = 36{\cdot}87°$

$$\text{Then}\quad \sin\alpha\cos\theta - \cos\alpha\sin\theta = \frac{1}{2}$$

$$\sin(\alpha - \theta) = \frac{1}{2}$$

$$\alpha° - \theta° = n(180°) + (-1)^n\,30°$$

$$\theta° = 36{\cdot}87° - n \times 180° - (-1)^n\,30°$$

$n = 0$ gives $\quad \theta = 6{\cdot}87°$
$n = 1$ gives $\quad \theta = 36{\cdot}87° - 180° + 30° = -113{\cdot}13°$

All other values of n give values outside the chosen range.

An alternative, but generally lengthier method, sometimes used for solving equations of this type, is to use the identities for $\sin\theta$ and $\cos\theta$ in terms of $t = \tan\frac{\theta}{2}$ as established in chapter 4.

$$\text{Thus}\quad \sin\theta = \frac{2t}{1+t^2} \quad\text{and}\quad \cos\theta = \frac{1-t^2}{1+t^2}$$

and the equation $a\cos\theta + b\sin\theta = c$ then becomes

$$\frac{a(1-t^2)}{(1+t^2)} + \frac{2bt}{(1+t^2)} = c$$

$$a(1-t^2) + 2bt = c(1+t^2)$$

$$t^2(a+c) - 2bt + (c-a) = 0$$

i.e. a quadratic equation in $t = \tan\frac{\theta}{2}$, which has solutions provided that

$$(2b)^2 \geqslant 4(a+c)(c-a)$$

$$4b^2 \geqslant 4(c^2 - a^2)$$

$$a^2 + b^2 \geqslant c^2, \quad \text{as before.}$$

The general solutions are developed through $\tan\frac{\theta}{2}$ to give $\frac{\theta}{2} = n\pi + \phi$, where ϕ is an angle satisfying the linear equation for $\tan\frac{\theta}{2}$.

Example 3
Solve, by this method, for $0° \leqslant \theta° \leqslant 90°$

$$2\cos\theta + 3\sin\theta + 4 = 0$$

$$\text{Using}\quad \cos\theta = \frac{1-t^2}{1+t^2},\quad \sin\theta = \frac{2t}{1+t^2}\quad \left(t = \tan\frac{\theta}{2}\right)$$

the given equation is

$$\frac{2(1-t^2)}{(1+t^2)}+\frac{6t}{1+t^2}+4=0$$

$$2-2t^2+6t+4(1+t^2)=0$$

$$2t^2+6t+6=0$$

or $t^2+3t+3=0$

Hence $t=\dfrac{-3\pm\sqrt{(-3)}}{2}$

and the task now becomes impossible since $\sqrt{(-3)}$ is meaningless within the real number system. Thus this equation does not have any solutions either within the given range or outside it.

Example 4
Solve, for $0° \leqslant \theta° \leqslant 360°$, the equation $3\sin\theta + 2\cos\theta = 1$

Replacing $\sin\theta$ by $\dfrac{2t}{1+t^2}$ and $\cos\theta$ by $\dfrac{1-t^2}{1+t^2}$ where $t=\tan\dfrac{\theta}{2}$, the equation becomes

$$\frac{6t}{1+t^2}+\frac{2(1-t^2)}{1+t^2}=1$$

$$6t+2(1-t^2)=1+t^2$$

Thus $3t^2-6t-1=0$

Hence $t=\dfrac{6\pm\sqrt{(36+12)}}{6}=\dfrac{3\pm\sqrt{12}}{3}$

$$= 2{\cdot}1547 \text{ or } -0{\cdot}1547$$

So $\tan\dfrac{\theta}{2}=2{\cdot}1547$

i.e. $\dfrac{\theta}{2}=n(180°)+65{\cdot}1°$

$$\theta = n(360°)+130{\cdot}2°$$

giving $\theta = 130{\cdot}2°$ or $490{\cdot}2°$ etc.

Or $\tan\dfrac{\theta}{2}=-0{\cdot}1547$

i.e. $\dfrac{\theta}{2}=n(180°)+171{\cdot}2°$

$$= n(360°)+342{\cdot}4°$$

and $\theta = 342{\cdot}4°$ or $702{\cdot}4°$

Therefore, in the given range $0 \leqslant \theta° \leqslant 360°$

$\theta = 130{\cdot}2$ and $342{\cdot}4$

Exercise D

In this exercise you are advised to try some examples by both methods, comparing the extent of the working needed to obtain complete solutions. Solve

1 $\sin\theta + 2\cos\theta = 1$ for $0° \leqslant \theta° \leqslant 360°$

2 $5\cos\theta - 12\sin\theta = 10$ for $-90° \leqslant \theta° \leqslant 90°$

3 $4\cos\theta - 6\sin\theta = 5$ for $0° \leqslant \theta° \leqslant 360°$

4 $3\cos\theta - 2\sin\theta = -1$ for $0° \leqslant \theta° \leqslant 360°$

5 $4\sin\theta + 3\cos\theta = 2$ for $-180° \leqslant \theta° \leqslant 180°$

6 $7\sin\theta - 24\cos\theta = 15$ giving the general solution.

7 $7\cos 2\theta + 6\sin 2\theta = 5$ for $0° \leqslant \theta° \leqslant 180°$

8 Express $\cos\theta + 3\sin\theta$ in the form $R\cos(\theta - \alpha)$ and write down the maximum and minimum values of the expression.

9 Does the equation $\cos 2\theta - \sin 2\theta = 2$ possess solutions? Illustrate your answer by drawing the graph of $y = \cos 2\theta - \sin 2\theta$ showing both maximum and minimum values.

Curve Sketching

CHAPTER SEVEN

7.1 Introduction

So far we have used various approaches to the process of sketching the curve of a given mathematical function. The curve, once obtained, has not only enabled us to illustrate and amplify the properties of the function, but has also allowed us to investigate the behaviour of the function under various circumstances.

It is clear, therefore, that the ability to represent a function by a sketched curve is significant in understanding and using mathematical functions, and in this chapter we seek to approach this task in a more systematic and exhaustive way.

It is not possible to study every type of curve, but there are certain special features which are common to most and these play a fundamental part in outlining the ultimate shape of the curve. We seek, by calculation (often approximate) or other methods, to answer the following seven questions.

1 Does the curve intersect the coordinate axes? If so, where?
2 Is the curve symmetrical about either the x-axis or the y-axis?
3 Has the curve stationary points (maximum, minimum or inflexion)?
4 Can we derive parts of the curve where y steadily increases or decreases as x increases?
5 Are there regions of the x-y plane in which the curve does not exist at all? (Or is the curve contained entirely within a particular region of the x-y plane?)
6 How does the curve behave when $x \to \pm\infty$?
7 And how does the curve behave when $y \to \pm\infty$?

Whenever we are considering the shape of the curve of a new function it may be necessary to include all of these features in our consideration. Nevertheless, until the shape of the curve has been completely defined it is advisable to seek to answer these questions in a methodical and systematic way, and to translate these answers into the relevant section(s) of the curve for which we are searching.

We now consider some of the more frequently occurring curves which have not yet been considered formally in this text.

Example 1

Consider the graph of the exponential function $y = 2^x$. We consider the questions:

1 Does the curve intersect the coordinate axes?

We see that when $x = 0$, $y = 2^0 = 1$, so that the curve cuts the y-axis at (0, 1).

However, the value of x when $y = 0$ (i.e. the solution of the equation $2^x = 0$) is not so apparent, nor possible to determine.

For positive values of x, 2^x is always positive; moreover as x increases, 2^x increases also.

For negative values of x, we can write $2^x = \frac{1}{2^X}$ where $X = -x$;

hence as $x \to -\infty$ i.e. $X \to \infty$ this expression $\to \frac{1}{\text{very large number}} \to 0$.

So strictly there is no point on the curve where the curve crosses the x-axis.

2 How does the curve behave when $x \to \pm\infty$?
We have seen that as $x \to +\infty$, $y \to +\infty$ also, and as $x \to -\infty$, $y \to 0$. (In effect, the curve approaches the line $y = 0$ as $x \to -\infty$.)

Such a line (a 'tangent at infinity') is called an **asymptote** and the curve is asymptotic to the x-axis in this case.

3 Can we derive sections of the curve where y steadily increases or decreases as x increases?
Starting from values of x close to $-\infty$, it is clear that y increases steadily, reaching the value 1 when $x = 0$, and increasing to infinity as $x \to \infty$.

Clearly we have now considered the behaviour of y as x traces out the set of values from $-\infty$ to $+\infty$ and the shape of the curve is as shown in figure 7.1. If we wished, we could consolidate the shape of the curve by plotting several specific points for a chosen set of values of x.

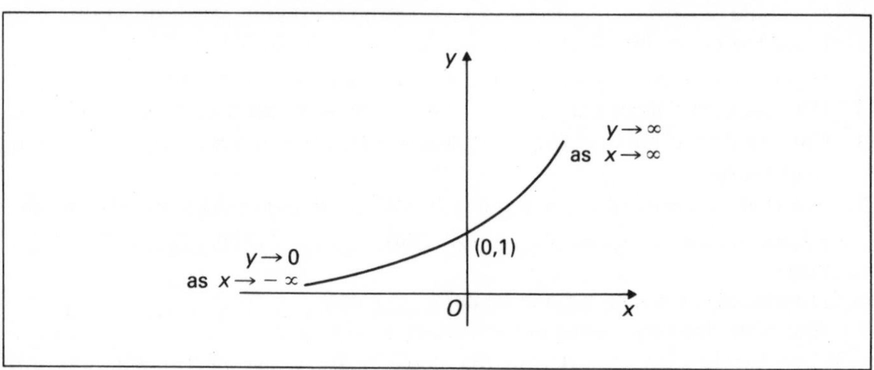

Figure 7.1

Example 2

Sketch the graph of $y = \frac{2}{x}$

Essential features

1 There are no points where the curve crosses the coordinate axes, for
   a) when $x \to 0$ through positive values, $y \to \infty$
   b) when $x \to 0$ through negative values, $y \to -\infty$
   c) when $x \to +\infty$, $y \to 0$ through positive values (we say $y \to +0$)
   d) when $x \to -\infty$, $y \to 0$ through negative values (we say $y \to -0$)

Thus both the x- and y-axes are asymptotes.

2 There are no turning points, for

$$\frac{dy}{dx} = \frac{-2}{x^2} = 0 \quad \text{has no finite solution.}$$

3 As x steadily increases from zero through positive values, $\frac{2}{x}$ steadily decreases, and as x steadily decreases from zero through negative values, $\frac{2}{x}$ steadily increases through negative values to zero.

These features indicate that the shape of the curve is as in figure 7.2. This curve is known as a **rectangular hyperbola.**

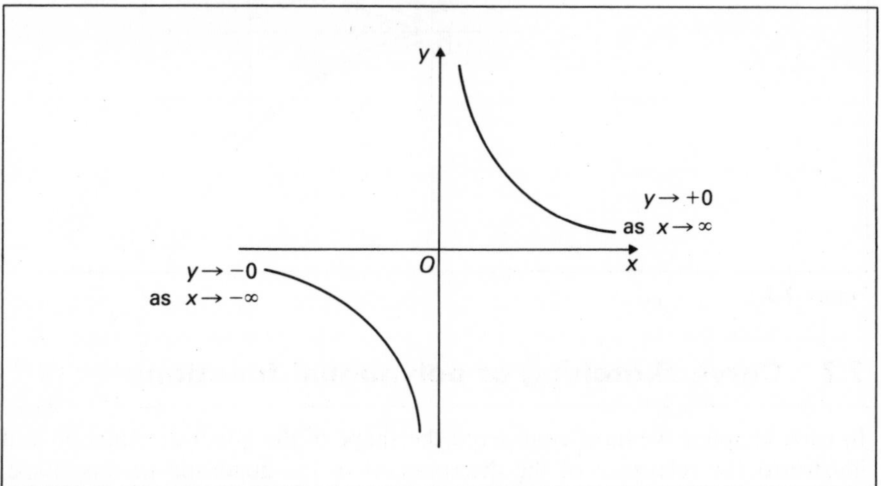

Figure 7.2

Example 3

Sketch the graph of $y^2 = x^3$.

Essential features

The first detail we note in this example is that if $x \geqslant 0$ the values of y are determined from the result $\pm\sqrt{(x^3)}$. That is, the curve consists of two **branches**, $y = +\sqrt{(x^3)}$, $y = -\sqrt{(x^3)}$.

Clearly for any value of x satisfying the condition $x \geqslant 0$ there are two values of y, equal in size, but opposite in sign. The curve is therefore **symmetrical** about the x-axis.

However, if $x < 0$, we cannot determine a real value of y, since y would be the square root of negative numbers for all such values of x. There is thus a region on the graph where no curve exists at all.

Clearly when $x = 0$, $y = 0$, and as $x \to \infty$,

$$y = \pm\sqrt{x^3} \to \pm\infty$$

Writing $y = \pm(x)^{3/2}$

$$\frac{dy}{dx} = \pm\tfrac{3}{2}(x)^{1/2} = 0 \text{ when } x = 0$$

The two branches therefore touch the x-axis at the origin. This curve is sketched in figure 7.3.

This particular curve is called a **semi-cubical parabola** and is distinguished from the ordinary parabola (described in Section 5.9) by the fact that, for the ordinary parabola, $x = 0$ is a tangent at (0, 0), whereas $y = 0$ is the tangent at (0, 0) to the semi-cubical one. (**Note** it is frequently useful to use the methods of calculus to derive not only points of maximum or minimum value on the curve, but also points of inflexion, given by the solutions of $\frac{d^2y}{dx^2} = 0$.)

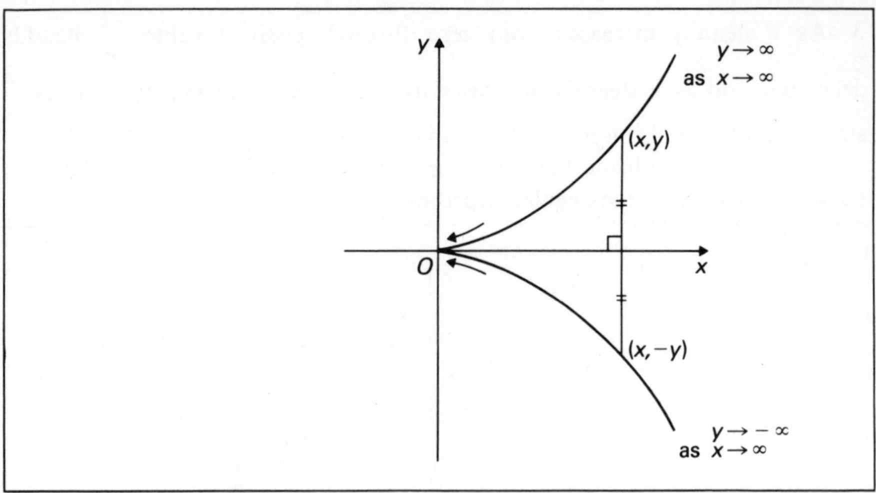

Figure 7.3

7.2 Curve sketching of polynomial functions

In early chapters we have considered the shape of the quadratic function and illustrated the relevance of the discriminant of the quadratic to this shape. The following examples are concerned with the shape of
a) the cubic polynomial function,
b) higher polynomial functions.

Example 1

Sketch the graph of the curve $y = x^3 - 3x + 5$.

Essential features

1 When $x = 0$, $y = 5$
When $y = 0$, $x^3 - 3x + 5 = 0 = f(x)$

The solutions of this equation, the roots not being whole numbers, may be derived using the method of section 11.7 or 13.1, but it is necessary only to recognise an approximate location of the solution(s) when curve sketching.

Since $f(-2) = -8 + 6 + 5 = +3$ and
$f(-3) = -27 + 9 + 5 = -13$

there is a root of the equation between $x = -2$ and $x = -3$.

2 $\dfrac{dy}{dx} = 3x^2 - 3 = 0$ when $x = \pm 1$

$$\frac{d^2y}{dx^2} = 6x \text{ and } \frac{d^2y}{dx^2} > 0 \text{ when } x = +1, < 0 \text{ when } x = -1$$

So the curve has a minimum point (1, 3), and a maximum point (−1, 7).

3 Now $y = x^3 - 3x + 5$

$$= x^3\left(1 - \frac{3}{x^2} + \frac{5}{x^3}\right)$$

When $x \to \pm\infty$ both $\dfrac{3}{x^2}$ and $\dfrac{5}{x^3} \to 0$

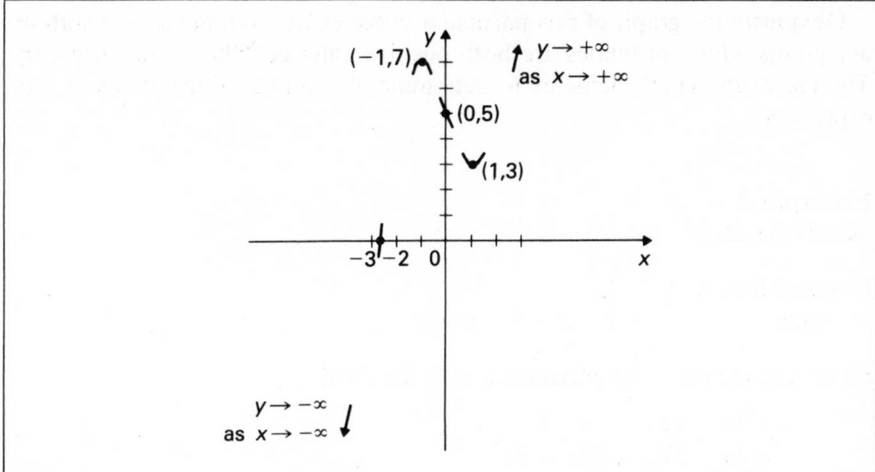

Figure 7.4

So $1 - \frac{3}{x^2} + \frac{5}{x^3} \to 1$

When $x \to +\infty$, $y \to \infty$, behaving as $x^3(1)$.
and when $x \to -\infty$, $y \to -\infty$, also behaving as $x^3(1)$.
Should we include only these features on the curve at this stage, we would have the outline given in figure 7.4.

4 A further observation of the gradient $\frac{dy}{dx} = 3x^2 - 3 = 3(x^2 - 1)$ shows:

a) if $x < -1$, $x^2 > 1$ and $\frac{dy}{dx} > 0$, i.e. the section of the curve for $x < -1$ is one for which y increases as x increases.

b) if $-1 < x < 1$, $x^2 < 1$ and $\frac{dy}{dx} < 0$, so the graph steadily decreases in this range.

c) if $x > 1$, $x^2 > 1$, $\frac{dy}{dx} > 0$ and the curve increases for all such values.

The essential features which we have already located may therefore simply be joined up to give the required shape, figure 7.5.

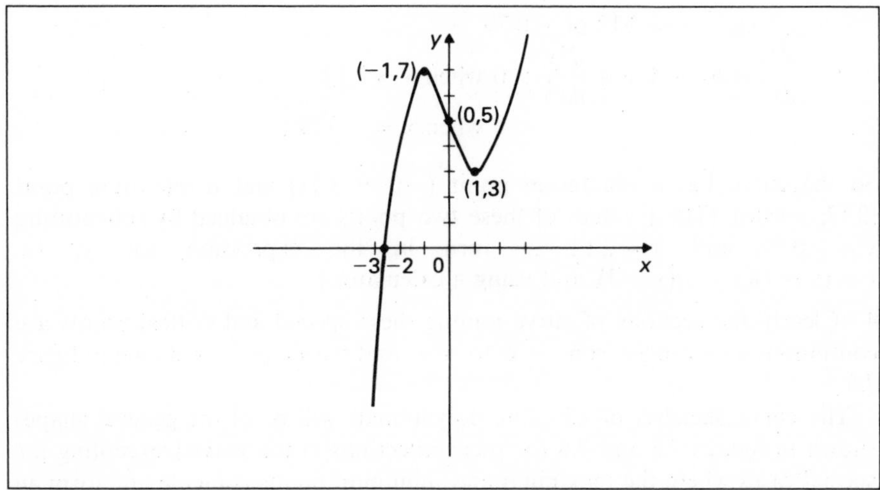

Figure 7.5

Obviously the graph of this particular cubic expression, possessing stationary points whose ordinates are both positive, intersects the x-axis once only. The use of the graph helps us to determine the number of real roots of such expressions.

Example 2

Sketch the curve $y = x^3 - 2x^2 - 5x + 6$

Essential features

1 When $x = 1$, $y = 1 - 2 - 5 + 6 = 0$

So we can factorise the polynomial into the form

$$y = (x - 1)(x^2 - x - 6)$$
$$= (x - 1)(x + 2)(x - 3)$$

Thus $y = 0$ when $x = 1, -2, 3$

The curve therefore crosses the x-axis at the three points

$$(1, 0), \quad (-2, 0), \quad (3, 0)$$

Moreover, when $x = 0$, $y = 6$, so that the curve crosses the y-axis at the point $(0, 6)$.

2 We may write $y = x^3\left(1 - \frac{2}{x} - \frac{5}{x^2} + \frac{6}{x^3}\right)$

So, since $\frac{2}{x}, \frac{5}{x^2}, \frac{6}{x^3}$ all tend to zero as $x \to \pm\infty$,

$y \to \infty$ behaving as x^3 when $x \to +\infty$

$y \to -\infty$ behaving as x^3 when $x \to -\infty$

3 $\frac{dy}{dx} = 3x^2 - 4x - 5$

$= 0$ when $x = \frac{1}{6}(4 \pm \sqrt{76})$

i.e. $x = 2{\cdot}12$ or $-0{\cdot}79$

$\frac{d^2y}{dx^2} = 6x - 4$ and $\frac{d^2y}{dx^2} > 0$ when $x = 2{\cdot}12$

< 0 when $x = -0{\cdot}79$

So the curve has a maximum point $(-0{\cdot}79, 8{\cdot}21)$ and a minimum point $(2{\cdot}12, -4{\cdot}06)$. (The y-values of these two points are obtained by substituting $x = -0{\cdot}79$ and $x = 2{\cdot}12$ in turn in the expression for y, i.e. $y = (x - 1)(x + 2)(x - 3)$, and using a calculator.)

4 Clearly the sections of curve joining these special and critical points are continuous as x ranges from $-\infty$ to $+\infty$ and the shape is as shown in figure 7.6.

The curve sketches of all cubic polynomials will be of the general shapes shown in figures 7.5 and 7.6 (or their reflections in the x-axis), excepting the special cases where the maximum and minimum points coincide and form an inflexional point.

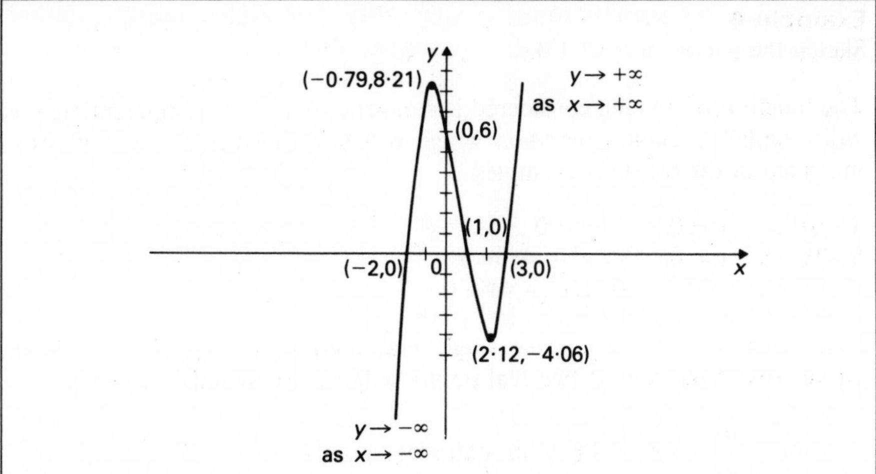

Figure 7.6

Example 3

Sketch the curve $y = x^3$

1 Obviously $x = 0$ when $y = 0$, and the curve passes through the origin.

2 $\dfrac{dy}{dx} = 3x^2, \qquad \dfrac{d^2x}{dx^2} = 6x,$

so that when $x = 0$ the origin is also a stationary point on the curve, this point being an inflexional one since

$$\frac{d^2y}{dx^2} = 0 \text{ when } x = 0.$$

3 When $x \to +\infty$, $y \to \infty$ also, behaving as x^3
When $x \to -\infty$, $y \to -\infty$ likewise

The shape of the curve is therefore as shown in figure 7.7.

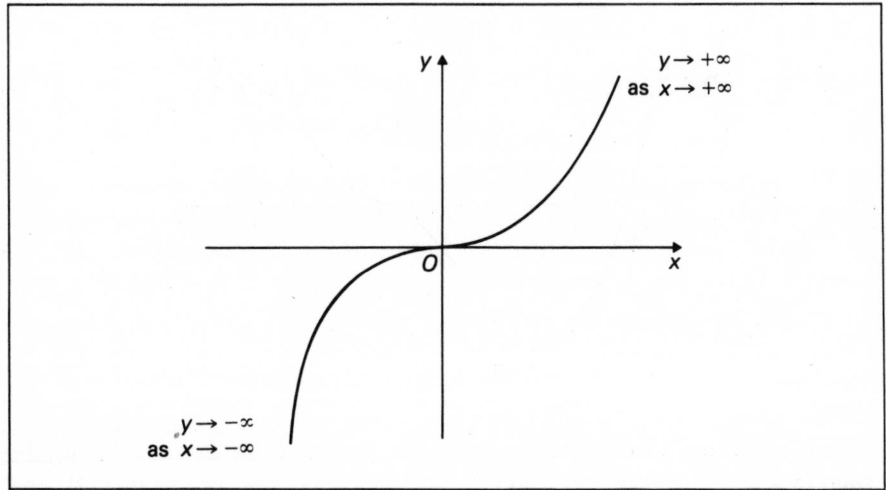

Figure 7.7

Example 4
Sketch the curve $y = x^4 + 4x$

The function of x being considered is a quartic. When this is differentiated we will establish a cubic expression which will have one or three real roots as indicated in the previous examples.

1 When $x = 0$, $\quad y = 0$
2 When $y = 0$, $\quad x^4 + 4x = 0$
$\Rightarrow x(x^3 + 4) = 0$

Now since $x^3 + 4 = 0$ has the single real solution $x = -\sqrt[3]{4} = -1{\cdot}59$, the curve crosses the axes at two real points only, i.e. $(0, 0)$ and $(-\sqrt[3]{4}, 0)$.

3 Now $\dfrac{dy}{dx} = 4x^3 + 4 = 0$ for stationary points.

But $x^3 + 1 \equiv (x + 1)(x^2 - x + 1) = 0$ for $x = -1$ only, since $x^2 - x + 1 = 0$ does not have real roots.

And $\dfrac{d^2y}{dx^2} = 12x^2 > 0$ when $x = -1$.

The curve has therefore a single stationary point $(-1, -3)$ which is a minimum point.

4 Since $y = x^4\left(1 + \dfrac{4}{x^3}\right)$ as $x \to \pm\infty$, $\dfrac{4}{x^3} \to 0$

Thus $y \to \infty$ behaving as x^4 as $x \to \pm\infty$

Incorporating these features in the shape of the curve, the form is that shown in figure 7.8.

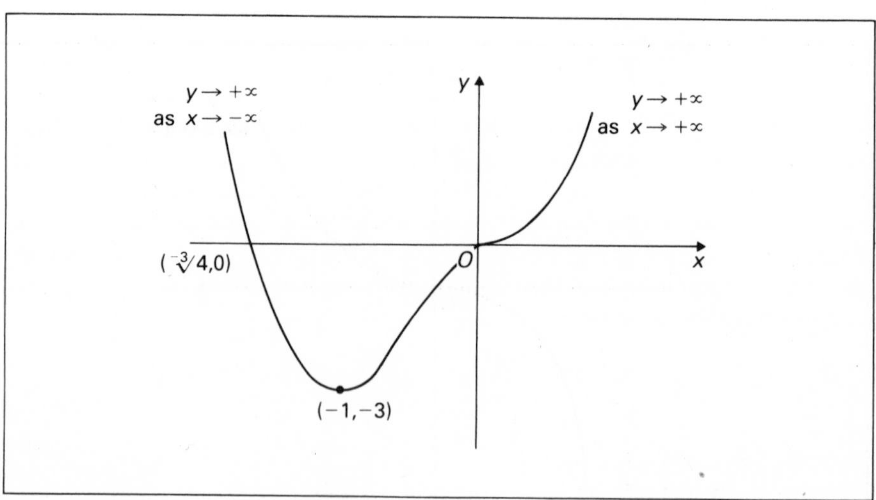

Figure 7.8

Exercise A

Sketch the graph of the curves defined by the following equations.

1 $y = 2^{-x}$
2 $y = 2^{x-1} + 2^{-x-1}$
3 $y = x(x-1)(x-2)$
4 $y = (x+1)(x^2+1)$
5 $y = 2x - x^3$
6 $y = x^3(x-2)$
7 $y^2 = x(x-2)^2$
8 $y^2 = x(x-2)(x-4)$

7.3 Asymptotes

So far, with the exception of the semi-cubical parabola and examples 7 and 8 of Exercise A, all of the curves considered have existed for all values of x in the infinite range $-\infty < x < \infty$. Furthermore, with the exception of the rectangular hyperbola, the curves have been **continuous in this range.** i.e. for each value of x, a finite value of y has been determinable from the given functional relationship.

By considering the graph of $y = \dfrac{f(x)}{g(x)}$, where $f(x)$ and $g(x)$ are quadratics, we shall now illustrate cases of curve sketching which

a) contain regions of the x-y plane in which the curve does not exist,

b) amplify the problem of discontinuity, such curves possessing asymptotes of the form $y = a$, $x = b$, or $y = mx + c$.

Moreover, in seeking to answer question 3 of the fundamental set of seven questions (posed earlier in this chapter, page 139), we shall also make use of an algebraic method of deriving stationary points for such types of curve.

Example 1

Sketch the graph of the curve $y = \dfrac{x^2 - x}{(x-2)(x+1)}$

Write down the equations of the asymptotes and the coordinates of the turning point(s), if any.

It is clear that we can write $y = \dfrac{x(x-1)}{(x-2)(x+1)}$

so that when $x = 0$, $y = 0$ and when $x = 1$, $y = 0$ follows by a consideration of the zeros of the numerator.

By considering likewise the zeros of the denominator, i.e. $x = 2$, $x = -1$ it is also clear that y is infinite for these values. Thus $x = 2$ and $x = -1$ are asymptotes to the curve and discontinuities exist for these values of x. A somewhat closer examination of the values of y in the neighborhood of these two critical values shows that

a) if $x = 2+$ i.e. a value just greater than 2,

$$y = \frac{(2+)(1+)}{(0+)(3+)} = \frac{(2+)}{(0+)}$$

that is, is an infinitely large **positive** number.

b) if $x = 2-$ i.e. a value just less than 2,

$$y = \frac{(2-)(1-)}{(0-)(3-)} = \frac{(2-)}{(0-)}$$

that is, is an infinitely large **negative** number.

Furthermore, if $x = -1+$, $y = \frac{(-1+)(-2+)}{(-3+)(0+)} \Rightarrow$ large $-$ve number

and if $x = -1-$, $y = \frac{(-1-)(-2-)}{(-3-)(0-)} \Rightarrow$ large $+$ve number

This analysis permits us to begin to 'shape' the curve as in figure 7.9(a). The asymptotes $x = 2$ and $x = -1$, and the parts of the curve as it approaches the asymptotes, show the discontinuity and beginnings of the formation of **branches** of the curve of the given function.

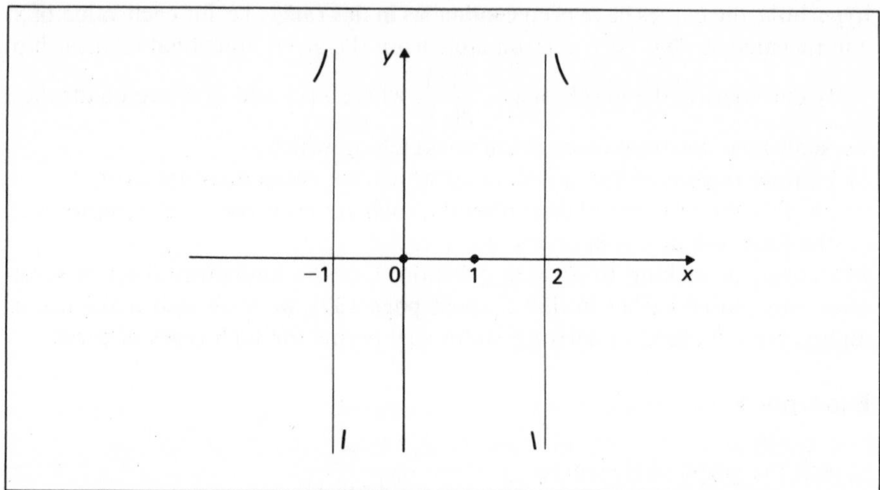

Figure 7.9 (a)

As the next step

since $y = \frac{x^2 - x}{x^2 - x - 2}$

we can also write

$$y = 1 + \frac{2}{x^2 - x - 2} \text{ by division.}$$

Now if $x \to +\infty$, $x^2 - x - 2 \to +\infty$ and $y \to 1 + \frac{2}{\text{large } +\text{ve number}}$

that is $y \to 1 + 0$

And if $x \to -\infty$, $x^2 - x - 2$ also $\to +\infty$ and $y \to 1 + 0$

We can then incorporate this feature in the graph of the curve of the function, and by choosing, say, $x = 3$, $x = -2$ derive the points (3, 1·5), (−2 . 1·5) which enable us to complete the two branches for $-\infty < x \leqslant -1$, $2 \leqslant x < +\infty$ (see figure 7.9(b)).

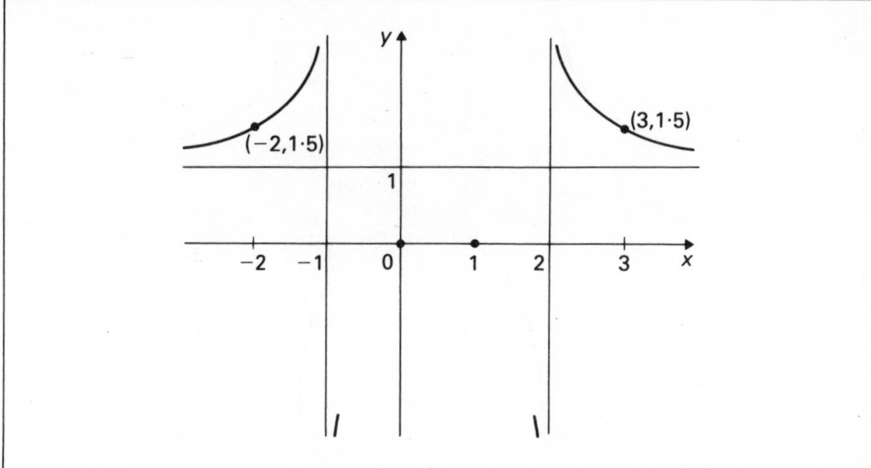

Figure 7.9 (b)

Now also from $y = \dfrac{x(x-1)}{(x-2)(x+1)}$, we can write

$y(x^2 - x - 2) = x^2 - x$

i.e. $x^2(y-1) + x(1-y) - 2y = 0$

Since we are concerned with values of $x \in \mathbb{R}$, for such values the discriminant of this quadratic in x must not be negative.

Thus $(1-y)^2 \geqslant 4(y-1)(-2y)$

or $(y-1)^2 + 8y(y-1) \geqslant 0$

i.e. $(y-1)(9y-1) \geqslant 0$

Consideration of the real number line for y immediately shows that for this inequality to be satisfied $y \geqslant 1$ or $\leqslant \frac{1}{9}$.

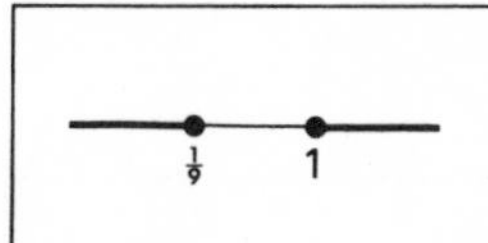

As a corollary, it follows that if $\frac{1}{9} < y < 1$, real values of x on the curve cannot be determined and thus such a region of the x-y plane does not contain any part of the curve. We have already seen what is happening when $y = 1$. When $y = \frac{1}{9}$ the values of x are determined from

$$x^2(\tfrac{1}{9} - 1) + x(1 - \tfrac{1}{9}) - \tfrac{2}{9} = 0$$
$$\Rightarrow \quad -8x^2 + 8x - 2 = 0$$
$$\Rightarrow \quad 4x^2 - 4x + 1 = 0$$
$$\Rightarrow \quad (2x-1)^2 = 0$$
$$\Rightarrow x = \tfrac{1}{2} \quad \text{(twice)}$$

So the point $(\frac{1}{2}, \frac{1}{9})$ lies on the curve, and, as is seen when the region of no curve is sketched in, is clearly a maximum point.

A final confirmation of the shape of the branch of the curve for $-1 \leqslant x \leqslant 2$ can be obtained by choosing two other suitable values of x to plot, on either side of the maximum. Thus when $x = -\frac{1}{2}$, $y = -0{\cdot}6$, and when $x = 1\frac{1}{2}$, $y = -0{\cdot}6$. The shape of the curve of this function is therefore as shown in figure 7.9(c).

Clearly all functions of the form $y = \dfrac{(x-\alpha)(x-\beta)}{(x-\gamma)(x-\delta)}$ where $\alpha, \beta, \gamma, \delta \in \mathbb{R}$ will have curve forms of this kind.

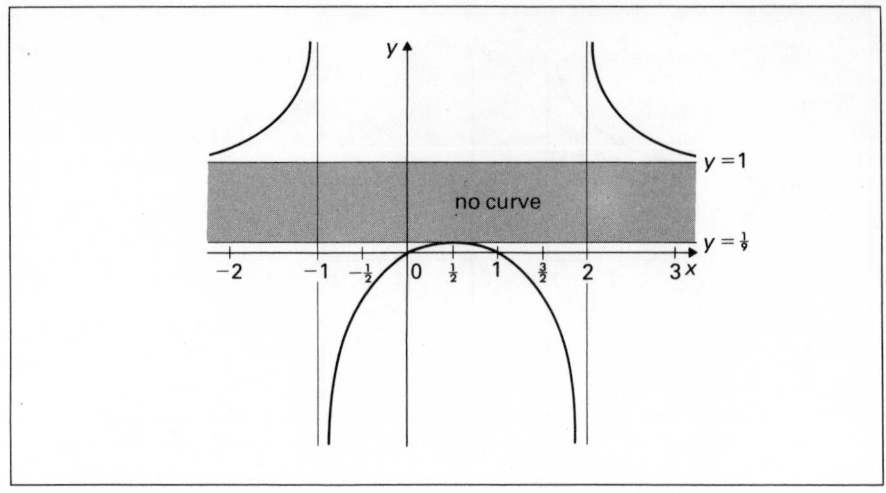

Figure 7.9 (c)

We now consider those cases in which the two quadratic expressions $f(x)$ and $g(x)$ either cannot be factorised or degenerate into binomial or monomial forms.

Example 2

Sketch the graph of $y = \dfrac{x}{x^2 + x + 1}$

We may write $y(x^2 + x + 1) = x$

$\Rightarrow \; x^2y + x(y - 1) + y = 0$

For $x \in \mathbb{R}$, $(y - 1)^2 \geqslant 4y^2$

i.e. $4y^2 - (y - 1)^2 \leqslant 0$

i.e. $\{2y - (y - 1)\}\{2y + (y - 1)\} \leqslant 0$

$$(y + 1)(3y - 1) \leqslant 0$$

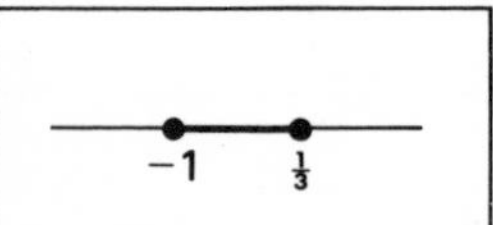

From the real number line for y the inequality is satisfied when $-1 \leqslant y \leqslant \frac{1}{3}$ only, so that for any other value of y, no real values of x exist for the curve.

The curve is therefore contained *entirely within* this narrow region of the x-y plane. Moreover, when $y = -1$,

$$-x^2 - 2x - 1 = 0 \quad \text{or} \quad (x + 1)^2 = 0 \quad \Rightarrow \quad x = -1$$

Hence $(-1, -1)$ is a point on the curve, which, when sketched, shows this point to be a minimum point.

When $y = \frac{1}{3}$, $\frac{1}{3}x^2 - \frac{2}{3}x + \frac{1}{3} = 0$ or $(x - 1)^2 = 0 \quad \Rightarrow \quad x = 1$

Hence $(1, \frac{1}{3})$ is also a point on the curve, which can be seen to be a maximum point.

Examination of the zeros of the numerator show that when $x = 0$, $y = 0$, and since $x^2 + x + 1 \neq 0$ for any real value of x there are no zeros of the denominator, and thus no asymptotes parallel to the y-axis. Since we can write

$$y = \frac{1}{x + 1 + \dfrac{1}{x}}, \text{ as } x \to +\infty, \; y \to 0+ \text{ and as } x \to -\infty, \; y \to -0.$$

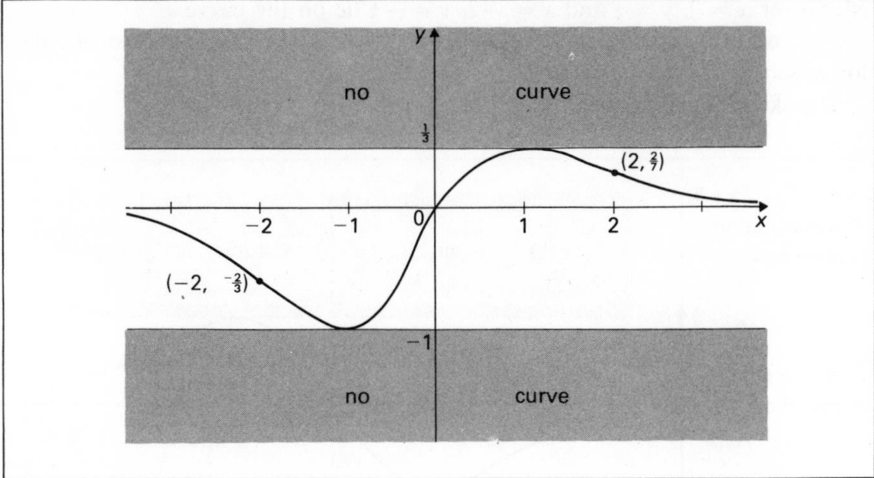

Figure 7.10

All of these features incorporated in the sketch of the curve of the function basically define the curve completely. The plotting of two extra points,

say $x = 2, \quad y = \frac{2}{7}$
$x = -2, \quad y = -\frac{2}{3}$
confirms the shape as shown in figure 7.10.

It is worth noting how important are the diagrams defining the values of y on the real number line for which y exists on the graph, since the information when translated to the sketch of the curve immediately rejects those regions of the x-y plane for which there is no curve. Also it is useful when considering an asymptote to judge the behaviour of the curve for $x \to \pm\infty$, $y \to \pm\infty$.

Example 3

Sketch the graph of $y = \dfrac{x}{x^2 + 1}$

We may write $\quad (x^2 + 1)y = x \Rightarrow x^2y - x + y = 0$

For $x \in \mathbb{R}, \quad 1 \geqslant 4y^2$
$\Rightarrow 4y^2 - 1 \leqslant 0$
$\Rightarrow (2y - 1)(2y + 1) \leqslant 0$

From the real number line for y the inequality is satisfied for $-\frac{1}{2} \leqslant y \leqslant \frac{1}{2}$ and in the x-y plane there is no curve for $y > \frac{1}{2}$ or $y < -\frac{1}{2}$.

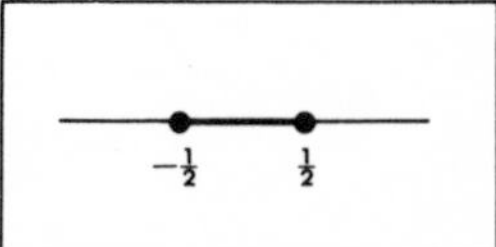

When $y = \frac{1}{2}, \quad x^2 + 1 = 2x \Rightarrow x = 1$ (twice)
When $y = -\frac{1}{2}, \quad x^2 + 1 = -2x \Rightarrow x = -1$ (twice)

The zeros of the numerator give $x = 0$, $y = 0$ and there are no zeros of the denominator.

When $x \to +\infty, \quad y = \dfrac{1}{x + \dfrac{1}{x}} \to 0^+$

and when $x \to -\infty$, $y \to 0^-$ (i.e. 'circle at infinity')

(Note the use of an alternative notation to $0+$, $0-$ used previously.)

Moreover $x = 2$, $y = \frac{2}{5}$ and $x = -2$, $y = -\frac{2}{5}$ lie on the curve.

The rotation of the part of the curve for which $x > 0$ to establish the part for which $x < 0$ is also clear in this example.

The sketch is shown in figure 7.11.

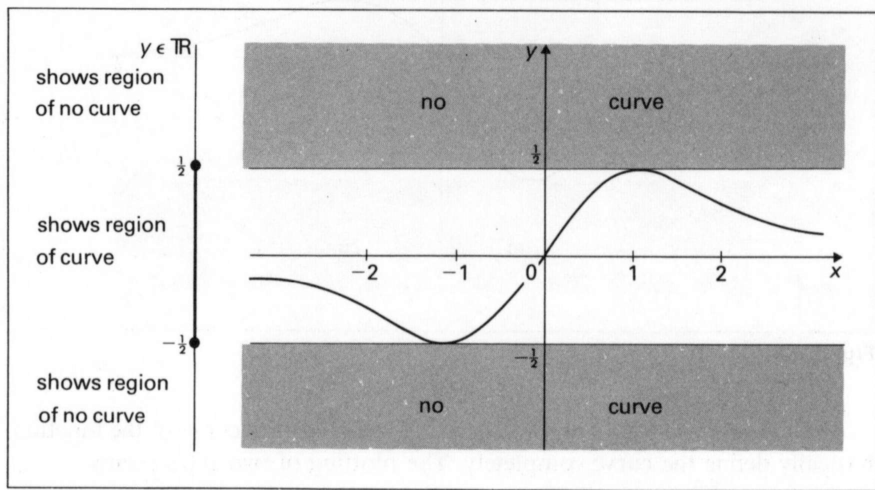

Figure 7.11

Example 4

Sketch the graph of $y = \dfrac{x-2}{x^2 - 2x + 1}$

Writing $y(x^2 - 2x + 1) = x - 2$

$\Rightarrow x^2y - x(2y + 1) + (y + 2) = 0$

For $x \in \mathbb{R}$, $(2y + 1)^2 \geqslant 4y(y + 2)$

$\Rightarrow 4y^2 + 4y + 1 \geqslant 4y^2 + 8y$

$\Rightarrow -4y + 1 \geqslant 0$

$y \leqslant \frac{1}{4}$

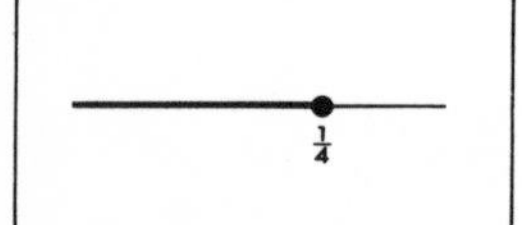

Hence, on the real number line for y, y can never be greater than $\frac{1}{4}$.

Also when $y = \frac{1}{4}$

$\frac{1}{4}x^2 - \frac{3}{2}x + \frac{9}{4} = 0$

$\Rightarrow x^2 - 6x + 9 = 0$

$\Rightarrow (x - 3)^2 = 0$

$x = 3$

Zeros of the numerator occur when $x = 2$, and of the denominator when $x = 1$ (twice).

And since $y = \dfrac{x-2}{(x-1)^2}$

when x is both $1+$ and $1-$, $y \to -\infty$. Since we may write

$$y = \frac{1 - \dfrac{2}{x}}{x - 2 + \dfrac{1}{x}},$$ (dividing numerator and denominator by x),

when $x \to +\infty$, $y \to 0^+$, and when $x \to -\infty$, $y \to 0^-$.

To confirm these features choose three points, namely

when $x = -1,\ y = -\frac{3}{4}$
$x = 0,\ y = -2$
$x = \frac{3}{2},\ y = -2$

The sketch of the curve is shown in figure 7.12.

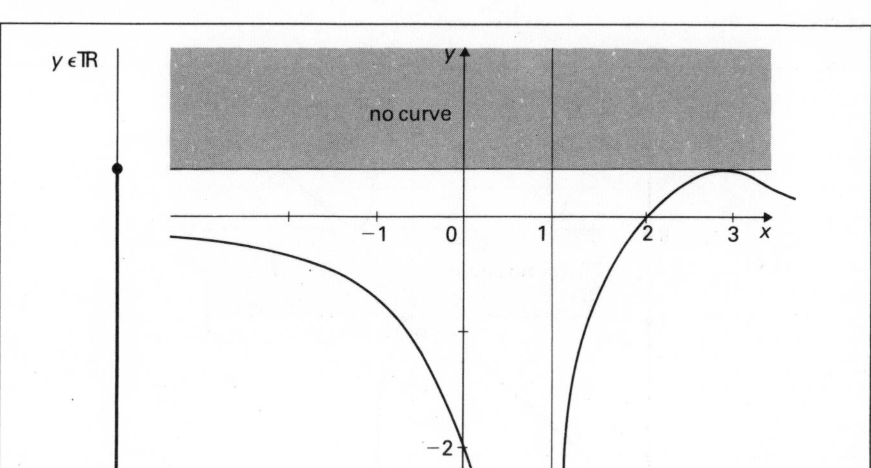

Figure 7.12

Example 5

Sketch the graph of $y = \dfrac{x^2}{x-2}$

In this example the linear denominator is considered to be a special degenerate case of the quadratic $g(x)$, in which the coefficient of the term x^2 is zero.

Rewriting as $(x-2)y = x^2$, the expression is $x^2 - xy + 2y = 0$

For $x \in \mathbb{R}$, $y^2 \geqslant 4(2y)$
i.e. $y(y-8) \geqslant 0$

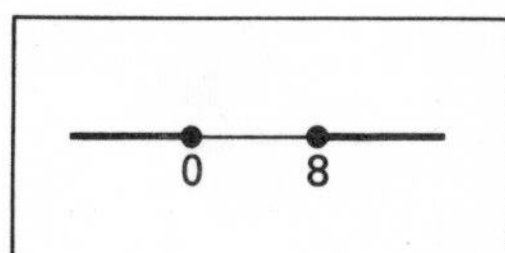

Consideration of the real number line for y shows that there is no curve in the x-y plane for $0 < y < 8$.

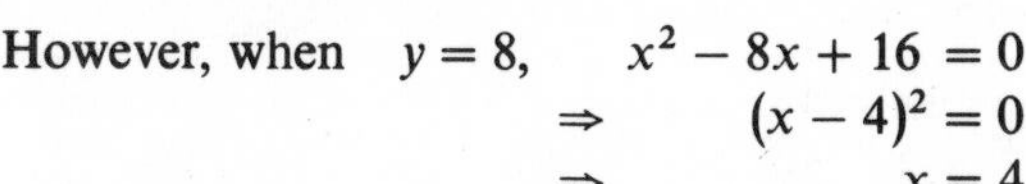

However, when $y = 8$,
$$x^2 - 8x + 16 = 0$$
$$\Rightarrow \quad (x-4)^2 = 0$$
$$\Rightarrow \quad x = 4$$

And when $y = 0$, $x^2 = 0 \Rightarrow x = 0$.

The numerator is zero when $x = 0$ (twice), and the curve passes through $(0, 0)$.

The denominator is zero when $x = 2$. Therefore this line is an asymptote to the curve. If $x = 2^-$, $y \to -\infty$ and if $x = 2^+$, $y \to +\infty$.

We may also write

$$y = x + 2 + \frac{4}{x-2}, \text{ by division,}$$

so that when $x \to +\infty$, $y \to x + 2+$, i.e. $y = x + 2$ is an **oblique asymptote** and the curve approaches it from above. However when $x \to -\infty$, $y \to x + 2-$, and the curve approaches the asymptote from below.

The curve, as shown in figure 7.13, is another form of the **hyperbola** possessing two asymptotes, one parallel to the y-axis and the other oblique.

It is not necessary in this example to plot further points, as the examination of the special features determines completely the shape of the branches of the curve.

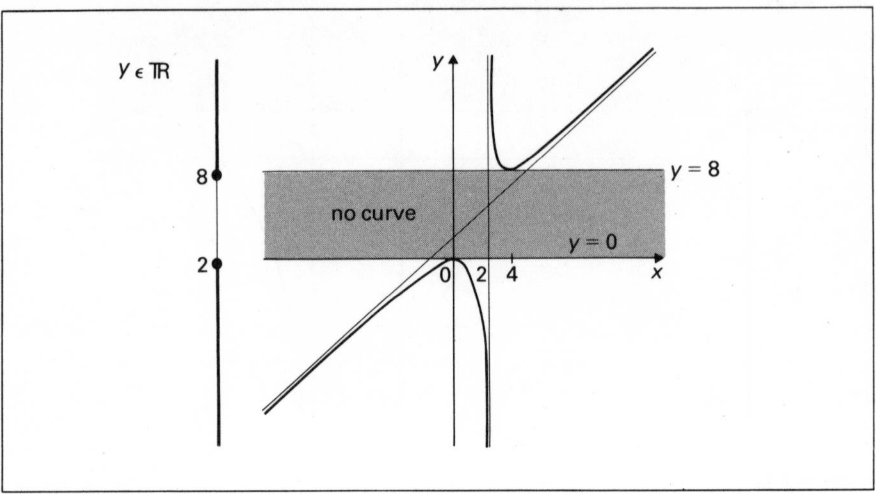

Figure 7.13

Exercise B

1 Analyse the special features of the shapes of the curves of the following functions, giving equations of their asymptotes and coordinates of turning points (whenever they exist); and sketch the curves.

a) $y = \dfrac{(x-1)(x+2)}{(x+1)(x-3)}$ f) $y = \dfrac{3(x-2)}{x^2+6x}$

b) $y = \dfrac{(x+2)(x-4)}{(x+1)(x-3)}$ g) $y = \dfrac{4}{(x-1)(x-3)}$

c) $y = \dfrac{(x-2)(x-4)}{x^2-x-6}$ h) $y = \dfrac{x-2}{x^2+2}$

d) $y = \dfrac{2x}{x^2+1}$ i) $y = \dfrac{3x-4}{2x^2-2x}$

e) $y = \dfrac{x^2-4x+1}{(x-2)^2}$ j) $y = \dfrac{2x^2-5x-3}{x^2+1}$

2 Using the methods of this chapter determine the turning points on the curve of

$$y = \frac{2x}{x^2+x+1}.$$

By drawing a sketch of the curve show that all other values of y lie between the two values of y of the turning points you have determined.

3 When, $x \in \mathbb{R}$, prove that if $y = \dfrac{\lambda x^2 + 3x - 4}{-4x^2 + 3x + \lambda}$, then $y \in \mathbb{R}$ if $1 \leqslant \lambda \leqslant 7$.
Sketch the graph of y against x in the special case of $\lambda = 1$.

7.4 Further functions

The notation $|x|$ is used in mathematics to denote the numerical (or absolute) value of the variable x. Thus $|-2{\cdot}5| = 2{\cdot}5$, $|-17| = 17$ and so on. We call this notation the **modulus** of the number x, and now consider the sketches of functions involving this form.

Example 1
Sketch the graph of $y = |x|$.

$y = x$ is, of course, a straight line passing through the origin and of unit gradient. The points on this line for $x < 0$ all have negative ordinate values. Thus the graph of $y = |x|$ for $x < 0$ simply consists of a section in which all of these ordinate values are positive, and is a reflection of the original section of the line in the x-axis. The part of the line for which $x > 0$ is not affected by the application of the modulus sign. The sketch is as shown in figure 7.14.

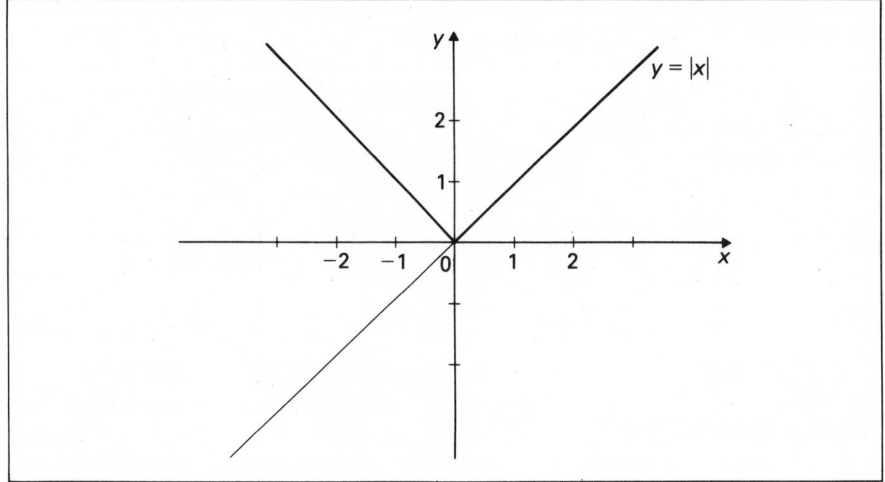

Figure 7.14

Again, if we consider the sketch of the graph of $y = |3x - 4|$ the technique of reflecting in the x-axis the section of the graph for which y is negative helps us to determine the shape of the graph.

For $y = 3x - 4$, the graph is a line of gradient 3, cutting the x-axis at $x = \frac{4}{3}$. For those values of x on the real number line such that $x < \frac{4}{3}$, y is negative. (The diagram used predominantly in the last section helps to identify those values of x satisfying this inequality.) Thus the shape of the graph is as shown in figure 7.15.

$\frac{4}{3}$

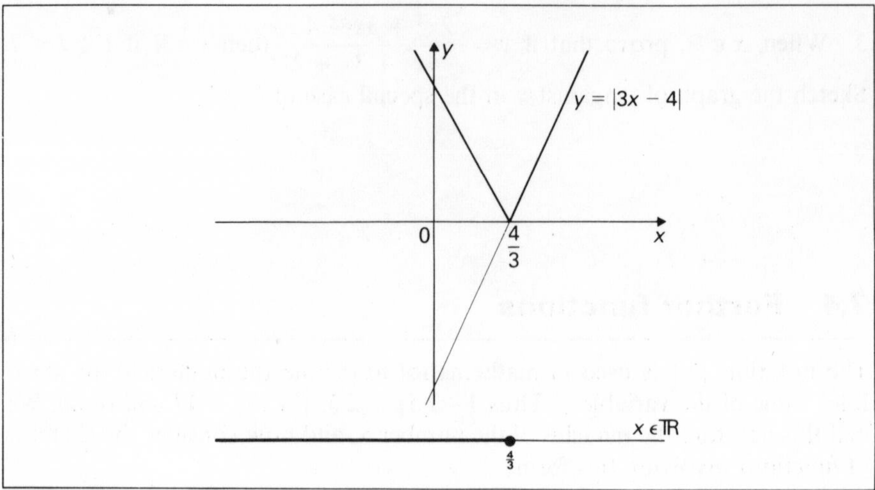

Figure 7.15

Further illustrations of the process of reflecting the sections for which y is negative on the graph are shown in figure 7.16, where the function being considered is not of the linear form.

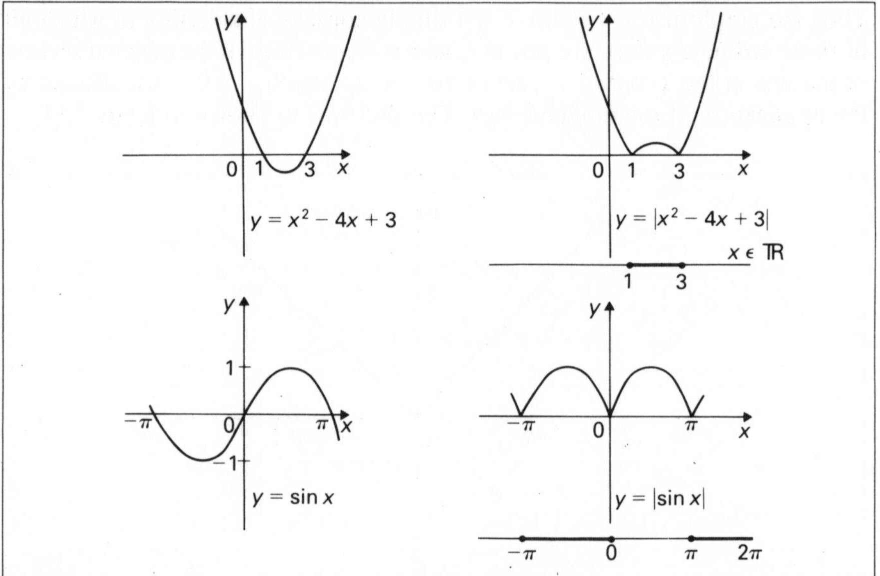

Figure 7.16

The process of curve sketching in which a modulus occurs becomes a little more complicated should the functional form for y involve more than one element.

Example 2

Sketch the graph of $y = |x + 1| + |x - 1|$

Dealing with the shapes of the two independent quantities involved in the moduli separately, their representation would be as shown in figure 7.17(a). The graph of the given form would therefore be derived by *adding* the ordinates on these separate sections for every value of x.

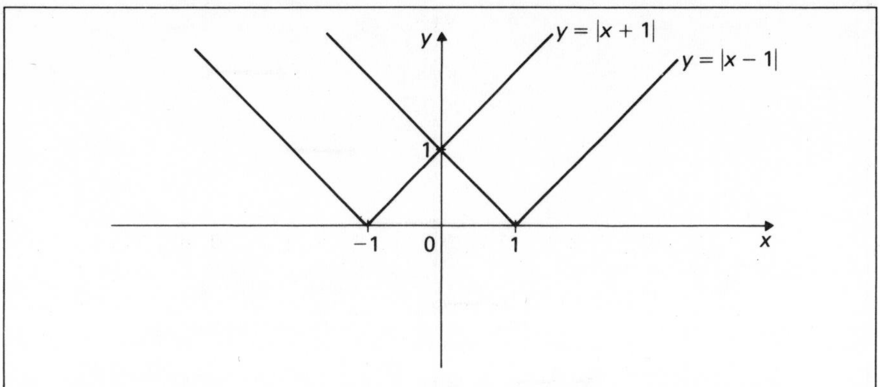

Figure 7.17 (a)

It is clear that for $0 < x \leqslant 1$ this would yield $y = x + 1 + 1 - x = 2$

for $x > 1$, $y = x + 1 + x - 1 = 2x$
for $-1 \leqslant x < 0$, $y = x + 1 + 1 - x = 2$
and for $x < -1$, $y = -1 - x + 1 - x = -2x$

The sketch is shown in figure 7.17(b).

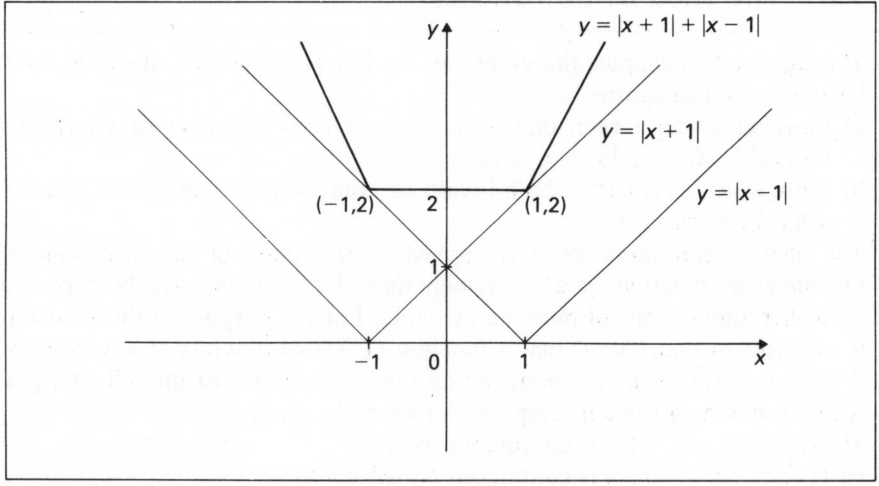

Figure 7.17 (b)

7.5 Step function

A further notation which is used, whose examination is relevant to this chapter, is the form $[x]$. This notation implies that we are considering the integral part of the variable x, i.e. the greatest integer not exceeding the value of x.

For example $[1{\cdot}8] = 1$ $[4] = 4$
$[-2{\cdot}4] = -3$ and so on.

Thus the graph of $y = [x]$ consists of a number of **steps,** each step representing the section of the graph from one integer to just short of the next, figure 7.18.

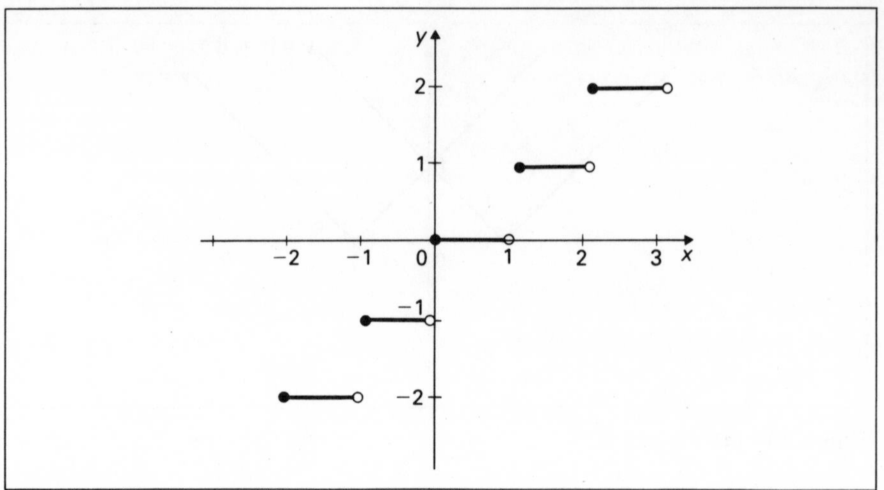

Figure 7.18

(It is important to understand that the line segments in the diagram must show inclusion of the value at one end and exclusion of the value at the other, by shaded, or unshaded rings.)

7.6 The idea of continuity

Throughout this chapter the graphs of the functions we have sketched have fallen into two categories,

a) those which have been unbroken or continuous as values of x have increased from $-\infty$ to $+\infty$, and

b) those which have had sudden breaks or jumps when some special value of x has been reached.

The idea of continuity as it is applied to the study of the behaviour of mathematical functions is an extremely important one and will be extended in further applications of pure mathematics. For the purpose of this course it is sufficient to understand that a function $f(x)$ is continuous at any point x, if we can move from this point along the curve, either to the left or right, without making a sudden jump. So, for example, given

a) $f(x) \equiv x^3$, $f(x)$ is continuous for all x

b) $f(x) \equiv \sqrt{x}$, $f(x)$ is continuous for all $x > 0$
(at $x = 0$, it is not possible to move to the left, on the curve)

c) $f(x) \equiv |x|$, $f(x)$ is continuous for all x,

d) $f(x) \equiv [x]$, $f(x)$ is continuous except when $x \in \mathbb{Z}$.

Exercise C

1 Sketch the graphs of the following curves.

a) $y = |x + 2|$
b) $y = |-x + 3|$
c) $y = |x^2 - 3x - 4|$
d) $y = |\cos x|$
e) $y = |2x + 1| - |2x - 1|$
f) $y = |2 \sin x| + |\cos x|$
g) $y = 3 + |x - 1|$

2 Sketch the graph of

a) $y = [2x]$ b) $y = [x^2]$ c) $y = x - [x]$

3 For what x value(s) (if any) are the functions defined by the following relationships discontinuous?

a) $f(x) \equiv 3x - 1$ c) $f(x) \equiv |x(x-1)(x-2)|$

b) $f(x) \equiv \dfrac{1}{x^2}$ d) $f(x) \equiv x - [x]$

7.7 Polar coordinates

In chapter 5, we discussed the idea that any point in two dimensions could be defined by two coordinates, each giving a distance from two fixed lines or axes. Much of the geometrical work at this level is based on such a cartesian system. However, we may also determine any point in a plane by **polar coordinates**. In this case, our reference is not to two axes, but to a fixed point O (called the **pole**) and a fixed line through the pole. This line is often called the **initial line**.

We fix any point P in the plane by giving its distance r from O, and the positive angle which OP makes with the initial line, so that in figure 7.19, the polar coordinates of P are (r, θ).

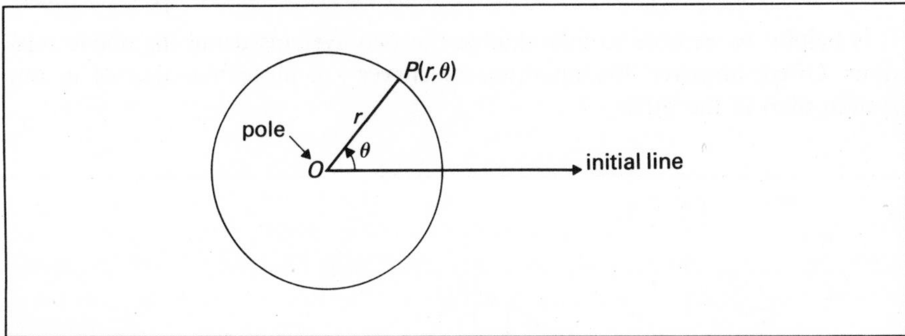

Figure 7.19

Points represented by polar coordinates may be plotted, just as points in the cartesian system can be plotted. It is possible to obtain polar graph paper, which consists of a series of concentric circles and lines at 1° intervals radiating from the centre of the circles. For most purposes, however a sketch on plain paper is adequate.

Exercise D

1 Plot the following points

$(1, 30°)$, $(2, 60°)$, $(1\frac{1}{2}, 45°)$, $(3, 160°)$.

2 Plot the following points, and state what they have in common, geometrically.

$(2, 10°)$, $(2, 70°)$, $(2, 210°)$, $(2, 316°)$, $(2, 108°)$, $(2, 170°)$.

3 Plot the following points and state what they have in common geometrically.

$(3, 30°), (4, 30°), (5, 30°), (\frac{1}{2}, 30°), (2\frac{1}{4}, 30°)$.

You will have noticed in question 2 above that all the points lie on the circle of radius 2. In fact, for each point there, the first coordinate is 2, and we would write $r = 2$ as the **polar** equation of the circle, centre O and radius 2.

Similarly in question 3, we see that all the points lie on the line through O which is at 30° to the initial line. We would give the polar equation of that line as $\theta = 30°$.

7.8 The relationship between polar and cartesian coordinates

By the nature of the two systems, we see that it is possible to change from a polar system to a cartesian system quite simply. In figure 7.20, if (r, θ) are the polar coordinates and (x, y) the cartesian coordinates of P, then

$$r^2 = x^2 + y^2 \quad \text{and} \quad \tan\theta = \frac{y}{x}$$

or

$$x = r\cos\theta \quad \text{and} \quad y = r\sin\theta$$

It is helpful to be able to interchange the two systems using the above relations. Often, however, the equations of curves are more manageable in one system than in the other.

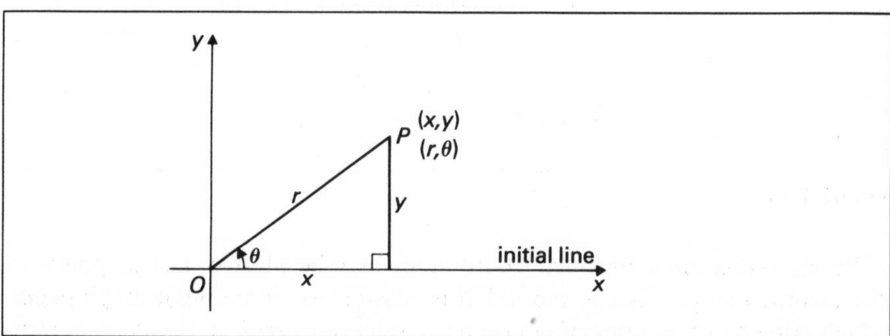

Figure 7.20

Example 1
Find the polar equation of the ellipse whose cartesian equation is

$$\frac{x^2}{4} + \frac{y^2}{9} = 1$$

From the relations above, we write

$$\frac{(r\cos\theta)^2}{4} + \frac{(r\sin\theta)^2}{9} = 1$$

$$\Rightarrow r^2(9\cos^2\theta + 4\sin^2\theta) = 36$$

or $\quad r^2(5\cos^2\theta + 4) = 36$

or $\quad r = \dfrac{6}{\sqrt{(5\cos^2\theta + 4)}}$

Example 2
Finding the cartesian equation of the curve whose polar equation is

$r = a(\cos\theta + 1)$

We use the relations $r = \sqrt{(x^2 + y^2)}$, $\tan\theta = \dfrac{y}{x}$.

$\left(\text{hence } \sin\theta = \dfrac{y}{\sqrt{(x^2+y^2)}},\quad \cos\theta = \dfrac{x}{\sqrt{(x^2+y^2)}}\right)$

The equation becomes

$$\sqrt{(x^2 + y^2)} = \frac{ay}{\sqrt{(x^2+y^2)}} + a$$

$$\Rightarrow x^2 + y^2 = ay + a\sqrt{(x^2 + y^2)}$$

In this case, it is clear that the polar equation is the more manageable.

7.9 Curve drawing

The sketching of curves given in polar forms is not as vital at this stage as curves given in cartesian form. It is probably more realistic to plot sufficient points in order to obtain the general shape of the curve.

We consider here some of the more 'usual' curves.

Example 1
Draw the curve whose polar equation is $r = \frac{1}{2}a(1 + \cos\theta)$

Our table of values for values of θ measured in degrees, is

θ	0	30	45	60	90	120	135	150	180
r	a	$0{\cdot}93a$	$0{\cdot}85a$	$0{\cdot}75a$	$0{\cdot}5a$	$0{\cdot}25a$	$0{\cdot}15a$	$0{\cdot}07a$	0

210	225	240	270	300	315	330	360
$0{\cdot}07a$	$0{\cdot}15a$	$0{\cdot}25a$	$0{\cdot}5a$	$0{\cdot}75a$	$0{\cdot}85a$	$0{\cdot}93a$	a

plotting these points, we obtain the curve known as a **cardioid** shown in figure 7.21.

Example 2
Draw the Archimedean spiral, whose polar equation is $r = \dfrac{\theta}{100}$.

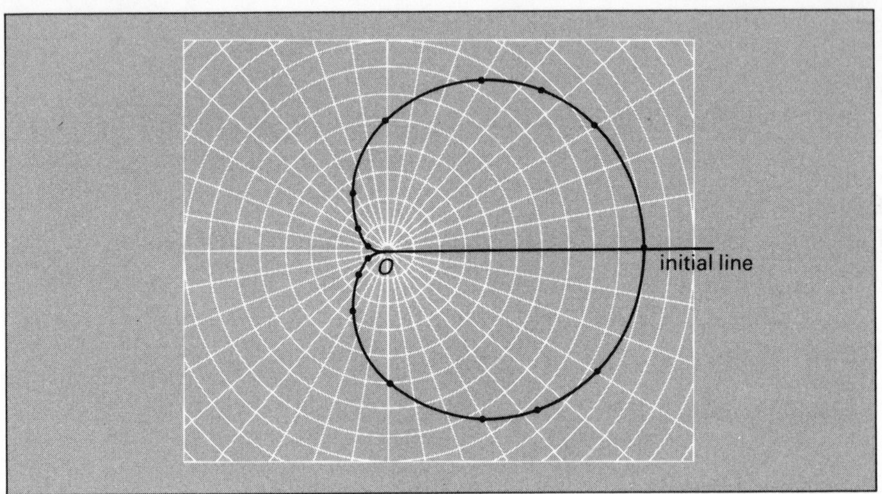

Figure 7.21

In this case, since the relationship connecting r and θ is linear, there is little to be gained by constructing a table of values. The points can be plotted directly, as in figure 7.22.

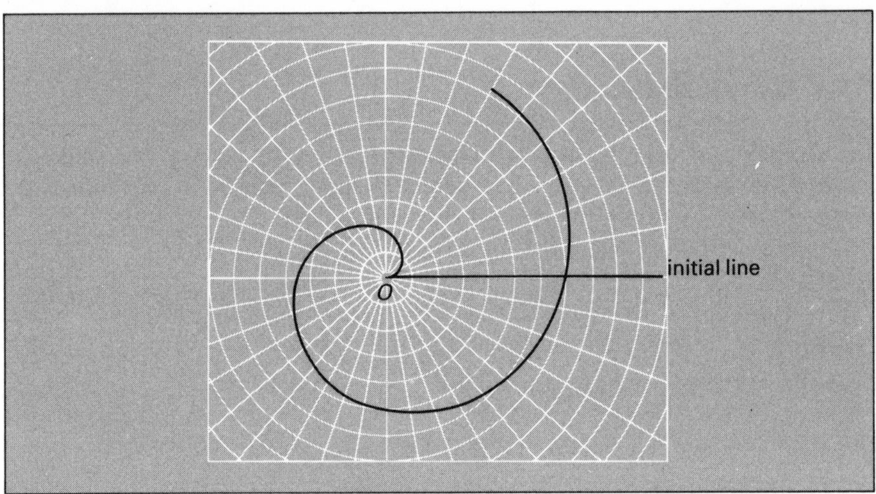

Figure 7.22

There are occasions when our table will yield values for r which are negative. How do we plot such points?

The point $P(2, 210)$ is clearly determined as in figure 7.23, but by considering that a negative r coordinate 'goes backwards', we can also write the coordinates of P as $(-2, 30°)$. Using this convention, we see that the polar coordinates for any point are not unique.

Check for yourself that the following pairs of coordinates represent the same point in each case.

a) $(4, 267°)$ $(-4, 87°)$

b) $(3, 30°)$ $(-3, 210°)$

c) $(-15, 345°)$ $(15, 165°)$

d) $(2, 0°)$ $(-2, 180°)$

e) $\left(6, \frac{7\pi}{5}\right)\left(-6, \frac{12\pi}{5}\right)$

Notice, too, that whilst it is usually convenient to use degrees in polar coordinates, it is equally correct to work in radians, (unless, of course, one particular type of angle measure is specified).

The following two examples are both cases where the r coordinate is negative for some values of θ.

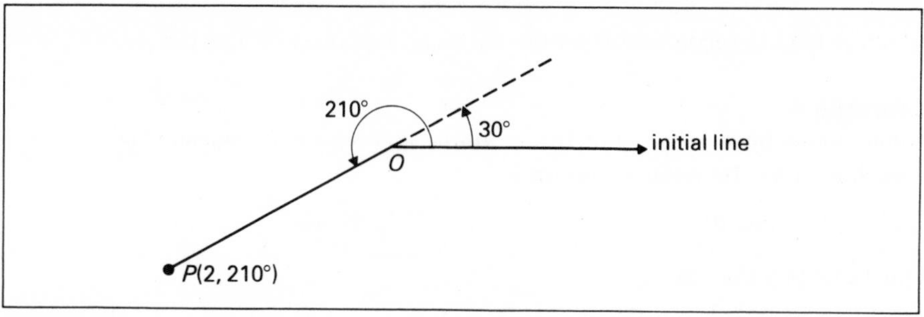

Figure 7.23

Example 3

Draw the curve whose polar equation is $r = a \cos \theta$.

We construct a table of values, as follows.

$\theta°$	0	30	45	60	90	120	135	150	180
r	$1a$	$0{\cdot}87a$	$0{\cdot}71a$	$0{\cdot}5a$	0	$-0{\cdot}5a$	$-0{\cdot}71a$	$-0{\cdot}87a$	$-1a$

210	225	240	270	300	315	330	360
$-0{\cdot}87a$	$-0{\cdot}71a$	$-0{\cdot}5a$	0	$0{\cdot}5a$	$0{\cdot}71a$	$0{\cdot}87a$	$1a$

Plotting the points, figure 7.24, we see that the curve is a circle of radius $\frac{1}{2}a$, whose centre is at $(\frac{1}{2}a, 0)$. In this example, it is relatively simple to convert the polar equation to its cartesian equivalent.

In $\quad r = a \cos \theta$, we write $r = \sqrt{(x^2 + y^2)}$ and $\cos \theta = \dfrac{x}{\sqrt{(x^2 + y^2)}}$

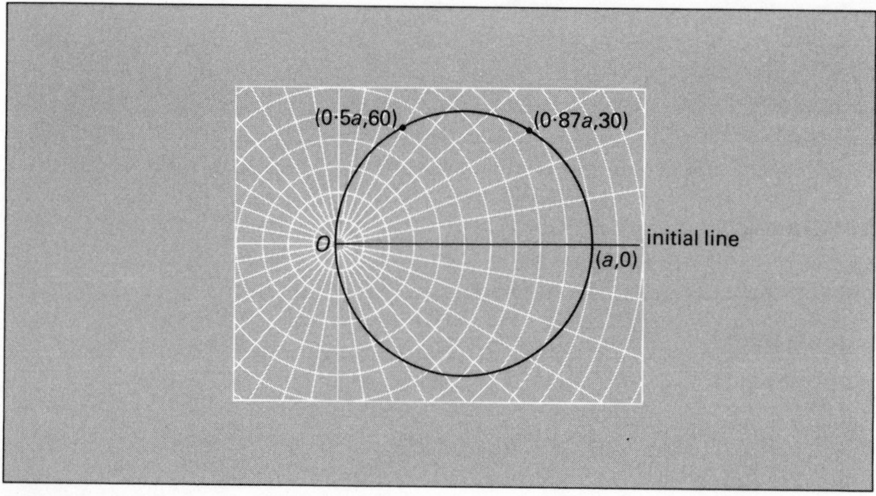

Figure 7.24

We obtain

$$\sqrt{(x^2 + y^2)} = \frac{ax}{\sqrt{(x^2 + y^2)}}$$

$$\Rightarrow x^2 + y^2 - ax = 0$$

which is readily recognisable as the cartesian equation of a circle.

Example 4

Some curves have loops, and this example, known as a **limaçon,** figure 7.25, is such a curve. Its polar equation is

$r^2 = a(\frac{1}{2} + \cos\theta)$

Our table of values is

$\theta°$	0	30	45	60	90	120	135	150	180
r	$1{\cdot}5a$	$1{\cdot}4a$	$1{\cdot}2a$	a	$0{\cdot}5a$	0	$-0{\cdot}21a$	$-0{\cdot}37a$	$-0{\cdot}5a$

210	225	240	270	300	315	330	360
$-0{\cdot}37a$	$-0{\cdot}21a$	0	$0{\cdot}5a$	a	$1{\cdot}2a$	$1{\cdot}4a$	$1{\cdot}5a$

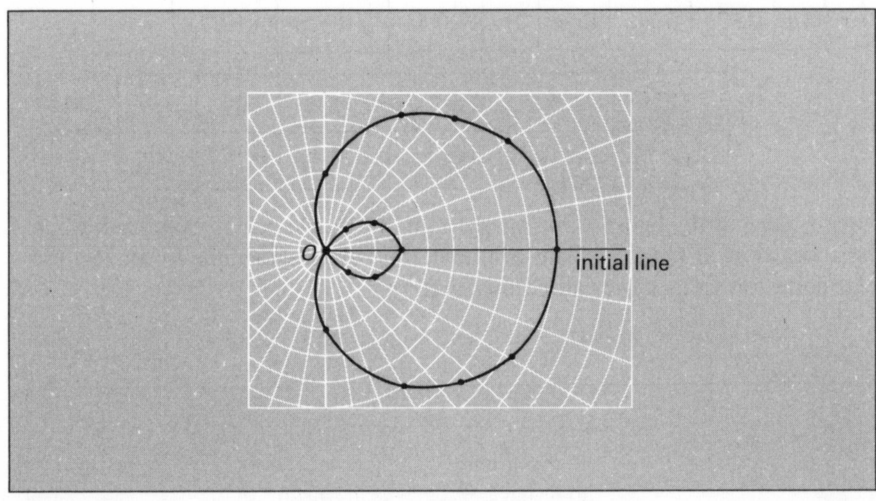

Figure 7.25

Exercise E

Draw the curves whose polar equations are

1 $r = a \sec\theta$

2 $r = \dfrac{1}{1 + \cos\theta}$

3 $r = a(1 + \frac{1}{2}\cos\theta)$

4 $r = a\cos 2\theta$

5 $r = \dfrac{a}{1 + \frac{1}{2}\cos\theta}$

6 $r = a\sqrt{(\cos 2\theta)}$

7 $r = a\sqrt{(\sec 2\theta)}$

8 $r = \dfrac{100}{\theta°}$

9 $r = \dfrac{\theta°}{100}$

10 $r = 1 + \cos 2\theta$

Sums, Areas and Integration

CHAPTER EIGHT

8.1 Introduction

There are many occasions in mathematics when we need to add together series of numbers. We are concerned here with series that have some clear pattern in their formation. We shall consider such series as

$$1 + 2 + 3 + 4 + \cdots$$
or $$1 + 4 + 9 + 16 + \cdots$$
or $$2 + 4 + 8 + 16 + \cdots$$
or $$5 + 9 + 13 + 17 + \cdots$$

We use the row of dots to indicate that the series continues indefinitely. Sometimes, we want the series to terminate at some point, and we would write, for example

$$1 + 2 + 3 + \cdots + 49 + 50$$

to indicate that we are trying to sum the first 50 positive integers, and here the dots indicate all the numbers we have omitted.

Or $1^2 + 2^2 + \cdots + 24^2 + 25^2$

represents the sum of the squares of the first 25 integers.

Notation

It is very cumbersome to write out these long strings of numbers each time we have a problem, and mathematicians use a shorthand notation.

Consider the series $1 + 3 + 5 + 7 + \cdots + 51$.

The first term, $T_1 = 1$
the second term $T_2 = 3$
the third term, $T_3 = 5$

Clearly we can generalise this and say that the rth term, $T_r = 2r - 1$. Check for yourself that, for different values of r, we do obtain the same series of numbers.

All of the terms, then, are of the form $(2r - 1)$ and we seek to add all the numbers like $(2r - 1)$ as r takes all the positive integer values from 1 to 26. Why 26? We could abbreviate this sentence into

$$S(2r - 1) \text{ as } r = 1, 2, \ldots, 26$$

where S represents the expression 'The sum of all the numbers like.' In practice, we use a Greek letter, $\sum$ (called **sigma**) for this notation, and we write

$$\sum_{r=1}^{26} (2r - 1).$$

The number at the bottom of the $\sum$ indicates the first value of r to be used, and the number at the top indicates the last. We would write

$\sum_{r=5}^{8} (2r - 1)$ for the series $9 + 11 + 13 + 15$,

since this is $(2 \times 5 - 1) + (2 \times 6 - 1) + (2 \times 7 - 1) + (2 \times 8 - 1)$.

Look now at the other series given at the beginning of this chapter.

Example 1

$1 + 2 + 3 + \cdots$ (N numbers)

The rth term, $T_r = r$ in this case and we write the series as $\sum_{r=1}^{N} (r)$. This gives gives us a notation for the sum of the first N integers.

Example 2

$1 + 4 + 9 + \cdots$

Here we rewrite the series as

$1^2 + 2^2 + 3^2 + \ldots$, and $T_r = r^2$.

Hence for the sum of the first N terms in this series, we write

$$\sum_{r=1}^{N} (r^2)$$

Example 3

Check for yourself that the first N terms of the series $2 + 4 + 8 + \cdots$ can be written as

$$\sum_{r=1}^{N} 2^r$$

Notice here that r occurs as an index. How would you write

$$1 + 2 + 4 + 8 + \cdots + 2^N?$$

How many terms are there? What is the value of 2^0? Check that we should write

$$\sum_{r=0}^{N} 2^r$$

and that there are $N + 1$ terms.

Example 4

$5 + 9 + 13 + 17 + \cdots$

In this case, the rth term could be written as $1 + 4r$, so that the sum of the first N numbers would be written as

$$\sum_{r=1}^{N} (1 + 4r)$$

We could also write this series as

$$\sum_{r=0}^{N-1} (5 + 4r)$$

Write out this series and check that the result is the same.

Exercise A

Write the following series using the $\sum$ notation.

1 $3 + 6 + 9 + \cdots + 42$

2 $1 + \frac{1}{2} + \frac{1}{4} + \frac{1}{8} + \frac{1}{16} + \frac{1}{32}$

3 $1 + 8 + 27 + \cdots + 343$

4 $11 + 8 + 5 + 2 - 1 - 4 - 7$

5 $5 + 8 + 11 + \cdots$ (N terms)

6 $1 - 2 + 4 - 8 + 16 - 32 + 64$

7 $\dfrac{1}{1 \times 2} + \dfrac{1}{2 \times 3} + \dfrac{1}{3 \times 4} + \cdots + \dfrac{1}{N(N+1)}$

8 $x + x^2 + x^3 + \cdots + x^N$

9 $1 + 12 + 144 + 1728 + \cdots$ (8 terms)

10 $1 + \frac{1}{2} + \frac{1}{3} + \frac{1}{4} + \frac{1}{5} + \frac{1}{6} + \frac{1}{7} + \frac{1}{8} + \frac{1}{9} + \frac{1}{10}$

Write out the following series without using the $\sum$ notation.

11 $\sum_{r=1}^{10} (r^2 - r)$

12 $\sum_{r=1}^{5} (3r + 2)$

13 $\sum_{r=1}^{7} \frac{1}{r^2}$

14 $\sum_{r=1}^{9} (-2r)$

15 $\sum_{r=1}^{10} (11 - r)$

16 $\sum_{r=1}^{10} r$ (is there any connection with question 15?)

17 $\sum_{r=1}^{5} r^3$

18 $\sum_{r=1}^{5} (6 - r)^3$ (is there any connection with question 17?)

19 $\sum_{r=1}^{4} (2r^2 - 3r - 5)$

20 $\sum_{r=1}^{6} (r^2 - 6)$

8.2 Some methods of summation

We need to be able to find the sums of many of these series, and there are many different methods and approaches that can be used. Often, a new technique was devised to fit each fresh series that occurred. We shall consider some of the more frequently occurring series here.

The arithmetic series

The series

$$1 + 2 + 3 + 4 + \cdots + N$$

and $3 + 5 + 7 + 9 + \cdots + (2N + 1)$

are both examples of **arithmetic** series in which each term has a constant amount added to make the next term. In the first series above, each term is 1 more than its predecessor, whereas in the second, each term is 2 more. We write a general arithmetic series as

$$a + (a + d) + (a + 2d) + \cdots + \{a + (N - 1)d\}$$

or

$$\sum_{r=1}^{N} \{a + (r - 1)d\}$$

To find the sum of this series, we write down the terms in ascending order

$$\text{i.e.} \quad \sum_{r=1}^{N} \{a + (r - 1)d\} = a + (a + d) + (a + 2d) + \cdots + \{a + (N - 1)d\}$$

and then in descending order

$$\text{i.e.} \quad \sum_{r=1}^{N} \{a + (r - 1)d\} = \{a + (N - 1)d\} + \{a + (N - 2)d\} + \{a + (N - 3)d\} + \cdots + a$$

Adding the two results

$$2\sum_{r=1}^{N} \{a + (r - 1)d\} = \{2a + (N - 1)d\} + \{2a + (N - 1)d\} + \{2a + (N - 1)d\} + \cdots + \{2a + (N - 1)d\}$$

$$\Rightarrow \quad 2\sum_{r=1}^{N} \{a + (r - 1)d\} = N\{2a + (N - 1)d\}$$

$$\Rightarrow \quad \sum_{r=1}^{N} \{a + (r - 1)d\} = \frac{N}{2}\{2a + (N - 1)d\}$$

In particular,

$$\sum_{r=1}^{N} (r), \quad \text{(where } a = 1,\ d = 1\text{) is given by}$$

$$\sum_{r=1}^{N} (r) = \frac{N}{2}\{2 + (N - 1)\} \qquad \text{i.e.}$$

$$\sum_{r=1}^{N} (r) = \frac{N}{2}(N + 1)$$

The geometric series

$\sum_{r=1}^{N} (ax^{r-1})$ is called a **geometric** series, in which the first term is a, and x is called the common ratio. To sum this series we write the terms in ascending order

$$\sum_{r=1}^{N} (ax^{r-1}) = a + ax + ax^2 + \cdots + ax^{N-1} \qquad \text{i}$$

Multiplying by x

$$x\sum_{r=1}^{N}(ax^{r-1}) = ax + ax^2 + \cdots + ax^{N-1} + ax^N \qquad \textbf{ii}$$

Subtracting **i** from **ii**

$$(x-1)\sum_{r=1}^{N}(ax^{r-1}) = -a + ax^N$$

i.e. $$\sum_{r=1}^{N}(ax^{r-1}) = \frac{ax^N - a}{x-1}$$ Hence

$$\sum_{r=1}^{N}(ax^{r-1}) = \frac{a(x^N - 1)}{x-1}.$$

Exercise B

1 Find the sum of the first 100 positive integers.

2 Find the sum of the first 50 odd integers.

3 Find the value of $1 + 5 + 9 + 13 + \cdots + 49$

4 Find the sum $1 + 4 + 16 + \cdots + 1024$

5 Find the sum $1 + \frac{1}{2} + \frac{1}{4} + \frac{1}{8} + \cdots + \frac{1}{2048}$

6 In an arithmetic series, the first term is a and the nth term is L. Prove that the sum of the first n terms is $\frac{n}{2}(a + L)$

7 In the geometric series

$$1 + \frac{1}{2} + \left(\frac{1}{2}\right)^2 + \left(\frac{1}{2}\right)^3 + \cdots + \left(\frac{1}{2}\right)^{N-1}$$

find the limiting value of the sum as $N \to \infty$.

8 Generalise the result of question 7 by considering the geometric series $1 + x + x^2 + \cdots + x^{N-1}$ where $0 < x < 1$, and find the limiting value of the sum as $N \to \infty$.

9 Investigate the sum of N terms of the series $1 - 1 + 1 - 1 + 1 - 1 \cdots$

10 Find $\sum_{r=1}^{18}(2r + 3)$

8.3 The natural numbers, their squares and cubes

We have already shown that

$$\sum_{r=1}^{N} r = \frac{N}{2}(N + 1).$$

We now consider two other important series, namely

$$\sum_{r=1}^{N} r^2 \text{ and } \sum_{r=1}^{N} r^3.$$

The methods used here are different again from those used earlier.

For $\sum_{r=1}^{N} r^2$, we consider the identity

$$(r+1)^3 - r^3 \equiv 3r^2 + 3r + 1$$

In this identity, we let r take the values 1, 2, 3, ..., N in turn, and obtain

$$\begin{array}{llllll}
2^3 & -1^3 & = 3 \times 1^2 & + 3 \times 1 & + 1 \\
3^3 & -2^3 & = 3 \times 2^2 & + 3 \times 2 & + 1 \\
4^3 & -3^3 & = 3 \times 3^2 & + 3 \times 3 & + 1 \\
5^3 & -4^3 & = 3 \times 4^2 & + 3 \times 4 & + 1 \\
\vdots & \vdots & \vdots & \vdots & \vdots \\
N^3 & -(N-1)^3 & = 3 \times (N-1)^2 & + 3(N-1) & + 1 \\
(N+1)^3 & -N^3 & = 3 \times N^2 & + 3N & + 1 \\
\hline
\end{array}$$

Adding, we obtain

$$(N+1)^3 - 1 = 3\sum_{r=1}^{N} r^2 + 3\sum_{r=1}^{N} r + N$$

$$\Rightarrow (N+1)^3 - 1 = 3\sum_{r=1}^{N} r^2 + 3\frac{N}{2}(N+1) + N$$

$$\Rightarrow \quad 3\sum_{r=1}^{N} r^2 = (N+1)^3 - 1 - 3\frac{N}{2}(N+1) - N$$

$$\Rightarrow \quad 6\sum_{r=1}^{N} r^2 = 2(N+1)^3 - 2 - 3N(N+1) - 2N$$

$$\Rightarrow \quad 6\sum_{r=1}^{N} r^2 = 2N^3 + 6N^2 + 6N + 2 - 2 - 3N^2 - 3N - 2N$$

$$\Rightarrow \quad 6\sum_{r=1}^{N} r^2 = 2N^3 + 3N^2 + N$$

$$\Rightarrow \quad 6\sum_{r=1}^{N} r^2 = N(N+1)(2N+1)$$

$$\Rightarrow \quad \sum_{r=1}^{N} r^2 = \frac{N(N+1)(2N+1)}{6}$$

Note that this method of obtaining an expression for $\sum_{r=1}^{N} r^2$ is not unique. Another identity which can be used to generate the sum is

$$\tfrac{1}{2}(r+1)^3 - \tfrac{1}{2}(r-1)^3 \equiv 3r^2 + 1$$

It is a worthwhile exercise, in practising the technique (sometimes called the 'method of differences'), to check that the same result can be obtained from this identity.

We use the same approach to find $\sum_{r=1}^{N} r^3$, and consider the identity

$$(r+1)^4 - r^4 \equiv 4r^3 + 6r^2 + 4r + 1$$

$$\begin{array}{llllll}
\text{let } r=1: & 2^4 - 1^4 & = 4\times 1^3 & + 6\times 1^2 & + 4\times 1 & + 1 \\
\text{let } r=2: & 3^4 - 2^4 & = 4\times 2^3 & + 6\times 2^2 & + 4\times 2 & + 1 \\
\text{let } r=3: & 4^4 - 3^4 & = 4\times 3^3 & + 6\times 3^2 & + 4\times 3 & + 1 \\
\text{let } r=4: & 5^4 - 4^4 & = 4\times 4^3 & + 6\times 4^2 & + 4\times 4 & + 1 \\
\vdots & \vdots & \vdots & \vdots & \vdots & \vdots \\
\text{let } r=N-1 & N^4 - (N-1)^4 & = 4\times (N-1)^3 & + 6\times (N-1)^2 & + 4\times (N-1) & + 1 \\
\text{let } r=N, & (N+1)^4 - N^4 & = 4\times N^3 & + 6\times N^2 & + 4\times N & + 1 \\
\hline
\text{Adding we get} & (N+1)^4 - 1 & = 4\sum_{r=1}^{N} r^3 & + 6\sum_{r=1}^{N} r^2 & + 4\sum_{r=1}^{N} r & + N
\end{array}$$

We have found, already, expressions for $\sum_{r=1}^{N} r$, $\sum_{r=1}^{N} r^2$, so we substitute these into our equation just obtained, and get

$$(N+1)^4 - 1 = 4\sum_{r=1}^{N} r^3 + N(N+1)(2N+1) + 2N(N+1) + N$$

$$\Rightarrow \quad 4\sum_{r=1}^{N} r^3 = (N+1)^4 - 1 - N(N+1)(2N+1) - 2N(N+1) - N$$

$$\Rightarrow \quad 4\sum_{r=1}^{N} r^3 = N^4 + 4N^3 + 6N^2 + 4N + 1 - 1 - 2N^3 - 3N^2 - N - 2N^2 - 2N - N$$

$$\Rightarrow \quad 4\sum_{r=1}^{N} r^3 = N^4 + 2N^3 + N^2$$

$$\Rightarrow \quad \sum_{r=1}^{N} r^3 = \tfrac{1}{4}(N^4 + 2N^3 + N^2)$$

$$\Rightarrow \quad \sum_{r=1}^{N} r^3 = \frac{N^2}{4}(N+1)^2$$

This is an interesting result, for we see that $\sum_{r=1}^{N} r^3 = \left(\sum_{r=1}^{N} r\right)^2$.

Note There is no similar simple relationship between $\sum_{r=1}^{N} r^2$ and $\sum_{r=1}^{N} r$.

Exercise C

1 Evaluate $\sum_{r=1}^{10} (r^2)$

2 Evaluate $\sum_{r=1}^{7} (r^3)$

3 Evaluate $\sum_{r=1}^{9} (3r^3 - 2r^2)$

4 The sum of the cubes of the first N positive integers is 14 400. Find N.

5 Show, by writing out the series, that

a) $3\sum_{r=1}^{10}(r^2) = \sum_{r=1}^{10}(3r^2)$ b) $16\sum_{r=1}^{10}(r^2) = \sum_{r=1}^{10}(4r)^2$

6 Evaluate $\sum_{r=6}^{15} r^2$, by calculating separately $\sum_{r=1}^{5} r^2$ and $\sum_{r=1}^{15} r^2$

7 Show that the sum of the first N odd numbers is always a perfect square, whatever the value of N.

8 Prove that there is only one positive value for N for which $\sum_{r=1}^{N} r^2 = \sum_{r=1}^{N} r^3$, and state this value.

8.4 Areas

It is often necessary to find areas between curves in a plane, and we try to find a method which will give us accurate values for areas. We consider first a particular example. Suppose we wish to calculate the area between the curve $y = x^2$, the x-axis and the line $x = a$, shown as the shaded area in figure 8.1.

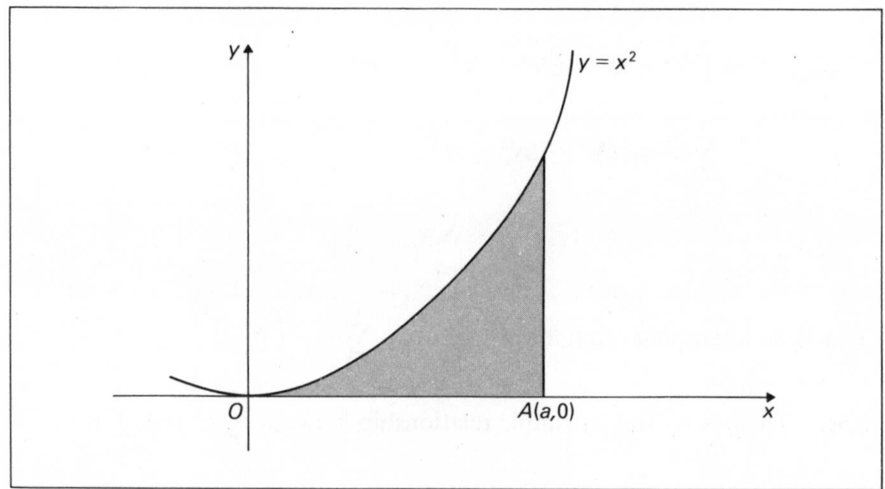

Figure 8.1

At an elementary level, we could draw the graph as accurately as possible on graph paper and count the squares to make the required area. We shall use a similar idea here and then apply limiting techniques to find an exact expression for the area.

First we divide the line OA into n equal divisions, each of length $\left(\frac{a}{n}\right)$, and then draw ordinates at each point, as in figure 8.2. We construct rectangles using these ordinates and obtain a kind of staircase effect. Clearly the area

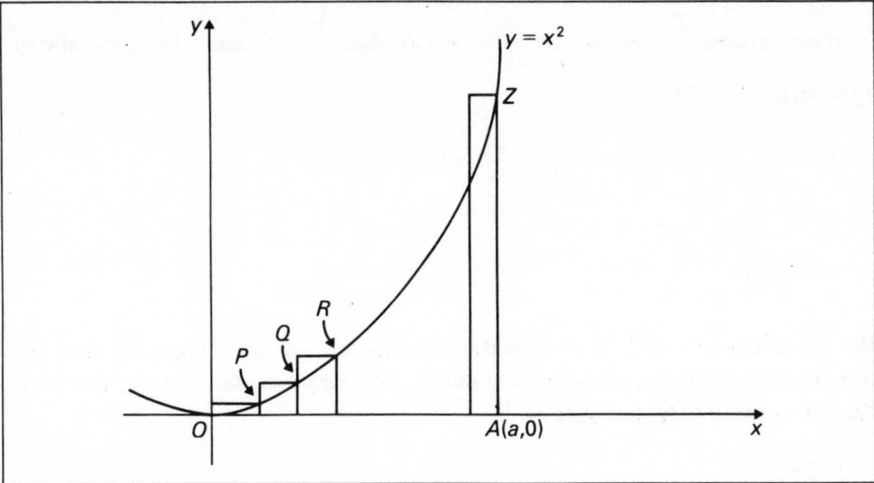

Figure 8.2

contained by all these rectangles is greater than the area we want. However, we will calculate this area to obtain an upper bound for the required area.

The coordinates of P are $\left\{\frac{a}{n}, \left(\frac{a}{n}\right)^2\right\}$,

of Q are $\left\{\frac{2a}{n}, \left(\frac{2a}{n}\right)^2\right\}$,

of R are $\left\{\frac{3a}{n}, \left(\frac{3a}{n}\right)^2\right\}$,

....

and of Z are $\left\{\frac{na}{n}, \left(\frac{na}{n}\right)^2\right\}$,

since each of these points lies on the curve $y = x^2$.

The total area of all the rectangles equals

$$\frac{a}{n} \times \left(\frac{a}{n}\right)^2 + \frac{a}{n} \times \left(\frac{2a}{n}\right)^2 + \frac{a}{n} \times \left(\frac{3a}{n}\right)^2 + \cdots + \frac{a}{n} \times \left(\frac{na}{n}\right)^2$$

$$= \left(\frac{a}{n}\right)^3 (1^2 + 2^2 + 3^2 + \cdots + n^2)$$

$$= \left(\frac{a}{n}\right)^3 \sum_{r=1}^{n} r^2$$

$$= \left(\frac{a}{n}\right)^3 \left\{\frac{n}{6}(n + 1)(2n + 1)\right\}$$

$$= \frac{a^3}{6n^3}(2n^3 + 3n^2 + n)$$

$$= \frac{a^3}{6}\left(2 + \frac{3}{n} + \frac{1}{n^2}\right)$$

If we now increase the number n of rectangles (and correspondingly reduce their width), we shall obtain a closer approximation to the required area, $\mathscr{A}$.

In more precise terms, we let $n \to \infty$ (so that $\frac{1}{n} \to 0$) and the area above approaches $\mathscr{A}$. That is,

$$\mathscr{A} = \lim_{n \to \infty} \left\{ \frac{a^3}{6} \left(2 + \frac{3}{n} + \frac{1}{n^2} \right) \right\} = \frac{a^3}{6} \times 2$$

$$\Rightarrow \mathscr{A} = \frac{a^3}{3}$$

We could equally well have considered a similar 'staircase' diagram with the tops of the rectangles all below the curve. You should check that, in this case, the lower bound for the area $\mathscr{A}$ is

$$\frac{a^3}{6} \left(2 - \frac{3}{n} + \frac{1}{n^2} \right)$$

Here again,

$$\lim_{n \to \infty} \left\{ \frac{a^3}{6} \left(2 - \frac{3}{n} + \frac{1}{n^2} \right) \right\}^* = \mathscr{A}$$

that is

$$\frac{a^3}{6} \times 2 = \mathscr{A}$$

$$\Rightarrow \mathscr{A} = \frac{a^3}{3}$$

It is worth carrying out a similar exercise using the curve $y = x^3$.

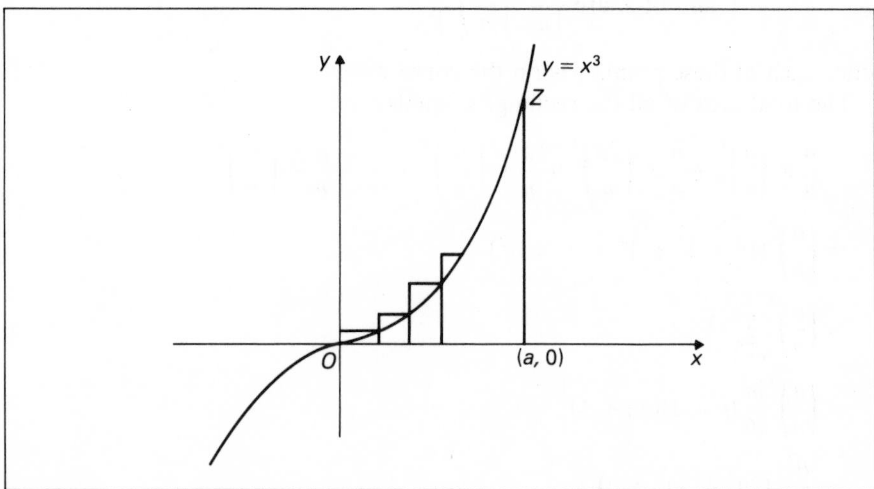

Figure 8.3

* If you had difficulty in obtaining this expression, ensure that you were counting only $(n - 1)$ rectangles, and that your sum at the end represents $\sum_{r=1}^{n-1} r^2$. Can you show that this is equal to $\frac{(n-1)n(2n-1)}{6}$?

Again, we draw n rectangles, each of width $\frac{a}{n}$, and taking the sum of the areas of these rectangles, we have

$$\frac{a}{n} \times \left(\frac{a}{n}\right)^3 + \frac{a}{n} \times \left(\frac{2a}{n}\right)^3 + \frac{a}{n} \times \left(\frac{3a}{n}\right)^3 + \cdots + \frac{a}{n} \times \left(\frac{na}{n}\right)^3$$

$$= \frac{a^4}{n^4}(1^3 + 2^3 + 3^3 + \cdots + n^3)$$

$$= \frac{a^4}{n^4}\sum_{r=1}^{n}(r^3)$$

and this is the upper bound area, and becomes

$$\frac{a^4}{n^4} \cdot \frac{n^2(n+1)^2}{4}$$

$$= \frac{a^4}{4}\left\{\frac{n^4 + 2n^3 + n^2}{n^4}\right\}$$

$$= \frac{a^4}{4}\left(1 + \frac{2}{n} + \frac{1}{n^2}\right)$$

We now let $n \to \infty$ $\left(\text{that is } \frac{1}{n} \to 0\right)$, and find that the required area is

$$\frac{a^4}{4}\left(1 + \frac{2}{n} + \frac{1}{n^2}\right) = \frac{a^4}{4}$$

Does the limiting value of the areas of the lower rectangles give the same answer in this case?

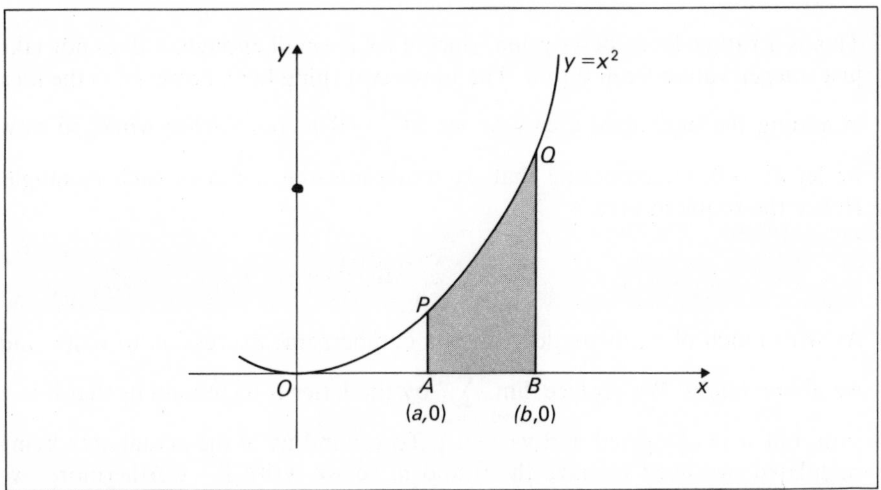

Figure 8.4

If we need to calculate the area shaded in figure 8.4, we see that we could calculate the area OQB and subtract the area OPA, and obtain

$$\tfrac{1}{3}b^3 - \tfrac{1}{3}a^3$$

We will now generalise the method by considering a general function $y = f(x)$, (figure 8.5). We want to find the area of the region between this curve, the x-axis, the y-axis and the line $x = a$.

We divide OA into a number of equal lengths, each being 'a little bit of x', that is δx – to use our earlier notation (see chapter 3). We have already seen that the upper bound area and the lower bound area both approach the same value when we take the limit. Here, we will take rectangles whose 'tops' cross

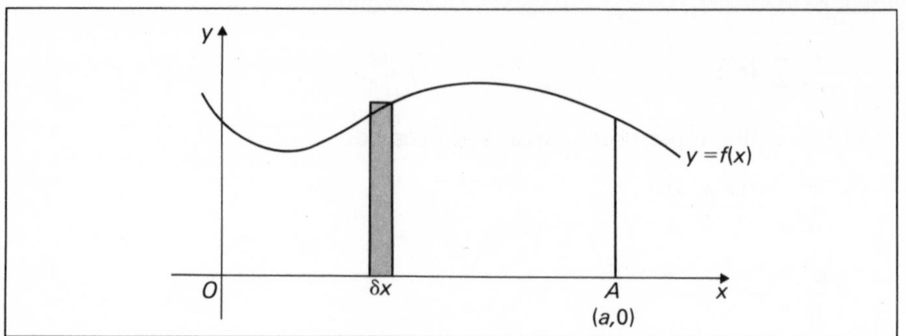

Figure 8.5

the curve. But it does not affect the result, since we increase the number of rectangles without limit, and hence reduce the width of each. The shaded area is one such rectangle, and we can use it as a representative of all the others. Suppose it is at $(x, 0)$, then its height is $f(x)$. The area of this rectangle is then $f(x)\,\delta x$.

In order to get the total area of all the rectangles, we add the areas of the individual rectangles, i.e.

$$\sum_{x=0}^{a} f(x)\,\delta x$$

This is a rather loose description, since if δx is small enough, x does not take just integer values from 0 to a. The important thing here, however, is the idea of adding the individual areas. As we let $\frac{1}{n} \to 0$ in our earlier work, so now we let $\delta x \to 0$, remembering that δx represents the width of each rectangle. Hence the required area is

$$= \lim_{\delta x \to 0} \sum_{x=0}^{a} f(x)\,\delta x$$

As with much of mathematics, this is a cumbersome expression to write, and we abbreviate it. We replace '$\lim\limits_{\delta x \to 0} \sum\limits_{x=0}^{a}$' by the letter S to remind us that it is a sum, but it is elongated and written $\int$. To remind us of the actual area being calculated, we need to have the 0 and a, so we write $\int_0^a$. Furthermore, we replace 'δx' by dx, just as we did in our differentiation work earlier. So our complete expression for the area described in Figure 8.5 is

$$\int_0^a f(x)\,dx$$

The '$\int$' is called an **integral sign**, and the expression above is read as 'the integral of $f(x)$ with respect to x as x varies from 0 to a'. The 'dx' reminds us that our rectangles were built up on the x-axis, and it is important to include it every time area expressions are used.

To revert to our first examples, we see that

$$\int_0^a x^2\,dx = \frac{a^3}{3}\ ;\ \int_0^a x^3\,dx = \frac{a^4}{4}\ ;\ \text{and} \int_a^b x^2\,dx = \frac{b^3}{3} - \frac{a^3}{3}$$

In general $\int_a^b f(x)\,dx$ represents the shaded area shown in figure 8.6.

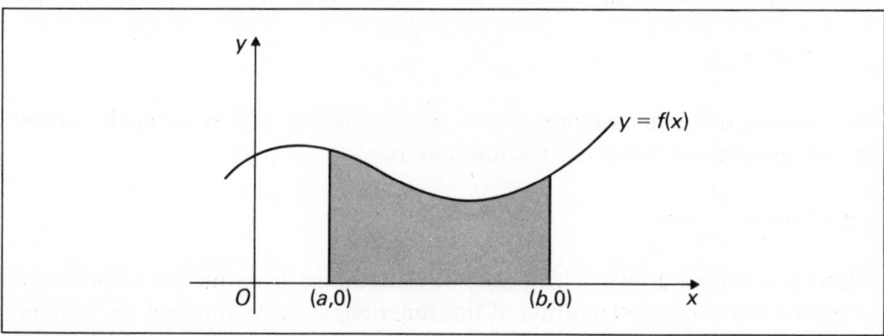

Figure 8.6

Exercise D

1 Find the shaded area in figure 8.6(a).

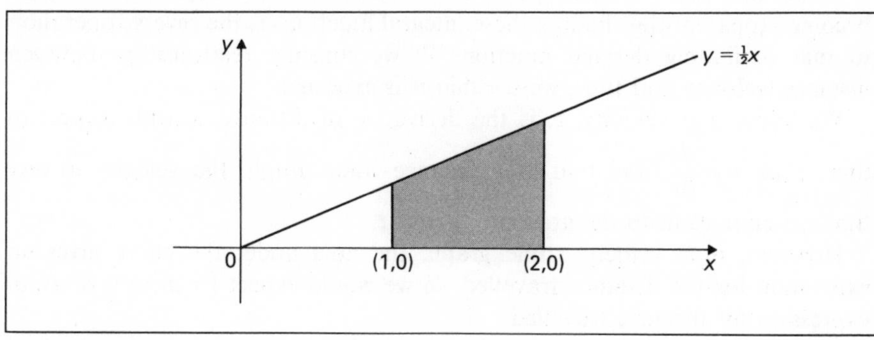

Figure 8.6 (a)

2 Calculate the area defined by $\int_1^3 x^2\,dx$, and sketch the graph to show the area you have found.

3 Sketch $y = 2x^3$ for $x \geqslant 0$, and show the area represented by $\int_1^3 2x^3\,dx$. Calculate this area and investigate its relationship with the area given by $2\int_1^3 x^3\,dx$.

4 Calculate, by referring to a sketch, $\int_2^4 2x\,dx$.

8.5 Generalisations

We have seen that

a) $\int_0^a x^2\,dx = \frac{1}{3}a^3$ b) $\int_1^a x^2\,dx = \frac{1}{3}a^3 - \frac{1}{3}$ c) $\int_2^a x^2\,dx = \frac{1}{3}a^3 - \frac{8}{3}$

and clearly there is a pattern emerging. These expressions differ only according to the position of the left hand starting point. In each case the term $\frac{a^3}{3}$ occurs, and we can, therefore define the **integral function** as

$$\int_0^a x^2\,dx = \frac{a^3}{3}$$

or if we use x as the upper end of the interval as well as the variable, we have

$$\int_0^x x^2\,dx = \frac{x^3}{3}$$

By choosing different starting points, we subtract or add some fixed number. So our generalised integral function here is

$$\int x^2\,dx = \frac{x^3}{3} + k$$

where k is some number which may be determined from further information.

Notice that for our definition of this function, we have omitted the 'starting' and 'finishing' points on the integral sign. In this way our integral function is perfectly general, and is then called an **indefinite integral**. In the same way, we define the integral function for $f(x) = x^3$ by

$$\int x^3\,dx = \frac{x^4}{4} + k$$

Whilst it is beyond the scope of this book to *prove* the necessary results, it becomes apparent that finding these integral functions is the reverse operation to that of finding derived functions. If we consider relationships between distance, velocity and time, we see that it is expected.

We know that velocity, v, is the derivative of distance, s, with respect to time, t, i.e. $v = \frac{ds}{dt}$, and that in a distance–time graph, the velocity at any time t is equivalent to the gradient at time t.

However, in a velocity–time graph, the area under the curve gives an expression for the distance travelled, so we would expect $\int v\,dt$ to give us an expression for distance travelled.

(**Note** This is one example which is used to justify our conjecture that finding integral functions – or integrating – is the reverse of differentiating. It is not a proof.)

We shall use, then, for our definition of integration, the following

$$\int f(x)\,dx = F(x) + k, \text{ where } F'(x) = f(x)$$

In much of the analysis, the constant is added just as a matter of course, but in the applications for which this analysis is fundamental, the constant is vital. It is a good idea, therefore, to establish the habit of including the constant whenever finding integral functions.

For definite integrals, the constant is not included, since we have seen that it is dependent only on our left-hand starting point in finding areas, and it is clear that

$$\int_c^b f(x)\,dx - \int_c^a f(x)\,dx = \int_a^b f(x)\,dx$$

and the position of c is irrelevant to the calculation. Note that in terms of our definition of integration above, we define a definite integral as

$$\int_a^b f(x)\,dx = F(b) - F(a)$$

It is convenient, too, to write this in terms of the integral function, as

$$\int_a^b f(x)\,dx = [F(x)]_a^b = F(b) - F(a)$$

Using the fact that integration is the reverse of differentiation, we have the following table of integral functions (the constants are omitted, for clarity).

$f(x)$	$\int f(x)\,dx$
k	kx
x	$\frac{1}{2}x^2$
x^2	$\frac{1}{3}x^3$
x^3	$\frac{1}{4}x^4$
$\sqrt{x} = x^{1/2}$	$\frac{2}{3}x^{3/2} = \frac{2}{3}x\sqrt{x}$

The general rule is

$$\int x^n\,dx = \frac{x^{n+1}}{n+1} + k$$

which is true for all values of n *except* $n = -1$. It is clear that the fraction $\dfrac{x^{n+1}}{n+1}$ is not defined when $n = -1$, and we need some further analysis to establish the integral function for $\int x^{-1}\,dx$ (see chapter 14).

In chapter 4, we showed that the derivative of sin x is cos x, so

$$\int \cos x\,dx = \sin x + k \quad \text{similarly} \quad \int \sin x\,dx = -\cos x + k$$

We recall that if $y = f(x) + g(x)$, then we can find $\dfrac{dy}{dx}$ by evaluating separately $f'(x)$ and $g'(x)$, so that $\dfrac{dy}{dx} = f'(x) + g'(x)$. It follows that

$$\int (f(x) + g(x))\,dx = F(x) + G(x) + k \text{ where } F'(x) = f(x) \text{ and } G'(x) = g(x)$$

Exercise E

1 Evaluate the following indefinite integrals.

a) $\int (x + x^2)\,dx$ b) $\int (3x^2 + 4x^3)\,dx$

c) $\int (2 + x)^2\,dx$ (evaluate $(2 + x)^2$ first, then integrate)

d) $\int (x^2 - 2x)\,dx$ e) $\int (x + \sin x)\,dx$

f) $\int (\cos x - \sin x)\, dx$ i) $\int (x^{1/2} + x^{3/2})\, dx$

g) $\int (1 + x + x^2)\, dx$ j) $\int \left(\sqrt{x} + \frac{1}{\sqrt{x}}\right) dx$

h) $\int (3x^2 + 2x + 1)\, dx$

2 Evaluate the following definite integrals.

a) $\int_0^2 (x^2 + 1)\, dx$ c) $\int_{-1}^1 x\, dx$

b) $\int_0^2 (x + 1)^2\, dx$ d) $\int_{-1}^1 x^3\, dx$

In each case, sketch graphs of the functions, and shade the areas you have calculated. How do you explain your answers to c) and d)?

3 Evaluate

a) $\int_{-\pi/2}^{\pi/2} \cos x\, dx$ b) $\int_0^{\pi} \sin x\, dx$

c) By means of sketch graphs, explain why $\int_0^{\pi} \cos x\, dx = 0$

and $\int_{-\pi/2}^{\pi/2} \sin x\, dx = 0$

4 Sketch the curve $y = x^2 - 4$, marking clearly the coordinates where it crosses the x-axis. Do you expect the area cut off below the x-axis to be positive or negative? Use integration to calculate this area. Is your previous expectation confirmed?

5 Evaluate

$$\int_0^{\pi/2} (\sin x + \cos x)\, dx,$$

and sketch a graph showing the area you have found.

6 Draw the graph of $y = x + 2$ and show on your graph the area represented by $\int_2^4 (x + 2)\, dx$. Evaluate this integral and show that you obtain the same answer by calculating the area of the appropriate trapezium in your diagram.

8.6 Volumes

We may calculate volumes of regular bodies using methods similar to those we have used in our area measurement. Integration is a particularly powerful tool when the volumes are obtainable by rotation of known curves. For example, we could trace out the surface of a cone by rotating around the x-axis the segment of the line $y = \left(\frac{r}{h}\right)x$ from $x = 0$ to $x = h$ as shown in Figure 8.7.

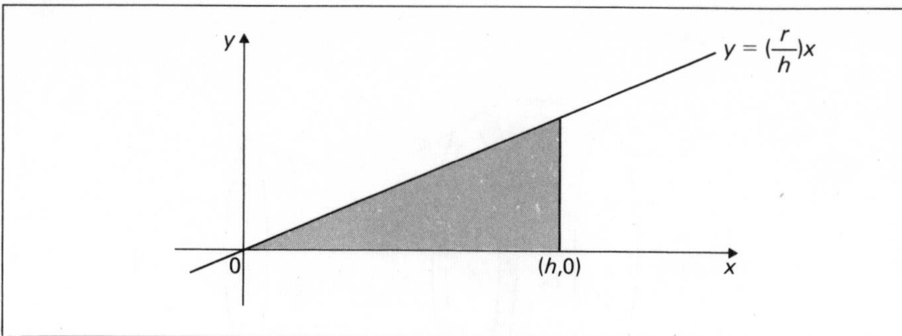

Figure 8.7

Or, we trace out a sphere by rotating the semicircle $y = +\sqrt{(a^2 - x^2)}$ around the x-axis through 2π radians, figure 8.8.

Consider the general case of the solid generated by rotation of $y = f(x)$, through 2π radians around the x-axis. The region enclosed by the curve $y = f(x)$, the segment of the x-axis from $x = a$ to $x = b$, and the ordinates drawn at these points.

Referring to figure 8.9, we see how our solid is formed. In order to find the volume, we divide the x-axis, from a to b, into equal intervals each of length δx. We see that we can consider the solid as being made up of a series of very 'thin' cylinders. One of these is shown at the point $(x, 0)$ in figure 8.9. The radius of the drawn cylinder is $f(x)$ so that its volume is

$$\pi[f(x)]^2 \, \delta x$$

Since the whole volume is made up of a series of such cylinders, our approximation to the total volume is

$$\sum_{x=a}^{b} \pi[f(x)]^2 \, \delta x$$

As we increase the number of such cylinders (and correspondingly decrease δx), we find that the volume equals

$$\lim_{\delta x \to 0} \left(\sum_{x=a}^{b} \pi[f(x)]^2 \, \delta x \right)$$

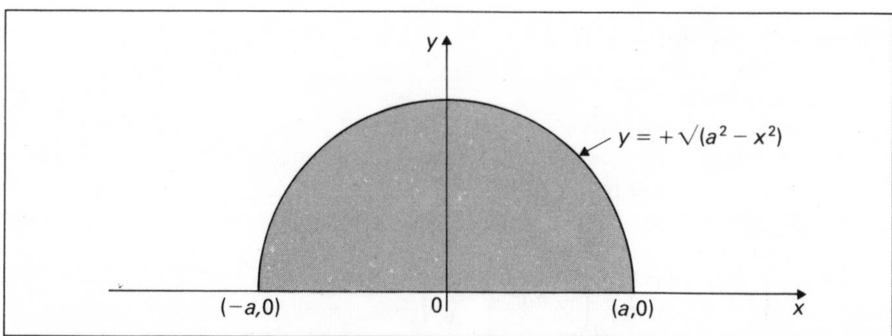

Figure 8.8

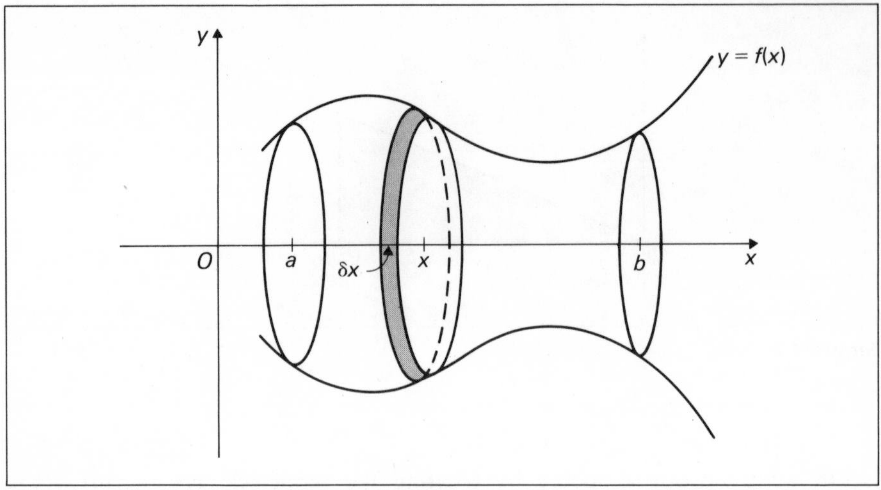

Figure 8.9

Simplifying this as before,

$$\text{Volume} = \int_a^b \pi[f(x)]^2\,dx$$

Volumes formed in this way are usually called **volumes of revolution**.

Example 1

Find the formula for the volume of a sphere of radius a.

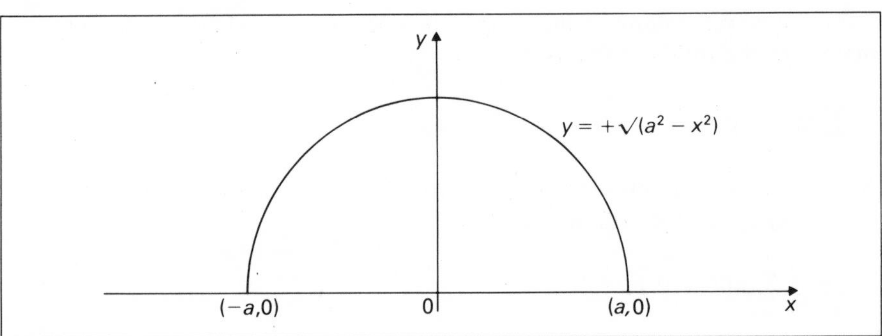

Figure 8.10

$$\begin{aligned}\text{Volume} &= \int_{-a}^{a} \pi(\sqrt{(a^2 - x^2)})^2\,dx \\ &= \int_{-a}^{a} \pi(a^2 - x^2)\,dx \\ &= \left[\pi a^2 x - \frac{\pi x^3}{3}\right]_{-a}^{a} \\ &= \left(\pi a^3 - \frac{\pi a^3}{3}\right) - \left(-\pi a^3 + \frac{\pi a^3}{3}\right) \\ &= \tfrac{4}{3}\pi a^3\end{aligned}$$

Example 2
Find the volume of revolution when $y^2 = x$ (for $0 \leqslant x \leqslant 9$) is rotated through 2π radians round the x-axis.

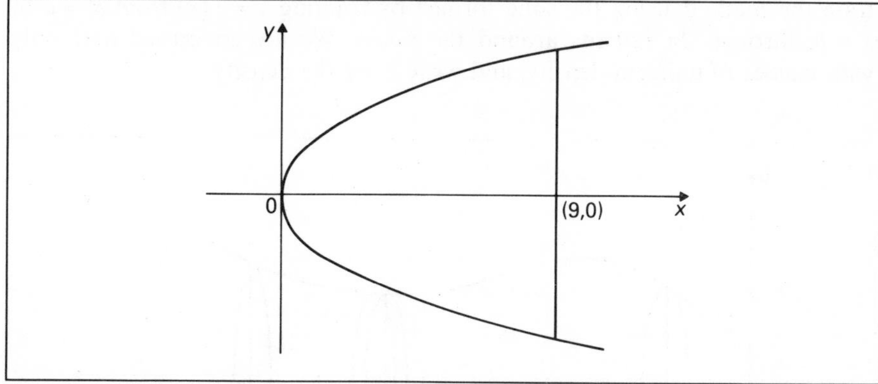

Figure 8.11

Here $\quad f(x) = \sqrt{x}$

$$[f(x)]^2 = x$$

and volume $= \displaystyle\int_0^9 \pi x \, dx$

$$= \left[\frac{\pi x^2}{2}\right]_0^9$$

$$= \frac{81\pi}{2}$$

Exercise F

1 Prove that the volume of the cone obtained by rotation in figure 8.7 is $\frac{1}{3}\pi r^2 h$.

2 Find the volume generated by rotation of $y = x^2 - 1$ from $x = -1$ to $x = +1$, through 2π radians around the x-axis.

3 An ellipsoid is formed by rotation of

$$y = +b\sqrt{\left(1 - \frac{x^2}{a^2}\right)}$$

from $x = -a$ to $x = +a$ through 2π radians around the x-axis. Find its volume.

4 The curve $y = x^3$ from $x = -2$ to $x = +2$ is rotated through 2π radians around the y-axis to form a solid which looks like an egg-cup. Find the volume of this solid.

5 Part of the semicircle $y = +\sqrt{(a^2 - x^2)}$ from $x = b$ $(b \geqslant 0)$ to $x = a$ is rotated through 2π radians around the x-axis. Find the volume so formed. (This formula is a standard result for the volume of a cap of a sphere.)

8.7 Centres of mass

We use integration as a limiting sum again when finding the positions of centres of mass – either of plane laminas or of solids of revolution. We illustrate the method using the solid formed by rotating $y = f(x)$ from $x = a$ to $x = b$, through 2π radians around the x-axis. We are concerned here only with masses of uniform density, and write ρ for the density.

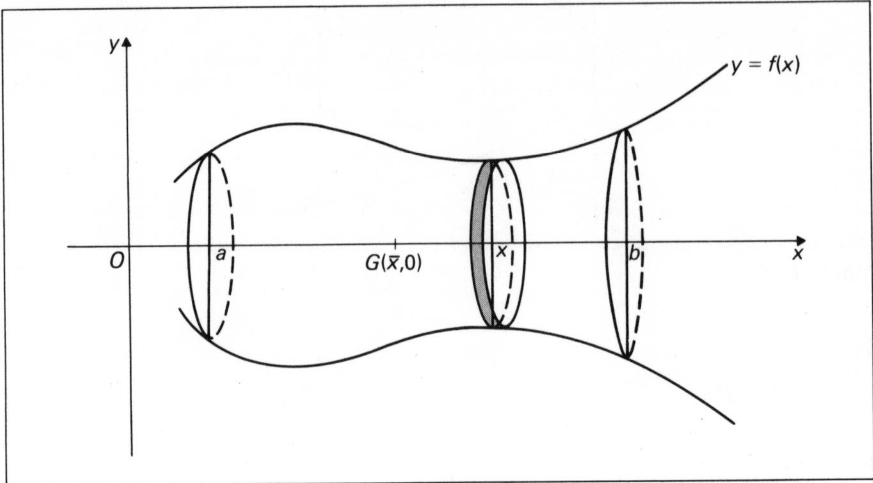

Figure 8.12

We know from previous work that

$$\text{volume} = \int_a^b \pi[f(x)]^2\, dx$$

$$\text{hence mass} = \rho \int_a^b \pi[f(x)]^2\, dx$$

Let the coordinates of the centre of mass, G, be $(\bar{x}, 0)$, then the total mass uniformly distributed over the solid has the same effect as the total mass at G. The **moment** of the total mass at G about O is then

$$\rho\bar{x} \int_a^b \pi[f(x)]^2\, dx \qquad \textbf{i}$$

If we consider the solid made up of a series of 'thin' cylinders, like the one drawn in figure 8.12, then each of these has mass $\rho\pi[f(x)]^2\, \delta x$ and each has its centre of mass at $(x, 0)$.

Consequently, the moment of each such cylinder about O is

$$x\rho\pi[f(x)]^2\, \delta x$$

We must add all such moments and allow $\delta x \to 0$ to make the result exact, i.e. we have

$$\lim_{\delta x \to 0} \sum_{x=a}^{b} \{x\rho\pi[f(x)]^2\, \delta x\} = \rho\pi \int_a^b x[f(x)]^2\, dx \qquad \textbf{ii}$$

Clearly expressions **i** and **ii** represent the same effect, and are equal.

Hence

$$\bar{x}\pi\rho \int_a^b [f(x)]^2\, dx = \pi\rho \int_a^b x[f(x)]^2\, dx$$

$$\bar{x} = \frac{\int_a^b x[f(x)]^2\, dx}{\int_a^b [f(x)]^2\, dx}$$

In the work above, the *y*-coordinate of *G* was zero, by the symmetry of the solid. However, if its values were different, the process of finding *y* would follow a similar pattern.

Example

Find the position of the centre of mass of a uniform solid hemisphere of radius *a*. Referring to figure 8.13.

$$\bar{x} = \frac{\int_0^a x(a^2 - x^2)\, dx}{\int_0^a (a^2 - x^2)\, dx}$$

$$\bar{x} = \frac{\left[\frac{a^2x^2}{2} - \frac{x^4}{4}\right]_0^a}{\left[a^2x - \frac{x^3}{3}\right]_0^a}$$

$$\bar{x} = \left(\frac{\frac{1}{4}a^4}{\frac{2}{3}a^3}\right) = \tfrac{3}{8}a$$

So that the position of the centre of mass of a uniform solid hemisphere is $\frac{3}{8}$ of the way up from the flat surface.

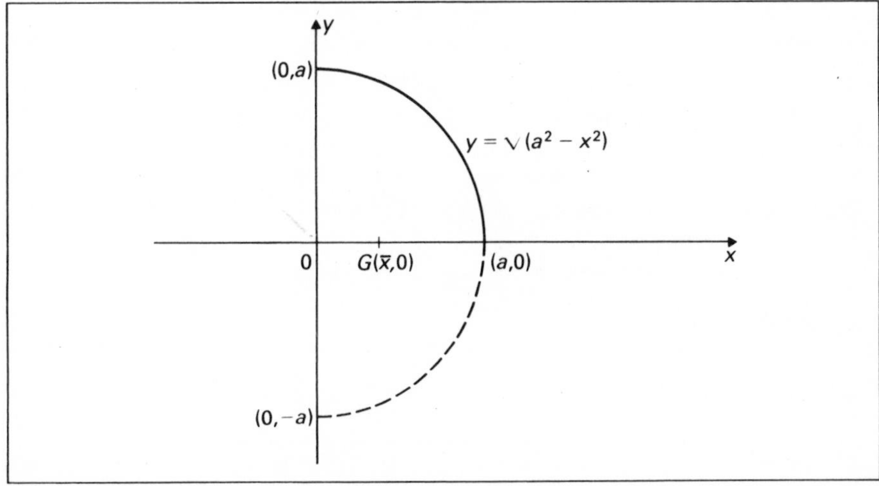

Figure 8.13

Exercise G

1 Find the position of the centre of mass of a uniform solid cone of height h and radius r. (See figure 8.7).

2 Find the coordinates of the centre of mass of the uniform solid formed by rotating the area bounded by $y^2 = x$ and $x = 4$ through 2π radians around the x-axis.

3 Find the position of the centre of mass of the uniform elliptical solid formed by rotating $y = +\sqrt{(1 - \frac{1}{4}x^2)}$, $(1 \leqslant x \leqslant 2)$ through 2π radians around the x-axis.

4 Find the position of the centre of mass of a cap of a uniform solid sphere of height h, as shown in figure 8.14.

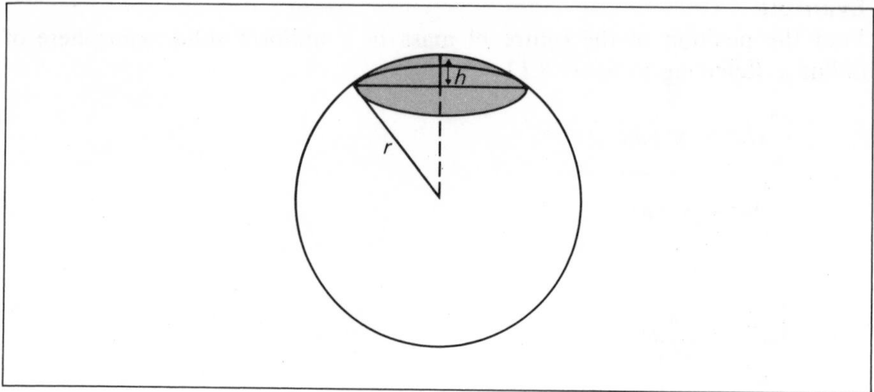

Figure 8.14

Vectors

CHAPTER NINE

9.1 Revision of basic ideas

Vectors were probably introduced to you in your previous work. They were defined as quantities which have a size, or magnitude, and a direction. Examples of vector quantities include velocity, force and momentum. Quantities which have size but no direction are called **scalars**. Examples of scalar quantities are speed, work, energy.

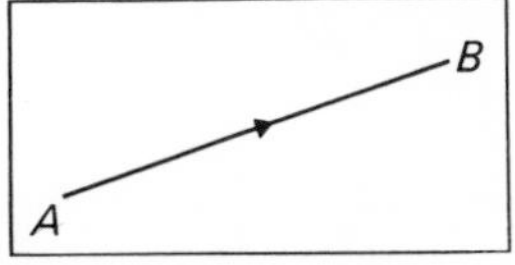

Figure 9.1

Figure 9.1 is a geometrical representation of a vector from A to B. We denote this by **AB**. Note that this is not the same as **BA**, but that they are related and we write **BA** = −**AB**. It is sometimes convenient to use a single lower case letter to denote a vector, and we write $\underset{\sim}{a}$ to denote the vector in figure 9.2. Note that in print $\underset{\sim}{a}$ appears as **a**.

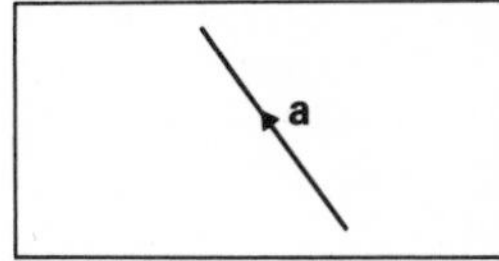

Figure 9.2

We define the addition of two vectors by the following method (see figure 9.3).

$$\mathbf{OA} + \mathbf{AB} = \mathbf{OB} \quad \text{or} \quad \mathbf{a} + \mathbf{b} = \mathbf{c}$$

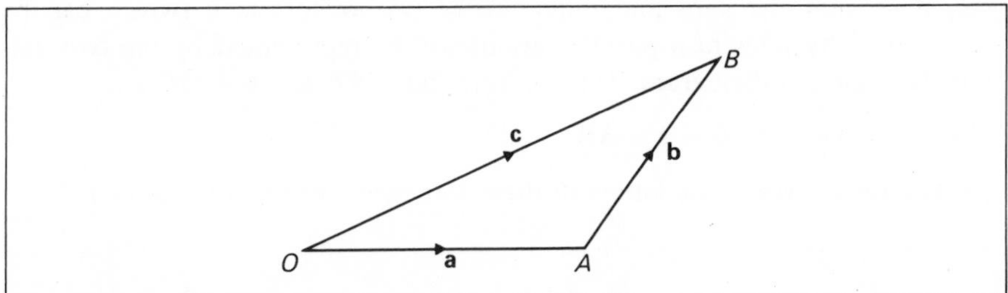

Figure 9.3

It is sometimes convenient to think of vector sums as alternative means of moving from O to B; i.e. we can either travel from O to A and from A to B, or directly from O to B.

We notice that vector addition is commutative, for, in figure 9.4

$$\mathbf{OA} + \mathbf{AB} = \mathbf{OB} \quad \text{or} \quad \mathbf{OC} + \mathbf{CB} = \mathbf{OB}$$

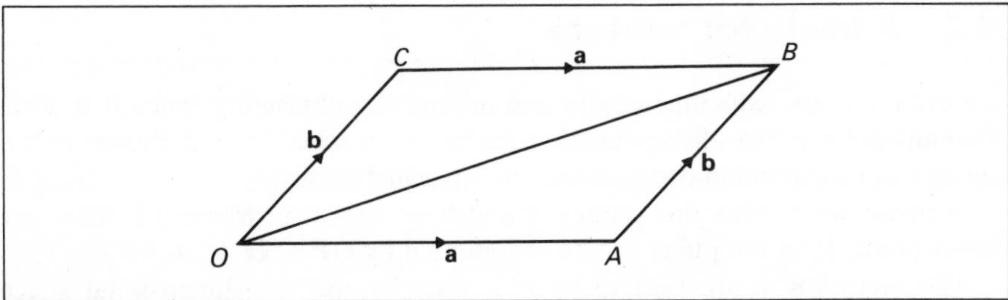

Figure 9.4

These two results can be written as

$$\mathbf{a} + \mathbf{b} = \mathbf{c} \quad \text{or} \quad \mathbf{b} + \mathbf{a} = \mathbf{c}.$$

Hence $\mathbf{a} + \mathbf{b} = \mathbf{b} + \mathbf{a}$.

We define a zero vector, $\mathbf{0}$, as the result of $\mathbf{a} + (-\mathbf{a})$ or $\mathbf{AB} + \mathbf{BA}$.

Following from this concept, we can also define what we mean by subtraction of vectors. We can write $\mathbf{a} - \mathbf{b}$ as $\mathbf{a} + (-\mathbf{b})$, so that we obtain for the resultant vector (see figure 9.5)

$$\mathbf{a} - \mathbf{b} = \mathbf{a} + (-\mathbf{b}) = \mathbf{OA} + \mathbf{OB'} = \mathbf{OC'}$$

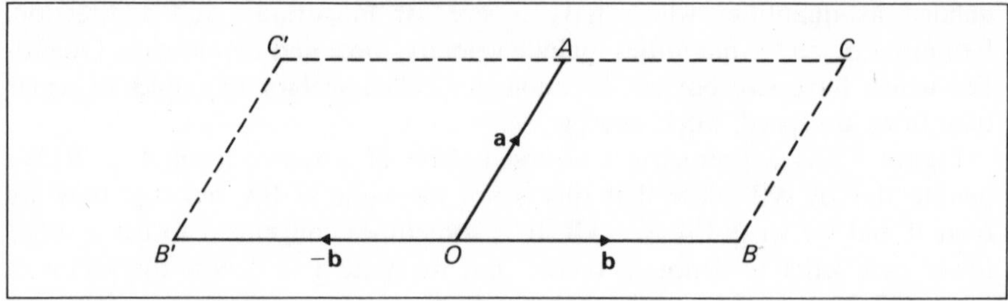

Figure 9.5

and we see that $\mathbf{OC'}$ is equal to $\mathbf{BA}$ (these vectors have the same magnitude and the same direction; it is only their point of application which differs). We see, then, that the sum and difference of two vectors $\mathbf{a}$, $\mathbf{b}$ (which can be represented by sides of a parallelogram) will be represented by the two diagonals of the parallelogram. That is, from figure 9.6, $\mathbf{a} + \mathbf{b} = \mathbf{OC}$ and

$$\mathbf{a} - \mathbf{b} = \mathbf{BA} \quad \text{or} \quad \mathbf{b} - \mathbf{a} = \mathbf{AB}$$

(Notice the orders of the letters in these statements of the differences.)

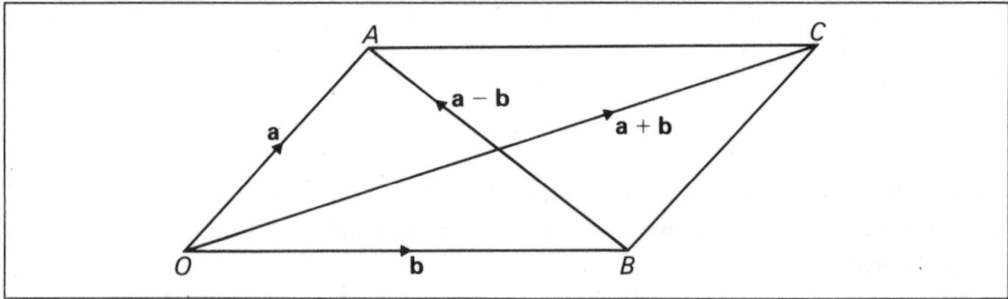

Figure 9.6

9.2 A basis for vectors

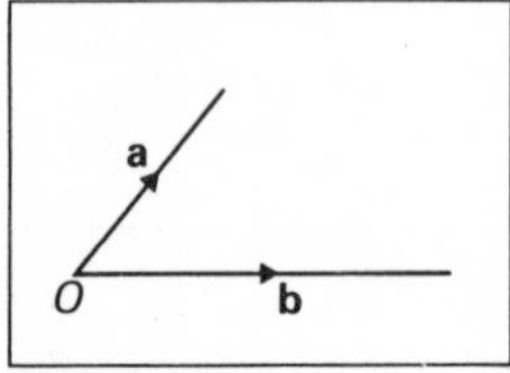

Figure 9.7

By extending or reducing vectors and adding or subtracting them, it is clear that any point in two-dimensional space can be reached from a chosen origin by appropriate combinations of two non-parallel vectors.

Suppose we choose the vectors $\mathbf{a}$ and $\mathbf{b}$, as shown in figure 9.7, then any other point, P, in the plane can be represented by $\mathbf{OP} = \lambda\mathbf{a} + \mu\mathbf{b}$, for $\lambda, \mu \in \mathbb{R}$.

The vectors $\mathbf{a}$, $\mathbf{b}$ are said to form a basis for the two-dimensional space (properly called a vector-space). In the same way we may find the vector of a point in three-dimensional space in terms of three non-parallel and non-coplanar vectors **a**, **b**, **c**. We see that

$$\mathbf{OP} = \lambda\mathbf{a} + \mu\mathbf{b} + \nu\mathbf{c}$$

It is often convenient to choose the basis for the space with the base vectors perpendicular to each other, and of unit length. There is then an immediate correspondence with the cartesian coordinate system.

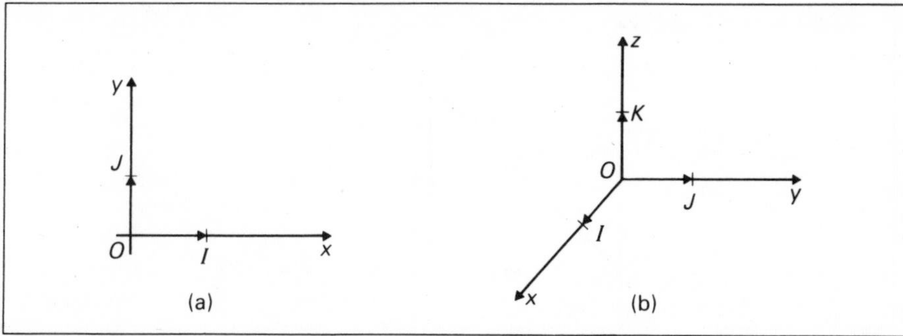

Figure 9.8

In figure 9.8(a) we take **OI**, **OJ** as unit base vectors along the x-axis, y-axis respectively. To correspond to the cartesian coordinate system, we write

$$\mathbf{OI} = \begin{pmatrix}1\\0\end{pmatrix} \text{ and } \mathbf{OJ} = \begin{pmatrix}0\\1\end{pmatrix}$$

Similarly in figure 9.8(b) we have the three dimensional situation with

$$\mathbf{OI} = \begin{pmatrix}1\\0\\0\end{pmatrix}, \quad \mathbf{OJ} = \begin{pmatrix}0\\1\\0\end{pmatrix}, \quad \mathbf{OK} = \begin{pmatrix}0\\0\\1\end{pmatrix}$$

For convenience, we sometimes write **OI** as $\hat{\mathbf{i}}$, **OJ** as $\hat{\mathbf{j}}$ and **OK** as $\hat{\mathbf{k}}$. The $\hat{}$ is used to remind us that the vectors are of unit length.

Using the base vectors, it is clear that a point, P, with coordinates (a, b, c) with respect to an origin O can be represented by the forms

$$\mathbf{OP} = a\begin{pmatrix}1\\0\\0\end{pmatrix} + b\begin{pmatrix}0\\1\\0\end{pmatrix} + c\begin{pmatrix}0\\0\\1\end{pmatrix}$$

or $$\mathbf{OP} = a\hat{\mathbf{i}} + b\hat{\mathbf{j}} + c\hat{\mathbf{k}}$$

or simply $$\mathbf{OP} = \begin{pmatrix}a\\b\\c\end{pmatrix}$$

9.3 Modulus of a vector

The modulus of a vector **a** is defined in 1, 2 or 3 dimensions as the length of the vector and is written as $|\mathbf{a}|$. (Compare the notation with that used for complex numbers in chapter 11.) When we use the cartesian system of coordinates, we see that we can use Pythagoras' Theorem to evaluate this length. For example, the vector

$$\mathbf{OA} = \begin{pmatrix} 3 \\ 4 \end{pmatrix} \text{ or } 3\hat{\mathbf{i}} + 4\hat{\mathbf{j}}$$

can be represented as shown in figure 9.9 and clearly the length of **OA** is 5 units. We say $|\mathbf{OA}| = 5$.

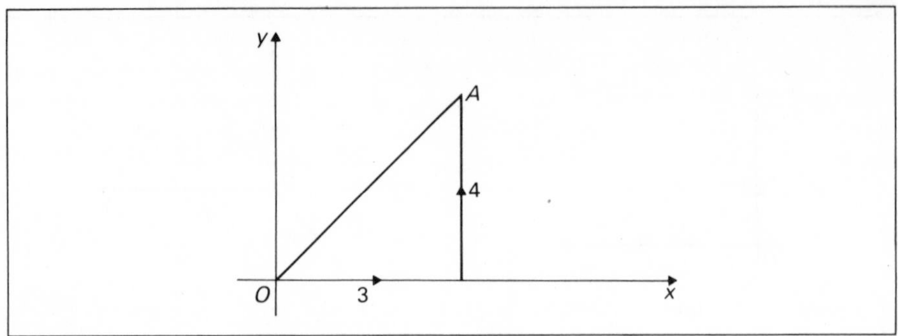

Figure 9.9

Example 1

Find the lengths of AB and AC, when A is (2, 1), B is (3, 7), and C is (5, 6).

We see from figure 9.10 that

$$\mathbf{AB} = \mathbf{OB} - \mathbf{OA} = \begin{pmatrix} 3 \\ 7 \end{pmatrix} - \begin{pmatrix} 2 \\ 1 \end{pmatrix} = \begin{pmatrix} 1 \\ 6 \end{pmatrix}$$

Hence

$$|\mathbf{AB}| = \sqrt{(1^2 + 6^2)} = \sqrt{37}$$

Similarly

$$\mathbf{AC} = \mathbf{OC} - \mathbf{OA} = \begin{pmatrix} 5 \\ 6 \end{pmatrix} - \begin{pmatrix} 2 \\ 1 \end{pmatrix} = \begin{pmatrix} 3 \\ 5 \end{pmatrix}$$

so that

$$|\mathbf{AC}| = \sqrt{(3^2 + 5^2)} = \sqrt{34}$$

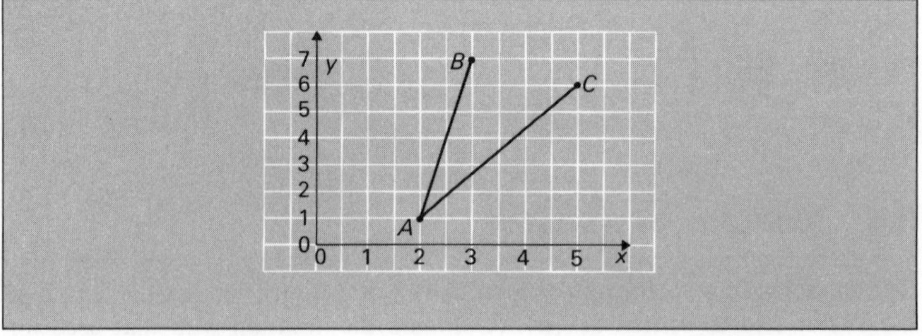

Figure 9.10

We apply the same principles to vectors in three dimensions, so that the length of

$$\begin{pmatrix} a \\ b \\ c \end{pmatrix} \text{ is } \sqrt{(a^2 + b^2 + c^2)}$$

as can be seen from figure 9.11.

$$(OB)^2 = a^2 + b^2, \text{ in } \triangle OAB$$

and $$(OC)^2 = (OB)^2 + (BC)^2, \text{ in } \triangle OBC$$

$$= a^2 + b^2 + c^2$$

Hence $|\mathbf{OC}| = \sqrt{(a^2 + b^2 + c^2)}$

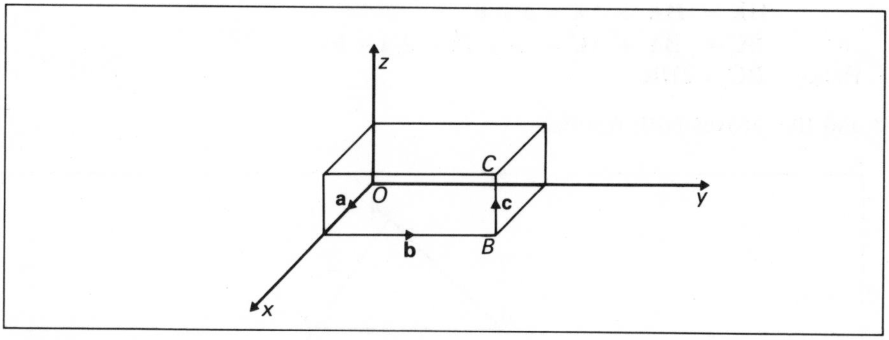

Figure 9.11

Example 2

Find the distance between A (1, 3, 7) and B (4, -1, -5).

$$\mathbf{AB} = \begin{pmatrix} 4 \\ -1 \\ -5 \end{pmatrix} - \begin{pmatrix} 1 \\ 3 \\ 7 \end{pmatrix} = \begin{pmatrix} 3 \\ -4 \\ -12 \end{pmatrix}$$

Hence $|\mathbf{AB}| = \sqrt{\{(3)^2 + (-4)^2 + (-12)^2\}}$

$$= \sqrt{\{9 + 16 + 144\}}$$
$$= \sqrt{169}$$
$$= 13$$

Equality of vectors

Two vectors are equal only when they have the same magnitude and the same direction. This is a helpful result in determining the nature of certain plane figures, as the following examples illustrate.

Example 3

Show that $OABC$ is a parallelogram where the coordinates of A, B, C are (5, 12), (17, 17), (12, 5) and O is the origin.

First we find the vectors **OA**, **CB**,

i.e. $\mathbf{OA} = \begin{pmatrix} 5 \\ 12 \end{pmatrix}$, $\mathbf{CB} = \begin{pmatrix} 17 \\ 17 \end{pmatrix} - \begin{pmatrix} 12 \\ 5 \end{pmatrix} = \begin{pmatrix} 5 \\ 12 \end{pmatrix}$

Hence **OA** = **CB**

i.e. $OA = CB$ and they are in the same direction.

Hence OA is equal and parallel to CB and $OABC$ is a parallelogram.

Example 4
In a triangle ABC, H and K are the midpoints of AB and AC respectively, figure 9.12. Prove that
a) HK is parallel to BC b) $HK = \frac{1}{2}BC$

Let $\mathbf{BA} = 2\mathbf{a}$ and $\mathbf{AC} = 2\mathbf{b}$
then $\mathbf{HA} = \mathbf{a}$ and $\mathbf{AK} = \mathbf{b}$

By vector addition,

$$\mathbf{HK} = \mathbf{HA} + \mathbf{AK} = \mathbf{a} + \mathbf{b}$$

and $\mathbf{BC} = \mathbf{BA} + \mathbf{AC} = 2\mathbf{a} + 2\mathbf{b} = 2(\mathbf{a} + \mathbf{b})$
Hence $\mathbf{BC} = 2\mathbf{HK}$

and this proves both results.

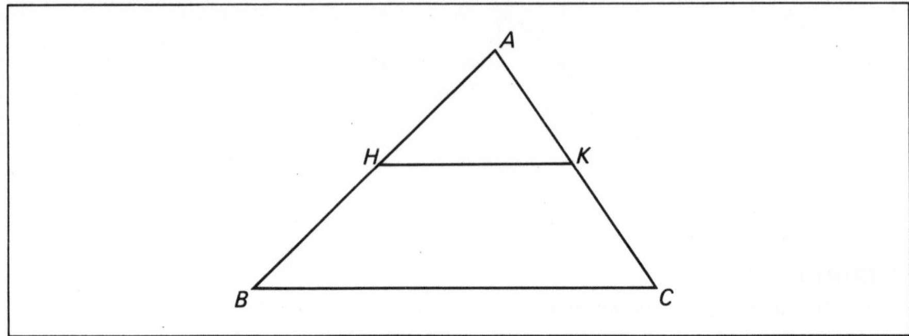

Figure 9.12

Exercise A

1 P, Q, R, S are the mid-points of the sides of any quadrilateral. Prove that $PQRS$ is a parallelogram. What is $PQRS$ if the quadrilateral is a rectangle?

2 $OABC$ is a parallelogram, where O is the origin, A is $(2, 1, -2)$ and B is $(-3, 2, 5)$. Find the coordinates of C, and find the lengths of the sides of the parallelogram.

3 Given that $\mathbf{a} = \hat{\mathbf{i}} + 2\hat{\mathbf{j}} - 3\hat{\mathbf{k}}$,
$\mathbf{b} = 2\hat{\mathbf{i}} - 3\hat{\mathbf{j}} + 4\hat{\mathbf{k}}$,
$\mathbf{c} = -\hat{\mathbf{i}} + 5\hat{\mathbf{j}} + \hat{\mathbf{k}}$.

a) Find in terms of $\hat{\mathbf{i}}, \hat{\mathbf{j}}, \hat{\mathbf{k}}$, the vectors
i) $\mathbf{a} + \mathbf{b} - \mathbf{c}$ iii) $3\mathbf{a} + 2\mathbf{b} - 5\mathbf{c}$
ii) $\mathbf{a} - \mathbf{b} - \mathbf{c}$

b) Find the lengths of each of the following vectors
i) $\mathbf{a}$ iv) $\mathbf{a} + \mathbf{b}$
ii) $\mathbf{b}$ v) $\mathbf{b} - \mathbf{c}$
iii) $\mathbf{c}$ vi) $\mathbf{a} - \mathbf{b} - \mathbf{c}$

4 Find the vector which is in a direction opposite to the sum of $\mathbf{F} = 2\hat{\mathbf{i}} - 3\hat{\mathbf{j}} + 5\hat{\mathbf{k}}$ and $\mathbf{G} = -\hat{\mathbf{i}} + \hat{\mathbf{j}} - 7\hat{\mathbf{k}}$, and find its size.

5 A ship P moves with velocity $2\hat{\mathbf{i}} + 3\hat{\mathbf{j}}$, and another ship Q moves with velocity $-4\hat{\mathbf{i}} - \hat{\mathbf{j}}$. What is the velocity of P relative to Q? Show the actual and relative velocities on a diagram.

6 Investigate whether $\begin{pmatrix} 3 \\ -4 \\ 5 \end{pmatrix}, \begin{pmatrix} \frac{1}{2} \\ 2 \\ 6 \end{pmatrix}, \begin{pmatrix} 4 \\ 6\frac{1}{2} \\ -9 \end{pmatrix}$ lie in a straight line.

7 Given that $\mathbf{a}$ and $\mathbf{b}$ are any two non-parallel vectors, draw diagrams to illustrate

i) $|\mathbf{a} + \mathbf{b}| \leqslant |\mathbf{a}| + |\mathbf{b}|$ ii) $|\mathbf{a} - \mathbf{b}| \geqslant |\mathbf{a}| - |\mathbf{b}|$

Generalise the first result with n vectors.

8 Given $\mathbf{PA} = \begin{pmatrix} 1 \\ 2 \\ 3 \end{pmatrix}$, $\mathbf{PB} = \begin{pmatrix} 3 \\ 1 \\ 9 \end{pmatrix}$

$\mathbf{QC} = \begin{pmatrix} -2 \\ 1 \\ 3 \end{pmatrix}$, $\mathbf{QD} = \begin{pmatrix} 4 \\ -2 \\ 21 \end{pmatrix}$

Find the size of the vectors $\mathbf{AB}$ and $\mathbf{CD}$, and state what relationship exists between them.

9.4 Ratio theorem

Often, we need to find the vector $\mathbf{OC}$, where C lies on AB, and C divides AB in a given ratio. (Compare this with the work done on ratios in chapter 5.)

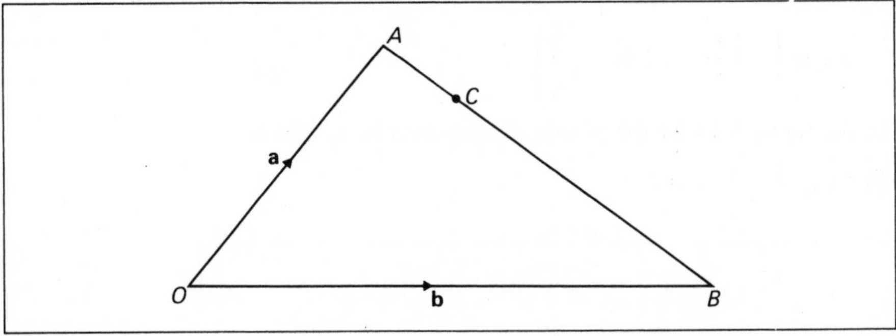

Figure 9.13

Suppose C divides AB in the ratio $\lambda : \mu$

that is $\dfrac{AC}{CB} = \dfrac{\lambda}{\mu}$, or $\dfrac{AC}{AB} = \dfrac{\lambda}{\lambda + \mu}$

Then $\quad \mathbf{OC} = \mathbf{OA} + \mathbf{AC}$

$$= \mathbf{OA} + \frac{\lambda}{\lambda+\mu}(\mathbf{AB})$$

$$= \mathbf{a} + \frac{\lambda}{\lambda+\mu}(\mathbf{b} - \mathbf{a})$$

$$= \frac{(\lambda+\mu)\mathbf{a} + \lambda\mathbf{b} - \lambda\mathbf{a}}{\lambda+\mu}$$

$$= \frac{\mu\mathbf{a} + \lambda\mathbf{b}}{\lambda+\mu}$$

and this is the required vector.

Example 1

Given that $\quad \mathbf{OA} = \begin{pmatrix} 1 \\ 3 \\ 7 \end{pmatrix}, \quad \mathbf{OB} = \begin{pmatrix} 2 \\ -1 \\ -2 \end{pmatrix}$

find **OC** where C divides AB in the ratio 1 : 4.

In this case $\lambda = 1$, $\mu = 4$, and we have

$$\mathbf{OC} = \frac{1\begin{pmatrix} 2 \\ -1 \\ -2 \end{pmatrix} + 4\begin{pmatrix} 1 \\ 3 \\ 7 \end{pmatrix}}{5}$$

$$= \frac{1}{5}\begin{pmatrix} 6 \\ 11 \\ 26 \end{pmatrix} \text{ which is the required vector.}$$

Example 2

Find vectors representing the points of trisection of the line AB, where

OA is $\begin{pmatrix} 2 \\ -1 \\ -3 \end{pmatrix}$ and **OB** is $\begin{pmatrix} 6 \\ 9 \\ 8 \end{pmatrix}$.

In the figure 9.14 let the points of trisection be C and D.

For C, $\quad \lambda = 1$, $\mu = 2$

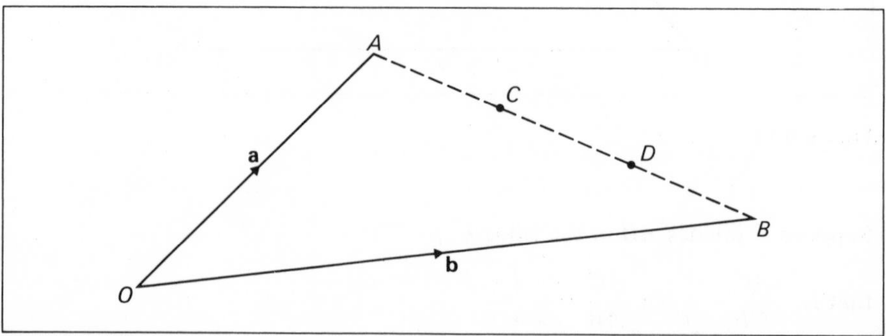

Figure 9.14

Hence $\mathbf{OC} = \dfrac{1\begin{pmatrix}6\\9\\8\end{pmatrix} + 2\begin{pmatrix}2\\-1\\-3\end{pmatrix}}{3} = \dfrac{1}{3}\begin{pmatrix}10\\7\\2\end{pmatrix}$

and, for D, $\lambda = 2$, $\mu = 1$

Hence $\mathbf{OD} = \dfrac{2\begin{pmatrix}6\\9\\8\end{pmatrix} + 1\begin{pmatrix}2\\-1\\-3\end{pmatrix}}{3} = \dfrac{1}{3}\begin{pmatrix}14\\17\\13\end{pmatrix}$

9.5 Equations of lines and planes

It follows from our work here that whatever the values of λ, μ, ($\in \mathbb{R}$), the point C given by

$$\mathbf{OC} = \frac{\mu\mathbf{a} + \lambda\mathbf{b}}{\lambda + \mu}$$

will always lie on the line AB. In establishing this result for OC, we saw that

$$\mathbf{OC} = \mathbf{a} + \frac{\lambda}{\lambda + \mu}(\mathbf{b} - \mathbf{a})$$

which we can re-write as

$$\mathbf{OC} = \mathbf{a} + k\mathbf{AB}$$

Hence the vector $\mathbf{r}$ of any point on the line AB can be written as

$$\mathbf{r} = \mathbf{a} + k\mathbf{AB}$$

We see that for different values of k, we obtain all the points on the line AB, and we call this the vector equation of the line AB.

Example 1
Find the equation of the line through $(2, 0, -3)$ which is in a direction $\begin{pmatrix}1\\-1\\4\end{pmatrix}$.

From our work,

$$\mathbf{a} = \begin{pmatrix}2\\0\\-3\end{pmatrix} \text{ and } \mathbf{AB} = \begin{pmatrix}1\\-1\\4\end{pmatrix}$$

Hence the vector equation of the line is

$$\mathbf{r} = \begin{pmatrix}2\\0\\-3\end{pmatrix} + k\begin{pmatrix}1\\-1\\4\end{pmatrix}$$

Note that k is a parameter, and this equation would be more properly called the vector parametric equation of the line. It may help to consider figure 9.15.

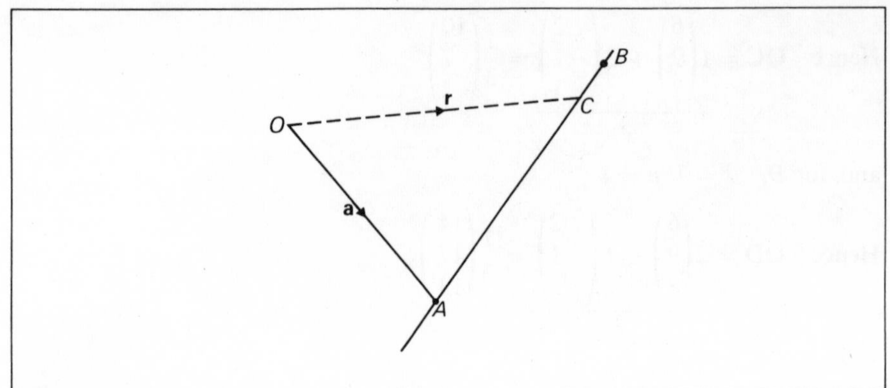

Figure 9.15

For the point C,

$$\begin{aligned} \mathbf{r} = \mathbf{OC} &= \mathbf{OA} + \mathbf{AC} \\ &= \mathbf{OA} + k(\mathbf{AB}) \\ &= \mathbf{a} \quad + k(\mathbf{AB}) \end{aligned}$$

In a similar way, we can obtain the vector parametric equation of a plane, figure 9.16. We know that any point in the plane can be represented by $s\mathbf{b} + t\mathbf{c}$ with reference to the point A in the plane. If we refer to an origin O, however, we have for any point in the plane,

$\mathbf{r} = \mathbf{a} + s\mathbf{b} + t\mathbf{c}$, where s, $t(\in \mathbb{R})$ are parameters.

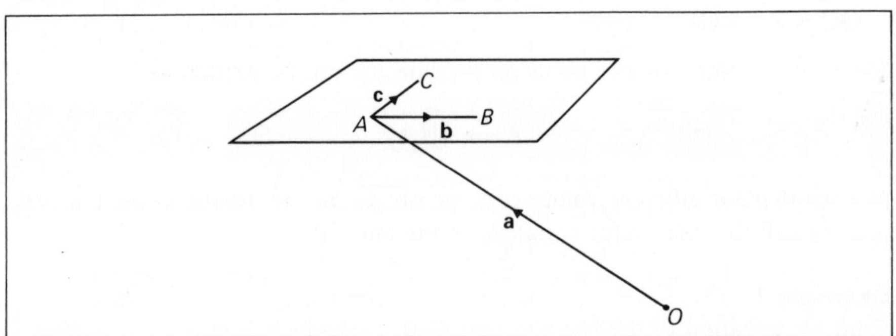

Figure 9.16

Example 2

Given that $A(1, 3, 5)$, $B(-2, 5, -1)$ and $C(-3, 3, 4)$ lie in a plane, find the vector parametric equation of the plane.

Here we are given $\mathbf{OA} = \begin{pmatrix} 1 \\ 3 \\ 5 \end{pmatrix}$, $\mathbf{OB} = \begin{pmatrix} -2 \\ 5 \\ -1 \end{pmatrix}$, $\mathbf{OC} = \begin{pmatrix} -3 \\ 3 \\ 4 \end{pmatrix}$

Hence $\mathbf{AB} = \begin{pmatrix} -2 \\ 5 \\ -1 \end{pmatrix} - \begin{pmatrix} 1 \\ 3 \\ 5 \end{pmatrix} = \begin{pmatrix} -3 \\ 2 \\ -6 \end{pmatrix}$

and $\mathbf{AC} = \begin{pmatrix} -3 \\ 3 \\ 4 \end{pmatrix} - \begin{pmatrix} 1 \\ 3 \\ 5 \end{pmatrix} = \begin{pmatrix} -4 \\ 0 \\ -1 \end{pmatrix}$

so that the equation of the plane is

$$\mathbf{r} = \mathbf{OA} + s\mathbf{AB} + t\mathbf{AC}$$

$$\Rightarrow \mathbf{r} = \begin{pmatrix} 1 \\ 3 \\ 5 \end{pmatrix} + s\begin{pmatrix} -3 \\ 2 \\ 6 \end{pmatrix} + t\begin{pmatrix} -4 \\ 0 \\ -1 \end{pmatrix}$$

9.6 Conversion of vector parametric equations of lines and planes to cartesian form

It is sometimes necessary to eliminate the parameters from a vector parametric equation. We consider first the two-dimensional vector equation of a line.

i.e. $\mathbf{r} = \begin{pmatrix} a \\ b \end{pmatrix} + t\begin{pmatrix} c \\ d \end{pmatrix}$

This can be written as

$$\begin{pmatrix} x \\ y \end{pmatrix} = \begin{pmatrix} a \\ b \end{pmatrix} + t\begin{pmatrix} c \\ d \end{pmatrix}$$

or $\begin{pmatrix} x \\ y \end{pmatrix} = \begin{pmatrix} a + tc \\ b + td \end{pmatrix}$

Hence $x = a + tc$ and $y = b + td$

or $t = \dfrac{x - a}{c}$ and $t = \dfrac{y - b}{d}$

so that $\dfrac{x - a}{c} = \dfrac{y - b}{d}$

which simplifies to

$$dx - ad = cy - bc$$

or $dx - cy + bc - ad = 0$

We see that the gradient is $\dfrac{d}{c}$ which is what we would have expected from the vector equation, since the line had a direction parallel to $\begin{pmatrix} c \\ d \end{pmatrix}$.

Example 1
Find the cartesian equation of the line which passes through (1, 2) and is parallel to $\begin{pmatrix} 1 \\ -2 \end{pmatrix}$.

The vector equation is

$$\mathbf{r} = \begin{pmatrix} 1 \\ 2 \end{pmatrix} + t\begin{pmatrix} 1 \\ -2 \end{pmatrix}$$

$$\Rightarrow \begin{pmatrix} x \\ y \end{pmatrix} = \begin{pmatrix} 1 + t \\ 2 - 2t \end{pmatrix}$$

$$\Rightarrow x = 1 + t, \quad y = 2 - 2t$$

$$\Rightarrow t = x - 1 \text{ and } t = \frac{2 - y}{2}$$

Hence $x - 1 = \dfrac{2 - y}{2}$

or $\quad 2x + y - 4 = 0$

We apply a similar method for the equation of a plane

$$\mathbf{r} = \begin{pmatrix} a_1 \\ a_2 \\ a_3 \end{pmatrix} + s\begin{pmatrix} b_1 \\ b_2 \\ b_3 \end{pmatrix} + t\begin{pmatrix} c_1 \\ c_2 \\ c_3 \end{pmatrix}$$

and eliminate s and t from the three equations

$$x = a_1 + sb_1 + tc_1$$
$$y = a_2 + sb_2 + tc_2$$
$$z = a_3 + sb_3 + tc_3$$

The method is more easily seen from an example.

Example 2

Find the cartesian equation of the plane whose vector equation is

$$\mathbf{r} = \begin{pmatrix} 1 \\ 0 \\ 2 \end{pmatrix} + s\begin{pmatrix} 1 \\ 1 \\ 1 \end{pmatrix} + t\begin{pmatrix} 2 \\ -1 \\ 3 \end{pmatrix}$$

Here we have

$$\begin{pmatrix} x \\ y \\ z \end{pmatrix} = \begin{pmatrix} 1 + s + 2t \\ 0 + s - t \\ 2 + s + 3t \end{pmatrix}$$

i.e. $x = 1 + s + 2t$ **i**

$y = s - t$ **ii**

$z = 2 + s + 3t$ **iii**

Subtracting equations **i** and **ii** gives

$$x - y = 1 + 3t$$

$$t = \frac{x - y - 1}{3}$$

Adding equation **i** to twice **ii** gives

$$x + 2y = 1 + 3s$$

$$s = \frac{x + 2y - 1}{3}$$

Substituting these values in **iii** gives

$$z = 2 + \frac{x + 2y - 1}{3} + 3\,\frac{x - y - 1}{3}$$

i.e. $3z = 6 + x + 2y - 1 + 3x - 3y - 3$

$$4x - y - 3z + 2 = 0$$

and this is the cartesian equation of the plane.

In general, any equation of the form

$ax + by + cz = d \quad (a, b, c, d \in \mathbb{R})$ represents a plane.

There is no such simple equation for a line in three dimensions. In order to give a cartesian representation of a line, we give the equations of planes which intersect in the required line. These equations are obtained by elimination of the parameter from the vector equation as the following example shows.

Example 3

Find the system of cartesian equations which represent the line

$$\mathbf{r} = \begin{pmatrix} -1 \\ 2 \\ -1 \end{pmatrix} + t\begin{pmatrix} 1 \\ 3 \\ -2 \end{pmatrix}$$

Here we have

$$x = -1 + t$$
$$y = 2 + 3t$$
$$z = -1 - 2t$$

and these can be rearranged, so that

$$t = x + 1$$

$$t = \frac{y - 2}{3}$$

$$t = \frac{1 - z}{2}$$

Hence $\dfrac{x + 1}{1} = \dfrac{y - 2}{3} = \dfrac{1 - z}{2}$, and this system of equations represents the given line.

In fact this system of equations represents a system of planes which have a common line, ℓ, as shown in figure 9.17

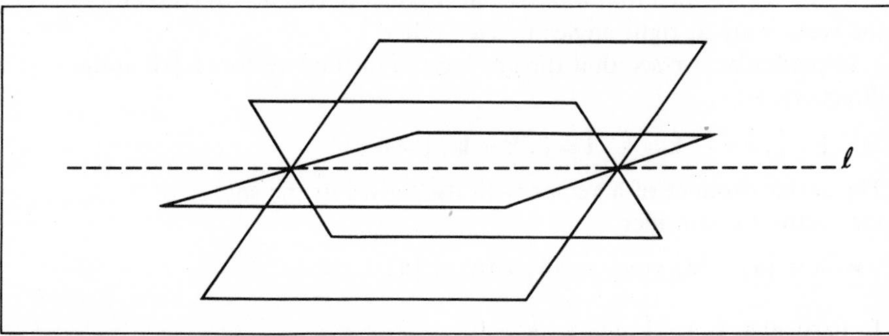

Figure 9.17

Exercise B

1 The centroid, G, of a triangle lies at the point of trisection of the three medians. If the vectors **OA**, **OB**, **OC** are represented by **a**, **b**, **c**, show that $\mathbf{OC} = \frac{1}{3}(\mathbf{a} + \mathbf{b} + \mathbf{c})$, by applying the ratio theorem twice.

Obtain a similar result for the tetrahedron $ABCD$.

2 $\mathbf{OA} = \begin{pmatrix} 2 \\ 1 \\ -3 \end{pmatrix}$ and $\mathbf{OB} = \begin{pmatrix} 4 \\ 2 \\ 1 \end{pmatrix}$.

The point P divides AB in the ratio $1:2$. Find the coordinates of P and obtain the parametric equation of the line PQ where Q is $(7, 2, 3)$.

3 Given that $5\mathbf{OC} = 3\mathbf{OB} + 2\mathbf{OA}$, what can you deduce about A, B and C? What is the value of the ratio $AC : CB$?

4 Points P, Q, R have coordinates $(-3, 5, 4)$, $(5, 9, 8)$, $(-1, 6, 5)$. Investigate whether they are collinear.

5 A is $(0, -1, 1)$, B is $(2, 4, 8)$, C is $(1, 1, 2)$ and D is $(3, 3, 6)$, find the vector parametric equation of the lines AB and CD.

6 A parallelepiped (a six-faced solid, each face of which is a parallelogram) has one vertex at the origin. The equations of three of its edges are $\mathbf{r} = \lambda\mathbf{a}$, $\mathbf{r} = \mu\mathbf{b}$, $\mathbf{r} = \nu\mathbf{c}$. Find, in terms of **a**, **b**, **c** and suitable parameters, the equations of the other edges.

9.7 Scalar product

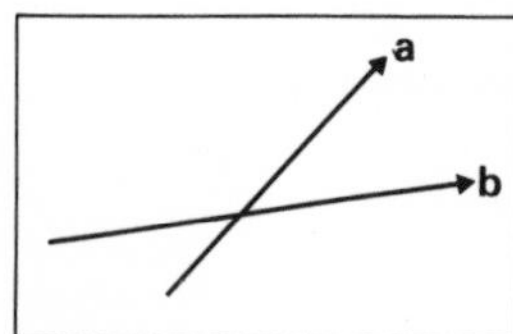

Figure 9.18

We consider now the product of two vectors **a** and **b**. There are two different methods of combining vectors by multiplication, one of which results in a scalar quantity and the other gives a vector quantity. We deal here only with the first kind, namely the **scalar product**.

We define the scalar (or dot) product $\mathbf{a} \cdot \mathbf{b}$ as $|\mathbf{a}||\mathbf{b}| \cos\theta$ where θ is the angle between **a** and **b**, figure 9.18. Clearly, there are two angles between the vectors **a** and **b** (in general one acute and one obtuse), so we need to specify in our definition which angle we mean. The commonly accepted convention is that we take the angle through which one vector would be rotated to be in the same direction as the other. So that the angle θ will be as shown in figure 9.19.

It follows from our definition that if $\mathbf{a} \cdot \mathbf{b} = 0$ then either $|\mathbf{a}| = 0$, $|\mathbf{b}| = 0$ or $\cos\theta = 0$, so that if **a** and **b** are non-zero vectors, then $\cos\theta = 0$ i.e. $\theta = 90°$. We see then that if the scalar product of two non-zero vectors is zero, the vectors are at right angles to each other.

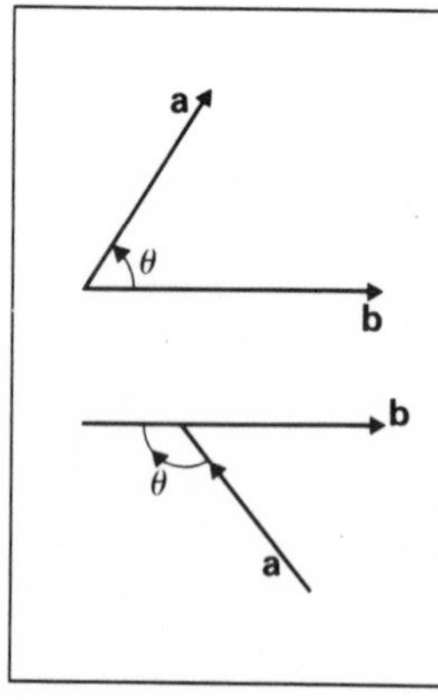

Figure 9.19

In particular, we see that the products of the unit vectors $\hat{\mathbf{i}}, \hat{\mathbf{j}}, \hat{\mathbf{k}}$ in the x, y, z directions give

$$\hat{\mathbf{i}} \cdot \hat{\mathbf{j}} = \hat{\mathbf{j}} \cdot \hat{\mathbf{i}} = \hat{\mathbf{i}} \cdot \hat{\mathbf{k}} = \hat{\mathbf{k}} \cdot \hat{\mathbf{i}} = \hat{\mathbf{j}} \cdot \hat{\mathbf{k}} = \hat{\mathbf{k}} \cdot \hat{\mathbf{j}} = 0$$

The scalar product of a vector with itself gives us the square of the length of the vector, for consider

$$\mathbf{a} \cdot \mathbf{a} = |\mathbf{a}| \cdot |\mathbf{a}| \cos 0 = |\mathbf{a}| \cdot |\mathbf{a}| = |\mathbf{a}|^2$$

In particular $\hat{\mathbf{i}} \cdot \hat{\mathbf{i}} = \hat{\mathbf{j}} \cdot \hat{\mathbf{j}} = \hat{\mathbf{k}} \cdot \hat{\mathbf{k}} = 1$

We notice also that $\mathbf{a} \cdot \mathbf{b} = \mathbf{b} \cdot \mathbf{a}$, since $\cos(-\theta) = \cos\theta$. Hence scalar multiplication of vectors is a commutative operation.

Scalar products of vectors in terms of cartesian components

Let $\mathbf{a} = a_1\hat{\mathbf{i}} + a_2\hat{\mathbf{j}} + a_3\hat{\mathbf{k}}$

and $\mathbf{b} = b_1\hat{\mathbf{i}} + b_2\hat{\mathbf{j}} + b_3\hat{\mathbf{k}}$

then
$$\begin{aligned}\mathbf{a} \cdot \mathbf{b} &= (a_1\hat{\mathbf{i}} + a_2\hat{\mathbf{j}} + a_3\hat{\mathbf{k}}) \cdot (b_1\hat{\mathbf{i}} + b_2\hat{\mathbf{j}} + b_3\hat{\mathbf{k}})\\ &= a_1b_1|\hat{\mathbf{i}}|^2 + a_2b_2|\hat{\mathbf{j}}|^2 + a_3b_3|\hat{\mathbf{k}}|^2 + a_1b_2\hat{\mathbf{i}} \cdot \hat{\mathbf{j}} + a_1b_3\hat{\mathbf{i}} \cdot \hat{\mathbf{k}}\\ &\quad + a_2b_1\hat{\mathbf{j}} \cdot \hat{\mathbf{i}} + a_2b_3\hat{\mathbf{j}} \cdot \hat{\mathbf{k}} + a_3b_1\hat{\mathbf{k}} \cdot \hat{\mathbf{i}} + a_3b_2\hat{\mathbf{k}} \cdot \hat{\mathbf{j}}\end{aligned}$$

and clearly the last six terms are all zero, so that the product simplifies to

$$a_1b_1|\hat{\mathbf{i}}|^2 + a_2b_2|\hat{\mathbf{j}}|^2 + a_3b_3|\hat{\mathbf{k}}|^2$$

but $|\hat{\mathbf{i}}|^2 = |\hat{\mathbf{j}}|^2 = |\hat{\mathbf{k}}|^2 = 1$

Hence we have

$$\mathbf{a} \cdot \mathbf{b} = a_1b_1 + a_2b_2 + a_3b_3$$

Example 1

Find the scalar product of the vectors

$$\begin{pmatrix}1\\3\\-2\end{pmatrix} \text{ and } \begin{pmatrix}-3\\4\\-1\end{pmatrix}$$

and hence find the angle between these vectors.

The product $\begin{pmatrix}1\\3\\-2\end{pmatrix} \cdot \begin{pmatrix}-3\\4\\-1\end{pmatrix} = (1)(-3) + (3)(4) + (-2)(-1) = 11$

Since the scalar product is defined as

$$\left|\begin{pmatrix}1\\3\\-2\end{pmatrix}\right|\left|\begin{pmatrix}-3\\4\\-1\end{pmatrix}\right| \cos\theta$$

we have

$$\left|\begin{pmatrix}1\\3\\-2\end{pmatrix}\right|\left|\begin{pmatrix}-3\\4\\-1\end{pmatrix}\right| \cos\theta = 11$$

$$\Rightarrow \cos\theta = \frac{11}{\left|\begin{pmatrix}1\\3\\-2\end{pmatrix}\right|\left|\begin{pmatrix}-3\\4\\-1\end{pmatrix}\right|}$$

$$= \frac{11}{\sqrt{14}\sqrt{26}}$$

Hence $\cos\theta = \dfrac{11}{19{\cdot}079} = 0{\cdot}5766$

i.e. $\theta = 54{\cdot}79°$

Further uses of scalar products occur in many applications where vectors are involved. In particular, when dealing with force-vectors it is often necessary to find the component of the forces in a given direction.

For example, in figure 9.20, if we require the component of **F** in the direction of x, we see that this is $|\mathbf{F}| \cos \theta$. Now, we know that $\mathbf{F} \cdot \mathbf{x} = |\mathbf{F}||\mathbf{x}| \cos \theta$ so that

$$|\mathbf{F}| \cos \theta = \frac{\mathbf{F} \cdot \mathbf{x}}{|\mathbf{x}|}$$

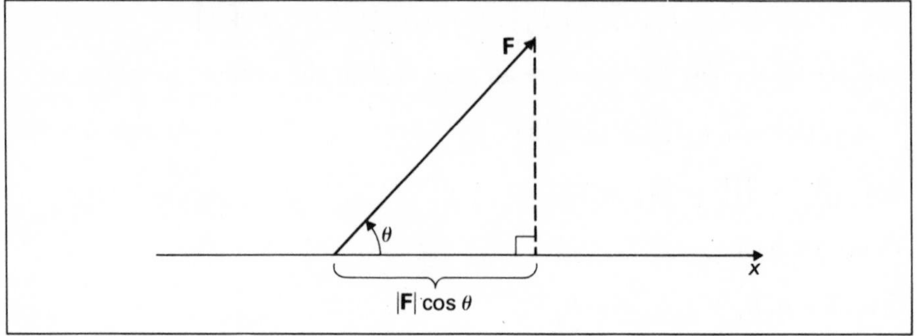

Figure 9.20

Example 2

Find the component of the vector $\begin{pmatrix} 7 \\ -3 \\ 2 \end{pmatrix}$ in the direction of the vector $\begin{pmatrix} 3 \\ 4 \\ 12 \end{pmatrix}$.

From the work just described, we see that the required component is

$$\frac{\begin{pmatrix} 7 \\ -3 \\ 2 \end{pmatrix} \cdot \begin{pmatrix} 3 \\ 4 \\ 12 \end{pmatrix}}{\left|\begin{pmatrix} 3 \\ 4 \\ 12 \end{pmatrix}\right|} = \frac{21 - 12 + 24}{13} = \frac{33}{13} = 2{\cdot}538$$

We look again at the equation of a plane, and consider first the general equation of a plane which passes through the origin. Such a plane has for its equation (as we saw in section 9.6),

$$ax + by + cz = 0$$

Now we can rewrite this equation as

$$\begin{pmatrix} x \\ y \\ z \end{pmatrix} \cdot \begin{pmatrix} a \\ b \\ c \end{pmatrix} = 0$$

or, more simply $\mathbf{r} \cdot \begin{pmatrix} a \\ b \\ c \end{pmatrix} = 0$

We know that if $\begin{pmatrix} a \\ b \\ c \end{pmatrix}$ is not a zero vector

then $\begin{pmatrix} a \\ b \\ c \end{pmatrix}$ is at right angles to the vector $\begin{pmatrix} x \\ y \\ z \end{pmatrix}$.

Hence $\begin{pmatrix} a \\ b \\ c \end{pmatrix}$ is at right angles to the plane

$$ax + by + cz = 0$$

and is usually referred to as the 'normal vector'.

We extend this idea and consider the plane $ax + by + cz = d$, which is parallel to $ax + by + cz = 0$ and hence $\begin{pmatrix} a \\ b \\ c \end{pmatrix}$ is also perpendicular to the plane $ax + by + cz = d$.

In view of this relationship between equations of planes and their normal vectors, we can use this to find the angle between two planes. From figure 9.21, we see that the angle between the two planes, $P\hat{R}Q$, is equal to one of the angles between the normal vectors – in this case $180° - P\hat{N}Q$. The following example will illustrate the method.

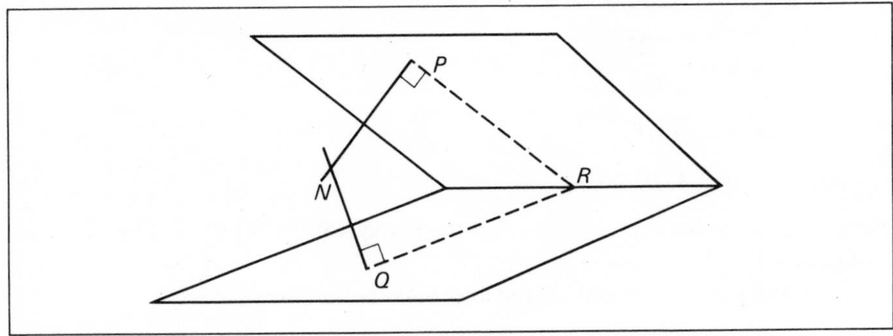

Figure 9.21

Example 3

Find the acute angle between the planes whose equations are

$$3x - y + 2z = 9 \text{ and } -x + 2y + z = 3$$

The two normal vectors are $\begin{pmatrix} 3 \\ -1 \\ 2 \end{pmatrix}$ and $\begin{pmatrix} -1 \\ 2 \\ 1 \end{pmatrix}$

Hence the scalar product gives

$$\begin{pmatrix} 3 \\ -1 \\ 2 \end{pmatrix} \cdot \begin{pmatrix} -1 \\ 2 \\ 1 \end{pmatrix} = \left|\begin{pmatrix} 3 \\ -1 \\ 2 \end{pmatrix}\right| \left|\begin{pmatrix} -1 \\ 2 \\ 1 \end{pmatrix}\right| \cos\theta$$

$$\Rightarrow \quad -3 - 2 + 2 = \sqrt{14}\sqrt{6}\cos\theta$$

So $\cos\theta = \dfrac{-3}{\sqrt{14}\sqrt{6}}$

$\cos\theta = -{\cdot}3273$

Hence $\theta = 109{\cdot}1°$.

Hence the acute angle between the plane is 70·9°.

Example 4

Find the angle made by the line $\mathbf{r} = \begin{pmatrix}1\\2\\-1\end{pmatrix} + t\begin{pmatrix}2\\-3\\5\end{pmatrix}$

with the plane $3x + y + 4z = 7$.

In this case, we first find the angle between the line and the normal vector to the plane.

The line has direction $\begin{pmatrix}2\\-3\\5\end{pmatrix}$ and the plane's normal vector is $\begin{pmatrix}3\\1\\4\end{pmatrix}$.

So we have $\begin{pmatrix}3\\1\\4\end{pmatrix} . \begin{pmatrix}2\\-3\\5\end{pmatrix} = \left|\begin{pmatrix}3\\1\\4\end{pmatrix}\right| \left|\begin{pmatrix}2\\-3\\5\end{pmatrix}\right| \cos\theta$

$$6 - 3 + 20 = \sqrt{26}\sqrt{38}\cos\theta$$

$$\Rightarrow \quad \cos\theta = \frac{23}{\sqrt{26}\sqrt{38}}$$

$$\cos\theta = {\cdot}7317$$

$$\Rightarrow \quad \theta = 43{\cdot}0°$$

Hence the angle between the line and the plane is $(90 - 43{\cdot}0)° = 47{\cdot}0°$, as seen in figure 9.22.

NA is the plane's normal vector and *BAC* is the line

$$\mathbf{r} = \begin{pmatrix}1\\2\\-1\end{pmatrix} + t\begin{pmatrix}2\\-3\\5\end{pmatrix}$$

so that our angle θ found above is $N\hat{A}C$. We wish to calculate $N\hat{C}A$, and from the figure $N\hat{C}A = 90° - N\hat{A}C = 47°$.

It is beyond the work of this book to consider physical scalar quantities which result from the product of vectors, but there are several definitions within applications of mathematics where such products occur. Some rather familiar results from mechanics are worthy of mention as illustrations of scalar products. For example

$$\mathbf{v} . \mathbf{v} = \mathbf{u} . \mathbf{u} - 2\mathbf{a} . \mathbf{s}$$

is a result obtainable by considering appropriate products from the vector relations

$$\mathbf{v} = \mathbf{u} + t\mathbf{a} \quad \text{and} \quad \mathbf{s} = \mathbf{u}t + \tfrac{1}{2}\mathbf{a}t^2$$

Also, 'work' may be defined by the expression

$$W = \int_0^T \mathbf{F} \,.\, \mathbf{v}\, dt$$

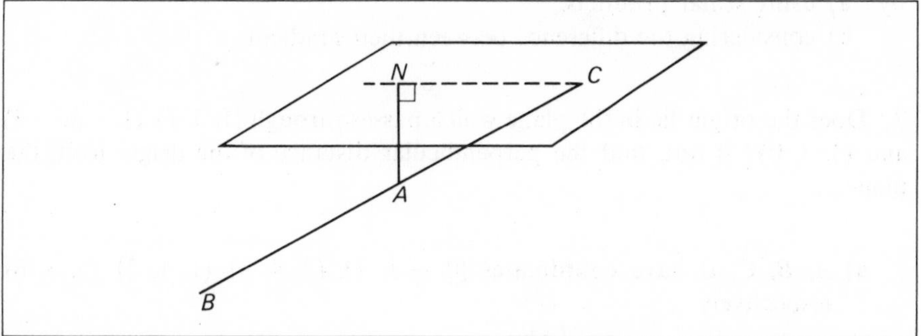

Figure 9.22

Exercise C

1 Show that the line $\mathbf{r} = \begin{pmatrix} 1 \\ 4 \\ -2 \end{pmatrix} + \lambda \begin{pmatrix} 1 \\ -2 \\ -1 \end{pmatrix}$

is parallel to the plane $\mathbf{r} \,.\, \begin{pmatrix} 1 \\ -3 \\ 7 \end{pmatrix} = 4.$

2 Find the vector-parametric equation of the line defined by

$$\frac{x-a}{m} = \frac{y-b}{n} = \frac{z-c}{p}.$$

3 In ΔABC, let $\mathbf{AB} = \mathbf{c}$, $\mathbf{BC} = \mathbf{a}$ and $\mathbf{CA} = \mathbf{b}$. Show that $\mathbf{a} + \mathbf{b} + \mathbf{c} = 0$. Rewrite this equation as $\mathbf{a} + \mathbf{b} = -\mathbf{c}$ and use scalar products to show that $(\mathbf{a} + \mathbf{b}) \,.\, (\mathbf{a} + \mathbf{b}) = \mathbf{c} \,.\, \mathbf{c}$. Hence prove the cosine rule for ΔABC.

4 Find, in parametric form, the equation of the plane which passes through $(0, 1, -1)$, $(2, 3, -2)$ and $(4, -3, -2)$. Eliminate the parameters to obtain the cartesian equation and hence write down a normal vector.

5 Show that the line $\mathbf{r} = \begin{pmatrix} 1 \\ 0 \\ -2 \end{pmatrix} + \lambda \begin{pmatrix} 3 \\ -3 \\ 1 \end{pmatrix}$

is at right angles to the plane $3x - 3y + z = 7$ and find the perpendicular distance of the point $(1, 0, -2)$ from the plane.

6 Find the acute angle between the lines

$$\mathbf{r} = \begin{pmatrix}1\\2\end{pmatrix} + t\begin{pmatrix}3\\4\end{pmatrix} \quad \text{and} \quad \mathbf{r} = \begin{pmatrix}3\\-2\end{pmatrix} + s\begin{pmatrix}2\\-1\end{pmatrix}$$

by a) using scalar products,
b) considering the difference between their gradients.

7 Does the origin lie in the plane which passes through (1, 3, 1), (1, −1, −5) and (3, 7, 0)? If not, find the perpendicular distance of the origin from the plane.

8 a) A, B, C, D have coordinates (0, −1, 1), (2, 4, 8), (1, 1, 2), (3, 3, 6) respectively.

Find λ, μ such that $\begin{pmatrix}\lambda\\\mu\\1\end{pmatrix}$ is perpendicular to both AC and BD.

b) Find the equation of the plane which contains the line AC and is perpendicular to $\begin{pmatrix}\lambda\\\mu\\1\end{pmatrix}$.

9 Of the following three equations, state which represent lines and which represent planes. State also how the lines are related to the planes.

a) $3x - 2y + z = 7$

b) $\dfrac{x-1}{3} = \dfrac{y+1}{-2} = \dfrac{z-2}{1}$

c) $\begin{pmatrix}x\\y\\z\end{pmatrix} = \begin{pmatrix}1\\-1\\2\end{pmatrix} + \lambda\begin{pmatrix}-2\\3\\12\end{pmatrix}$

10 Find the length of the perpendicular drawn from the origin to the plane

$8x + 9y - 12z = 17$

Hence write down the equation of the parallel plane which is twice this distance from the origin.

11 Find the component of the force $\mathbf{F} = -3\hat{\mathbf{i}} + 6\hat{\mathbf{j}} + 2\hat{\mathbf{k}}$ in the direction of $-4\hat{\mathbf{i}} + 4\hat{\mathbf{j}} + 7\hat{\mathbf{k}}$.

12 The sides of a triangle ABC are represented by vectors **a**, **b** and **c**, where

$$\mathbf{a} = \hat{\mathbf{i}} + 4\hat{\mathbf{j}} + 5\hat{\mathbf{k}}$$
$$\mathbf{b} = 2\hat{\mathbf{i}} + 2\hat{\mathbf{j}} + 8\hat{\mathbf{k}}$$
$$\mathbf{c} = -\hat{\mathbf{i}} + 2\hat{\mathbf{j}} - 3\hat{\mathbf{k}}$$

Find the angles of the triangle.

13 Three sides of a unit cube have vectors $\hat{\mathbf{i}}, \hat{\mathbf{j}}, \hat{\mathbf{k}}$. One corner of the cube is at the origin. Find an expression for the acute angle between the diagonals of the cube.

14 Using the facts that $\mathbf{b} = \mathbf{c} - \mathbf{a}$ and $\mathbf{h} \cdot \mathbf{b} = 0$, in figure 9.23, prove that $c \cos \beta = a \cos \alpha$.

Hence prove the sine formula, $\dfrac{a}{\sin A} = \dfrac{c}{\sin C}$.

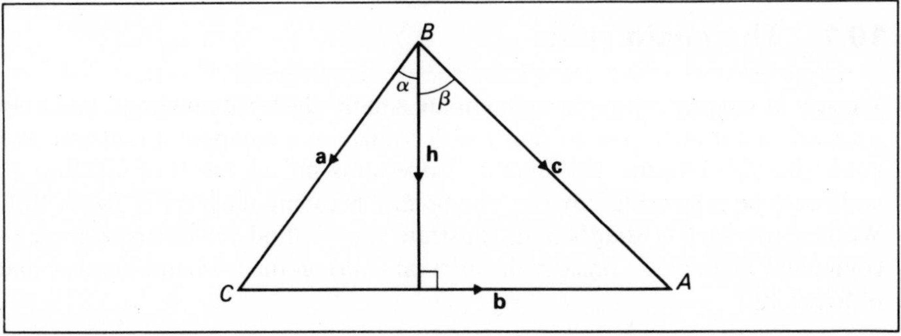

Figure 9.23

CHAPTER TEN

Further Differentiation

10.1 The chain rule

We saw in chapter 3 how to differentiate simple algebraic functions, and now we need to consider how to find the derivative of a compound function. We know that $fg(x)$ means 'the function f operating on the result of function g', and could be represented by the 'compound' mapping diagram in figure 10.1. We can use such a diagram to illustrate the method for differentiation of compound functions. Consider the interval (a, b) on the left hand number-line of figure 10.2.

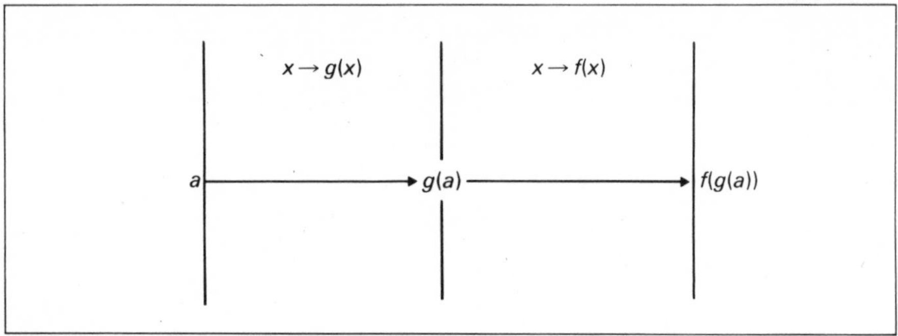

Figure 10.1

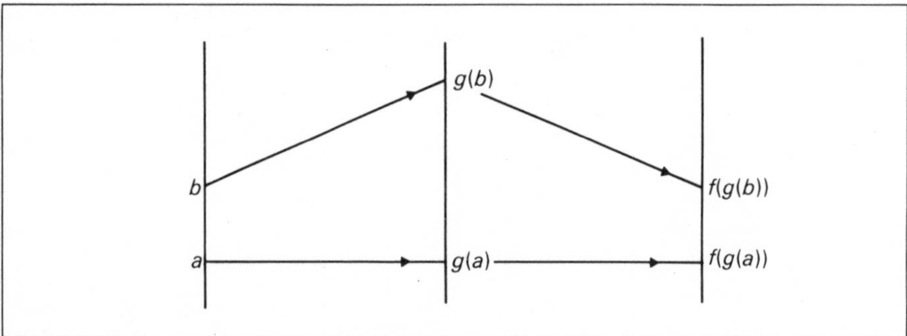

Figure 10.2

Our average scale factor for the function g is

$$\left|\frac{g(b) - g(a)}{b - a}\right| \quad \text{i}$$

and the average scale factor for the function f

$$\left|\frac{f(g(b)) - f(g(a))}{g(b) - g(a)}\right| \quad \text{ii}$$

To obtain the derivative of $fg(x)$, we need to consider the scale factor

$$\left\{\frac{f(g(b) - f(g(a))}{b - a}\right\}$$

which is the product of the scale factors **i** and **ii**. We assume here that

$$\lim_{b\to a}\left\{\frac{f(g(b)) - f(g(a))}{g(b) - g(a)}\right\} \lim_{b\to a}\left\{\frac{g(b) - g(a)}{b - a}\right\}$$

$$= \lim_{b\to a}\left\{\frac{f(g(b)) - f(g(a))}{g(b) - g(a)} \cdot \frac{g(b) - g(a)}{b - a}\right\}$$

and we see that this simplifies to

$$\lim_{b\to a}\left\{\frac{f(g(b)) - f(g(a))}{b - a}\right\}$$

This is our definition (see chapter 3) of the derivative of $fg(x)$ at $x = a$.

Hence, if we need to differentiate a compound, or composite function $y = fg(x)$, we differentiate the function $g(x)$ with respect to x and the function $f(g(x))$ with respect to $g(x)$.

It is more usual to simplify the working by saying in $y = fg(x)$,

let $\quad y = f(z) \qquad$ where $z = g(x)$

then $\quad \dfrac{dy}{dz} = \dfrac{df(z)}{dz}, \qquad \dfrac{dz}{dx} = g'(x)$

hence $\quad \dfrac{dy}{dx} = \dfrac{dy}{dz} \cdot \dfrac{dz}{dx} = \left(\dfrac{df(z)}{dz}\right) \cdot (g'(x))$

This process is often called the **Chain Rule**, and the method will be more easily seen by means of some examples.

Example 1

To find the derivative of $(x^2 + 2)^3$,

write $\quad y = (x^2 + 2)^3$

let $\quad y = z^3 \qquad$ where $z = x^2 + 2$

then $\quad \dfrac{dy}{dz} = 3z^2, \qquad \dfrac{dz}{dx} = 2x$

$$\frac{dy}{dx} = \frac{dy}{dz} \cdot \frac{dz}{dx} = 6xz^2$$

$$= 6x(x^2 + 2)^2$$

Example 2

Find $\dfrac{dy}{dx}$ when $y = \dfrac{1}{x^3 + 3x^2}$

let $\quad y = \dfrac{1}{z} \qquad$ where $z = x^3 + 3x^2$

then $\frac{dy}{dz} = -\frac{1}{z^2}$, $\frac{dz}{dx} = 3x^2 + 6x$

$$\frac{dy}{dx} = \frac{dy}{dz} \cdot \frac{dz}{dx} = \left(-\frac{1}{z^2}\right)(3x^2 + 6x)$$

$$= \frac{-(3x^2 + 6x)}{(x^3 + 3x^2)^2}$$

$$= \frac{-3(x + 2)}{x^3(x + 3)^2}$$

Example 3

Find $\frac{dy}{dx}$ when $y = \sin^3 x$

let $y = z^3$ where $z = \sin x$

$$\frac{dy}{dz} = 3z^2 \qquad \frac{dz}{dx} = \cos x$$

$$\frac{dy}{dx} = \frac{dy}{dz} \cdot \frac{dz}{dx} = 3z^2 \cos x$$

$$= 3 \sin^2 x \cos x$$

Exercise A

Find $\frac{dy}{dx}$ in each of the following.

1 $y = (3x^2 + 1)^2$

2 $y = \frac{1}{3x + x^2}$

3 $y = \sin(x^2 + 2)$

4 $y = \cos(3x - 1)$

5 $y = (x^4 + 5x^3)^3$

6 $y = (x^2 + 2x + 1)^3$

7 $y = (x - 2)^2(x + 2)^2$

8 $y = (3x^3 + 2x)^{-2}$

9 $y = \cos^2(x)$

10 $y = \frac{1}{\sin x}$

Question 10 in this exercise will have shown you a method for differentiating csc x. The result of differentiating csc x is usually written as $-\csc x \cot x$. Can you change your answer to question 10 into that form?

By writing sec x as $\frac{1}{\cos x}$, show that the derivative of sec x is sec x tan x.

This chain process may be extended to deal with even more complex functions.

Consider $y = \sin^2(3x^2 + 2)$
let $y = z^2$, where $z = \sin \theta$, $\theta = 3x^2 + 2$

then $\dfrac{dy}{dz} = 2z, \quad \dfrac{dz}{d\theta} = \cos\theta, \quad \dfrac{d\theta}{dx} = 6x$

$$\frac{dy}{dx} = \frac{dy}{dz} \cdot \frac{dz}{d\theta} \cdot \frac{d\theta}{dx} = (2z)(\cos\theta)(6x)$$

$$= 12x \sin\theta \cos\theta$$

$$= 12x \sin(3x^2 + 2)\cos(3x^2 + 2)$$

How could you use the results of chapter 4 to simplify this answer? Investigate why it could be written as

$$\frac{dy}{dx} = 6x \sin(6x^2 + 4)$$

Often we use the chain rule when dealing with functions which are described parametrically. For example, for the curve described by the equations

$$x = at^2, \qquad y = at^3$$

we have

$$\frac{dx}{dt} = 2at \qquad \frac{dy}{dt} = 3at^2$$

hence

$$\frac{dy}{dx} = \frac{dy}{dt} \cdot \frac{dt}{dx} \quad \text{or} \quad \frac{dy}{dt} \Big/ \frac{dx}{dt}$$

i.e. $\dfrac{dy}{dx} = \dfrac{3at^2}{2at} = \dfrac{3t}{2}$ and the *gradient* of the curve defined by these parametric equations is given by $\dfrac{3t}{2}$ for the point (at^2, at^3).

Of course, we could have eliminated t in the first place and obtained y in terms of x, so that

$$y = a^{-1/2} \,.\, x^{3/2}$$

from which we obtain,

$$\frac{dy}{dx} = \frac{3}{2} a^{-1/2} x^{1/2}$$

Is this answer consistent with $\dfrac{dy}{dx} = \dfrac{3t}{2}$ just obtained?

Use of the chain rule in physical situations

Consider an example:

A perfectly spherical balloon is blown up so that its volume increases at a rate of 5 cm^3/sec.

Find a formula in terms of radius r for the rate at which r increases.

We introduce variables r and V to represent the radius and the volume at any time t.

We are told in the first sentence that the balloon is a sphere, hence at any time t, V and r are related by the equation

$$V = \tfrac{4}{3}\pi r^3$$

We are also told the rate of increase of the volume, i.e. $\dfrac{dV}{dt}$, is 5.

Hence, $\dfrac{dV}{dt} = 5$.

We are asked to calculate $\dfrac{dr}{dt}$, the rate of increase of the radius. Mathematically, we may summarise the question as:

Given $V = \tfrac{4}{3}\pi r^3$ **i**, $\dfrac{dV}{dt} = 5$ **ii**

Find $\dfrac{dr}{dt}$.

From **i** we have $\dfrac{dV}{dr} = 4\pi r^2$;

from **ii**, we may write

$$\frac{dV}{dt} = \frac{dV}{dr} \cdot \frac{dr}{dt} = 5 \quad \text{(using the chain rule).}$$

Hence, $\dfrac{dr}{dt} = \dfrac{5}{\left(\dfrac{dV}{dr}\right)} = \dfrac{5}{4\pi r^2}$,

and this is the required formula.

Let us look at a second example:
Water is poured into a cylindrical tank of radius 40 cm and height h cm at a rate of 7 litres per minute. Let x be the depth of water at time t; find the rate of increase of the depth of water when $x = \tfrac{1}{2}h$. Explain why this rate is constant.

The problem can be summarised as follows:

At time t, $V = 1600\pi x$, $\dfrac{dV}{dt} = 7000$, find $\dfrac{dx}{dt}$ when $x = \tfrac{1}{2}h$.

We have $\dfrac{dV}{dx} = 1600\pi$, and, by the chain rule,

$$\Rightarrow \quad \frac{dV}{dt} = \frac{dV}{dx} \cdot \frac{dx}{dt}$$

$$\Rightarrow \quad 7000 = 1600\pi \frac{dx}{dt}$$

$$\Rightarrow \quad \frac{dx}{dt} = \frac{7000}{1600\pi} \approx 1{\cdot}39 \text{ cm/minute}$$

This rate is constant since the volume of water is a linear function of the depth of water. This would not be the case, for example, had the tank been a cone, when the volume would be a function of the cube of the depth.

Exercise B

1 Water is poured at a rate of 7 litres per minute into a conical container in which the height $2r$ cm, is equal to the diameter, as shown in figure 10.2(a). Let x cm be the depth of water at time t min; calculate the rate of increase of the depth when $x = r$.

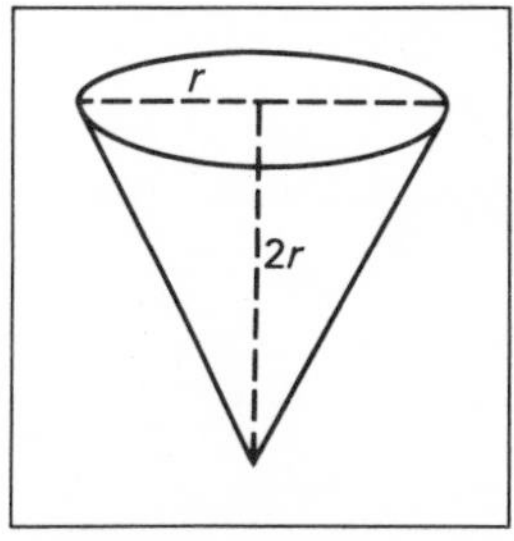

Figure 10.2 (a)

2 A lady making a circular rug of diameter 2 m has worked out that starting from the centre, the area increases at a constant rate k m^2/hr where r m is the radius at any time. Find the rate of increase of the radius when half the area has been completed.

3 In shading a triangular area a boy starts in one corner and maintains an isoceles triangle of shading, as shown in figure 10.2(b). His rate of working is such that the shaded area increases at 2cm^2/min. Find a formula, in terms of x, for the rate of increase of his distance from O (shown as OL in the diagram).

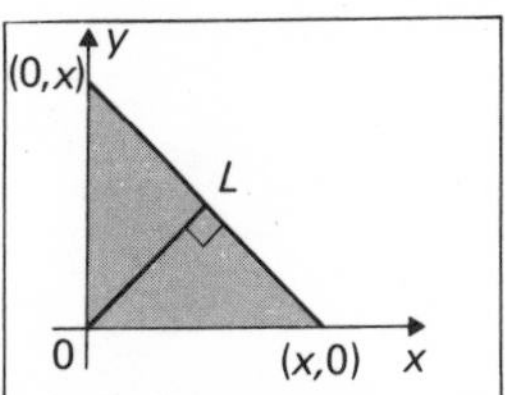

Figure 10.2 (b)

4 A piece of metal in the shape of a cube of side 10 cm is heated uniformly so that its volume increases at a rate proportional to the area of one face. It is known that when one side is 10·5 cm, the rate of increase of the volume is 11·025 cm^3/sec. Find, in terms of x, the rate at which the length x metres of each edge increases.

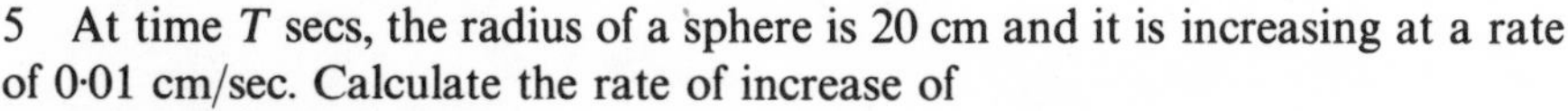

5 At time T secs, the radius of a sphere is 20 cm and it is increasing at a rate of 0·01 cm/sec. Calculate the rate of increase of
a) the surface area, and
b) the volume at time T.
(Leave π in your answers.)

10.2 The inverse trigonometric functions

Definition: If $y = \tan x$, we define x to be **'the inverse tangent of y'**, and we write $x = \tan^{-1} y$ (sometimes written arc tan y).
Similar definitions are applied to $\sin^{-1} y$, $\cos^{-1} y$ and so on, (also written arc sin y, arc cos y).

It is clear that there is an infinite number of values of x which equal $\tan^{-1} y$, for any one value of y. For this reason, we restrict the range of x so that

$$-\frac{\pi}{2} \leqslant x \leqslant \frac{\pi}{2}.$$

The value of x so obtained is called the **principal** value of x. For $x = \sin^{-1} y$, the range for x is restricted so that

$$-\frac{\pi}{2} \leqslant x \leqslant \frac{\pi}{2}$$

and for $x = \cos^{-1} y$, the relevant range is $0 \leqslant x \leqslant \pi$.

For a sound understanding of the relationship between trigonometric functions and their inverses, the graphs of $y = \tan^{-1} x$, $y = \sin^{-1} x$ and $y = \cos^{-1} x$ should be drawn for the principal values of the angle. Can you see why the given ranges for each inverse function were chosen?

The derivatives of the inverse trigonometric functions

We use the chain rule here by seeing that

$$\frac{dy}{dx} \cdot \frac{dx}{dy} = 1 \Rightarrow \frac{dy}{dx} = \frac{1}{\dfrac{dx}{dy}}$$

1 The inverse tangent

$$y = \tan^{-1} x$$

$$x = \tan y$$

$$\frac{dx}{dy} = \sec^2 y$$

$$\frac{dy}{dx} = \frac{1}{\sec^2 y}$$

$$\frac{dy}{dx} = \frac{1}{1 + \tan^2 y} = \frac{1}{1 + x^2}$$

2 The inverse sine

$$y = \sin^{-1} x$$

$$x = \sin y$$

$$\frac{dx}{dy} = \cos y$$

$$\frac{dy}{dx} = \frac{1}{\cos y}$$

$$\frac{dy}{dx} = \frac{1}{\sqrt{(1 - \sin^2 y)}} = \frac{1}{\sqrt{(1 - x^2)}}$$

3 The inverse cosine

$$y = \cos^{-1} x$$

$$x = \cos y$$

$$\frac{dx}{dy} = -\sin y$$

$$\frac{dy}{dx} = -\frac{1}{\sin y}$$

$$\frac{dy}{dx} = -\frac{1}{\sqrt{(1-\cos^2 y)}}$$

$$\frac{dy}{dx} = -\frac{1}{\sqrt{(1-x^2)}}$$

10.3 The product rule for differentiation

Sometimes we shall meet functions which are compounded from two simpler functions by multiplication. For example, $x \to (2x+3)(x^2+2)$ is the product of the two functions $x \to 2x+3$ and $x \to x^2+2$. We consider now how we find the derivatives of such functions.

The product rule

Suppose $y = uv$ where u and v are some functions of x. We recall that one of our approaches to finding derived functions was to take nearby points on a curve (see chapter 3). If P is the point that has coordinates (x, uv) then our nearby point Q is obtained by increasing x by a small amount δx. Since u and v are functions of x, then there will be corresponding increases δu in u and δv in v. So the coordinates of Q will be $\{(x+\delta x), (u+\delta u)(v+\delta v)\}$. Figure 10.3 shows this. The gradient of the chord PQ then is $\frac{QN}{PN}$, that is

$$\frac{\delta y}{\delta x} = \frac{(u+\delta u)(v+\delta v) - uv}{\delta x}$$

$$\frac{\delta y}{\delta x} = \frac{uv + u\,\delta v + v\,\delta u + \delta u\,\delta v - uv}{\delta x}$$

$$\frac{\delta y}{\delta x} = u\frac{\delta v}{\delta x} + v\frac{\delta u}{\delta x} + \frac{\delta u\,\delta v}{\delta x}$$

Now, as $\delta x \to 0$ (so that the chord PQ approaches the tangent at P), $\delta u \to 0$ and $\delta v \to 0$. We have

$$\lim_{\delta x \to 0}\left(\frac{\delta y}{\delta x}\right) = \lim_{\delta x \to 0}\left(u\frac{\delta v}{\delta x}\right) + \lim_{\delta x \to 0}\left(v\frac{\delta u}{\delta x}\right) + \lim_{\delta x \to 0}\left(\frac{\delta u\,\delta v}{\delta x}\right)$$

$$\frac{dy}{dx} = u\frac{dv}{dx} + v\frac{du}{dx} + 0$$

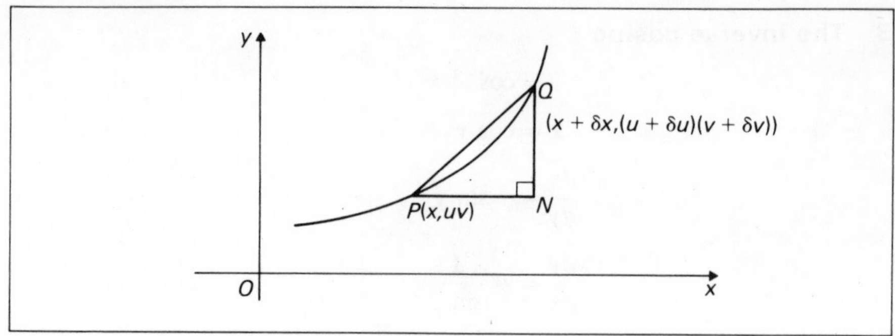

Figure 10.3

The last term, $\dfrac{\delta u\,\delta v}{\delta x}$, tends to zero because if we write

$$\frac{\delta u\,\delta v}{\delta x} \quad \text{as} \quad \frac{\delta u\,\delta v\,\delta x}{\delta x\,\delta x}, \quad \text{this is the same as}$$

$$\frac{\delta u}{\delta x}\cdot\frac{\delta v}{\delta x}\,.\,\delta x$$

In the limit, this tends to

$$\left(\frac{du}{dx}\right)\left(\frac{dv}{dx}\right)(0) \quad \text{i.e.} = 0$$

We have, then, the rule for differentiating a product, namely

$$\frac{d}{dx}(uv) = u\frac{dv}{dx} + v\frac{du}{dx}$$

Example 1

Given $\quad y = (7x^2 + 3)(11x^4 + 5x)$, find $\dfrac{dy}{dx}$.

Here $\quad u = 7x^2 + 3 \quad$ and $\quad v = 11x^4 + 5x$

then $\quad \dfrac{du}{dx} = 14x \quad$ and $\quad \dfrac{dv}{dx} = 44x^3 + 5$

So that $\quad \dfrac{dy}{dx} = (7x^2 + 3)(44x^3 + 5) + 14x(11x^4 + 5x)$

$$= 462x^5 + 132x^3 + 105x^2 + 15$$

In this case we can check our result by multiplying out the original function, i.e. $y = 77x^6 + 33x^4 + 35x^3 + 15x$, and differentiating term by term.

Example 2

To find the derived function for $f(x) = (x + 1)^4 \,.\, x^{3/2}$ (for which it is not so simple to multiply out.)

We write $u = (x + 1)^4$ and $v = x^{3/2}$.

Then $\dfrac{du}{dx} = 4(x+1)^3 \qquad \dfrac{dv}{dx} = \dfrac{3x^{1/2}}{2}$

and $f'(x) = (x+1)^4 \,.\, \dfrac{3x^{1/2}}{2} + 4(x+1)^3 \,.\, x^{3/2}$

which can be written more neatly by taking out the common factors and simplifying, as follows,

$$f'(x) = \frac{(x+1)^3x^{1/2}}{2}((x+1)3 + 8x)$$

$$= \frac{(x+1)^3x^{1/2}(11x+3)}{2}$$

It is often important that we simplify our expression for the derivative, especially if we are looking for the stationary points when we put $\dfrac{dy}{dx} = 0$.

Example 3

Find the turning points for the function

$$y = (x-1)^2(x-2)^2$$

and determine for each whether it is a maximum, minimum or a point of inflexion.

Let $\quad u = (x-1)^2 \qquad v = (x-2)^2$

then $\dfrac{du}{dx} = 2(x-1) \qquad \dfrac{dv}{dx} = 2(x-2)$

and $\dfrac{dy}{dx} = 2(x-1)^2(x-2) + 2(x-1)(x-2)^2$

$$= 2(x-1)(x-2)\{(x-1)+(x-2)\}$$

$$= 2(x-1)(x-2)(2x-3)$$

so $\dfrac{dy}{dx} = 0 \quad \text{when} \quad x = 1, 2 \text{ or } \tfrac{3}{2}$

For the second derivative, we could multiply out all three brackets, or multiply out one pair and use the product rule,

thus $\dfrac{dy}{dx} = 2(x^2 - 3x + 2)(2x-3)$

Let $\quad u = x^2 - 3x + 2 \qquad v = 2x - 3$

and $\dfrac{du}{dx} = 2x - 3 \qquad \dfrac{dv}{dx} = 2$

so $\dfrac{d^2y}{dx^2} = 2\{2(x^2 - 3x + 2) + (2x-3)(2x-3)\}$

$$\frac{d^2y}{dx^2} = 2(2x^2 - 6x + 4 + 4x^2 - 12x + 9)$$

$$\frac{d^2y}{dx^2} = 12x^2 - 36x + 26$$

When $x = 1$, $\dfrac{d^2y}{dx^2} = 12 - 36 + 26 = 2 > 0$ i.e. minimum at $x = 1$

when $x = \frac{3}{2}$, $\dfrac{d^2y}{dx^2} = 27 - 54 + 26 = -1 < 0$ i.e. maximum at $x = \frac{3}{2}$

when $x = 2$, $\dfrac{d^2y}{dx^2} = 48 - 72 + 26 = 2 > 0$ i.e. minimum at $x = 2$

Hence substituting these x-values in our original equation we have: **minima** at (1, 0), (2, 0) and a **maximum** at $(\frac{3}{2}, \frac{1}{16})$.

Exercise C

Differentiate the following products with respect to x.

1 $(x^2 + 2)(3x + 4)$
2 $(x + 1)^3(x - 2)^2$
3 $x^2 \sin x$
4 $(2x + 1)\cos x$
5 $\sin x \cos x$
6 $\cos x \cos x$
7 $(7x + 3)(x + 5)^3$
8 $(x + 8)^2 \sin x$
9 $(x^2 + 2)(x^2 - 2)$
10 $(x^3 + x^2 + x)(x + 1)^5$

In each part of the product in this exercise, the function was simple. It often occurs that the separate parts are compound functions when we need to use the chain rule within the product rule.

Example 1

Suppose we wish to differentiate $y = (x^2 + 3)^3 \sin x$. We write $u = (x^2 + 3)^3$ and $v = \sin x$.

To find $\dfrac{du}{dx}$, by the chain rule, we write $u = t^3$, $t = x^2 + 3$

$$\Rightarrow \frac{du}{dt} = 3t^2 \qquad \frac{dt}{dx} = 2x$$

$$\Rightarrow \frac{du}{dx} = 6xt^2 = 6x(x^2 + 3)^2$$

We can now use the product rule, and obtain

$$\frac{dy}{dx} = (x^2 + 3)^3 \cos x + 6x(x^2 + 3)^2 \sin x$$

and this expression can be simplified to give

$$\frac{dy}{dx} = (x^2 + 3)^2\{(x^2 + 3)\cos x + 6x \sin x\}$$

Great care is needed in writing out solutions where both the chain rule and the product rule are used together. It is important, too, to state clearly which part of the question is being dealt with at each stage.

Example 2

$$y = (\sin 2x)^2(x^3 + 2x)^4$$

Let $u = (\sin 2x)^2$ and $v = (x^3 + 2x)^4$.

Then $u = t^2, t = \sin\theta, \theta = 2x$ and $v = z^4, z = x^3 + 2x$

$$\Rightarrow \frac{du}{dt} = 2t, \frac{dt}{d\theta} = \cos\theta, \frac{d\theta}{dx} = 2 \text{ and } \frac{dv}{dz} = 4z^3, \frac{dz}{dx} = 3x^2 + 2$$

$$\Rightarrow \frac{du}{dx} = 4t\cos\theta \quad \text{and } \frac{dv}{dx} = 4z^3(3x^2+2)$$

$$\Rightarrow \frac{du}{dx} = 4\sin 2x\cos 2x \quad \text{and } \frac{dv}{dx} = 4(x^3+2x)^3(3x^2+2)$$

Using the product rule, we obtain finally,

$$\frac{dy}{dx} = (\sin 2x)^2 4(x^3+2x)^3(3x^2+2) + 4\sin 2x\cos 2x(x^3+2x)^4$$
$$= 4(x^3+2x)^3\sin 2x\{(3x^2+2)\sin 2x + (x^3+2x)\cos 2x\}$$

Exercise D

Differentiate the following products with respect to x.

1 $(x^2+1)^2(2x+3)$
2 $(2x+5)^3(x^2+1)$
3 $(2x+1)\sin^2 x$
4 $\sin^2 x\cos^2 x$
5 $(x^2+2)^3\sin 2x$
6 $\sin x\sec x$
7 $\cos x\csc x$
8 $(11x+13)^2(13x-11)^2$
9 $(14x^2+1)^3(14x^2-1)^3$
10 $x^2(\sin^3 2x)$

11 Find the derivatives of $\sin^2 x$; $\cos^2 x$; $\sin 2x$; $\cos 2x$ and show that

a) $\frac{d}{dx}(\sin^2 x + \cos^2 x) = 0$ b) $\frac{d}{dx}(\cos^2 x - \sin^2 x) = -2\sin 2x$

12 Writing $\sin\theta = \frac{2t}{1+t^2}$, where $t = \tan\frac{\theta}{2}$,

show that $\frac{d}{d\theta}(\sin\theta) = \frac{1-t^2}{1+t^2}$.

How does this answer relate to the work done in chapter 4?

10.4 The quotient rule for differentiation

Derivatives of functions which are made up from the quotient of two simpler functions, such as $x \to \frac{3x^4+6x}{2x-1}$

may be considered in a way similar to products. We will find a result for the general quotient function

$y = \frac{u}{v}$ where u and v are both functions of x

We have

$$\frac{\delta y}{\delta x} = \frac{\left(\dfrac{u + \delta u}{v + \delta v} - \dfrac{u}{v}\right)}{\delta x}$$

$$\Rightarrow \frac{\delta y}{\delta x} = \frac{uv + v\,\delta u - uv - u\,\delta v}{v(v + \delta v)\,\delta x}$$

$$\Rightarrow \frac{\delta y}{\delta x} = \frac{v\,\delta u - u\,\delta v}{v(v + \delta v)\,\delta x}$$

$$\Rightarrow \frac{\delta y}{\delta x} = \frac{v\dfrac{\delta u}{\delta x} - u\dfrac{\delta v}{\delta x}}{v(v + \delta v)}$$

and taking the limit as $\delta x \to 0$, we obtain

$$\frac{dy}{dx} = \lim_{\delta x \to 0} \left\{ \frac{v\dfrac{\delta u}{\delta x} - u\dfrac{\delta v}{\delta x}}{v(v + \delta v)} \right\}$$

$$\Rightarrow \frac{dy}{dx} = \frac{v\dfrac{du}{dx} - u\dfrac{dv}{dx}}{v^2}$$

An alternative method of establishing this result is to use the product rule for the product (uv^{-1}). We note that the chain rule will also be involved with the derivative of (v^{-1}) with respect to x.

The product rule gives

$$\frac{d}{dx}(uv^{-1}) = u\frac{d}{dx}(v^{-1}) + v^{-1}\frac{du}{dx}$$

$$= -uv^{-2}\frac{dv}{dx} + v^{-1}\frac{du}{dx}$$

$$= v^{-1}\frac{du}{dx} - uv^{-2}\frac{dv}{dx}$$

$$= \frac{v\dfrac{du}{dx} - u\dfrac{dv}{dx}}{v^2} \quad \text{as before.}$$

Example 1

Find $f'(x)$ for $f(x) = \dfrac{x^2 + 1}{2x + 3}$.

Here $\quad u = x^2 + 1 \qquad v = 2x + 3$

and $\quad \dfrac{du}{dx} = 2x \qquad \dfrac{dv}{dx} = 2$

Hence $f'(x) = \dfrac{(2x+3)(2x) - 2(x^2+1)}{(2x+3)^2}$

$$= \frac{4x^2 + 6x - 2x^2 - 2}{(2x+3)^2}$$

$$= \frac{2x^2 + 6x - 2}{(2x+3)^2}$$

Example 2

Write $y = \tan x$ as $\dfrac{\sin x}{\cos x}$ and hence find the derivative of tan x.

$u = \sin x \qquad v = \cos x$

and $\dfrac{du}{dx} = \cos x \qquad \dfrac{dv}{dx} = -\sin x$

Thus $\dfrac{dy}{dx} = \dfrac{\cos x(\cos x) - (\sin x)(-\sin x)}{\cos^2 x}$

$$\frac{dy}{dx} = \frac{\cos^2 x + \sin^2 x}{\cos^2 x}$$

$$\frac{dy}{dx} = \frac{1}{\cos^2 x} \quad (\text{since } \cos^2 x + \sin^2 x = 1)$$

$$\frac{dy}{dx} = \sec^2 x$$

Exercise E

1 Find $\dfrac{dy}{dx}$ for each of the following, simplifying your answers as much as possible.

a) $y = \dfrac{x}{x^2+1}$ d) $y = \dfrac{1}{x} + \dfrac{1}{x^2}$

b) $y = \dfrac{2x+1}{x+1}$ e) $y = \dfrac{x(x+1)^2}{(2x-1)}$

c) $y = \dfrac{(x+1)^3}{x}$ f) $y = \dfrac{\sin x}{\tan x}$

2 Use the fact that $\cot x = \dfrac{\cos x}{\sin x}$ to find the derivative of cot x.

3 a) Use the quotient rule to differentiate $\dfrac{1}{x^2+1}$

b) Write $\dfrac{1}{x^2+1}$ as $(x^2+1)^{-1}$

and use the chain rule to find its derivative. Do your two answers agree?

4 Differentiate each of the following functions, with respect to x. (Notice that in some of them, the chain rule must be used when dealing with the numerator or the denominator.)

a) $\dfrac{2x}{(x^2+1)^2}$ f) $\tan^2 x + \sec^2 x$

b) $\dfrac{\sin^2 x}{\cos x}$ g) $\dfrac{4x - 3x^2}{(x^2-4)^2}$

c) $\dfrac{\sin 2x}{\cos 2x}$ h) $\dfrac{3x+2}{2x-3}$

d) $\dfrac{(x^3+x^2+1)^2}{(x+1)^3}$ i) $\dfrac{\sin x}{x}$

e) $\dfrac{(1+\sin x)^2}{1-\sin x}$ j) $\dfrac{x}{\cos x}$

5 Investigate the turning points on the graph of

$$y = \frac{x}{(x-2)(3-x)}$$

Sketch the graph, marking clearly these turning points and the asymptotes.

As a result of these exercises, you have established the derivatives of all the trigonometric functions. It is helpful if these are remembered, and they are listed here as a summary.

y	$\dfrac{dy}{dx}$
$\sin x$	$\cos x$
$\cos x$	$-\sin x$
$\tan x$	$\sec^2 x$
$\sec x$	$\sec x \tan x$
$\csc x$	$-\csc x \cot x$
$\cot x$	$-\csc^2 x$

10.5 Implicit differentiation

There are occasions when we need to find $\dfrac{dy}{dx}$ for some expression which cannot easily be expressed in the form $y = f(x)$. For example, the equation $xy^2 - yx^2 = 3$ cannot be written in the form $y = f(x)$. We seek an alternative method of finding $\dfrac{dy}{dx}$ as a corollary to the use of the chain rule. We treat the two products on the L.H.S. separately and differentiate each by the product rule. We must differentiate the R.H.S. also, and we must remember that we are differentiating with respect to x. We consider $xy^2 - yx^2 = 3$ as our first example.

Example 1

The derivative of xy^2 is $(1)y^2 + x(2y)\dfrac{dy}{dx}$.

Similarly, the derivative of yx^2 is $x^2\dfrac{dy}{dx} + y(2x)$.

Hence for $xy^2 - yx^2 = 3$, we have the derived equation

$$\left(y^2 + x(2y)\frac{dy}{dx}\right) - \left(x^2\frac{dy}{dx} + y(2x)\right) = 0$$

$$y^2 + 2xy\frac{dy}{dx} - x^2\frac{dy}{dx} - 2xy = 0$$

$$\frac{dy}{dx}(2xy - x^2) = 2xy - y^2$$

i.e. $$\frac{dy}{dx} = \frac{2xy - y^2}{2xy - x^2}$$

or, simplifying,

$$\frac{dy}{dx} = \frac{y(2x - y)}{x(2y - x)}$$

and so the derivative is a function of both x and y. A further example will illustrate the method again.

Example 2

Find the gradient of the curve whose equation is

$xy + y^2x^2 = 2(x + y)$ at the point (1, 2).

The curve is $xy + y^2x^2 = 2x + 2y$.

The derived equation is $x\dfrac{dy}{dx} + y + 2y\dfrac{dy}{dx}x^2 + 2xy^2 = 2 + 2\dfrac{dy}{dx}$

$$\Rightarrow x\frac{dy}{dx} + 2yx^2\frac{dy}{dx} - 2\frac{dy}{dx} = 2 - y - 2xy^2$$

$$\Rightarrow \frac{dy}{dx}(x + 2yx^2 - 2) = 2 - y - 2xy^2$$

$$\Rightarrow \frac{dy}{dx} = \frac{2 - y - 2xy^2}{x + 2yx^2 - 2}$$

At the point (1, 2)

$$\frac{dy}{dx} = \frac{2 - 2 - 2(1)(2^2)}{1 + 2(2)(1^2) - 2}$$

$$= -\tfrac{8}{3}$$

$$= -2\tfrac{2}{3}$$

Example 3

It might be helpful to a fuller understanding if we consider the curve whose equation is $xy = 1$, since this can be differentiated both by implicit methods (as here) and by the more 'usual' method.

Implicit method

$$xy = 1$$

$$\Rightarrow x\frac{dy}{dx} + y(1) = 0$$

$$\Rightarrow \frac{dy}{dx} = -\frac{y}{x} \qquad \textbf{i}$$

Direct method

$$xy = 1, \quad \text{i.e. } y = \frac{1}{x}$$

Hence $\dfrac{dy}{dx} = -\dfrac{1}{x^2}$ **ii**

Why do the two answers **i** and **ii** differ? Can you see that they are effectively the same? Remember that you have the relation $xy = 1$ as the equation of the curve and you should be able to transform equation **i** into equation **ii**.

Exercise F

---

1 Find $\dfrac{dy}{dx}$ at $(3, -4)$ on the circle $x^2 + y^2 = 25$

a) by implicit differentiation
b) by rewriting the equation in the form $y = \sqrt{(25 - x^2)}$
c) by applying the chain rule to the parametric forms

$$x = 5\cos\theta, \qquad y = 5\sin\theta.$$

2 Find $\dfrac{dy}{dx}$ for the function given by $xy = x^2 \sin y$.

3 Find the gradient of the curve $y = \dfrac{x-1}{x+2}$ at $(2, \frac{1}{4})$,

a) by using the quotient rule and
b) by rewriting the equation as $(x + 2)y = x - 1$ and using the method of implicit differentiation.

4 Find $\dfrac{dy}{dx}$ for the ellipse $x^2 + xy + y^2 = 4$.

5 If $x^p y^q = 4$, show that

$$\frac{dy}{dx} = -\frac{py}{qx}$$

Complex Numbers

CHAPTER ELEVEN

11.1 Imaginary numbers

In the earlier chapters of this book we have discussed the use, when solving certain algebraic problems, of the set of real numbers, $\mathbb{R}$; also the sets of rational numbers $\mathbb{Q}$, and irrational numbers. Such sets of numbers may be shown either as points on the real number line or displacements from the zero position on that line. Figure 11.1 shows the position of a few typical members of these sets.

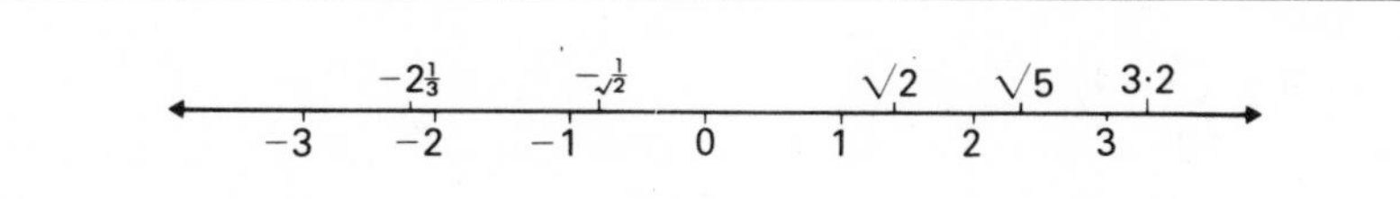

Figure 11.1

We now show the existence of a further set of numbers whose identity may be defined using the following example.

Example
Solve the equation $x^2 + 1 = 0$.

Following normal practice we would write

$$x^2 = -1$$
$$\Rightarrow x = \pm\sqrt{(-1)}$$

and then be forced to stop because of a lack of a ready made value for $\sqrt{(-1)}$. Whereas $\sqrt{1} = \pm 1$, $\sqrt{4} = \pm 2$, $\sqrt{9} = \pm 3\ldots$ we cannot find a real number which fulfils the condition that its square is a negative real number. However, if we define $\sqrt{(-1)}$ to be an **imaginary number,** and label it as the number i, the solution of our equation would be

$$x = \pm i$$

Similarly, given

$$x^2 + 4 = 0 \quad \text{we could write} \quad x = \pm 2i$$
$$x^2 + 5 = 0 \qquad x = \pm i\sqrt{5}$$
$$3x^2 + 11 = 0 \qquad x = \pm i\sqrt{\tfrac{11}{3}}$$

and so on.

We therefore introduce a new number system, consisting of numbers which are real multiples of i. All such numbers are **imaginary**. We may draw an **imaginary number line**, but this is covered fully in section 11.4. The following properties should be noted

$$i^2 = -1, \quad i^3 = i^2 i = -i, \quad i^4 = i^3 i = -i^2 = +1 \text{ and so on.}$$

11.2 Complex numbers

If we combine together, by addition or subtraction, any real number x and any purely imaginary number iy, we form as a result a further kind of number $x + iy$, called a **complex number** (x and y are both real). The set of all complex numbers is denoted by $\mathbb{C}$. Such numbers arise naturally from the solution of quadratic equations in those cases where the discriminant $b^2 - 4ac < 0$.

Example 1
Solve $x^2 - 4x + 5 = 0$

From the formula

$$x = \frac{-b \pm \sqrt{(b^2 - 4ac)}}{2a}$$

$$= \frac{4 \pm \sqrt{(16 - 20)}}{2}$$

$$= \frac{4 \pm \sqrt{(-4)}}{2}$$

$$= \frac{4 \pm 2i}{2}$$

$$\Rightarrow x = 2 + i \quad \text{or} \quad 2 - i$$

Example 2
Solve $2x^2 - 6x + 11 = 0$

$$x = \frac{6 \pm \sqrt{(36 - 88)}}{4}$$

$$= \frac{6 \pm \sqrt{(-52)}}{4}$$

$$= \frac{6 \pm 2i\sqrt{(13)}}{4}$$

$$\Rightarrow x = \tfrac{3}{2} + \tfrac{1}{2}i\sqrt{(13)} \quad \text{or} \quad \tfrac{3}{2} - \tfrac{1}{2}i\sqrt{(13)}$$

Note that the solutions of quadratics of this type consist of one root of the form $p + iq$ and the other of the form $p - iq$. Such complex numbers are said to be conjugate to each other and it will be seen in the subsequent work that both the sum and product of complex conjugate pairs $\in \mathbb{R}$.

Exercise A

1 Solve the following equations giving answers as imaginary numbers.

a) $x^2 + 9 = 0$ c) $49x^2 + 1 = 0$ e) $5x^2 + 17 = 0$

b) $4x^2 + 25 = 0$ d) $x^2 + 11 = 0$ f) $\dfrac{3x^2}{2} + 15 = 0$

2 Solve the following equations giving answers in $\mathbb{C}$ involving, where they occur, surds in their simplest forms.

a) $x^2 + 2x + 2 = 0$ d) $x^2 + x + 1 = 0$
b) $2x^2 - 2x + 5 = 0$ e) $2x^2 - 6x + 7 = 0$
c) $x^2 + 14x + 50 = 0$ f) $5x^2 - 8x + 6 = 0$.

11.3 Simplification of expressions involving complex numbers using the four basic rules

Addition and subtraction

Since every complex number consists of a real part (in special cases zero), and an imaginary part, the real parts may be collected together independently, and imaginary parts collected together independently.

Thus, for example,

$$(2 + 3i) + (5 - 2i) = (2 + 5) + (3i - 2i) = 7 + i$$

$$(2 + 3i) - (5 - 2i) = (2 - 5) + (3i + 2i) = -3 + 5i$$

and $$(2 + 3i) - (3 + 4i) + (5 - 6i) = (2 - 3 + 5) + (3i - 4i - 6i) = 4 - 7i$$

In general, $(a + bi) \pm (c + di) = (a \pm c) + i(b \pm d)$ where $a, b, c, d \in \mathbb{R}$.

Multiplication

To multiply complex numbers together we apply the normal distributive law,

$$\begin{aligned}(2 + 3i)(5 - 2i) &= (2 + 3i)5 + (2 + 3i)(-2i)\\ &= 10 + 15i - 4i - 6i^2\\ &= 10 + 11i + 6 \text{ (since } i^2 = -1)\\ &= 16 + 11i\end{aligned}$$

and $$\begin{aligned}(2 + 3i)(2 - 3i) &= (2 + 3i)2 + (2 + 3i)(-3i)\\ &= 4 + 6i - 6i - 9i^2\\ &= 13\end{aligned}$$

This second example shows that when the complex number $2 + 3i$ is multiplied by its conjugate $2 - 3i$, the result is the real number 13.

Generally $$\begin{aligned}(x + iy)(x - iy) &= x^2 + ixy - ixy - i^2y^2\\ &= x^2 + y^2\end{aligned}$$

and the product of two complex conjugate numbers is simply equal to the sum of the squares of their real and imaginary parts.

Thus $$(-3 + 4i)(-3 - 4i) = (-3)^2 + 4^2 = 25$$
$$(\sqrt{3} + 2i)(\sqrt{3} - 2i) = (\sqrt{3})^2 + 2^2 = 7$$

Repeated products are worked out in stages, for example,

$$\begin{aligned}(2+3i)(-1+2i)(3-i) &= (-2-3i+4i-6)(3-i)\\ &= (-8+i)(3-i)\\ &= -24+3i+8i+1\\ &= -23+11i\end{aligned}$$

and binomial expansions follow the methods for real numbers.

$$\begin{aligned}(1-i\sqrt{2})^3 &= 1+3(-i\sqrt{2})+3(-i\sqrt{2})^2+(-i\sqrt{2})^3\\ &= 1-i\,3\sqrt{2}-6+i\,2\sqrt{2} \quad (\text{since } i^2=-1 \text{ and } i^3=-i)\\ &= -5-i\sqrt{2}\end{aligned}$$

Division

Division of one complex number by a second is, in effect, converted to one of multiplication using the property of multiplication of complex conjugates.

Thus for example,

$$\frac{2+3i}{5-2i} = \frac{(2+2i)(5+2i)}{(5-2i)(5+2i)} \quad \text{(multiplying numerator and denominator by } 5+2i)$$

$$= \frac{(2+3i)(5+2i)}{5^2+2^2}$$

$$= \tfrac{1}{29}\{10+15i+4i-6\}$$

$$= \tfrac{4}{29}+\tfrac{19}{29}\,i$$

This process of creating a real denominator from a complex one is comparable to the process of rationalising the denominator in the division of surd expressions. Division of polynomials in which the coefficients are real, imaginary, or a mixture of both, may be performed by using the remainder theorem and the methods described in chapter 1.

Example 1

Determine the remainder when $f(x) \equiv x^4 - 3$ is divided by $x + 2i$.

$$\begin{aligned}R \equiv f(-2i) &= (-2i)^4 - 3\\ &= 16i^4 - 3 = 16(+1) - 3 = 13\end{aligned}$$

Example 2

Find the quotient and remainder when

$$f(x) \equiv x^3 + (2i-1)x^2 + x + (i-3)$$

is divided by $x - 1 + i$.

Using detached coefficients, synthetic division and $\lambda = 1 - i$

1	$2i-1$	1	$i-3$
	$1(1-i)$	$i(1-i)$	$(i+2)(1-i)$
1	i	$i+2$	R

The quotient is $x^2 + ix + (i + 2)$ and the remainder is

$$\begin{aligned} R &= i - 3 + (i + 2)(1 - i) \\ &= i - 3 + i + 2 + 1 - 2i \\ &= 0 \end{aligned}$$

Two further general results concerning complex numbers are also important.

The equality of complex numbers

Let $\quad x_1 + iy_1 = x_2 + iy_2 \quad (x_1, x_2, y_1, y_2 \in \mathbb{R})$

Then $\quad x_1 - x_2 = -i(y_1 - y_2)$

So, squaring,

$$\begin{aligned} (x_1 - x_2)^2 &= i^2(y_1 - y_2)^2 \\ &= -(y_1 - y_2)^2 \end{aligned}$$

Hence within the set $\mathbb{R}$, such a relationship is true if, and only if, both $(x_1 - x_2)^2$ and $(y_1 - y_2)^2$ are zero.

i.e. $\quad x_1 - x_2 = 0 \Rightarrow x_1 = x_2$

and $\quad y_1 - y_2 = 0 \Rightarrow y_1 = y_2$

Thus the complex numbers $x_1 + iy_1$ and $x_2 + iy_2$ are equal only if the real parts of each are equal and the imaginary parts are equal.

We write $\mathcal{R}e(x + iy)$ when we wish to isolate the real part, x, of the complex number; and $\mathcal{I}m(x + iy)$ when referring to the imaginary part y.

Thus

$$x_1 + iy_1 = x_2 + iy_2 \Leftrightarrow \begin{cases} \mathcal{R}e(x_1 + iy_1) = \mathcal{R}e(x_2 + iy_2) \\ \quad \text{i.e.} \quad x_1 = x_2 \\ \mathcal{I}m(x_1 + iy_1) = \mathcal{I}m(x_2 + iy_2) \\ \quad \text{i.e.} \quad y_1 = y_2 \end{cases}$$

The complex number zero

The result $x + iy = 0 = 0 + 0i$ from the above, is true if and only if,

$$\begin{aligned} \mathcal{R}e(x + iy) = 0 &\Rightarrow x = 0 \\ \mathcal{I}m(x + iy) = 0 &\Rightarrow y = 0 \end{aligned}$$

These ideas lead to a further alternative method for dealing with division.

Example 3
Divide $3 + 2i$ by $3 - 2i$

Let $\quad \dfrac{3 + 2i}{3 - 2i} = x + iy$

Then $\quad \begin{aligned} 3 + 2i &= (x + iy)(3 - 2i) \\ &= 3x + 3iy - 2ix + 2y \\ &= (3x + 2y) + i(3y - 2x) \end{aligned}$

Hence comparing real and imaginary parts of these equal complex numbers

$$\left.\begin{aligned} 3 &= 3x + 2y \\ 2 &= -2x + 3y \end{aligned}\right\} \Rightarrow x = \tfrac{5}{13} \quad y = \tfrac{12}{13}$$

So that $\dfrac{3+2i}{3-2i} = \tfrac{5}{13} + \tfrac{12}{13}i$

Example 4

Evaluate $\dfrac{(4-3i)(2+3i)}{(5-3i)}$

Let $\quad x + iy = \dfrac{(4-3i)(2+3i)}{(5-3i)}$

Then $\quad (x+iy)(5-3i) = (4-3i)(2+3i)$

$\Rightarrow \mathcal{R}e\{(x+iy)(5-3i)\} = \mathcal{R}e\{(4-3i)(2+3i)\}$

$\Rightarrow 5x + 3y = 8 + 9 = 17$

and

$\mathcal{I}m\{(x+iy)(5-3i)\} = \mathcal{I}m\{(4-3i)(2+3i)\}$

$\Rightarrow -3x + 5y = 12 - 6 = 6$

(The associated elements are shown by the curved linking lines.)

Hence $\quad 34y = 51 + 30 \quad y = \tfrac{81}{34}$

and $\quad 34x = 85 - 18 \quad x = \tfrac{67}{34}$

so that $\dfrac{(4-3i)(2+3i)}{(5-3i)} = \tfrac{1}{34}(67 + 81i)$

The process of extracting square roots of complex numbers may also be carried out this way.

Example 5

Calculate the square root of $-5 + 12i$

Let $\quad x + iy = \sqrt{(-5+12i)} \quad (x, y \in \mathbb{R})$

Square $\quad (x+iy)^2 = -5 + 12i$

So comparing real and imaginary parts

$$x^2 - y^2 = -5$$

$$2xy = 12$$

Thus $\quad x^2 - \left(\dfrac{6}{x}\right)^2 = -5$

$$x^4 + 5x^2 - 36 = 0$$

$$(x^2 + 9)(x^2 - 4) = 0$$

$\Rightarrow x = \pm 2$ or $x = \pm 3i$

$\Rightarrow y = \pm 3$ or $y = \pm 2i$

Hence $\sqrt{(-5+12i)} = 2 + 3i$ or $-2 - 3i$

(**Note** these roots are repeated by this method.)

Example 6
Calculate $\sqrt{(-7-24i)}$

Let $x + iy = \sqrt{(-7-24i)} \quad (x, y \in \mathbb{R})$

Then $(x+iy)^2 = -7-24i$

i.e. $x^2 - y^2 + 2ixy = -7-24i$

Hence, comparing real and imaginary parts

$$x^2 - y^2 = -7 \quad 2xy = -24$$

$$\Rightarrow y = \frac{-12}{x} \text{ and } x^2 - \frac{144}{x^2} = -7$$

$$\Rightarrow x^4 + 7x^2 - 144 = 0$$
$$\Rightarrow (x^2+16)(x^2-9) = 0$$
$$\Rightarrow x = \pm 3 \text{ or } x = \pm 4i$$

When $x = 3 \quad y = -4$; when $x = -3 \quad y = 4$

(Values obtained with $x = \pm 4i$ may be ignored since x is known to be real.)

Hence the roots of $\sqrt{(-7-24i)}$ are $3-4i$ and $-3+4i$
or $\pm(3-4i)$

Exercise B

1 Evaluate
a) $(3+2i)+(2-3i)$
b) $(5-2i)-(-3+i)$
c) $(1+i)+(2-i)-(-3-i)$

2 Evaluate the products
a) $(4-3i)(-3+4i)$
b) $(2+i\sqrt{3})(2-i\sqrt{3})$
c) $(1+2i)(3-7i)$
d) $i(2+i)+(1-i)(1+i)$
e) $(1-i)^4$
f) $2(1-i)^2-(1-i)^3$

3 By making the denominators real, reduce to the form $x+iy$,

a) $\dfrac{1}{3+4i}$

b) $\dfrac{5}{1+2i}$

c) $\dfrac{2-i}{3+7i}$

d) $\dfrac{(2+i)(3+i)}{(1+i)}$

e) $\dfrac{3-4i}{(2+i)(1+3i)}$

f) $\dfrac{1+i}{-1+2i}+\dfrac{1-i}{1+2i}$

g) $\dfrac{3+i}{1+2i}-\dfrac{2}{3+i}$

h) $\dfrac{1}{(1+i)^2}+\dfrac{1}{(1-i)^2}$

4 By equating to $x+iy$ and comparing real and imaginary parts, work out

a) $\dfrac{4}{2+3i}$

b) $\dfrac{1}{3-i\sqrt{2}}$

c) $\dfrac{3}{(4-i)(1+3i)}$

d) $\sqrt{(2i)}$

e) $\sqrt{(-5+12i)}$

f) $\sqrt{(p^2-1+2pi)}$

11.4 Geometrical representation; the Argand diagram

We can represent real numbers on a real number line, and imaginary numbers on an imaginary number line. It is convenient to draw these two number lines at right angles, as we did the axes for real graphs. We choose the x-axis to be the real number line and the y-axis to be the imaginary number line.

It is then clear that any complex number $z = x + iy$ (z is commonly used to signify complex numbers) may be represented in the plane of the axes by either

a) the point P whose coordinates are (x, y), or by

b) the vector **OP** written in column form as $\begin{pmatrix} x \\ y \end{pmatrix}$.

This is shown in figure 11.2.

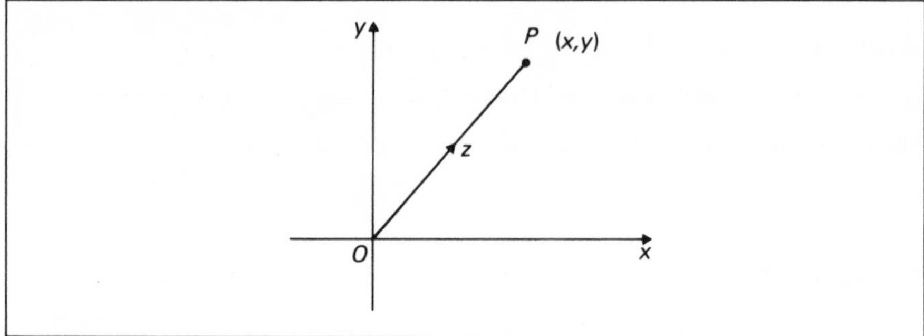

Figure 11.2

This diagrammatic way of helping us to visualise the idea of a complex number was due to a Swiss mathematician, Jean Robert Argand (1768–1822) and is referred to as the **Argand diagram**.

Such diagrams should enable us to understand more easily some of the processes involving complex numbers. It must be clearly understood that the Ox axis is the **real axis of the system** and Oy **the imaginary axis**.

Whenever complex numbers are represented by **free vectors**, it must be understood that any other vector of equal magnitude and direction may be used to represent that complex number. Thus in the diagram of the complex number, figure 11.3, z is represented by **OP**, so **AB**, **CD** will also represent z provided that $OP = AB = CD$ in length and AB, CD, OP are parallel.

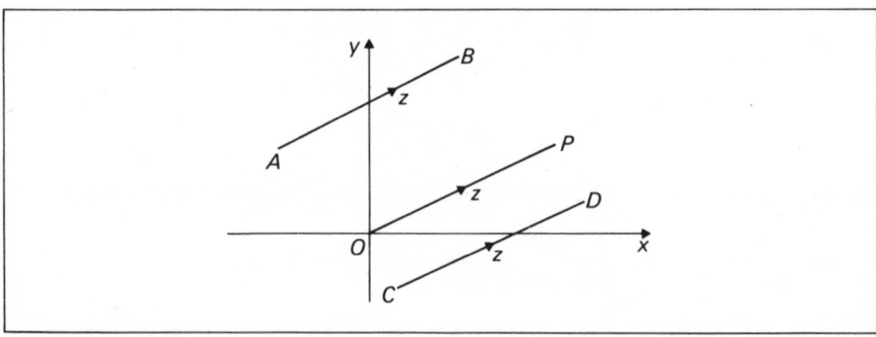

Figure 11.3

This principle may be used immediately to illustrate the addition and subtraction of two complex numbers. For if, in figure 11.4, $\mathbf{z}_1 = \mathbf{OP}$ and $\mathbf{z}_2 = \mathbf{OQ}$ we can immediately introduce for $\mathbf{z}_2$ an equal vector **PR** where **PR** = **OQ**. By the vector law of addition we then have

$$\mathbf{OP} + \mathbf{PR} = \mathbf{OR}$$

or $\mathbf{z}_1 + \mathbf{z}_2 = \mathbf{OR}$

Likewise $\mathbf{OQ} + \mathbf{QP} = \mathbf{OP}$

or

$$\begin{aligned}\mathbf{QP} &= \mathbf{OP} - \mathbf{OQ} \\ &= \mathbf{z}_1 - \mathbf{z}_2\end{aligned}$$

It is clear that if we take such a form of representation for z_1 and z_2 then the diagonals OR, QP of the parallelogram $OPRQ$ represent in turn the sum and the difference of the complex numbers.

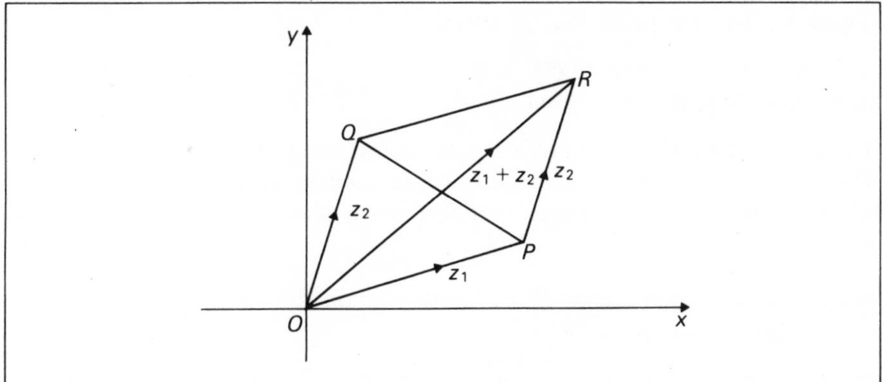

Figure 11.4

Should we choose however to follow the coordinate representation, the addition and subtraction of the two complex numbers $z_1 = x_1 + iy_1$ and $z_2 = x_2 + iy_2$ would be derived from figures 11.5 and 11.6 as follows.

In figure 11.5, clearly P being $(x_1 y_1)$ R is $(x_1 + x_2,\ y_1 + y_2)$ and $z = (x_1 + x_2) + i(y_1 + y_2)$ is represented by the point R.

In figure 11.6, PR' represents $-z_2$ and R' is $(x_1 - x_2,\ y_1 - y_2)$. So $z = (x_1 - x_2) + i(y_1 - y_2)$ is represented by the point R'.

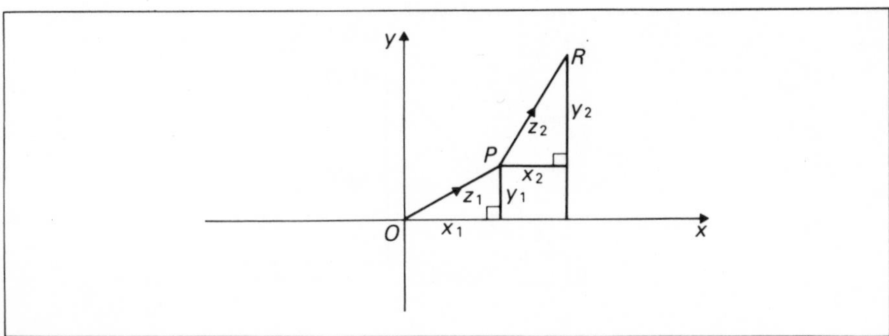

Figure 11.5

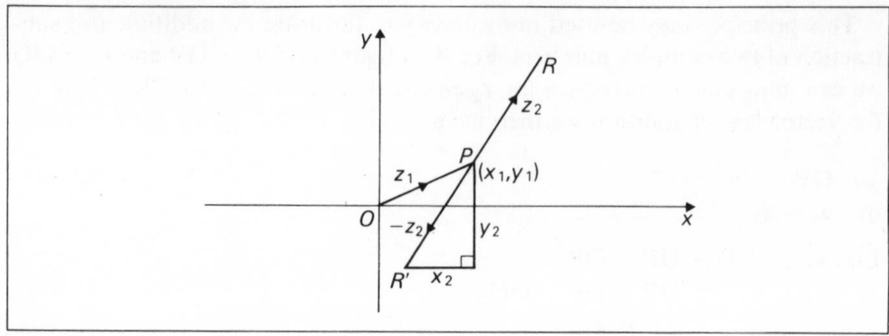

Figure 11.6

11.5 Modulus and argument

If the complex number $z = x + iy$ is represented in the Argand diagram (figure 11.7) by the point $P(x, y)$, letting

$$OP = r \quad \text{and} \quad P\hat{O}X = \theta$$

then $\quad x = r\cos\theta \quad$ and $\quad y = r\sin\theta$

The coordinates (r, θ) of P are the polar coordinates of this point.
The length OP i.e. r is now defined to be the **modulus of the complex number** z which we write as $|z|$. Notice that this is the same as the definition of the modulus of a vector.

Since $\quad |z| = r \quad$ and $\quad x^2 + y^2 = r^2(\cos^2\theta + \sin^2\theta) = r^2$

$$|z| = r = \sqrt{(x^2 + y^2)}$$

The angle θ is called the argument of z and is written Arg z.

Now since $\quad \tan\theta = \dfrac{y}{x}, \quad \theta = \left\{\tan^{-1}\left(\dfrac{y}{x}\right) + n\pi\right\}$

Hence $\qquad \text{Arg } z = \tan^{-1}\left(\dfrac{y}{x}\right) + n\pi$, where $n \in \mathbb{Z}$

and hence there are an infinite number of values for Arg z. But for a given value of z, OP is unique in its direction and will therefore correspond to a single value of θ in the range $-\pi < \theta \leqslant \pi$. We call this value of θ its principal value and write arg z (using a small letter 'a') for the single value of θ concerned.

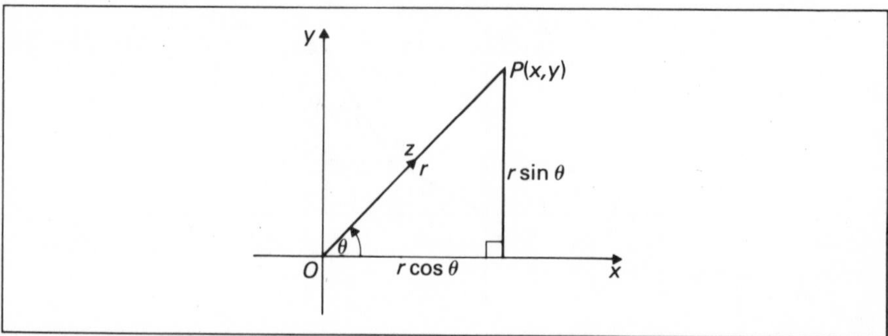

Figure 11.7

The representation of the complex number z using the polar coordinates of P also shows that z can be expressed in the form

$$z = r\cos\theta + ir\sin\theta$$
$$= r(\cos\theta + i\sin\theta)$$

Thus $|z| = |r(\cos\theta + i\sin\theta)|$

$$\Rightarrow |z| = |r|\sqrt{(\cos^2\theta + \sin^2\theta)}$$
$$\Rightarrow |z| = |r|$$

In the following examples we use the Argand diagram to help in identifying the principal value of the argument of the complex number z.

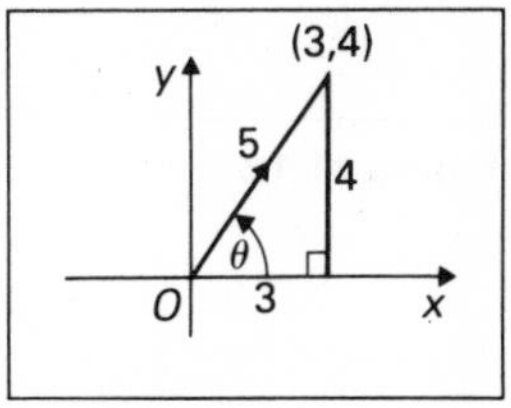

Example 1

$z = 3 + 4i$

We can see from figure 11.8 that $|z| = \sqrt{(3^2 + 4^2)} = 5$

and $\arg z = \tan^{-1}\frac{4}{3}$
$= 0{\cdot}927$

i.e. the value of arg z corresponds to an angle in the range $-\pi < \theta \leqslant \pi$ and so is the principal value of the argument.

Figure 11.8

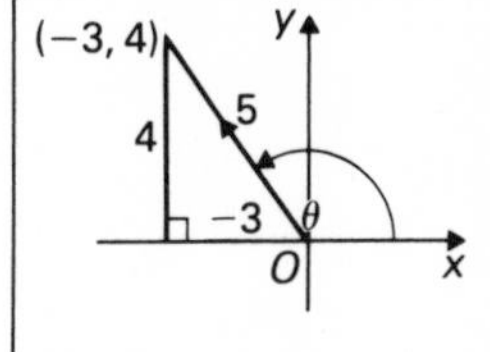

Figure 11.9

Example 2

$z = -3 + 4i$

From figure 11.9, again $|z| = 5$

and $\arg z = \tan^{-1}\left(\frac{-4}{3}\right)$
$= 2{\cdot}214$

Again the value of arg z corresponds to an angle in the range $-\pi < \theta \leqslant \pi$ and so is the principal value.

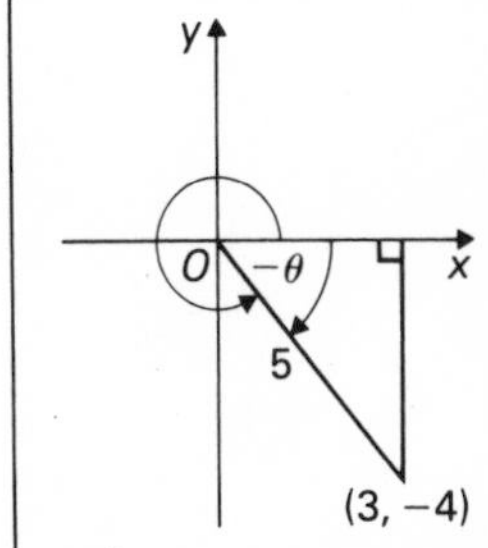

Figure 11.10

Example 3

$z = 3 - 4i$

Once more, from figure 11.10, $|z| = 5$

and $\operatorname{Arg} z = \tan^{-1}\left(\frac{-3}{4}\right)$
$= 5{\cdot}356$

But in this case θ does not lie in the range $-\pi < \theta \leqslant \pi$, i.e. Arg $z = 5{\cdot}356$ is not the principal value. We overcome this by using $-\theta$ as in the diagram where $\theta = 0{\cdot}927$. Hence arg $z = -0{\cdot}927$.

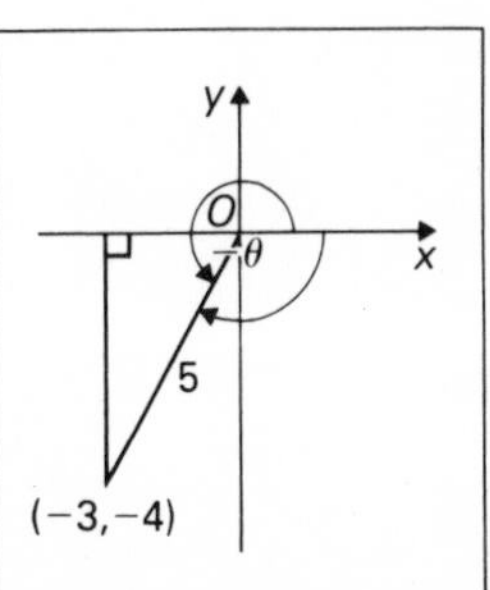

Figure 11.11

Example 4

$z = -3 - 4i$

Finally, figure 11.11 shows that again, $|z| = 5$ and we have the same difficulties as before.

$$\operatorname{Arg} z = \tan^{-1}\left(\frac{-4}{-3}\right) = 4{\cdot}069$$

but $\arg z = -2{\cdot}215$

Note that, in figure 11.12, for $z = \sqrt{3} + i$, $|z| = = \sqrt{(3 + 1)} = 2$

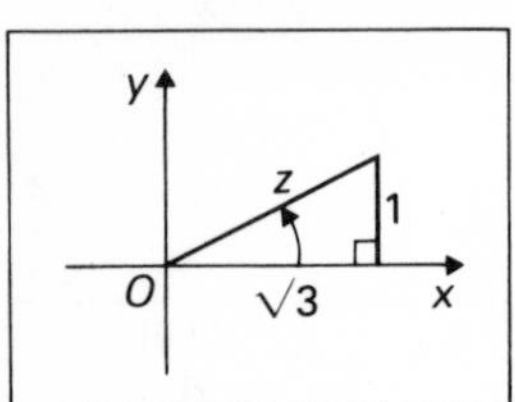

Figure 11.12

$$\arg z = \frac{\pi}{6} = 0{\cdot}524$$

and we can write

$$z = 2\left(\cos\frac{\pi}{6} + i\sin\frac{\pi}{6}\right)$$

Or, when $z = 5\sqrt{2}\left(\cos\frac{\pi}{4} - i\sin\frac{\pi}{4}\right)$, since $\cos\frac{\pi}{4} = \cos\left(\frac{-\pi}{4}\right)$

and $-\sin\frac{\pi}{4} = \sin\left(\frac{-\pi}{4}\right)$, we can write

$$z = 5\sqrt{2}\left\{\cos\left(\frac{-\pi}{4}\right) + i\sin\left(\frac{-\pi}{4}\right)\right\}$$

$$\Rightarrow |z| = 5\sqrt{2} \text{ and } \arg z = \frac{-\pi}{4}$$

and
$$z = 5\sqrt{2}\left(\frac{1}{\sqrt{2}} - \frac{i}{\sqrt{2}}\right)$$
$$= 5 - 5i$$

The idea of writing a complex number in the form

$$z = r\cos\theta + i\sin\theta$$

is associated with Abraham de Moivre (1667–1754) and is sometimes referred to as the de Moivre form of a complex number. We now use this form to achieve a diagrammatic representation of the product $z_1 z_2$ of two complex numbers, and also of their quotient $\frac{z_1}{z_2}$.

Multiplication

Let the de Moivre forms be
$$z_1 = r_1(\cos\theta_1 + i\sin\theta_1)$$
$$z_2 = r_2(\cos\theta_2 + i\sin\theta_2)$$

Then

$$\begin{aligned} z_1 z_2 &= r_1 r_2(\cos\theta_1 + i\sin\theta_1)(\cos\theta_2 + i\sin\theta_2) \\ &= r_1 r_2\{\cos\theta_2\cos\theta_2 - \sin\theta_1\sin\theta_2 + i(\sin\theta_1\cos\theta_2 + \cos\theta_1\sin\theta_2)\} \\ &= r_1 r_2\{\cos(\theta_1 + \theta_2) + i\sin(\theta_1 + \theta_2)\} \end{aligned}$$

Hence $|z_1 z_2| = r_1 r_2$

and $\arg z_1 z_2 = \theta_1 + \theta_2 = \arg z_1 + \arg z_2$

Thus the modulus of the product of z_1 and z_2 is the product of their moduli and the argument of the product is the sum of their arguments.

To see how this result is represented on the Argand diagram let $\mathbf{OP}_1 = z_1$ and $\mathbf{OP}_2 = z_2$ in figure 11.13

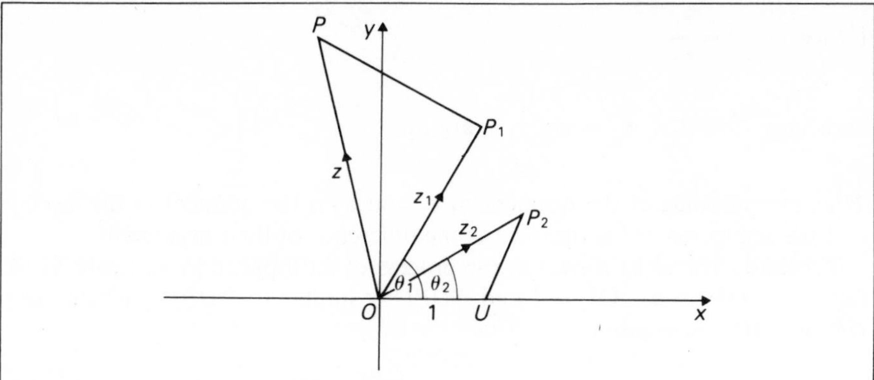

Figure 11.13

Also let **OU** = 1 i.e. U is the point (1, 0) on the real axis.

Construct $\triangle OP_1P$ similar to $\triangle OUP$ with OP_1 and OU a corresponding pair of sides.

Then $P\hat{O}P_1 = P_2\hat{O}U = \theta_2$ by construction,
and therefore $P\hat{O}U = \theta_1 + \theta_2$

If **OP** represents z, arg $z = \theta_1 + \theta_2$

Moreover $\dfrac{OP}{OP_2} = \dfrac{OP_1}{OU}$

i.e. $\dfrac{OP}{r_2} = \dfrac{r_1}{1}$

$\Rightarrow OP = r_1 r_2$

So that $|z| = r_1 r_2$

Thus **OP** $= z$ has $|z| = r_1 r_2$

and arg z = arg z_1 + arg z_2

i.e. z represents $z_1 z_2$

Division

Using the same de Moivre forms

$$\frac{z_1}{z_2} = \frac{r_1(\cos\theta_1 + i\sin\theta_1)}{r_2(\cos\theta_2 + i\sin\theta_2)}$$

$$= \frac{r_1}{r_2}\frac{(\cos\theta_1 + i\sin\theta_1)(\cos\theta_2 - i\sin\theta_2)}{(\cos\theta_2 + i\sin\theta_2)(\cos\theta_2 - i\sin\theta_2)}$$

adopting the standard method to 'realise' the denominator of the quotient.

Then

$$\frac{z_1}{z_2} = \frac{r_1}{r_2}\left\{\frac{\cos\theta_1\cos\theta_2 + \sin\theta_1\sin\theta_2 + i(\sin\theta_1\cos\theta_2 - \cos\theta_1\sin\theta_2)}{\cos^2\theta_2 + \sin^2\theta_2}\right\}$$

$$= \frac{r_1}{r_2}\{\cos(\theta_1 - \theta_2) + i\sin(\theta_1 - \theta_2)\}$$

Hence $\left|\dfrac{z_1}{z_2}\right| = \dfrac{r_1}{r_2}$

and $\arg \dfrac{z_1}{z_2} = \theta_1 - \theta_2 = \arg z_1 - \arg z_2$

Thus the modulus of the quotient of z_1 and z_2 is the quotient of the moduli and the argument of the quotient is the difference of their arguments.

Repeating the ideas shown in the process of multiplication, in figure 11.14, $\mathbf{OP}_1 = z_1$, $\mathbf{OP}_2 = z_2$, $\mathbf{OU} = 1$ and $\triangle OPU$ is similar to $\triangle OP_1P_2$ where sides OP_1 and OP correspond.

Then $P_1\hat{O}P_2 = P\hat{O}U = \theta_1 - \theta_2$ and $\dfrac{OP_1}{OP} = \dfrac{OP_2}{OU}$

i.e. $OP = \dfrac{r_1}{r_2}$

Thus if $\mathbf{OP} = z$, $|z| = OP_1 = \dfrac{r_1}{r_2}$

and $\arg z = \theta_1 - \theta_2$
$= \arg z_1 - \arg z_2$

Hence $\mathbf{OP} = z$ represents $\dfrac{z_1}{z_2}$

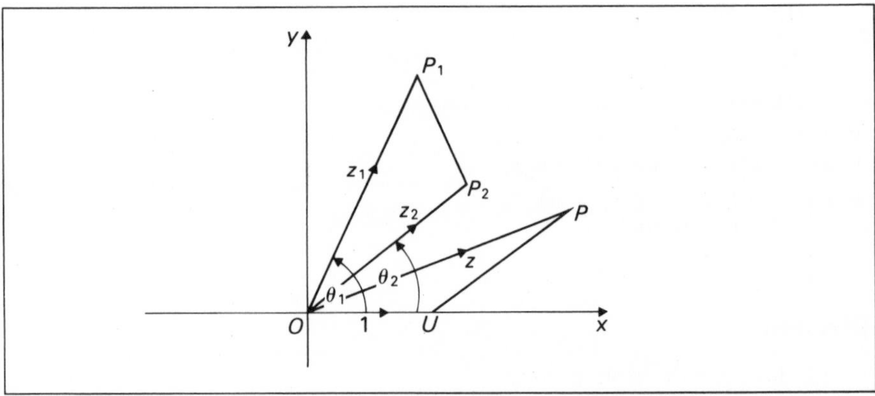

Figure 11.14

Example 5

If $z_1 = 3 - 4i$ and $z_2 = 1 + i$, write down $|z_1 z_2|$ and $\arg z_1 z_2$

From figure 11.15 we can write

$$z_1 = 5(\tfrac{3}{5} - \tfrac{4}{5}i)$$
$$= 5(\cos \theta_1 + i \sin \theta_1)$$

where $\cos \theta_1 = \tfrac{3}{5}$ and $\sin \theta_1 = -\dfrac{4}{5}$

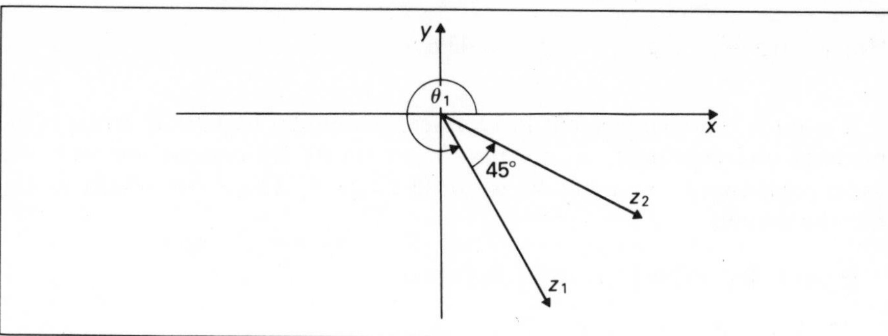

Figure 11.15

$$\text{and} \quad z_2 = \sqrt{2}\left(\frac{1}{\sqrt{2}} + \frac{1}{\sqrt{2}}i\right)$$

$$= \sqrt{2}\left(\cos\frac{\pi}{4} + i\sin\frac{\pi}{4}\right)$$

Hence $|z_1 z_2| = 5\sqrt{2}$ (product of moduli).

Working in degrees $\theta_1 = 306{\cdot}87°$

$\theta_1 + 45° = 351{\cdot}87°$

$\Rightarrow \text{Arg } z_1 z_2 = 351{\cdot}87°$

$\Rightarrow \arg z_1 z_2 =$ principal value of Arg $351{\cdot}87°$

$= -8{\cdot}13°$

Example 6

If $z_1 = 5 + 2i$, $z_2 = 5 - 2i$, determine $\left|\frac{z_1}{z_2}\right|$ and $\arg\frac{z_1}{z_2}$.

From figure 11.16,

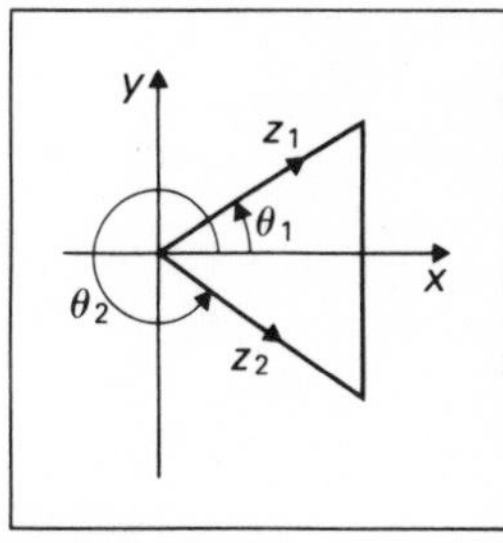

Figure 11.16

$$z_1 = \sqrt{29}\left\{\frac{5}{\sqrt{29}} + \frac{2}{\sqrt{29}}i\right\}$$

$$= \sqrt{29}\{\cos\theta_1 + i\sin\theta_1\}$$

where $\cos\theta_1 = \frac{5}{\sqrt{29}}$, $\sin\theta_1 = \frac{2}{\sqrt{29}}$

$$z_2 = \sqrt{29}\left\{\frac{5}{\sqrt{29}} - \frac{2}{\sqrt{29}}i\right\}$$

$$= \sqrt{29}\{\cos\theta_2 + i\sin\theta_2\}$$

where $\cos\theta_2 = \frac{5}{\sqrt{29}}$, $\sin\theta_2 = -\frac{2}{\sqrt{29}}$

Hence $\left|\frac{z_1}{z_2}\right| = \frac{\sqrt{29}}{\sqrt{29}} = 1$

$\tan\theta_1 = \frac{2}{5}$

$\Rightarrow \theta_1 = 21{\cdot}8° = \arg z_1$

$\tan\theta_2 = -\frac{2}{5}$

$\Rightarrow \theta_2 = 360 - 21{\cdot}8° = 338{\cdot}2°$

so $\arg z_2 = -21{\cdot}8°$

Hence $\arg \dfrac{z_1}{z_2} = \arg z_1 - \arg z_2 = 43{\cdot}6°$

It is often more convenient to represent a complex number in terms of its modulus and argument, i.e. $z = r(\cos \theta + i \sin \theta)$. By comparison with the usual polar coordinate notation, we write $z = [r, \theta]$. The above results would then be written

$$[r_1, \theta_1] . [r_2, \theta_2] = [r_1 r_2, \theta_1 + \theta_2] \quad \text{and}$$

$$\frac{[r_1, \theta_1]}{[r_2, \theta_2]} = \left[\frac{r_1}{r_2}, \theta_1 - \theta_2\right]$$

Exercise C

1 Display the following complex numbers on an Argand diagram. Also determine their moduli and the principal values of their arguments.

a) 2 f) $1 - i$

b) $3i$ g) $-2 + 5i$

c) $-0{\cdot}5i$ h) $1 + i\sqrt{3}$

d) $12 + 5i$ i) $2 - i$

e) $3 + 2i$

2 Express, in the form $z = x + iy$, the complex numbers

$$z_1 = 2\left(\cos \frac{\pi}{3} - i \sin \frac{\pi}{6}\right) \quad \text{and}$$

$$z_2 = 1{\cdot}5\left(\cos \frac{\pi}{6} + i \sin \frac{\pi}{6}\right)$$

Determine also, the moduli and principal values of the arguments of $z_1 z_2$ and $\dfrac{z_1}{z_2}$.

3 Prove, by direct multiplication, that

a) $(\cos \theta + i \sin \theta)^2 = \cos 2\theta + i \sin 2\theta$

b) $\dfrac{1}{(\cos \theta + i \sin \theta)^2} = \cos 2\theta - i \sin 2\theta$

Determine the value of k if

$$(\cos \theta + i \sin \theta)^3 + \frac{1}{(\cos \theta + i \sin \theta)^3} = k \cos 3\theta$$

4 Show that if $r = \sqrt{2}$ and $\theta = -\frac{3}{4}\pi$, then the complex number $z = r(\cos \theta + i \sin \theta)$ may also be represented as

$$z = \frac{2}{-1 + i}.$$

5 Given that $z_1 = 1 + 2i$ and $z_2 = 2 + i$, determine

a) $|z_1 z_2|$ and $\left|\dfrac{z_1}{z_2}\right|$ b) arg $z_1 z_2$ and arg $\dfrac{z_1}{z_2}$

6 Represent on an Argand diagram the complex number $z = 2 + 3i$ and its conjugate $z^* = 2 - 3i$. Construct the complex number $Z = zz^*$. Write down the value of $|Z^*|$ and the value of arg (Z^*).

7 Given that $z = r(\cos\theta + i\sin\theta)$, determine in terms of r and θ,
a) $z + z^{-1}$ b) $z^2 + z^{-2}$
Calculate $|z + (z^*)^{-1}|$ and arg$\{z + (z^*)^{-1}\}$.

11.6 Solution of the equation $z^3 = 1$

We know that $z^3 - 1 \equiv (z - 1)(z^2 + z + 1)$. Therefore the solutions of $z^3 = 1$ are $z = 1$ and the solutions of $z^2 + z + 1 = 0$

i.e. $z = \dfrac{-1 \pm \sqrt{(1-4)}}{2}$ or $z = -\dfrac{1}{2} \pm i\dfrac{\sqrt{3}}{2}$

Denoting these roots by z_1, z_2 and z_3

$$z_1 = 1$$

$$z_2 = -\frac{1}{2} + i\frac{\sqrt{3}}{2} = \cos\frac{2}{3}\pi + i\sin\frac{2}{3}\pi$$

$$z_3 = -\frac{1}{2} - i\frac{\sqrt{3}}{2} = \cos\frac{4}{3}\pi + i\sin\frac{4}{3}\pi$$

which when represented on the Argand diagram, figure 11.17, are seen to be positioned symmetrically at points on the circumference of a unit circle.

Now $z_2^2 = (\cos\frac{2}{3}\pi + i\sin\frac{2}{3}\pi)(\cos\frac{2}{3}\pi + i\sin\frac{2}{3}\pi)$
$= \cos(\frac{2}{3}\pi + \frac{2}{3}\pi) + i\sin(\frac{2}{3}\pi + \frac{2}{3}\pi)$
$= \cos\frac{4}{3}\pi + i\sin\frac{4}{3}\pi$
$= z_3$

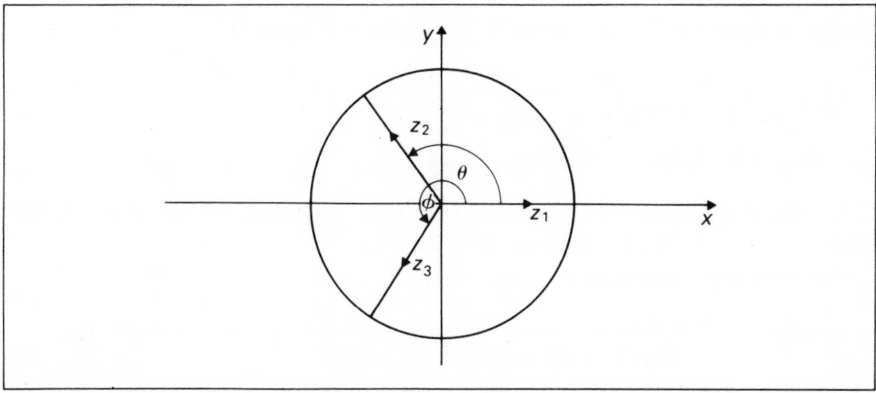

Figure 11.17

$$\begin{aligned} \text{And} \quad z_3{}^2 &= \cos \tfrac{8}{3}\pi + i \sin \tfrac{8}{3}\pi \quad \text{similarly} \\ &= \cos (2\pi + \tfrac{2}{3}\pi) + i \sin (2\pi + \tfrac{2}{3}\pi) \\ &= \cos \tfrac{2}{3}\pi + i \sin \tfrac{2}{3}\pi \\ &= z_2 \end{aligned}$$

Therefore the square of either of the complex roots of the equation is equal to the other. In this particular study it is commonly accepted that ω be written for one of these roots. Then ω^2 is the other, the third root being the real root.

Thus ω, ω^2 satisfy $z^3 - 1 = 0$

and also satisfy $z^2 + z + 1 = 0$

So $\omega^2 + \omega + 1 = 0$ and $\omega^4 + \omega^2 + 1 = 0$

and $\omega^3 = \omega^6 = 1$

Notice that the three roots 1, ω, ω^2 when marked on an Argand diagram lie on the circle $|z| = 1$ and are symmetrically placed.

Example 1

If ω is one of the complex roots of unity, evaluate

a) $(1 + \omega^2)^3$ b) $\dfrac{\omega}{(1 - \omega)^2}$

$$\text{a)}\ (1 + \omega^2)^3 = 1 + 3\omega^2 + 3\omega^4 + \omega^6 = 1 + 3(\omega^4 + \omega^2) + \omega^6$$
$$= 1 + 3(-1) + 1 = -1 \quad (\text{using } \omega^4 + \omega^2 + 1 = 0,\ \omega^6 = 1)$$

$$\text{or} \quad (1 + \omega^2)^3 = (1 + \omega + \omega^2 - \omega)^3 = (0 - \omega)^3 \quad (\text{using } 1 + \omega + \omega^2 = 0)$$
$$= -\omega^3 = -1 \quad (\text{using } \omega^3 = 1)$$

$$\text{b)}\ \frac{\omega}{(1 - \omega)^2} = \frac{\omega}{1 - 2\omega + \omega^2} = \frac{\omega}{1 + \omega + \omega^2 - 3\omega}$$

$$= \frac{\omega}{0 - 3\omega} \quad \text{since } 1 + \omega + \omega^2 = 0$$

$$= -\tfrac{1}{3}$$

Example 2

Express $\dfrac{3}{z^3 - 1}$ in partial fractions with linear denominators.

Since the roots of $z^3 - 1 = 0$ are 1, ω, ω^2 we can write

$$\frac{3}{z^3 - 1} \equiv \frac{A}{z - 1} + \frac{B}{z - \omega} + \frac{C}{z - \omega^2}$$

i.e. $3 \equiv A(z - \omega)(z - \omega^2) + B(z - 1)(z - \omega^2) + C(z - 1)(z - \omega)$

The following method of evaluating A, B, C is chosen to illustrate further manipulative exercises involving $\omega^2 + \omega + 1 = 0$.

Compare coefficients of z^2, z, 1

$$\begin{aligned} z^2\!: \quad & 0 = A + B + C \\ z\!: \quad & 0 = -(\omega + \omega^2)A - (1 + \omega^2)B - (1 - \omega)C \\ \text{i.e.} \quad & 0 = A + \omega B + \omega^2 C \quad (\text{using } 1 + \omega + \omega^2 = 0) \\ 1\!: \quad & 3 = \omega^3 A + \omega^2 B + \omega C \end{aligned}$$

Adding these three equations

$$3 = (2 + \omega^3)A + (1 + \omega + \omega^2)B + (1 + \omega + \omega^2)C$$
$$= (2 + 1)A + 0(B) + 0(C) \quad \Rightarrow \quad A = 1$$

So $\quad B + C = -1, \quad \omega B + \omega^2 C = -1$

i.e. $(1 - \omega)B = (\omega^2 - 1)C \quad \Rightarrow \quad B = -(\omega + 1)C$ since $\omega \neq 1$ being one of the complex roots of unity.

So $\qquad B = \omega^2 C$

$\Rightarrow \qquad (\omega^2 + 1)C = -1$ i.e. $-\omega C = -1 \Rightarrow C = \dfrac{1}{\omega}$

And $\qquad B = \omega^2\left(\dfrac{1}{\omega}\right) = \omega$

Hence $\qquad \dfrac{3}{z^3 - 1} \equiv \dfrac{1}{z - 1} + \dfrac{\omega}{z - \omega} + \dfrac{1}{\omega(z - \omega^2)}$

Example 3

Prove that $E = (p + q + r)(p + \omega q + \omega^2 r)(p + \omega^2 q + \omega r)$
$= p^3 + q^3 + r^3 - 3pqr$

$$\begin{aligned} E &= (p + q + r)\{(p^2 + \omega^3 q^2 + \omega^3 r^2 + pq(\omega + \omega^2) + qr(\omega^4 + \omega^2) + pr)(\omega + \omega^2)\} \\ &= (p + q + r)(p^2 + q^2 + r^2 - pq - qr - rp) \\ &= p^3 + q^3 + r^3 + (p^2 q - p^2 q) + (p^2 r - p^2 r) + (q^2 r - q^2 r) \\ &\quad + (q^2 p - q^2 p) + (r^2 p - r^2 p) + (r^2 q - r^2 q) - pqr - pqr - pqr \\ &= p^3 + q^3 + r^3 - 3pqr \end{aligned}$$

Exercise D

1 If ω is one of the complex cube roots of unity, evaluate
 a) $(1 + \omega)^3$ b) $(1 - \omega)^2(1 + \omega)$ c) $(1 - \omega + \omega^2)(1 + \omega + \omega^2)$

2 Prove that

 a) $\omega^2(1 + \omega)^2 + \omega^2(1 - \omega)^2 = -2$

 b) $\dfrac{\omega^2(\omega - 1)^2}{(\omega + 1)^3} = 3$

 c) $(\omega p + \omega^2 q)(\omega^2 p + \omega q) = p^2 - pq + q^2$

3 If $\dfrac{x^2}{x^3 + 1} \equiv \dfrac{A}{x + 1} + \dfrac{B}{x + \omega} + \dfrac{C}{x + \omega^2}$,
determine the values of A, B and C.

4 Simplify $(p + \omega q + \omega^2 r)(p + \omega^2 q + \omega r)$

11.7 Solution of other cubic and quartic equations

Example 1

The equation $z^3 = -1$ is $(z + 1)(z^2 - z + 1) = 0$

with roots $z_1 = -1$ and complex roots $z_2, z_3 = \dfrac{1 \pm \sqrt{(1 - 4)}}{2}$

$$= \frac{1}{2} \pm i\frac{\sqrt{3}}{2}$$

So $z_2 = \frac{1}{2} + i\frac{\sqrt{3}}{2} = \cos\frac{\pi}{3} + i\sin\frac{\pi}{3}$

$$z_3 = \frac{1}{2} - i\frac{\sqrt{3}}{2} = \cos\frac{5\pi}{3} + i\sin\frac{5\pi}{3}$$

$$\text{or} \quad \cos\left(\frac{-\pi}{3}\right) + i\sin\left(\frac{-\pi}{3}\right)$$

Example 2

To solve $z^3 = -i$ we write $z^3 = i^3$ or $\left(\frac{z}{i}\right)^3 = 1$

The roots of the equation are therefore $z_1 = i$ and z_2, z_3 where

$$z_2 = i\left(\cos\frac{2}{3}\pi + i\sin\frac{2}{3}\pi\right) \text{ and}$$

$$z_3 = i\left(\cos\frac{4}{3}\pi + i\sin\frac{4}{3}\pi\right)$$

(from i multiplied by the roots of $z^2 + z + 1 = 0$)

Hence $z_2 = -\sin\frac{2}{3}\pi + i\cos\frac{2}{3}\pi = \sin\left(-\frac{2}{3}\pi\right) + i\cos\left(-\frac{2}{3}\pi\right)$

$$= \cos\left(\frac{\pi}{2} - \frac{2}{3}\pi\right) + i\sin\left(\frac{\pi}{2} - \frac{2}{3}\pi\right)$$

$$= \cos\left(\frac{-\pi}{6}\right) + i\sin\left(\frac{-\pi}{6}\right)$$

and $z_3 = -\sin\frac{4}{3}\pi + i\cos\frac{4}{3}\pi$

$$= \cos\left(\frac{\pi}{2} - \frac{4}{3}\pi\right) + i\sin\left(\frac{\pi}{2} - \frac{4}{3}\pi\right)$$

$$= \cos\left(\frac{-5\pi}{6}\right) + i\sin\left(\frac{-5\pi}{6}\right)$$

Example 3

The quartic equation $z^4 = 1$ is $(z^2 - 1)(z^2 + 1) = 0$ with two real roots ± 1 and two purely imaginary roots $\pm i$.

i.e. $z_1 = 1$, $z_2 = -1$, $z_3 = i = \cos\frac{\pi}{2} + i\sin\frac{\pi}{2}$,

$$z_4 = -i = \cos\frac{3\pi}{2} + i\sin\frac{3\pi}{2}$$

The next example illustrates a method of solving a cubic equation of the form $z^3 + pz + q = 0$ where p, q are real. Should it be possible to find a root of the equation by, for example, using the remainder theorem, the other two roots will follow from the solution of the equation obtained from the quadratic quotient. It may not be possible however to find this root so simply nor will the root necessarily be of a simple rational form.

We first consider the simplification of the expression

$$E \equiv (x - \lambda - \mu)(x - \omega\lambda - \omega^2\mu)(x - \omega^2\lambda - \omega\mu)$$

where ω is one of the complex roots of unity and λ, μ are constants. Multiplying the second and third brackets together,

$$\begin{aligned}E &\equiv (x-\lambda-\mu)\{x^2 - x[\lambda(\omega+\omega^2)+\mu(\omega+\omega^2)] \\ &\quad +\omega^3\lambda^2 + (\omega^4+\omega^2)\lambda\mu + \omega^3\mu^2\} \\ &\equiv (x-\lambda-\mu)\{x^2 - x(\omega+\omega^2)(\lambda+\mu) + \omega^3(\lambda^2+\mu^2) \\ &\quad +(\omega^4+\omega^2)\lambda\mu\} \\ &\equiv (x-\lambda-\mu)\{x^2 + x(\lambda+\mu) + (\lambda^2 - \lambda\mu + \mu^2)\}\end{aligned}$$

since $\omega^3 = 1$, $\omega^2 + \omega = -1$ and $\omega^4 + \omega^2 = -1$.

So
$$\begin{aligned}E &\equiv x^3 + x^2\{(\lambda+\mu)-(\lambda+\mu)\} + x\{\lambda^2 - \lambda\mu + \mu^2 - (\lambda+\mu)^2\} \\ &\quad -(\lambda+\mu)(\lambda^2 - \lambda\mu + \mu^2) \\ &\equiv x^3 - 3\lambda\mu x - (\lambda^3 + \mu^3)\end{aligned}$$

Thus the roots of the equation $E = 0$ i.e. $x^3 - 3\lambda\mu x - (\lambda^3 + \mu^3) = 0$ are $\lambda + \mu$, $\omega\lambda + \omega^2\mu$ and $\omega^2\lambda + \omega\mu$.

We now use this information in the following numerical example.

Example 4

Determine the roots of $x^3 - 2x + 3 = 0$.

By comparison with $x^3 - 3\lambda\mu x - (\lambda^3 + \mu^3) = 0$ we choose λ, μ so that $3\lambda\mu = 2$ and $\lambda^3 + \mu^3 = -3$.

Then eliminating μ and λ in turn

$$\lambda \text{ satisfies } \lambda^3 + \left(\frac{2}{3\lambda}\right)^3 = -3 \quad \text{i.e. } \lambda^6 + 3\lambda^3 + \left(\frac{2}{3}\right)^3 = 0$$

$$\text{and} \quad \mu \text{ satisfies } \mu^3 + \left(\frac{2}{3\mu}\right)^3 = -3 \quad \text{i.e. } \mu^6 + 3\mu^3 + \left(\frac{2}{3}\right)^3 = 0$$

Hence λ^3, μ^3 are the roots of the quadratic $t^2 + 3t + (\frac{2}{3})^3 = 0$

$$\text{i.e.} \quad \lambda^3, \mu^3 = \frac{-3 \pm \sqrt{\{9 - 4(\frac{2}{3})^3\}}}{2}$$

which when evaluated gives

$$\begin{aligned}&\lambda^3 = -0{\cdot}102 \text{ or } \lambda = \sqrt[3]{(-0{\cdot}102)} = -0{\cdot}468 \\ \text{and} \quad &\mu^3 = -2{\cdot}898 \text{ or } \mu = \sqrt[3]{(-2{\cdot}898)} = -1{\cdot}426\end{aligned}$$

Therefore the three roots of the given equation are

$$\begin{aligned}&x_1 = \lambda + \mu = -0{\cdot}468 - 1{\cdot}426 = -1{\cdot}894 \\ &x_2 = \omega\lambda + \omega^2\mu = -0{\cdot}468\omega - 1{\cdot}426\omega^2 \\ \text{and} \quad &x_3 = \omega^2\lambda + \omega\mu = -0{\cdot}468\omega^2 - 1{\cdot}426\omega\end{aligned}$$

It was Argand in his publication ESSAI (1806) who justified that every algebraic polynomial equation has a root. Generally, cubic equations with real coefficients have either 3 real roots (corresponding to figure 11.18(a)) or 1 real and 2 complex roots (corresponding to figure 11.18(b)).

Quartic equations likewise may be expected to have four real, two real and two complex or four complex roots, corresponding to figures 11.19(a), (b), (c), respectively.

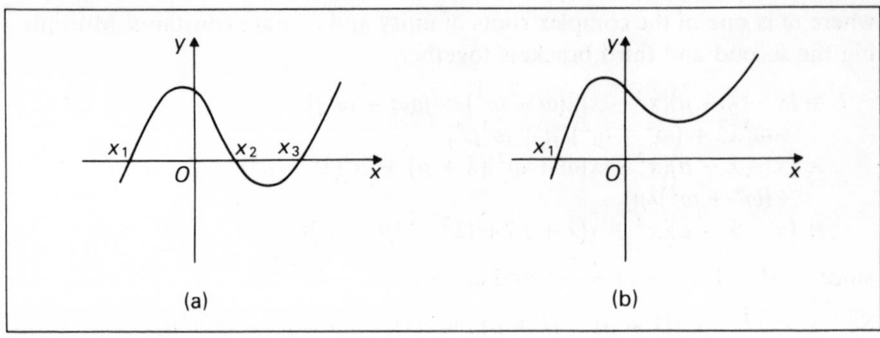

Figure 11.18

If the quartic can be factorised by the remainder theorem, or some other process, it will take up the biquadratic form

$$(x^2 + px + q)(x^2 + Px + Q) = 0$$

Should the quartic have four real roots x_1, x_2, x_3, x_4 as in figure 11.19(a), both of the quadratic factors, when equated to zero, will yield real roots. However, if only two roots are real (figure 11.19(b)) one of these quadratics will yield real roots and the other complex conjugate roots, whereas if all four roots are complex (figure 11.19(c)) both quadratics will then yield complex conjugate roots.

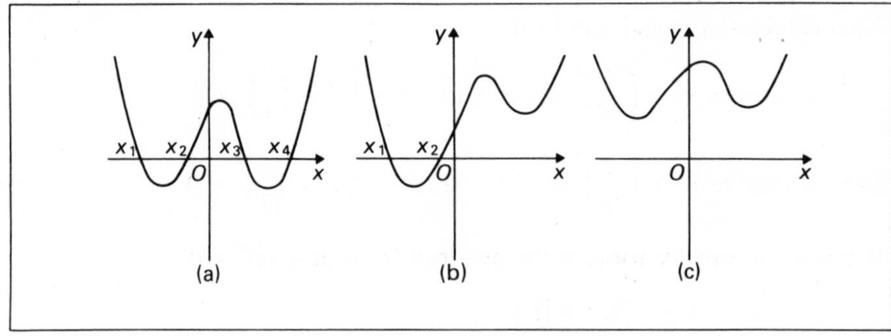

Figure 11.19

Finally in this chapter, we illustrate a general, but rather complicated, method of writing a quartic in biquadratic form and utilise this process in a numerical case to evaluate the four roots of a quartic with given numerical coefficients.

Let the quartic be $x^4 + ax^3 + bx^2 + cx + d = 0$, where a, b, c, d are all real.

We seek to express the quartic as the difference of two squares by introducing a number λ which we first use to turn the terms $x^4 + ax^3$ into a perfect square, i.e. we write $x^4 + ax^3$ as

$$\left(x^2 + \frac{a}{2}x + \lambda\right)^2 - \frac{a^2x^2}{4} - 2\lambda x^2 - a\lambda x - \lambda^2$$

(Check that this is so by multiplying out the bracket.)

Then the given quartic equation may be written as

$$\left(x^2 + \frac{a}{2}x + \lambda\right)^2 - \frac{a^2x^2}{4} - 2\lambda x^2 - a\lambda x - \lambda^2 + bx^2 + cx + d = 0$$

or

$$\left(x^2+\frac{a}{2}x+\lambda\right)^2 - x^2\left(\frac{a^2}{4}+2\lambda-b\right) - x(a\lambda-c) - (\lambda^2-d) = 0$$

i.e.

$$\left(x^2+\frac{a}{2}x+\lambda\right)^2 - \left\{x^2\left(\frac{a^2}{4}+2\lambda-b\right) + x(a\lambda-c) + (\lambda^2-d)\right\} = 0$$

Now if λ is chosen so that

$$x^2\left(\frac{a^2}{4}+2\lambda-b\right) + x(a\lambda-c) + (\lambda^2-d)$$

is a perfect square of the form $(Ax+B)^2$, the given equation will then reduce to the two quadratic factors,

$$\left(x^2+\frac{a}{2}x+\lambda+Ax+B\right)\left(x^2+\frac{a}{2}x+\lambda-Ax-B\right) = 0$$

and the four roots will be determinable.

For the perfect square to be developed λ must satisfy the equation

$$(a\lambda-c)^2 = 4\left(\frac{a^2}{4}+2\lambda-b\right)(\lambda^2-d)$$

which is clearly a cubic possessing either one or three real roots. A value of λ is found from this cubic and the solution obtained.

Example 5

Solve the equation $x^4 - 2x^3 + x^2 - 4x + 4 = 0$.

Write the equation in the form

$$(x^2-x+\lambda)^2 - x^2 - 2\lambda x^2 + 2\lambda x - \lambda^2 + x^2 - 4x + 4 = 0$$

i.e. $(x^2-x+\lambda)^2 - \{2\lambda x^2 + (4-2\lambda)x + (\lambda^2-4)\} = 0$

Choose λ so that $2\lambda x^2 + (4-2\lambda)x + (\lambda^2-4)$ is a perfect square.

Then $(4-2\lambda)^2 = 4 \times 2\lambda(\lambda^2-4)$

i.e. $4(\lambda-2)^2 = 4 \times 2\lambda(\lambda-2)(\lambda+2)$

Clearly $\lambda = 2$ is a root of this equation and the above quadratic becomes

$$4x^2 + 0x + 0$$

The original equation then becomes $(x^2-x+2)^2 - 4x^2 = 0$

i.e. $(x^2-x+2+2x)(x^2-x+2-2x) = 0$

$(x^2+x+2)(x^2-3x+2) = 0$

From $x^2-3x+2=0 \quad x = 1 \text{ or } 2$

From $x^2+x+2=0 \quad x = \dfrac{-1\pm\sqrt{(1-8)}}{2}$

$\Rightarrow \quad x = -\dfrac{1}{2} \pm i\dfrac{\sqrt{7}}{2}$

Hence the four roots are 1, 2, $-\frac{1}{2}+i\frac{\sqrt{7}}{2}$ and $-\frac{1}{2}-i\frac{\sqrt{7}}{2}$.

(It is obvious that if the subsidiary cubic equation in λ does not have a root which can be found immediately, then this process becomes very complicated indeed. This course of study will not demand such extensive processes.)

Exercise E

1 Solve the equation $z^4 + 1 = 0$ and show the position of the roots on an Argand diagram.

2 Solve $x^7 + x^4 - x^3 - 1 = 0$ by first factorising the equation.

3 Express $x^6 + 1$ as the product of three real quadratic factors.

4 Solve
a) $x^3 + x - 10 = 0$ b) $x^3 - x - 12 = 0$ c) $x^3 - 9x - 31 = 0$

Check the validity of your real root by substituting back into the given cubic equation.

5 By expressing each of the following equations in biquadratic form, solve them
a) $x^4 - 4x^3 + 4x^2 - 4x + 3 = 0$ c) $x^4 + 4x^2 + 4x + 15 = 0$
b) $x^4 + x^2 + 1 = 0$

Further Integration

CHAPTER TWELVE

12.1 Algebraic substitutions

We have already learnt how to deal with integrals of simple functions, both polynomial functions and simple trigonometric functions. We consider now more complex functions.

Example 1

Suppose we wish to find the indefinite integral

$$\int (2x+3)(x^2+3x)^2\, dx$$

We could, of course, multiply out the brackets and integrate term by term. This is laborious, and we seek an alternative method. We substitute u for the expression (x^2+3x). What then happens to $(2x+3)$ and 'dx'? We recall that we may have only one variable in our integral, so all the x's must be changed to u's. Clearly,

$$u = x^2 + 3x \quad \Rightarrow \quad \frac{du}{dx} = 2x + 3$$

and, if we regard $\left(\frac{du}{dx}\right)$ as behaving like a fraction,* we obtain

$$du = (2x+3)\, dx$$

this is the remaining part of the integral.

Hence, $\int (2x+3)(x^2+3x)^2\, dx$ becomes $\int u^2\, du$

and, on performing the integration, we obtain $\frac{u^3}{3} + k$

that is, $\frac{(x^2+3x)^3}{3} + k$

Example 2

Evaluate $\int \sin^3 x \cos x\, dx$

Here we write $u = \sin x$, $du = \cos x\, dx$ and our integral becomes

$$\int u^3\, du = \frac{u^4}{4} + k = \tfrac{1}{4}\sin^4 x + k$$

---

* The notation $\frac{du}{dx}$ is used mainly because of the similarity of behaviour of derived functions to fractions. The proofs of the validity of these similarities will be found in more advanced works.

Exercise A

1 Find the following indefinite integrals using the substitutions suggested.

a) $\int (x^2 + 1)^2 2x \, dx$ $\quad u = x^2 + 1$

b) $\int -\cos^2 x \sin x \, dx$ $\quad u = \cos x$

c) $\int (x^3 + 7x)^2(3x^2 + 7) \, dx$ $\quad u = x^3 + 7x$

d) $\int \frac{2x}{(x^2 + 1)^4} \, dx$ $\quad u = x^2 + 1$

e) $\int (x^6 + x^7 + x^8)^{10}(6x^5 + 7x^6 + 8x^7) \, dx$ $\quad u = x^6 + x^7 + x^8$

2 Find the following indefinite integrals, by choosing suitable substitutions.

a) $\int (x^3 + x^2)(3x^2 + 2x) \, dx$

b) $\int \sin^2 x \cos x \, dx$

c) $\int \frac{3x^2}{(x^3 + 1)^2} \, dx$

d) $\int 2(x + 1)(x^2 + 2x + 3)^2 \, dx$

e) $\int 6x^2(x^3 + 5)^3 \, dx$

f) $\int (\cos x - \sin x)(\cos x + \sin x)^3 \, dx$

g) $\int \left(x + \frac{1}{x}\right)^2\left(1 - \frac{1}{x^2}\right) dx$

These questions were very conveniently chosen so that when the substitution was made, the new integral in terms of u was very simple. In practice, this seldom happens, and although the substitutions may be straightforward, the resulting integral will not always be as simple.

Example 3

Consider $\int (x + 1)(7x + 5)^2 \, dx$

We write $u = 7x + 5$, so that $du = 7 \, dx$

that is $\quad x = \frac{u - 5}{7}$ and $dx = \frac{1}{7} \, du$

Then the integral becomes

$$\int \left(\frac{u - 5}{7} + 1\right)(u^2)\left(\frac{1}{7} \, du\right) = \int \left(\frac{u + 2}{7}\right)(u^2)\left(\frac{1}{7}\right) du$$

$$= \frac{1}{49} \int (u^3 + 2u^2) \, du$$

$$= \frac{1}{49}\left(\frac{u^4}{4} + \frac{2u^3}{3}\right) + k$$

$$= \frac{u^4}{196} + \frac{2u^3}{147} + k$$

$$= \frac{(7x+5)^4}{196} + \frac{2(7x+5)^3}{147} + k$$

The method, then, is very similar to that used in the previous set of questions. The main difference is in the amount of algebraic manipulation involved in simplifying the new integral. It is not always easy to decide what to substitute, and much practice is essential to get a real grip of this aspect of integration.

A further example will illustrate the method again.

Example 4

Evaluate $\int (3x - 7)(2x + 3)^3 \, dx$

Let $u = 2x + 3$, then $du = 2 \, dx$

$$x = \frac{u-3}{2} \quad \text{and} \quad dx = \tfrac{1}{2} \, du$$

So, the integral becomes

$$\int \left\{3\left(\frac{u-3}{2}\right) - 7\right\}(u^3) \,.\, \tfrac{1}{2} \, du = \frac{1}{4}\int (3u - 9 - 14)u^3 \, du$$

$$= \frac{1}{4}\int (3u^4 - 23u^3) \, du$$

$$= \frac{1}{4}\left(\frac{3u^5}{5} - \frac{23u^4}{4}\right) + k$$

$$= \tfrac{3}{20}(2x+3)^5 - \tfrac{23}{16}(2x+3)^4 + k$$

Exercise B

1 Find the following indefinite integrals using the suggested substitutions.

a) $\int (x + 1)(4x + 3)^2 \, dx$ $\qquad u = 4x + 3$

b) $\int (7x - 2)(3x + 5)^3 \, dx$ $\qquad u = 3x + 5$

c) $\int (3x + 2)(3x - 2)^2 \, dx$ $\qquad u = 3x - 2$

d) $\int x \sin (x^2) \, dx$ $\qquad u = x^2$

e) $\int 2 \sin^2 (2x + 5) \cos (2x + 5)\, dx \qquad u = \sin (2x + 5)$

f) $\int \dfrac{(11x^2 - 3)x}{(x^2 + 1)^3}\, dx \qquad u = x^2 + 1$

2 Find the following indefinite integrals, using appropriate substitutions.

a) $\int (3x + 5)(2x + 1)^2\, dx$

b) $\int (2x + 1)(3x - 5)^3\, dx$

c) $\int -\cos^3 (3x + 4) \sin (3x + 4)\, dx$

d) $\int 2 \tan (2x - 1) \sec^2 (2x - 1)\, dx$

e) $\int \dfrac{14x}{(2x^2 - 1)^4}\, dx$

12.2 Definite integration

When we make a substitution in a definite integral, we also need to make the same substitution in the limits of integration. The method is best illustrated by examples.

Example 1

Evaluate $\int_3^4 (x + 1)(2x - 5)^4\, dx$

let $u = 2x - 5, \qquad du = 2\, dx$

$$x = \frac{u + 5}{2} \quad \text{and } dx = \tfrac{1}{2}\, du$$

Hence the integral becomes

$$\int \left(\frac{u + 5}{2} + 1\right) u^4 \,.\, \tfrac{1}{2}\, du$$

Now, our relationship connecting u and x is

$u = 2x - 5$

the lower limit, 3, for $x \Rightarrow 2 \times 3 - 5 = 1$ as the lower limit for u, and the upper limit 1 for $x \Rightarrow 2 \times 4 - 5 = 3$ as the upper limit for u.

Hence the definite integral in terms of u becomes

$$\int_1^3 \left(\frac{u+5}{2}+1\right)u^4 \cdot \tfrac{1}{2}\, du = \frac{1}{4}\int_1^3 (u+7)u^4\, du$$
$$= \frac{1}{4}\int_1^3 (u^5+7u^4)\, du$$
$$= \frac{1}{4}\left[\left(\frac{u^6}{6}+\frac{7u^5}{5}\right)\right]_1^3$$
$$= \frac{1}{4}\left\{\left(\frac{3^6}{6}+\frac{7(3)^5}{5}\right)-\left(\frac{1}{6}+\frac{7}{5}\right)\right\}$$
$$= \tfrac{1}{4}\{(121{\cdot}5+340{\cdot}2)-(0{\cdot}167+1{\cdot}4)\}$$
$$= \tfrac{1}{4}\{(461{\cdot}7)-(1{\cdot}567)\}$$
$$= 115{\cdot}03$$

Example 2

Evaluate $\int_0^1 2\sin^2(2x+3)\cos(2x+3)\,dx$

Let $u=\sin(2x+3)$, then $du = 2\cos(2x+3)\,dx$

so the integral becomes $\int u^2\,du$

Lower limit: $x=0 \Rightarrow u=\sin 3 = 0{\cdot}141$

Upper limit: $x=1 \Rightarrow u=\sin 5 = -0{\cdot}959$

Hence

$$\int_{0\cdot141}^{-0\cdot959} u^2\,du = \left[\tfrac{1}{3}u^3\right]_{0\cdot141}^{-0\cdot959}$$
$$= -0{\cdot}295 \text{ (correct to 3 decimal places)}$$

Exercise C

Evaluate the following:

1 $\int_1^{10} (x-1)^{1/2}\,dx$

2 $\int_0^{0\cdot6} x(1-x^2)^3\,dx$

3 $\int_{-1/4}^{0} (4x+1)^4\,dx$

4 $\int_0^1 \frac{x\,dx}{(2x+1)^3}$

5 $\int_0^{\pi/4} (\sin x+\cos x)(\cos x-\sin x)\,dx$

6 $\int_{-1}^{+1} (3x+2)(2x+3)^3\,dx$

7 $\int_{\pi/12}^{7\pi/12} \sin\left(2x+\frac{\pi}{12}\right)dx$

8 $\int_{\pi/2}^{\pi} (\sin 3x+\cos\tfrac{1}{2}x)\,dx$

9 $\int_0^3 x\sqrt{(x^2+16)}\,dx$

10 $\int_0^3 \frac{x}{\sqrt{(x^2+16)}}\,dx$

11 $\int_0^{\pi/3} \frac{\sin x}{\cos^2 x}\,dx$

12 $\int_0^{1/3} x(2+3x)^5\,dx$

13 Show that $\int \sin^2 x\,dx$ can be written as $\int \frac{1}{2}(1-\cos 2x)\,dx$,

and hence evaluate $\int_0^{\pi/2} \sin^2 x\,dx$

14 Use a similar method to evaluate $\int_{-\pi/2}^{0} \cos^2 x\,dx$

15 Evaluate $\int_0^4 x\sqrt{(x^2+9)}\,dx$ and explain why you cannot use a similar method to evaluate $\int_0^4 x^2\sqrt{(x^2+9)}\,dx$

16 Evaluate $\int_0^1 x^5(x^3+1)^4\,dx$

17 Sketch the graph of $y = \frac{x}{1+x^2}$

and use this to find the value of $\int_{-1}^{+1} \frac{x}{1+x^2}\,dx$

18 Evaluate a) $\int_{-\pi/2}^{0} \sin^3 x \cos x\,dx$ b) $\int_{-\pi/2}^{\pi/2} \cos^3 x \sin x\,dx$

19 Evaluate

$$\int_3^5 (x+1)(x^2+2x)\,dx$$

a) by multiplying out and integrating term by term and
b) by substituting $u = x^2 + 2x$

20 Evaluate $\int_0^{\pi/2} \frac{\cos x\,dx}{2\sqrt{(1+\sin x)}}$

12.3 Trigonometric substitutions

In all the questions so far, we have made algebraic substitutions, sometimes replacing an algebraic term by a trigonometric term. There are times, however, when integrals can be simplified considerably by substituting a trigonometric function for an algebraic one.

The main reasons are that sine and cosine functions are related by their derivatives and the relation $\sin^2\theta + \cos^2\theta = 1$. Secant and tangent are similarly related through their derivatives and the relation $\tan^2\theta + 1 = \sec^2\theta$. Let us consider two specific examples and then their generalisations.

Example 1

$$\int \frac{1}{\sqrt{(1-x^2)}}\,dx$$

We know that $\sqrt{(1-\sin^2\theta)} = \cos\theta$, so that if in our integral we write $x = \sin\theta$, the denominator becomes simply $\cos\theta$, and we shall have removed the difficulty that existed there.

Let $x = \sin\theta$ then $dx = \cos\theta\, d\theta$

so the integral becomes

$$\int \frac{1}{\sqrt{(1-\sin^2\theta)}} \cdot \cos\theta\, d\theta = \int \frac{1}{\cos\theta} \cdot \cos\theta\, d\theta$$

$$= \int 1\, d\theta$$

$$= \theta + k$$

$$= \sin^{-1} x + k$$

(see chapter 10 for definition of $\sin^{-1} x$)

We can make this result more general if we now consider

$$\int \frac{1}{\sqrt{(a^2 - b^2x^2)}}\,dx \quad \text{where } a \text{ and } b \text{ are constants.}$$

We need to compare $(a^2 - b^2x^2)$ with $(1 - \sin^2\theta)$ in order to find the appropriate substitution. We can write

$$a^2 - b^2x^2 = a^2\left(1 - \frac{b^2}{a^2}x^2\right)$$

Hence, we need $\dfrac{b^2x^2}{a^2}$ to be equal to $\sin^2\theta$

that is, the required substitution is $x = \dfrac{a}{b}\sin\theta$

and $dx = \dfrac{a}{b}\cos\theta\, d\theta$.

Hence the integral becomes

$$\int \frac{1}{\sqrt{(a^2 - a^2 \sin^2 \theta)}} \cdot \frac{a}{b} \cos \theta \, d\theta = \int \frac{1}{a\sqrt{(1 - \sin^2 \theta)}} \cdot \frac{a}{b} \cos \theta \, d\theta$$

$$= \frac{1}{b} \int \frac{\cos \theta}{\cos \theta} \, d\theta$$

$$= \frac{1}{b} \int 1 \, d\theta$$

$$= \frac{1}{b} \theta + k$$

$$= \frac{1}{b} \sin^{-1}\left(\frac{bx}{a}\right) + k$$

(Note that in both cases above a cosine substitution could have given an equally straightforward integral.)

Example 2

Here we consider integrals where $\tan \theta$ is an appropriate function to use in our substitution.

Consider $\int \frac{1}{1 + x^2} \, dx$

We compare the $(1 + x^2)$ of the denominator with the relationship $1 + \tan^2 \theta = \sec^2 \theta$, and substitute $x = \tan \theta$ so that $dx = \sec^2 \theta \, d\theta$.

The integral becomes

$$\int \frac{1}{(1 + \tan^2 \theta)} \sec^2 \theta \, d\theta = \int \frac{1}{\sec^2 \theta} \cdot \sec^2 \theta \, d\theta$$

$$= \int 1 \, d\theta$$

$$= \theta + k$$

$$= \tan^{-1}(x) + k$$

Our generalisation here is obtained by considering

$$\int \frac{1}{(a^2 + b^2x^2)} \, dx \quad (a, b \text{ constants})$$

i.e. $$\int \frac{1}{a^2\left(1 + \frac{b^2x^2}{a^2}\right)} \, dx$$

and we see that we must write $\frac{b^2x^2}{a^2} = \tan^2 \theta$

So the substitution is

$$x = \frac{a}{b} \tan \theta \quad \text{and} \quad dx = \frac{a}{b} \sec^2 \theta \, d\theta$$

The integral becomes

$$\int \frac{1}{a^2(1+\tan^2\theta)} \cdot \frac{a}{b}\sec^2\theta\, d\theta = \frac{1}{ab}\int \frac{1}{\sec^2\theta} \cdot \sec^2\theta\, d\theta$$

$$= \frac{1}{ab}\int 1\, d\theta$$

$$= \frac{1}{ab}\theta + k$$

$$= \frac{1}{ab}\tan^{-1}\left(\frac{bx}{a}\right) + k$$

The results obtained should be compared with those used when differentiating the inverse trigonometric functions in chapter 10 (p. 214).

In dealing with individual questions, we should evaluate each integral using the constants given in the question. It is insufficient to substitute the numbers into the results shown here.

Exercise D

1 Find the following indefinite integrals, using an appropriate $\sin\theta$ substitution.

a) $\int \frac{1}{\sqrt{(9-x^2)}}\,dx$ d) $\int \frac{1}{\sqrt{(\frac{1}{4}-x^2)}}\,dx$

b) $\int \frac{1}{\sqrt{(4-x^2)}}\,dx$ e) $\int \frac{1}{\sqrt{(9-4x^2)}}\,dx$

c) $\int \frac{1}{\sqrt{(2-x^2)}}\,dx$ f) $\int \frac{1}{\sqrt{(\frac{1}{9}-\frac{1}{4}x^2)}}\,dx$

2 Find the following indefinite integrals using an appropriate $\cos\theta$ substitution.

a) $\int \frac{1}{\sqrt{(16-x^2)}}\,dx$ d) $\int \frac{3}{\sqrt{(9-x^2)}}\,dx$

b) $\int \frac{1}{\sqrt{(1-4x^2)}}\,dx$ e) $\int \frac{1}{\sqrt{(12-3x^2)}}\,dx$

c) $\int \frac{1}{\sqrt{(1-\frac{1}{4}x^2)}}\,dx$ f) $\int \frac{1}{\sqrt{(2-8x^2)}}\,dx$

3 Use appropriate $\tan\theta$ substitutions to find

a) $\int \frac{1}{1+4x^2}\,dx$ d) $\int \frac{1}{12+3x^2}\,dx$

b) $\int \frac{1}{4+9x^2}\,dx$ e) $\int \frac{1}{\frac{1}{4}+\frac{1}{16}x^2}\,dx$

c) $\int \frac{1}{25+x^2}\,dx$ f) $\int \frac{3}{8+4x^2}\,dx$

4 Use similar methods to find the following indefinite integrals

a) $\int \frac{1}{1+(x-1)^2}\,dx$ d) $\int \frac{1}{x^2+4x+5}\,dx$

b) $\int \frac{1}{\sqrt{(4-(x+2)^2)}}\,dx$ e) $\int \frac{1}{\sqrt{(5-4x-x^2)}}\,dx$

c) $\int \frac{1}{\sqrt{(1-(2-x)^2)}}\,dx$ f) $\int \frac{1}{4x^2+12x+10}\,dx$

5 Evaluate the following integrals

a) $\int_0^1 \frac{1}{4+x^2}\,dx$ d) $\int_0^{2/3} \frac{3}{4+9x^2}\,dx$

b) $\int_0^1 \frac{1}{\sqrt{(4-x^2)}}\,dx$ e) $\int_{-1}^0 \frac{1}{\sqrt{(4-(3x+1)^2)}}\,dx$

c) $\int_2^5 \frac{1}{\sqrt{(9-(x-2)^2)}}\,dx$ f) $\int_0^1 \frac{1}{4+(3x-1)^2}\,dx$

12.4 Integration by parts

We saw in chapter 10 how to deal with the problem of differentiating the product of two functions. It is reasonable, then, to ask the question, 'Is there a formula for integrating a product?' The immediate answer is 'No', since there is not an equivalent formula. However, we can transform the formula for differentiation of a product to help us with the integration of some products.

Our formula for differentiation of a product is

$$\frac{d}{dx}(uv) = u\frac{dv}{dx} + v\frac{du}{dx}$$

We now integrate each term of this formula with respect to x, and obtain

$$\int \frac{d}{dx}(uv)\ dx = \int u\frac{dv}{dx}\,dx + \int v\frac{du}{dx}\,dx$$

$$\Rightarrow \int d(uv) = \int u\,dv + \int v\,du$$

$$\Rightarrow uv = \int u\,dv + \int v\,du$$

which we can rearrange to obtain

$$\int u\,dv = uv - \int v\,du$$

Clearly, this is not a formula that immediately works out an integral for us, since there is another integral, still to be found, on the right hand side of the equation. Usually, when applied, this formula (known as the formula for '**integration by parts**') alters a difficult or awkward integral into a more simple one. Some examples will show how the formula is used.

Example 1

$$\int x \sin x \, dx$$

Comparing this with the formula, we write

$$u = x \quad \text{and } dv = \sin x \, dx$$

so that we obtain $du = dx$ and $v = -\cos x$

and applying the formula, we have

$$\int x \sin x \, dx = x(-\cos x) - \int (-\cos x) \, dx$$

i.e. $$= -x \cos x + \int \cos x \, dx$$

and we see that the new integral on the R.H.S. is more simple, so the final answer is

$$\int x \sin x \, dx = -x \cos x + \sin x + k$$

Example 2

$$\int (x+1)(3x+8)^8 \, dx$$

Here, let $u = x + 1$ and $dv = (3x+8)^8 \, dx$

then $du = dx$ and $v = \dfrac{(3x+8)^9}{27}$ (using reverse of Chain Rule)

So that, applying the formula for integration by parts, we have

$$\int (x+1)(3x+8)^8 \, dx = (x+1)\,\frac{(3x+8)^9}{27} - \int \tfrac{1}{27}(3x+8)^9 \, dx$$

$$= \frac{(x+1)(3x+8)^9}{27} - \tfrac{1}{810}(3x+8)^{10} + k$$

(**Note** This integral could also have been found by substitution, by writing $z = 3x + 8$. Try it by this method and compare your answer with that just obtained.)

Example 3

In this example, we have to apply the formula for integration by parts twice before the resulting integral is simplified enough.

$$\int x^2 \cos x \, dx$$

let $u = x^2$ and $dv = \cos x \, dx$

then $du = 2x \, dx$ and $v = \sin x$

So the given integral becomes

$$x^2 \sin x - \int 2x \sin x \, dx$$

and applying the formula to the remaining integral, with $u = 2x$, $dv = \sin x\, dx$, we obtain

$$x^2 \sin x - \left\{-2x \cos x - \int -2 \cos x\, dx\right\}$$

$$= x^2 \sin x + 2x \cos x - 2 \int \cos x\, dx$$

$$= x^2 \sin x + 2x \cos x - 2 \sin x + k$$

In each of these examples, and indeed with the integration of any product, we have a choice of terms to be represented by u and dv. Look back at these three examples, and investigate why choosing the alternative terms would have given us a less simple final integral.

Exercise E

Find the following indefinite integrals, using the method of integration by parts

1 $\int x(x + 10)^7\, dx$

2 $\int (2x + 3)(3x + 2)^5\, dx$

3 $\int x \tan^{-1} x\, dx$ $\left(\text{In the resulting integral, write } \dfrac{x^2}{1 + x^2} \text{ as } 1 - \dfrac{1}{1 + x^2}\right)$

4 $\int \sec^3 x \tan x\, dx$ $\left(\text{write this integral as } \int \sec^2 x(\sec x \tan x)\, dx\right)$

5 $\int x \sin^{-1} x\, dx$

6 $\int (x + 3)^2 \cos x\, dx$

7 $\int (2x + 7)^3(x - 1)\, dx$

8 $\int \frac{1}{4}x^3(x + 4)^3\, dx$

9 $\int \cot x \sin x\, dx$ (compare your answer with that obtained by simplifying $\cot x \sin x$ first)

10 $\int x^3 \sin x\, dx$

It should be remembered that the method of integration by parts is not always the easiest way of dealing with integrals of products. Often a simple substitution will result in a more straightforward method. Numbers 1, 2, 7, 8 and 9 in the above exercise are such questions. Check in some of these cases that you can do them by an alternative method and that you can obtain the same answers.

We can apply the method of integration by parts to the two integrals

$$\int \sin^n x\,dx \quad \text{and} \quad \int \cos^n x\,dx \quad (\text{for } n \neq 1\ n \in \mathbb{Z}^+)$$

Although these particular results are beyond the level expected within the core of mathematics at this level, the technique can be used to advantage in many other situations. They are included here merely to illustrate the technique.

Let $I_n = \int_0^{\pi/2} \sin^n x\,dx$

i.e. $I_n = \int_0^{\pi/2} \sin^{n-1} x \sin x\,dx$

Integrating by parts with

$$u = \sin^{n-1} x \qquad \text{and} \quad dv = \sin x\,dx$$

i.e. $du = (n-1)\sin^{n-2} x \cos x\,dx$ and $v = -\cos x$

we obtain

$$I_n = [-\cos x \sin^{n-1} x]_0^{\pi/2} - \int_0^{\pi/2} (-\cos x)(n-1)\sin^{n-2} x \cos x\,dx$$

Now the first term is zero at both limits, hence,

$$I_n = \int_0^{\pi/2} (\cos x)(n-1)\sin^{n-2} x \cos x\,dx$$

$$= (n-1)\int_0^{\pi/2} \sin^{n-2} x \cos^2 x\,dx$$

$$= (n-1)\int_0^{\pi/2} \sin^{n-2} x(1-\sin^2 x)\,dx$$

$$= (n-1)\int_0^{\pi/2} (\sin^{n-2} x - \sin^n x)\,dx$$

$$= (n-1)\int_0^{\pi/2} \sin^{n-2} x\,dx - (n-1)\int_0^{\pi/2} \sin^n x\,dx$$

At this stage, we are tempted to abandon the approach since both integrals on the right-hand side present the same difficulties as the original integral. However, since we denoted $\int_0^{\pi/2} \sin^n x\,dx$ by I_n, it is logical to write $\int_0^{\pi/2} \sin^{n-2} x\,dx$ as I_{n-2}.

Hence, we have

$$I_n = (n-1)I_{n-2} - (n-1)I_n$$

which we rearrange to give

$$nI_n = (n-1)I_{n-2}$$

$$\text{or} \quad I_n = \left(\frac{n-1}{n}\right)I_{n-2}$$

We see, therefore, that we can reduce the problem of finding $\int_0^{\pi/2} \sin^n x\,dx$ to that of finding $\int_0^{\pi/2} \sin^{n-2} x\,dx$. Clearly, we may continue this process until we finish up with either $\int_0^{\pi/2} \sin x\,dx$ or $\int_0^{\pi/2} 1\,dx$, depending whether n was odd or even. An example will illustrate how the result is used.

Example 1

Evaluate $\qquad I_6 = \int_0^{\pi/2} \sin^6 x\,dx$

Applying our previous result $\qquad I_6 = \frac{5}{6} I_4$

and $\qquad I_4 = \frac{3}{4} I_2$

and $\qquad I_2 = \frac{1}{2} I_0$

when $\qquad I_0 = \int_0^{\pi/2} 1\,dx = \frac{\pi}{2}$

so that $\qquad I_6 = \frac{5}{6} \times \frac{3}{4} \times \frac{1}{2} \times \frac{\pi}{2} = \frac{5\pi}{32}$

We follow a similar procedure for $\int_0^{\pi/2} \cos^n x\,dx$ $(n \neq 1, n \in \mathbb{Z}^+)$

Let $I_n = \int_0^{\pi/2} \cos^n x\,dx$

i.e. $I_n = \int_0^{\pi/2} \cos^{n-1} x \cos x\,dx$

$$\Rightarrow \quad I_n = [\cos^{n-1} x \sin x]_0^{\pi/2} - \int_0^{\pi/2} (n-1) \cos^{n-2} x(-\sin x) \sin x\,dx$$

$$\Rightarrow \quad I_n = (n-1) \int_0^{\pi/2} \cos^{n-2} x \sin^2 x\,dx$$

$$\Rightarrow \quad I_n = (n-1) \int_0^{\pi/2} \cos^{n-2} x(1 - \cos^2 x)\,dx$$

$$\Rightarrow \quad I_n = (n-1) \int_0^{\pi/2} (\cos^{n-2} x - \cos^n x)\,dx$$

$$\Rightarrow \quad I_n = (n-1)(I_{n-2} - I_n)$$

By rearrangement, this gives the formula

$$I_n = \frac{n-1}{n} I_{n-2}$$

and we see that this is exactly the same result which we obtained for

$$\int_0^{\pi/2} \sin^n x\,dx$$

Example 2

$$I_7 = \int_0^{\pi/2} \cos^7 x\, dx$$

$$I_7 = \tfrac{6}{7} I_5$$

and $\quad I_5 = \tfrac{4}{5} I_3$

and $\quad I_3 = \tfrac{2}{3} I_1$

where $\quad I_1 = \int_0^{\pi/2} \cos x\, dx = [\sin x]_0^{\pi/2} = 1$

Hence $\quad I_7 = \tfrac{6}{7} \cdot \tfrac{4}{5} \cdot \tfrac{2}{3} \cdot 1 = \tfrac{16}{35}$

Exercise F

Miscellaneous

In this exercise, choose the most appropriate method for working out each integral. There may be several methods which will simplify the problem and some may be better than others.

1 a) $\int \sin^2 x\, dx$ b) $\int \cos^2 x\, dx$

(Hint: use the formulae $\cos 2\theta = 2\cos^2\theta - 1 = 1 - 2\sin^2\theta$ to rewrite both integrals in terms of $\cos 2x$.)

2 a) $\int \sin^3 x\, dx$ b) $\int \cos^3 x\, dx$

(Use the identity $\cos^2\theta + \sin^2\theta = 1$ before finding an appropriate substitution.)

3 a) $\int x\sqrt{(x^2 + 1)}\, dx$ b) $\int \dfrac{x}{\sqrt{(x^2 + 1)}}\, dx$

4 a) $\int_{-1}^{+1} x \cos x\, dx$ b) $\int_{-1}^{+1} x \sin x\, dx$

Note that in one of these integrals, the symmetry should make the evaluation very simple.

5 Use the substitution $x = a\cos^2\theta + b\sin^2\theta$ to evaluate

$$\int_a^b \frac{dx}{\sqrt{\{(x - a)(b - x)\}}}, \; b > a$$

and hence evaluate

a) $\int_1^2 \dfrac{dx}{\sqrt{\{(2 - x)(x - 1)\}}}$ b) $\int_3^5 \dfrac{dx}{\sqrt{\{(5 - x)(x - 3)\}}}$

6 $\int \dfrac{1 - x}{\sqrt{(1 - x^2)}}\, dx$ (Hint: use the formulae for $\cos 2\theta$)

7 a) $\int_0^{\pi} x^2 \cos x \, dx$ b) $\int_{-\pi/2}^{0} \sin x \cos x \, dx$

8 Sketch graphs to show why, for some convenient values of n ($n \in \mathbb{Z}^+$),

$$\int_0^{\pi/2} \sin^n x \, dx = \int_0^{\pi/2} \cos^n x \, dx$$

(**Note** The same reduction formula was established for both integrals, but graphical reasoning is required here.)

9 a) $\int 2x(x^2 - 3)^2 \, dx$ b) $\int \tan x \sec^2 x \, dx$

10 a) $\int x\sqrt{(x + 3)} \, dx$ b) $\int_0^{\pi/2} \sin 7x \cos x \, dx$

11 Show that $\int_0^{\pi} \frac{\sin x}{4 - \sin^2 x} dx = \frac{\pi}{3\sqrt{3}}$

12 Make use of the relation $\sin x = \frac{2t}{1 + t^2}$ when substituting $t = \tan \frac{x}{2}$ to evaluate $\int_0^{\pi/2} \frac{1}{1 + \sin x} dx$

Approximate Methods

CHAPTER THIRTEEN

There are many occasions in mathematics when exact methods prove difficult or even impossible. For this reason, various techniques have been developed for finding approximate numerical solutions. These have become used to a great extent with the advent of pocket calculators and microprocessors; we shall consider here those numerical methods which are relevant to this level of work.

13.1 The approximate solution of equations

We seek solutions of equations which can be generalised by $f(x) = 0$. From our graphical work, we know that the solutions of $f(x) = 0$ occur wherever $y = f(x)$ crosses the x-axis. We take this as our approach, and consider the graph of $y = f(x)$, figure 13.1.

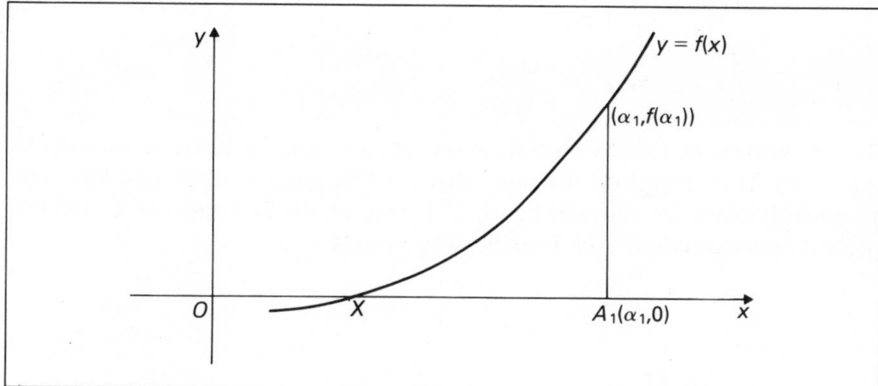

Figure 13.1

Our method is to guess an approximate solution at A_1, where $x = \alpha_1$, which is 'reasonably close' to the point X representing the exact solution. We construct the tangent to $y = f(x)$ at the point $(\alpha_1, f(\alpha_1))$ and produce it to meet the x-axis at A_2, with coordinates $(\alpha_2, 0)$. Then from figure 13.2, it is clear that α_2 is a better approximation to the solution than α_1.

We then repeat the process to find $A_3(\alpha_3, 0)$ which is yet closer to X. Clearly we may repeat this process as many times as we like to achieve the necessary accuracy.

Consider, in figure 13.2, $\triangle A_1 A_2 P_1$,

$$\tan P_1\hat{A}_2A_1 = \frac{P_1A_1}{A_1A_2} = \frac{f(\alpha_1)}{\alpha_1 - \alpha_2}$$

Now, $\tan P_1\hat{A}_2A_1$ is the gradient of the curve at P_1,

hence $\tan P_1\hat{A}_2A_1 = f'(\alpha_1)$

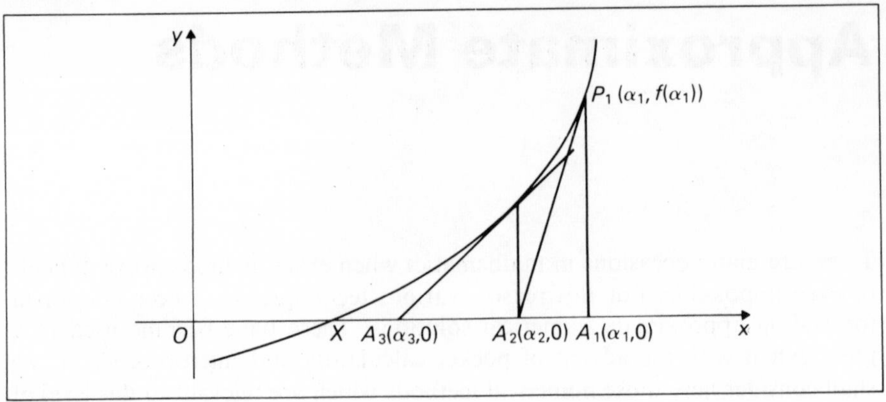

Figure 13.2

So we have

$$f'(\alpha_1) = \frac{f(\alpha_1)}{\alpha_1 - \alpha_2}$$

$$\Rightarrow (\alpha_1 - \alpha_2)f'(\alpha_1) = f(\alpha_1)$$

$$\Rightarrow \quad \alpha_1 - \alpha_2 = \frac{f(\alpha_1)}{f'(\alpha_1)}$$

$$\Rightarrow \quad \alpha_2 = \alpha_1 - \frac{f(\alpha_1)}{f'(\alpha_1)}$$

This is known as the Newton-Raphson process for the iterative solution of equations. It is important to realise that the first guess at a solution has to be reasonably close, for consider figure 13.3. Making the first guess at A_1 yields a second approximation even further away from X.

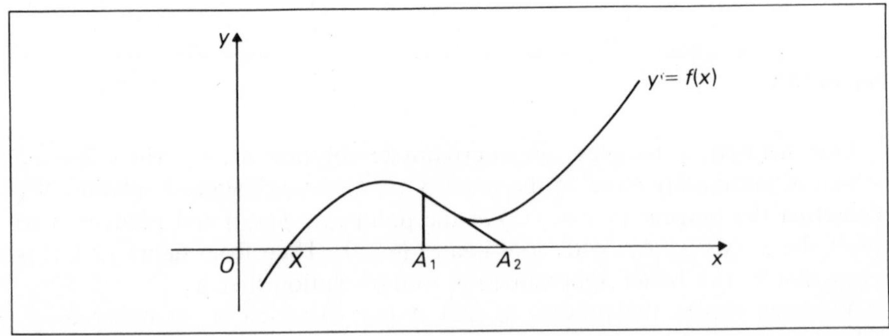

Figure 13.3

It is also important to be aware of any discontinuities within the graph, when making the first guess. Clearly in figure 13.4, a first guess at A_1 will not give a better approximation to X which is 'on the other side' of the discontinuity. Some examples will illustrate the method.

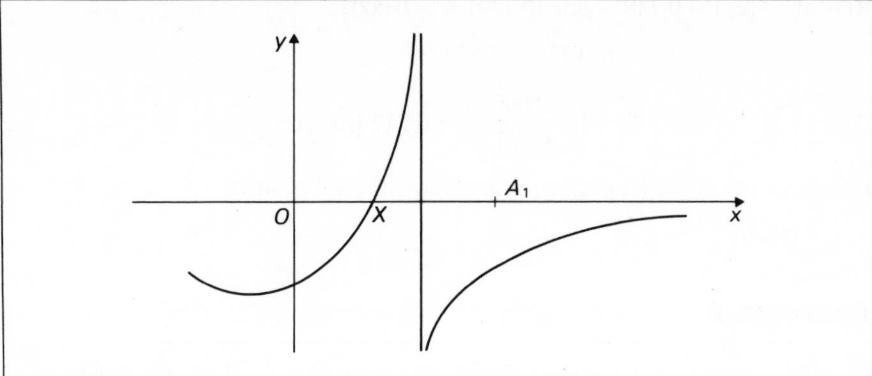

Figure 13.4

Example 1

Find an approximate solution of the equation $x^2 - 27 = 0$, taking $x = 5$ as first guess.

We have $\quad f(x) = x^2 - 27, \quad f'(x) = 2x, \quad \alpha_1 = 5$

Hence $\quad f(\alpha_1) = -2, \qquad f'(\alpha_1) = 10.$

Thus $$\alpha_2 = 5 - \left(\frac{-2}{10}\right) = 5{\cdot}2$$

Repeating the process with $\alpha_2 = 5{\cdot}2$, we have

$$f(\alpha_2) = (5{\cdot}2)^2 - 27 = 0{\cdot}04, \quad f'(\alpha_2) = 10{\cdot}4$$

So $$\alpha_3 = 5{\cdot}2 - \frac{0{\cdot}04}{10{\cdot}4} = 5{\cdot}196 \text{ (to 3 d.p.)}$$

Repeating the process with $\alpha_3 = 5{\cdot}196$, we have

$$f(\alpha_3) = (5{\cdot}196)^2 - 27 = -0{\cdot}002, \quad f'(\alpha_3) = 10{\cdot}392$$

And $$\alpha_4 = 5{\cdot}196 - \left(\frac{-0{\cdot}002}{10{\cdot}392}\right)$$

$$\alpha_4 = 5{\cdot}196 \quad \text{(to 3 d.p.)}$$

Since α_3 and α_4 are the same to an accuracy of three decimal place, there is no value in continuing further; we have found a reasonable approximation to the solution.

Example 2

Solve $x - \cos x = 0$ giving the answer correct to 3 decimal places, using $x = 0{\cdot}7$ as the first guess.

$$f(x) = x - \cos x, \quad f'(x) = 1 + \sin x, \quad \alpha_1 = 0{\cdot}7$$

Then $\quad f(\alpha_1) = 0{\cdot}7 - \cos(0{\cdot}7) = -0{\cdot}065, \quad f'(\alpha_1) = 1{\cdot}644$?

$$\alpha_2 = 0{\cdot}7 - \left(\frac{-0{\cdot}065}{1{\cdot}664}\right) = 0{\cdot}739$$

Repeat $f(\alpha_2) = 0{\cdot}739 - \cos(0{\cdot}739) = -0{\cdot}0001,$

$$f'(\alpha_2) = 1 + \sin 0{\cdot}739 = 1{\cdot}674$$

And $$\alpha_3 = 0{\cdot}739 - \left(\frac{-0{\cdot}0001}{1{\cdot}673}\right) = 0{\cdot}739 \text{ (to 3 d.p.)}$$

Hence, to the required degree of accuracy, the solution is

$$x = 0{\cdot}739$$

Exercise A

1 Find, correct to 3 decimal places, the root of $x^3 - 9x + 1 = 0$ which is near to $x = 3$.

2 Solve $2 \sin x = x$, by first sketching the two graphs $y = 2 \sin x$ and $y = x$ to obtain a reasonable guess for your first estimate. Find the solution correct to 3 decimal places.

3 The ancient Greek mathematician, Hero, used the following method to find the square roots of a number. To find $\sqrt{N}$, guess a root P, then a better approximation is $\frac{1}{2}\left(P + \frac{N}{P}\right)$. Show that the Newton-Raphson method (applied to the function $f(x) = x^2 - N$) yields the same solution.

4 Find both roots of $x^2 + x - 1 = 0$
a) by the Newton-Raphson process,
b) using the quadratic formula.

5 Solve the equation $2 \sin x - \cos x = 0$ using the Newton-Raphson method with a first estimate of $x = 0{\cdot}5$. Give your answer accurate to 3 decimal places. Check your solution by finding the exact solution of the equation.

6 Find, correct to 3 decimal places, the solution of $t^4 - t - 3 = 0$, near to $t = 1{\cdot}4$.

13.2 Polynomial approximations to functions

In chapter 3, when we started to develop our method of differentiation, we established that

$$f'(a) = \lim_{h \to 0}\left\{\frac{f(a+h) - f(a)}{h}\right\}$$

It follows, then, that *if h is small*, we can approximate and say

$$f'(a) \approx \frac{f(a+h) - f(a)}{h}$$

$$hf'(a) \approx f(a+h) - f(a)$$

i.e. $f(a+h) \approx f(a) + hf'(a)$

This means that if we know the value of the function at a particular point $x = a$ we can determine an approximate value for the function at a nearby point $x = a + h$.

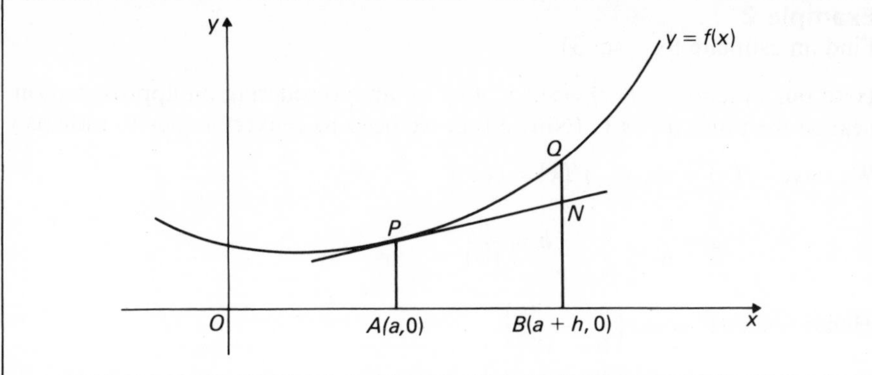

Figure 13.5

Graphically, we see that we are approximating to the curve $y = f(x)$ by using the tangent to the curve at $(a, f(a))$, as shown in figure 13.5. We are effectively saying that if the distance AB (i.e. h) is small enough, then the ordinate of N is approximately the same as the ordinate of Q. We can generalise this result by replacing $a + h$ by x, so that we obtain the approximate relation

$$f(x) \approx f(a) + (x - a)f'(a)$$

or $$y \approx f(a) + (x - a)f'(a)$$

This is a linear relationship and is the equation of the tangent at $x = a$ to the curve $y = f(x)$.

We can use the linear approximation to estimate values of many functions near to some known point, as the following examples illustrate.

Example 1

Estimate $\sqrt{50}$

Here we consider the function $f(x) = +\sqrt{x}$ near to the point $a = 49$.

We know that $f(a + h) \approx f(a) + hf'(a)$

i.e. $$\sqrt{(49 + 1)} \approx \sqrt{49} + 1\left(\frac{1}{2\sqrt{49}}\right), \text{ since } f'(x) = \frac{1}{2\sqrt{x}}$$

$$= 7 + \tfrac{1}{14}$$

$$= 7{\cdot}0714$$

Notice that without any further detailed calculation we could also find $\sqrt{48}$, by considering $\sqrt{(49 - 1)}$

So $$\sqrt{48} = \sqrt{(49 - 1)} \approx \sqrt{49} - \frac{1}{2\sqrt{49}}$$

$$= 7 - \tfrac{1}{14}$$

$$= 6{\cdot}9286$$

Compare these results with calculator values for $\sqrt{50}$ and $\sqrt{48}$.

Example 2

Find an estimate for sin 31°

Here our function is $f(x) = \sin x$, and we are considering an approximation near to the point $a = \pi/6$. (Notice that we need to convert angles to radians.)

We have $f(x) = \sin x$, $f'(x) = \cos x$

$$a = \frac{\pi}{6} \qquad h = \frac{\pi}{180}$$

Hence
$$\sin 31° = \sin\left(\frac{\pi}{6} + \frac{\pi}{180}\right)$$
$$\approx \sin\frac{\pi}{6} + \frac{\pi}{180}\left(\cos\frac{\pi}{6}\right)$$
$$= 0{\cdot}5 + (0{\cdot}0175)(0{\cdot}866)$$
$$= 0{\cdot}5151$$

Checking this result against tables or a calculator, we see that it gives quite reasonable accuracy.

Exercise B

1 Using the fact that $\sqrt[3]{125} = 5$, find estimates for $\sqrt[3]{126}$, $\sqrt[3]{127}$, $\sqrt[3]{124}$, $\sqrt[3]{123}$. How well do your answers compare with tables or calculator values?

2 Find estimates for tan 46° and tan 44°. (Remember to change to radians before using the formula.)

3 Find the equation of the tangent to the curve $y = x^3 + x$ at (2, 10). Draw the graph and its tangent and estimate from your graph how close to $x = 2$ the linear approximation is valid.

4 Find a linear approximation for $\sin^{-1}(a + h)$ for small values of h. Draw graphs of $y = \sin^{-1} x$ and the graph of your approximation to show how close they are.

5 Prove that an alternative linear approximation for $f(a + h)$ is given by

$$f(a + h) \approx f(a - h) + 2hf'(a)$$

Show on a diagram how this fits with the curve $y = f(x)$ near to $x = a$.

Better approximations

We have seen that the linear approximation

$$f(a + h) \approx f(a) + hf'(a)$$

is not very accurate, because a straight line takes no account of the curvature of the curve away from the point where $x = a$. We would obtain a better approximation by attempting to find a quadratic expression in h. Suppose such an approximation is

$$f(a + h) \approx f(a) + hf'(a) + h^2 \,.\, (P), \qquad \text{i}$$

where we have to find (P), which is a function of a only.

In **i**, our variable is h, and we see that, by differentiation with respect to h, (a is constant).

$$f'(a+h) \approx f'(a) + 2h(P)$$

(where dashes here denote differentiation with respect to h).

We differentiate again, and obtain

$$f''(a+h) \approx 2(P)$$

so that for small h, $P = \frac{1}{2}f''(a)$, and our quadratic approximation becomes

$$f(a+h) \approx f(a) + hf'(a) + \frac{h^2}{2} f''(a)$$

We can continue this process as many times as we like to obtain an nth order approximation

$$f(x+h) \approx f(a) + hf'(a) + \frac{h^2}{2!} f''(a) + \frac{h^3}{3!} f'''(a) + \frac{h^4}{4!} f^{(4)}(a) \cdots + \frac{h^n}{n!} f^{(n)}(a).$$

It is helpful if the reader verifies this result at least for the cubic approximation.

All of these approximations are known as **Taylor approximations**, and no attempt has been made here to give a rigorous proof of the result for the nth order approximation. It should be noted, too, that some consideration must be given to terms which follow the term in h^n. The analysis required is well beyond the coverage required for this course, and is omitted.

Example 3

Find a cubic approximation for

$\tan\left(\frac{\pi}{4} + h\right)$, and evaluate $\tan\left(\frac{\pi}{4} + 0{\cdot}01\right)$.

$$f(x) = \tan x \qquad \Rightarrow f\left(\frac{\pi}{4}\right) = 1$$

$$f'(x) = \sec^2 x \qquad \Rightarrow f'\left(\frac{\pi}{4}\right) = 2$$

$$f''(x) = 2\sec^2 x \tan x \qquad \Rightarrow f''\left(\frac{\pi}{4}\right) = 4$$

$$f'''(x) = 4\sec^2 x \tan^2 x + 2\sec^4 x \Rightarrow f'''\left(\frac{\pi}{4}\right) = 16$$

Hence
$$\tan\left(\frac{\pi}{4} + h\right) \approx 1 + 2h + \frac{4h^2}{2!} + \frac{16h^3}{3!}$$

$$= 1 + 2h + 2h^2 + \frac{8h^3}{3}$$

$$\tan\left(\frac{\pi}{4} + 0{\cdot}01\right) = 1 + 2(0{\cdot}01) + 2(0{\cdot}01)^2 + \tfrac{8}{3}(0{\cdot}01)^3$$

$$= 1 + 0{\cdot}02 + 0{\cdot}0002 + 0{\cdot}000\,002\,7$$

$$= 1{\cdot}020\,202\,7$$

and this compares very favourably with tabulated values.

Maclaurin approximations

There are many occasions when it is helpful to find approximate polynomial expressions for functions near to the origin. In the general Taylor approximation, we write $a = 0$, and obtain the **Maclaurin expansion**

$$f(h) \approx f(0) + hf'(0) + \frac{h^2}{2!}f''(0) + \frac{h^3}{3!}f'''(0) + \cdots$$

$$= \sum_{r=0}^{\infty} \frac{h^r f^{(r)}(0)}{r!} \qquad \textbf{(Note } 0! = 1\textbf{)}$$

We sometimes need to restrict the values which h can take for the series to be convergent.

Using the above, we can establish series for the trigonometric functions $\sin h$ and $\cos h$, as follows.

Example 4

$$\begin{aligned} f(x) &= \sin x &\Rightarrow\ & f(0) = 0 \\ f'(x) &= \cos x &\Rightarrow\ & f'(0) = 1 \\ f''(x) &= -\sin x &\Rightarrow\ & f''(0) = 0 \\ f'''(x) &= -\cos x &\Rightarrow\ & f'''(0) = -1 \\ f^{(4)}(x) &= \sin x &\Rightarrow\ & f^{(4)}(0) = 0 \\ f^{(5)}(x) &= \cos x &\Rightarrow\ & f^{(5)}(0) = 1 \end{aligned}$$

and so on. We obtain the series

$$\sin h \approx 0 + h(1) + \frac{h^2}{2!}(0) + \frac{h^3}{3!}(-1) + \cdots$$

$$\text{i.e.}\quad \sin h \approx h - \frac{h^3}{3!} + \frac{h^5}{5!} - \cdots$$

This series converges quite rapidly, as can be verified by comparing values of $\sin h$ with values obtained from the first three non-zero terms of the series. For example, consider $\sin(0{\cdot}2) = 0{\cdot}1987$ to 4 decimal places and

$$h - \frac{h^3}{3!} + \frac{h^5}{5!} = (0{\cdot}2) - \frac{(0{\cdot}2)^3}{3!} + \frac{(0{\cdot}2)^5}{5!}$$

$$= 0{\cdot}2 - 0{\cdot}001\,333 + 0{\cdot}000\,002\,7$$

$$= 0{\cdot}1987 \text{ (to 4 d.p.)}$$

It is a useful exercise to consider how large h may be taken before the first three non-zero terms fail to give a good approximation. This is left to the reader.

Exercise C

1 Use Maclaurin's expansion to find the first four non-zero terms in the series for $\cos h$.

2 Verify that the derivative of the series for $\sin h$ does yield the series for $\cos h$.

3 Find the general Taylor expansion for $(a + h)^n$ where $n \in \mathbb{R}$. How does your answer compare with the Binomial Theorem for $(a + h)^n$ where $n \in \mathbb{Z}^+$?

4 If $f(x) = \tan^{-1} x$, find the first five derivatives of $f(x)$, and hence use Maclaurin's expansion to find a series for $\tan^{-1} x$. (This series is known as 'Gregory's series'.)

5 By writing $(a + h)$ as x in the quadratic Taylor approximation, find the equation of the curve which approximates to $y = \sin x + \cos x$ near to $x = \frac{\pi}{4}$. Sketch the curve of $y = \sin x + \cos x$ and the quadratic curve on the same axes. Hence estimate a range of values round $\frac{\pi}{4}$ for which the quadratic curve is a reasonable approximation.

13.3 Binomial expansions

Particularly important at this stage is the result you obtained (did you?) to question 3 in the above exercise. Let us look more closely at this example.

$$f(x) = x^n$$
$$f'(x) = nx^{n-1}$$
$$f''(x) = n(n-1)x^{n-2}$$
$$f'''(x) = n(n-1)(n-2)x^{n-3}$$

and so on. We obtain

$$(a + h)^n = a^n + a^{n-1}\ \ nh + a^{n-2}\frac{n(n-1)h^2}{2!} + a^{n-3}\frac{n(n-1)(n-2)h^3}{3!} + \cdots$$

which we could write as $(a + h)^n = \sum_{r=0}^{N} \frac{a^{n-r}h^r n(n-1)\cdots(n-r+1)}{r!}$

where the upper limit in the summation is dependent on the number of terms required in the expansion. When $n \in \mathbb{Z}^+$, clearly we can simplify $n(n-1)\cdots(n-r+1)$ into the form $\frac{n!}{(n-r)!}$ but this expression is, in general, meaningless when $n \in \mathbb{R}$. We consider some special cases.

Example 1

$$(1 + h)^{-1} = 1 - h + h^2 - h^3 + h^4 - \cdots$$

Here we obtain the series quickly, by substitution of appropriate values in our general expression above. Notice, however, that we have a geometric series on the right-hand side, with a first term of 1 and a common ratio of $(-h)$. If we sum such a series (see chapter 8) to N terms, we obtain

$$\left\{\frac{1 - (-h)^N}{1 + h}\right\}$$

If we wish to find the sum as $N \to \infty$, we note that if $|h| < 1$, then $(-h)^N \to 0$, and the sum becomes $\frac{1}{1 + h}$ or $(1 + h)^{-1}$.

Notice that we have needed here to restrict values that h may take, so we must treat our binomial expansions with caution, and ensure that they do converge. Let us look again at our original expansion

$$(1 + h)^{-1} = 1 - h + h^2 - h^3 + h^4 - \cdots$$

One test for convergence is to compare two successive terms in the expansion, and we see that a series will converge if the ratio of the absolute values of the $(n+1)$th term and the nth term, written as $\left|\dfrac{(n+1)\text{th term}}{n\text{th term}}\right|$, tends to a limit (as $n \to \infty$) which is less than 1.

In our example, we have convergence if

$$\left|\frac{(-h)^n}{(-h)^{n-1}}\right| < 1$$

$$\Rightarrow \quad |h| < 1 \text{ as before.}$$

Example 2
Find the binomial expansion for $(1+h)^{1/2}$, and find the range of values of h for which the series converges. Use your result to calculate to 4 decimal places $\sqrt{(1{\cdot}03)}$.

$$(1+h)^{1/2} = 1 + \tfrac{1}{2}h + \frac{\frac{1}{2}(-\frac{1}{2})}{2!}h^2 + \frac{\frac{1}{2}(-\frac{1}{2})(-\frac{3}{2})}{3!}h^3 + \frac{\frac{1}{2}(-\frac{1}{2})(-\frac{3}{2})(-\frac{5}{2})}{4!}h^4 + \cdots$$

i.e. $(1+h)^{1/2} = 1 + \tfrac{1}{2}h - \tfrac{1}{8}h^2 + \tfrac{1}{16}h^3 - \tfrac{5}{128}h^4 + \cdots$

Convergence occurs when

$$\left|\frac{\frac{1}{2}(-\frac{1}{2})(-\frac{3}{2})\cdots(1\frac{1}{2}-n)h^n}{n!} \Big/ \frac{\frac{1}{2}(-\frac{1}{2})(-\frac{3}{2})\cdots(2\frac{1}{2}-n)h^{n-1}}{(n-1)!}\right| < 1$$

$$\Rightarrow \left|\frac{(1\frac{1}{2}-n)h}{n}\right| < 1$$

$$\Rightarrow |h| < 1, \text{ since } \left|\frac{1\frac{1}{2}-n}{n}\right| \text{ is clearly } < 1$$

To find $\sqrt{(1{\cdot}03)}$, $h = 0{\cdot}03$, and we have

$$(1+0{\cdot}03)^{1/2} = 1 + \tfrac{1}{2}(0{\cdot}03) - \tfrac{1}{8}(0{\cdot}03)^2$$

(remaining terms cannot affect 4th decimal place)

$$= 1 + 0{\cdot}015 - 0{\cdot}0001$$
$$= 1{\cdot}0149$$

Exercise D

1 Find the binomial expansion for $(1+h)^{-2}$, and find the range of values of h for which the series converges.

2 Find $(1+h)^{-1/2}$ and use your result to estimate $(1{\cdot}03)^{-1/2}$. Is your answer a reasonable estimate to the reciprocal of $(1{\cdot}03)^{1/2}$ calculated in Example 2 above?

3 Expand $(1+h)^{-3}$ as far as the term in h^4, and investigate the range of values for h for which the series converges.

4 $(1 + h)^n$ when expanded has its term in h^2 given by $-\dfrac{3}{32}h^2$

Find two possible values for n. If the following term is $\frac{7}{128}h^3$ state which of your two values of n is now valid.

5 Find the binomial expansion for $(1 + h^2)^{-1}$. Integrate your series between the limits 0 and x and compare the result with that obtained for Gregory's series (see Exercise C question 4). Explain the connection.

6 Find an expansion for $(1 - h^2)^{1/2}$ and integrate your series between the limits 0 and x. Assuming that this method of integration of a series is valid, state the functon for which this is the series expansion.

13.4 Approximate integration

In order to develop methods for the approximate evaluation of definite integrals, we recall that we began our work on integration by considering areas under curves. Clearly, we could draw a curve on graph paper and count the squares within the required area to give an estimate for the area. As a refinement of that, we shall look at the problem as follows (see figure 13.6).

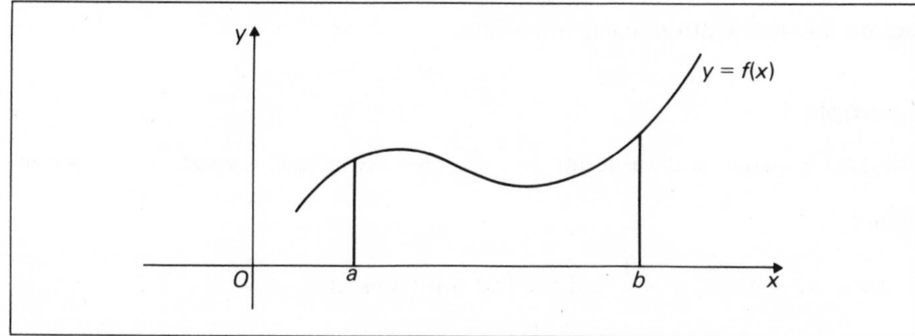

Figure 13.6

We are attempting to evaluate $\int_a^b f(x)\,dx$ to obtain the area under $y = f(x)$ from $x = a$ to $x = b$. We divide the interval $[a, b]$ of the domain into n equally wide subintervals; the width of each is then $\left(\dfrac{b - a}{n}\right)$ as shown in figure 13.7.

Let the length of the ordinates be $y_0, y_1, y_2, \ldots, y_n$. We join the tops of

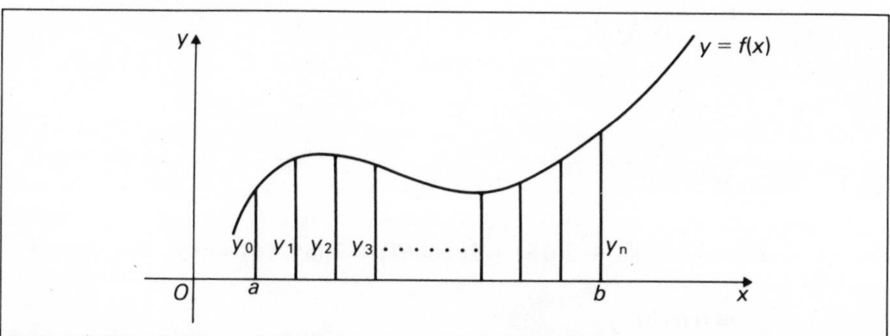

Figure 13.7

these ordinates by straight lines so that we have n trapezia. The total area of these n trapezia will give us an approximation to the area under the curve.

The area of the first trapezium is

$$\left(\frac{b-a}{n}\right)\left(\frac{y_0+y_1}{2}\right)$$

The area of the second trapezium is

$$\left(\frac{b-a}{n}\right)\left(\frac{y_1+y_2}{2}\right) \quad \text{and so on}$$

Hence the total area of the n trapezia is

$$\left(\frac{b-a}{n}\right)\left[\frac{y_0+y_1}{2}+\frac{y_1+y_2}{2}+\frac{y_2+y_3}{2}+\cdots+\frac{y_{n-1}+y_n}{2}\right]$$

$$=\frac{1}{2}\left(\frac{b-a}{n}\right)(y_0+2y_1+2y_2+\cdots+2y_{n-1}+y_n)$$

This is known as the **Trapezium Rule** for approximate evaluation of definite integrals. It is more usual to write this as

$$\tfrac{1}{2}h(y_0+2y_1+2y_2+\cdots+2y_{n-1}+y_n)$$

where h is the width of each trapezium.

Example 1

Find an approximate value for $\int_1^2 \frac{1}{x}\,dx$, by taking four trapezia as shown in figure 13.8.

Here we see that $y=\frac{1}{x}$ and the five ordinates are

$$y_0=\frac{1}{1}=1$$

$$y_1=\frac{1}{1{\cdot}25}=0{\cdot}8$$

$$y_2=\frac{1}{1{\cdot}5}=0{\cdot}6667$$

$$y_3=\frac{1}{1{\cdot}75}=0{\cdot}5714$$

$$y_4=\frac{1}{2}=0{\cdot}5$$

and $h=0{\cdot}25$

So $\int_1^2 \frac{1}{x}\,dx \approx \tfrac{1}{2}(0{\cdot}25)[1+2(0{\cdot}8)+2(0{\cdot}6667)+2(0{\cdot}5714)+0{\cdot}5]$

$$=0{\cdot}6970$$

(**Note** The exact value of this integral will be found in chapter 14.)

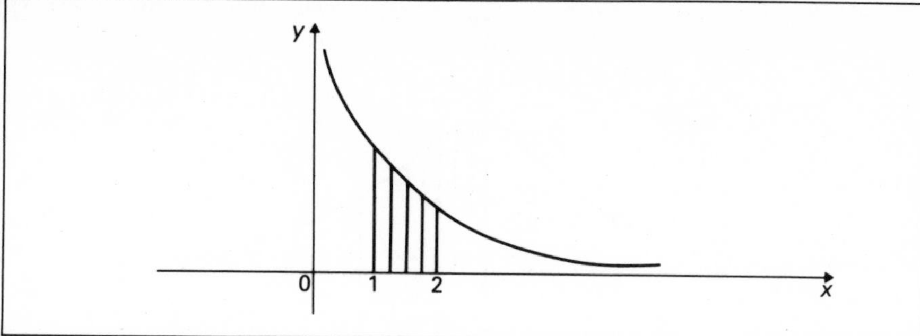

Figure 13.8

Example 2

Use the trapezium rule with eight strips (figure 13.9) to find an approximation for $\int_0^1 \frac{1}{1+x^2}\,dx$

Compare your answer with the exact value of the integral to find an estimate for π, using the value of the integral found in chapter 12.

In this case, the width of each strip is 0·125 and the lengths of the ordinates are

$$y_0 = \frac{1}{1+0^2} = 1$$

$$y_1 = \frac{1}{1+(0{\cdot}125)^2} = 0{\cdot}9846$$

$$y_2 = \frac{1}{1+(0{\cdot}25)^2} = 0{\cdot}9412$$

$$y_3 = \frac{1}{1+(0{\cdot}375)^2} = 0{\cdot}8767$$

$$y_4 = \frac{1}{1+(0{\cdot}5)^2} = 0{\cdot}8$$

$$y_5 = \frac{1}{1+(0{\cdot}625)^2} = 0{\cdot}7191$$

$$y_6 = \frac{1}{1+(0{\cdot}75)^2} = 0{\cdot}64$$

$$y_7 = \frac{1}{1+(0{\cdot}875)^2} = 0{\cdot}5664$$

$$y_8 = \frac{1}{1+1^2} = 0{\cdot}5$$

Hence

$$\int_0^1 \frac{1}{1+x^2}\,dx \approx$$

$$\tfrac{1}{2}(0{\cdot}125)[1 + 2(0{\cdot}9846 + 0{\cdot}9412 + 0{\cdot}8767 + 0{\cdot}8 + 0{\cdot}7191 + 0{\cdot}64 + 0{\cdot}5664) + 0{\cdot}5]$$
$$= 0{\cdot}7847$$

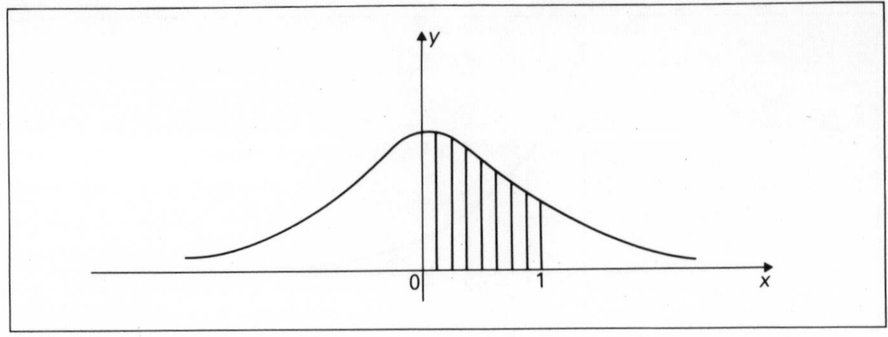

Figure 13.9

Now, the integral may be evaluated exactly by means of a trigonometric substitution (see chapter 12) and we obtain

$$\int_0^1 \frac{1}{1+x^2}\,dx = [\tan^{-1} x]_0^1 = \frac{\pi}{4}$$

Hence, we may deduce that

$$0{\cdot}7847 \approx \frac{\pi}{4}$$

$$\Rightarrow \qquad \pi \approx 3{\cdot}139$$

And we see that this is a reasonable estimate for π, but is only accurate to the first decimal place. Clearly we could improve on the accuracy by taking a greater number of strips, or, alternatively, develop a more accurate method.

Exercise E

1 Evaluate, by the trapezium rule with four strips, $\int_0^3 (2x+3)\,dx$. Compare your answer with the value obtained by integration. Explain the results.

2 Use six strips in the trapezium rule to estimate the value of $\int_0^{0{\cdot}6} \sin x\,dx$, and compare your answer with the value obtained by integration.

3 Use four strips to estimate the value of

$$\int_{-2}^{-1} \frac{1}{x}\,dx$$

and by sketching appropriate graphs, explain how your answer is related to that obtained in Example 1 on page 276.

4 Evaluate $\int_0^2 \frac{1}{1+x^3}\,dx$ using the trapezium rule with

a) four strips,
b) eight strips in.

5 Use a) four b) six c) eight strips to estimate the value of $\int_0^{\pi/4} \tan x\,dx$.
(**Note** The integration of tan x will be explained in chapter 14.)

You will have realised from this exercise and the previous examples, that approximations to integrals by the trapezium rule do not, in general, give very good results. The reason is that straight line segments are not good approximations to curves. We seek, therefore, a curved approximation as we did when trying to improve on the linear Taylor approximation.

We know that a parabola can always be drawn through any three non-collinear points, and we can use this fact to obtain a better estimation of area. Consider the curve $y = f(x)$ and the ordinates y_0, y_1, y_2, equally spaced as shown in figure 13.10. Let h be the width of the two strips, as before.

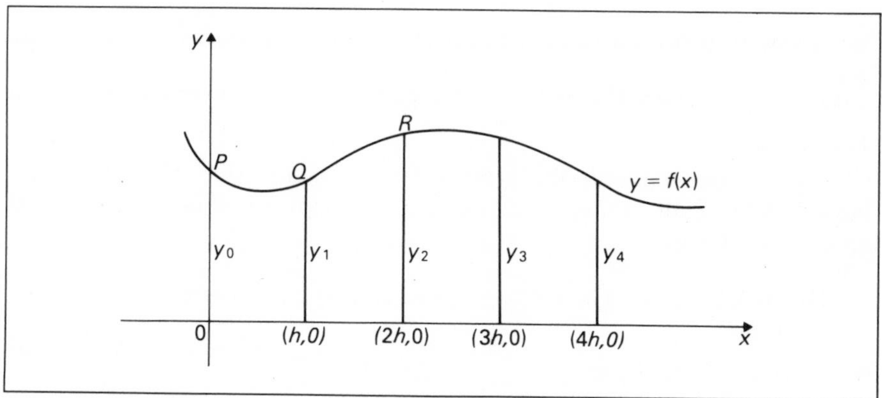

Figure 13.10

We wish to find a parabola, whose general equation we may take as $y = ax^2 + bx + c$, which passes through P, Q, R. Note that we do not lose any generality of the result by taking P on the y-axis.

Hence $y_0 = c$
$y_1 = ah^2 + bh + c$
$y_2 = 4ah^2 + 2bh + c$

and solving these equations for a, b, c we obtain

$$a = \frac{y_2 - 2y_1 + y_0}{2h^2}$$

$$b = \frac{4y_1 - y_2 - 3y_0}{2h}$$

$$c = y_0$$

Now, we can find the area of the region under $y = ax^2 + bx + c$ and between $x = 0$ and $x = 2h$ by integration:

$$\int_0^{2h} (ax^2 + bx + c)\, dx = \left[\frac{ax^3}{3} + \frac{bx^2}{2} + cx\right]_0^{2h}$$

$$= \frac{8ah^3}{3} + 2bh^2 + 2ch$$

By substituting the values of a, b and c just found, we see that the area becomes

$$\frac{8h^3}{3}\left(\frac{y_2 - 2y_1 + y_0}{2h^2}\right) + 2h^2\left(\frac{4y_1 - y_2 - 3y_0}{2h}\right) + 2hy_0$$

$$= \frac{4h}{3}(y_2 - 2y_1 + y_0) + h(4y_1 - y_2 - 3y_0) + 2hy_0$$

$$= \frac{h}{3}(4y_2 - 8y_1 + 4y_0 + 12y_1 - 3y_2 - 9y_0 + 6y_0)$$

$$= \frac{h}{3}(y_0 + 4y_1 + y_2)$$

We repeat the process with the next two strips, and, in a similar way, we obtain $\frac{h}{3}(y_2 + 4y_3 + y_4)$ for the area of the region under the curve containing the two strips.

Clearly, we can continue the process over as many pairs of strips as are needed. Adding all the separate results for n strips, we obtain, for the total area, the expression

$$\frac{h}{3}[(y_0 + 4y_1 + y_2) + (y_2 + 4y_3 + y_4) + (y_4 + 4y_5 + y_6) + \cdots + (y_{n-2} + 4y_{n-1} + y_n)]$$

which we can simplify, as

$$\frac{h}{3}\{y_0 y_n + 4(y_1 + y_3 + y_5 + \cdots + y_{n-1}) + 2(y_2 + y_4 + y_6 + \cdots + y_{n-2})\}$$

This is known as **Simpson's Rule** for the approximate evaluation of definite integrals. Note that, by the nature of the result, we must have an *even* number of strips.

Example 3

Evaluate, approximately, $\int_0^4 \frac{x}{1 + x^2}\,dx$ using eight strips

The width of each strip, h, is 0·5 units. Tabulating the calculations as follows eases the final evaluation.

$y_0 = \frac{0}{1 + 0} = 0$ $\quad y_1 = \frac{0{\cdot}5}{1 + (0{\cdot}5)^2} = 0{\cdot}4$ $\quad y_2 = \frac{1}{1 + 1^2} = 0{\cdot}5$

$y_3 = \frac{1{\cdot}5}{1 + (1{\cdot}5)^2} = 0{\cdot}4615$ $\quad y_4 = \frac{2}{1 + 2^2} = 0{\cdot}4$

$y_5 = \frac{2{\cdot}5}{1 + (2{\cdot}5)^2} = 0{\cdot}3448$ $\quad y_6 = \frac{3}{1 + 3^2} = 0{\cdot}3$

$y_7 = \frac{3{\cdot}5}{1 + (3{\cdot}5)^2} = 0{\cdot}2642$

$y_8 = \frac{4}{1 + 4^2} = 0{\cdot}2353$

$y_0 + y_8 = 0{\cdot}2353$	odd ordinates $= 1{\cdot}4705$	even ordinates $= 1{\cdot}2$

Hence area $\approx \frac{1}{3} \times (0{\cdot}5)\{(0{\cdot}2353) + 4(1{\cdot}4705) + 2(1{\cdot}2)\} = 1{\cdot}4196$

Example 4

Evaluate, approximately, $\int_0^8 (x^3 + x)\,dx$ using Simpson's Rule with eight strips.

$y_0 = 0$	$y_1 = 2$	$y_2 = 10$
	$y_3 = 30$	$y_4 = 68$
	$y_5 = 130$	$y_6 = 222$
	$y_7 = 350$	
$y_8 = 520$		
$y_0 + y_8 = 520$	odd ordinates $= 512$	even ordinates $= 300$

Hence $\int_0^8 (x^3 + x)\,dx \approx \frac{1}{3} \times 1\{520 + 4(512) + 2(300)\}$

$= 1056$

We compare this with the value obtained by integration:

$$\int_0^8 (x^3 + x)\,dx = \left[\frac{x^4}{4} + \frac{x^2}{2}\right]_0^8$$

$$= \frac{8^4}{4} + \frac{8^2}{2} = 1024 + 32$$

$$= 1056$$

You should sketch the curve $y = x^3 + x$ for $x = 0$ to $x = 8$, and sketch some typical parabolas to see why the approximate method yields an exact answer.

Exercise F

1 Use Simpson's Rule with a) 4 strips b) 8 strips to find an approximate value for

$$\int_0^{\pi/4} \tan x\,dx$$

2 a) Compare the results of using i) trapezium rule and ii) Simpson's rule, with eight strips in each case, to evaluate

$$\int_0^2 \frac{1}{1 + x^2}\,dx, \text{ correct to 4 decimal places.}$$

b) Evaluate the integral by means of the substitution $x = \tan\theta$, again correct to 4 decimal places. Compare this answer with the two obtained in a).

3 Explain why Simpson's rule will give an exact answer for $\int_a^b p(x)\,dx$ where $p(x)$ is a polynomial of degree 1 or 2. Check this by evaluating

$$\int_1^2 (2x + 3)\,dx \text{ and } \int_0^1 (x^2 + x)\,dx$$

both by Simpson's Rule and by direct integration.

4 Evaluate $\int_1^2 \frac{1}{x^2 + 2x}\, dx$, using Simpson's rule with a) four strips and b) eight strips. What improvement would you expect in the accuracy of the result if the rule were applied with 16 strips?

5 Evaluate $\int_0^2 x\, dx$ directly. Use Simpson's rule with two strips to show that the result is exact. Explain why Simpson's rule with four strips is also accurate.

6 Evaluate $\int_1^4 \log x\, dx$ by taking six strips in Simpson's rule.
(**Note** Log x denotes the **common logarithm** of x and should not be confused with ln x, which denotes the **natural logarithm** of x – see also chapter 14.)

7 Use Simpson's rule with an appropriate number of strips to evaluate

$$\int_0^{\pi/2} \frac{1}{1 + \cos\theta}\, d\theta \text{ correct to 3 decimal places.}$$

How can you be sure that your answer is as accurate as this?

8 We know that the equation of a circle of radius a, centre $(0, 0)$ is $x^2 + y^2 = a^2$. Write down an integral which represents the area of the circle and use Simpson's rule with an appropriate number of strips to calculate an approximate value for the area. Hence find an estimate for π.

9 The equation of an ellipse may be written in the form

$$y = \pm\sqrt{\left(9 - \frac{9x^2}{4}\right)}$$

Sketch the curve and use Simpson's rule to calculate its area. The exact value of the area is 6π. How accurate is your estimate?

10 Prove that $\int_{-a}^{a} x^3\, dx$ has the same value whether evaluated directly, by the trapezium rule or by Simpson's rule. Sketch the curve to show why this is so. Consider $\int_{-a}^{a} x^2\, dx$ by all three methods, and compare the results obtained in this case.

Investigate $\int_{a-h}^{a+h} x^3\, dx$ evaluated both directly and by Simpson's rule with 2 strips.

Two Special Functions

CHAPTER FOURTEEN

14.1 The exponential function

We have seen that an exponential function of the form a^x is one in which a given base a is raised to a variable exponent (i.e. power or index). Typical of such functions, for differing values of a, are 2^x, 3^x, 4^x, 2^{-x}, 3^{2x}, ... A common property of these functions is that for varying values of x the graphs of all of them, when drawn on cartesian axes, will pass through the fixed point $Q(0, 1)$. This is due to the fact that $a^0 = 1$, for all a. Figure 14.1 shows sketches of the graphs of some of these functions. Note also that the graphs are all asymptotic to the x-axis.

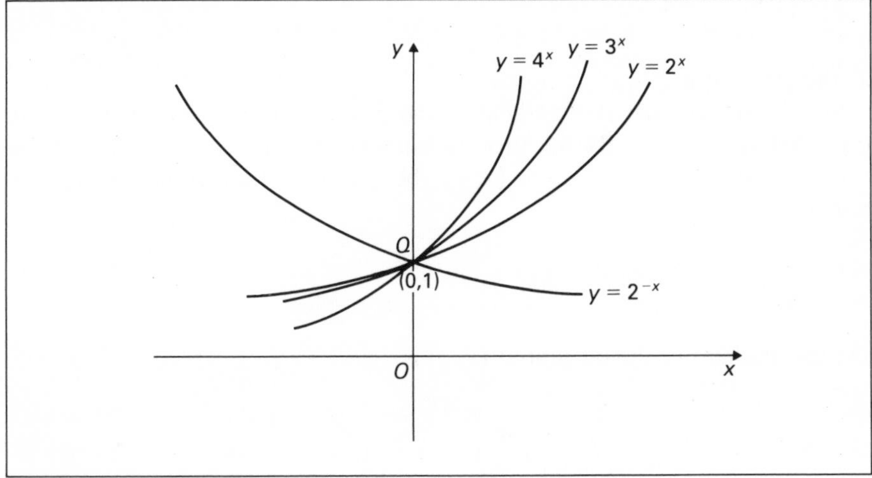

Figure 14.1

When $a > 0$, we can seek to derive the gradient of the tangent at the point $Q(0, 1)$ to these curves by the usual method of finding the limiting value of a successive set of values of the gradients of chords PQ. In this process, if we take P to be the point $(\delta x, y)$, the gradient of the chord PQ is $m = \dfrac{y-1}{\delta x}$ (figure 14.2), and δx would tend to zero.

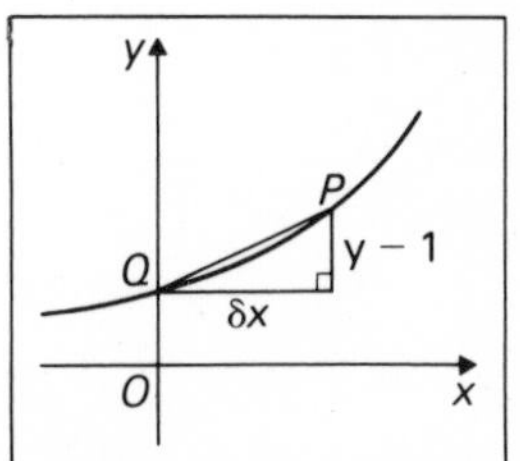

Figure 14.2

The approximate numerical values of the gradients of chords PQ, for the exponential functions $y = a^x$ where $a = 4$, 3, 2 and 1·5, are shown in the following table.

	δx	0·5	0·25	0·1	0·01	0·001	0·0001
For $y = 4^x$	m	2·00	1·66	1·49	1·40	1·39	1·39
For $y = 3^x$	m	1·46	1·26	1·16	1·10	1·10	1·10
For $y = 2^x$	m	0·83	0·76	0·72	0·70	0·69	0·69
For $y = 1{\cdot}5^x$	m	0·45	0·43	0·41	0·41	0·41	0·41

It is a worthwhile exercise to check these values on your calculator.

From these values it is clear that, as the base a takes decreasing values, the values of the gradient m decrease likewise, and so, ultimately, will give the gradients of the tangents at (0, 1). Also as δx becomes smaller and smaller the table shows the limit to which the gradients of the chords tend, as $\delta x \to 0$.

For $a = 4$ the gradient of the tangent approximates to 1·39
$a = 3$ the gradient of the tangent approximates to 1·10
$a = 2$ the gradient of the tangent approximates to 0·69
$a = 1{\cdot}5$ the gradient of the tangent approximates to 0·41

It is apparent therefore, that there will be, somewhere within the set of values of a, a particular value of a for which the gradient of the tangent at (0, 1) is unity. Also the values calculated suggest that this value is between $a = 2$ and $a = 3$, closer to the value 3 than to 2. We call this special number e and the corresponding function **the exponential function**, which we write either as e^x or $\exp(x)$.

Thus e is defined by the fact that

$$\left[\frac{d}{dx}(e^x)\right]_{x=0} = 1$$

i.e. the gradient of $y = e^x$, at $x = 0$, is 1.

Now let us consider the gradient of the tangent to the graph of $y = e^x$ at any other point (x, y). By the process outlined in chapter 3, if $x + h$ is the x-coordinate of a point close to (x, y), the gradient of the chord joining the points is

$$\frac{e^{x+h} - e^x}{h}$$

and the gradient of the tangent at (x, y) is

$$\lim_{h\to 0}\left\{\frac{e^{x+h} - e^x}{h}\right\}$$

Now to evaluate this limit, we write $e^{x+h} - e^x = e^x(e^h - 1)$ and, using Maclaurin's theorem, in the form

$$f(h) \equiv e^h = f(0) + \frac{f'(0)}{1!}h + \frac{f''(0)}{2!}h^2 + \cdots$$

$$f(h) = e^h \qquad \Rightarrow f(0) = e^0 = 1$$

and $f'(0) = e^0 = \dfrac{d}{dh}[e^h]_{h=0} = 1$

Thus $e^h = 1 + h +$ terms in $h^2, h^3, \ldots$

So $\dfrac{e^h - 1}{h} = 1 + h \times \text{(other terms)}$

And when $h \to 0$, $\displaystyle\lim_{h\to 0}\left\{\frac{e^h - 1}{h}\right\} = 1 + 0 \times \text{(other terms)}$

$$= 1$$

Hence the gradient of the tangent at (x, y) is

$$\lim_{h\to 0}\left\{\frac{e^{x+h}-e^x}{h}\right\} = \lim_{h\to 0}\left\{\frac{e^x(e^h-1)}{h}\right\}$$

$$= e^x \lim_{h\to 0}\left\{\frac{e^h-1}{h}\right\}$$

$$= e^x,$$

using the value for the limit as established.

This result can be written in the form

$$\frac{d}{dx}(e^x) = e^x$$

and it follows immediately that

$$e^x = \frac{d}{dx}(e^x) = \frac{d^2}{dx^2}(e^x) = \frac{d^3}{dx^3}(e^x) = \cdots = \frac{d^n}{dx^n}(e^x) = \cdots$$

for all $n \in \mathbb{R}^+$.

14.2 Exponential series

We have used Maclaurin's Theorem to establish the form of the expansion of a function $f(x)$ as an infinite series of ascending powers of x, namely

$$e^x \equiv 1 + \frac{x}{1!} + \frac{x^2}{2!} + \frac{x^3}{3!} + \cdots + \frac{x^n}{n!} + \cdots$$

This series helps us to compute an approximate value for e more closely than the gradient method at the beginning of this chapter.

For putting $x = 1$

$$e^1 \approx 1 + 1 + 0{\cdot}5 + 0{\cdot}1\dot{6} + 0{\cdot}041\dot{6} + 0{\cdot}0083 + 0{\cdot}001382 + \cdots$$
$$\approx 2{\cdot}718281 \ldots$$

or, corrected to three places of decimals, $e = 2{\cdot}718$.

Replacing x by $-x$ in the series expansion for e^x leads to the series for

$$e^{-x} \equiv 1 - \frac{x}{1!} + \frac{x^2}{2!} - \frac{x^3}{3!} + \cdots + (-1)^n\frac{x^n}{n!} + \cdots$$

Further, by considering the graphs of $y = e^x$ and $y = e^{-x}$, it is clear that $y = e^{-x}$ is a reflection of $y = e^x$ in the y-axis (figure 14.3).

The curve shown by the dotted line in figure 14.3 is that of

$$y = \tfrac{1}{2}(e^x + e^{-x})$$

and is achieved by determining, for values of x on both curves, the average value of the corresponding ordinates. Such a curve is called a **catenary** and is the shape taken up by a heavy chain hanging freely under gravity.

Other basic exponential graphs which should be recognised are $y = -e^x$ (obtained by reflecting $y = e^x$ in the x-axis) and $y = -e^{-x}$ (obtained by reflecting $y = e^{-x}$ in the x-axis). A worthwhile exercise now is to draw the

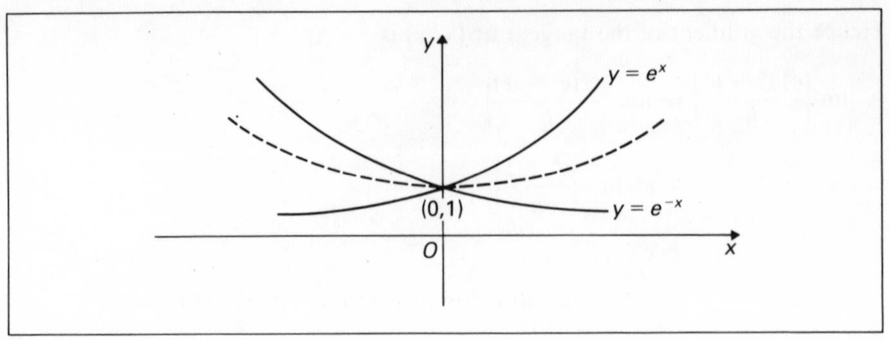

Figure 14.3

graphs of these functions for values of x, such that $-2 \leqslant x \leqslant 2$. Many standard sets of mathematical tables include tables of values of e^x and e^{-x}, as do electronic calculators. You will possibly find the e^x 'key' associated as an inverse operation with the ln 'key'. We relate these two operations later in this chapter.

Example 1

Expand $(x + 1)e^x$ as a series of ascending powers of x, writing down the coefficient of the x^n term in the expansion.

$$(x + 1)e^x$$

$$\equiv (x+1)\left(1 + \frac{x}{1!} + \frac{x^2}{2!} + \frac{x^3}{3!} + \cdots + \frac{x^n}{n!} + \cdots\right)$$

$$= 1 + \frac{x}{1!} + \frac{x^2}{2!} + \frac{x^3}{3!} + \cdots + \frac{x^n}{n!} + \cdots$$

$$+ x + \frac{x^2}{1!} + \frac{x^3}{2!} + \cdots \frac{x^n}{(n-1)!} + \cdots$$

$$= 1 + \left(1 + \frac{1}{1!}\right)x + \left(\frac{1}{1!} + \frac{1}{2!}\right)x^2 + \left(\frac{1}{2!} + \frac{1}{3!}\right)x^3 + \cdots + \left(\frac{1}{(n-1)!} + \frac{1}{n!}\right)x^n + \cdots$$

$$= 1 + \frac{2}{1!}x + \left(\frac{2}{2!} + \frac{1}{2!}\right)x^2 + \left(\frac{3}{3!} + \frac{1}{3!}\right)x^3 + \cdots + \left(\frac{n}{n!} + \frac{1}{n!}\right)x^n + \cdots$$

$$= 1 + \frac{2}{1!}x + \frac{3}{2!}x^2 + \frac{4}{3!}x^3 + \cdots + \frac{(n+1)}{n!}x^n + \cdots$$

Hence the coefficient of x^n is $\dfrac{n+1}{n!}$

Example 2

Using the series expansion for e^x derive, to three decimal places, the value of $\sqrt{e}$ and check your value from tables.

If $N = \sqrt{e}$, then $N = e^{1/2} = 1{\cdot}6487$ from tables

i.e. $N = 1{\cdot}649$ to 3 decimal places

Now $e^{1/2} = e^{0.5} = 1 + \frac{0.5}{1!} + \frac{(0.5)^2}{2!} + \frac{(0.5)^3}{3!} + \frac{(0.5)^4}{4!} + \cdots$

$$= 1 + 0.5 + 0.125 + 0.020\,83 + 0.002\,60 + 0.000\,26 + \cdots$$

$$\approx 1.648\,69$$

$$= 1.649 \text{ (to 3 d.p.)}$$

Example 3
Find the equation of the tangent at the point (0, 1) on the curve $y = e^x$.

The equation of the tangent is

$$y - 1 = m(x - 0)$$

where m is the gradient of the tangent at the point (0, 1) on the curve.
Now when $y = e^x$

$$\frac{dy}{dx} = e^x$$

$$\Rightarrow \quad m = \left[\frac{dy}{dx}\right]_{x=0} = e^0 = 1$$

$\Rightarrow$ the equation of the tangent is $y - 1 = x$
i.e. $y = x + 1$

This example illustrates a method of approach which will be used later in this section. We knew of course, that at (0, 1) the gradient of the tangent was 1, and also that the curve passed through this point.

Example 4
The parametric equations of a given curve are

$$x = e^t, \qquad y = e^{2t} + 1$$

Determine the cartesian equation of the locus.

Since $y = (e^t)^2 + 1$ (using **Rule 3** of the rules of indices)
$y = x^2 + 1$
$\Rightarrow$ $x = \pm\sqrt{(y - 1)}$ is the equation of the locus

Exercise A

1 Use the appropriate exponential series to determine the value of

a) $\frac{1}{e}$ b) $e^{0.1}$ c) $e^{-1/2}$

Give your answers correct to three places of decimals and check them against values obtained from a set of tables.

2 Sketch the graph of the function $f(x) = \frac{e^x - e^{-x}}{2}$ for values of x between -2 and 2 inclusive.

3 Expand in the form of a series of ascending powers of x, up to and including the term in x^4

a) $e^{x/2}$ c) e^{-2x}

b) e^{2x} d) $\dfrac{e^x + e^{-x}}{2}$

4 Prove that the sum of the series

$$1 + \frac{1}{2!} + \frac{1}{4!} + \frac{1}{6!} + \cdots \quad \text{is } \frac{1}{2}\left(e + \frac{1}{e}\right)$$

Derive a similar answer for the sum of the series

$$\frac{1}{1!} + \frac{1}{3!} + \frac{1}{5!} + \frac{1}{7!} + \cdots$$

5 Determine the first four terms of the series expansions of

a) $\dfrac{e^{2x} + e^x}{e^{3x}}$

b) $(x - 1)e^{-x} + (x + 1)e^x$

c) $e^x \sin x$ (using the expansion of $\sin x$ as a series of powers of x also)

6 Prove that

$$(e^x + e^{-x})(e^y + e^{-y}) + (e^x - e^{-x})(e^y - e^{-y}) \quad = \quad 2(e^{x+y} + e^{-x-y})$$

7 The tangent and normal at the point (0, 2) on the curve $y = 2e^x$ meet the x-axis at T and N respectively. Determine the length of TN.

8 If $\tan \theta = \frac{1}{2}(e^x - e^{-x})$ determine, in terms of x, the value of $\cos \theta$.

9 Evaluate

$$\frac{2^2}{1!} + \frac{2^4}{3!} + \frac{2^6}{5!} + \cdots$$

10 Determine the cartesian equations of the curves whose equations given in parametric form are

a) $x = e^t + e^{-t}, \quad y = e^t - e^{-t}$ b) $x = \dfrac{e^t + e^{-t}}{e^t - e^{-t}}, \quad y = e^{2t} - 1$

14.5 Properties of the exponential function

Since e is only a special value of the general base a, the exponential functions will obey the general rules of indices. Particularly

$$e^x \times e^y = e^{x+y}$$
$$e^x \div e^y = e^{x-y}$$
$$(e^x)^y = e^{xy}$$

and so on.

Furthermore, since $\dfrac{d}{dx}(e^x) = e^x, \quad \int e^x \, dx = e^x + K$

Now let us consider the derivatives of more complicated expressions, such as $e^{f(x)}$ where $f(x)$ is of any forms already met in this text.

Let $\quad y = e^{f(x)}$ and write $u = f(x)$

Then $\quad y = e^u$

So $\quad \dfrac{dy}{du} = e^u \quad$ and $\quad \dfrac{du}{dx} = f'(x)$

By the chain rule,

$$\frac{dy}{dx} = \frac{dy}{du} \cdot \frac{du}{dx}$$

Hence $\quad \dfrac{dy}{dx} = e^u f'(x)$

Hence $\quad \dfrac{dy}{dx} = f'(x)e^{f(x)}$, replacing u by $f(x)$

We now have a straightforward method which can generally be applied in a single line of writing.

Thus, for example, if

$y = e^{2x} \qquad \dfrac{dy}{dx} = \left\{\dfrac{d}{dx}(2x)\right\}e^{2x} = 2e^{2x}$

$y = e^{-3x} \qquad \dfrac{dy}{dx} = -3e^{-3x}$

$y = e^{x^2} \qquad \dfrac{dy}{dx} = 2xe^{x^2}$

$y = e^{\tan x} \qquad \dfrac{dy}{dx} = \sec^2 x \; e^{\tan x}$

$y = e^{\sqrt{(x^2+1)}} \qquad \dfrac{dy}{dx} = \frac{1}{2}(x+1)^{-1/2} \,.\, 2x \,.\, e^{\sqrt{(x^2+1)}}$

(the chain rule is also applied to $\dfrac{d}{dx}\{\sqrt{(x^2+1)}\}$)

So $\quad \dfrac{dy}{dx} = \dfrac{x}{\sqrt{(x^2+1)}} \cdot e^{\sqrt{(x^2+1)}}$

Moreover, the exponential functions may also be associated with other functions whose differential properties have already been investigated. The following examples illustrate the techniques which are used in the cases when the compound function is either a product or a quotient.

Example 1

Find $\dfrac{dy}{dx}$ if $y = e^x(x^2+1)$

We write $\quad y = e^x(x^2+1) = uv \quad$ where $u = e^x$, $v = x^2 + 1$

By the product rule

$$\frac{dy}{dx} = u\frac{dv}{dx} + v\frac{du}{dx}$$

$$= e^x(2x) + (x^2 + 1)e^x$$

$$= e^x(2x + x^2 + 1) = e^x(x^2 + 2x + 1)$$

$$= (x + 1)^2e^x$$

Example 2

Find $\dfrac{dy}{dx}$ if $y = \dfrac{e^{2x} + 1}{e^{2x} - 1}$

Here $y = \dfrac{u}{v}$ where $u = e^{2x} + 1$ and $v = e^{2x} - 1$

By the quotient rule

$$\frac{dy}{dx} = \frac{v\dfrac{du}{dx} - u\dfrac{du}{dx}}{v^2}$$

$$= \frac{(e^{2x} - 1)2e^{2x} - (e^{2x} + 1)2e^{2x}}{(e^{2x} - 1)^2}$$

$$= \frac{2e^{2x}\{(e^{2x} - 1) - (e^{2x} + 1)\}}{(e^{2x} - 1)^2}$$

$$= \frac{-4e^{2x}}{(e^{2x} - 1)^2}$$

Example 3

Given that $y = e^{-x}\cos 2x$, prove that

$$\frac{d^2y}{dx^2} + 2\frac{dy}{dx} + 5y = 0$$

We can answer this problem by differentiating twice, either explicitly or implicitly. Both methods are shown.

Method 1 (Explicit differentiation)

$$y = e^{-x}\cos 2x = uv$$

Hence $$\frac{dy}{dx} = -e^{-x}2\sin 2x - e^{-x}\cos 2x$$

$$= -e^{-x}(2\sin 2x + \cos 2x)$$

$$\frac{d^2y}{dx^2} = e^{-x}(2\sin 2x + \cos 2x) - e^{-x}(4\cos 2x - 2\sin 2x)$$

$$= e^{-x}(4\sin 2x - 3\cos 2x)$$

Hence L.H.S.

$$= \frac{d^2y}{dx^2} + 2\frac{dy}{dx} + 5y$$

$$= e^{-x}(4 \sin 2x - 3 \cos 2x) - 2e^{-x}(2 \sin 2x + \cos 2x) + 5e^{-x} \cos 2x$$

$$= e^{-x}\{\sin 2x(4 - 4) + \cos 2x(-3 - 2 + 5)\}$$

$$= 0$$

Method 2 (Implicit differentiation)

$$y = e^{-x} \cos 2x = \frac{1}{e^x} \cos 2x$$

Hence $e^x y = \cos 2x$

In this case the L.H.S. is a product and we differentiate with respect to x, *implicitly*.

$$e^x \frac{dy}{dx} + e^x y = -2 \sin 2x$$

Differentiating again,

$$\left\{e^x \frac{d^2y}{dx^2} + e^x \frac{dy}{dx}\right\} + \left\{e^x \frac{dy}{dx} + ye^x\right\} = -4 \cos 2x$$

$$= -4e^x y$$

$$\Rightarrow \qquad e^x \frac{d^2y}{dx^2} + 2e^x \frac{dy}{dx} + 5e^x y = 0$$

$$e^x\left(\frac{d^2y}{dx^2} + 2\frac{dy}{dx} + 5y\right) = 0$$

But for finite x, $e^x \neq 0$ (the curve of $y = e^x$ is asymptotic to the x-axis).

$$\Rightarrow \frac{d^2y}{dx^2} + 2\frac{dy}{dx} + 5y = 0$$

Example 4

Determine the coordinates of, and the nature of the turning points on the graph of $y = xe^x$ for finite values of x. Sketch the graph.

$$y = xe^x = uv$$

So $\dfrac{dy}{dx} = xe^x + 1 \,.\, e^x$

$$= (x + 1)e^x$$

For turning points on the graph $\dfrac{dy}{dx} = 0$, so $x = -1 (e^x \neq 0)$

Also $\qquad \dfrac{d^2y}{dx^2} = (x + 1)e^x + e^x = (x + 2)e^x$

and when $x = -1, \dfrac{d^2y}{dx^2} > 0$

Thus when $x = -1$, the turning point is a minimum.

Substituting in $y = xe^x$, $y = -e^{-1} = -\dfrac{1}{e} \approx -0{\cdot}37$

The graph, therefore, has a minimum point at $\left(-1, -\dfrac{1}{e}\right)$.

When $x = 0$, $y = 0$ and the graph passes through the origin. When x is large and negative, $xe^x \to -0$, i.e. tends to zero through negative values, for y is then $-h/e^h$ (where $x = -h$ and $h \to \infty$). When x is positive, y is positive; and as $x \to \infty$, $y \to \infty$. The sketch of the graph is as shown in figure 14.4.

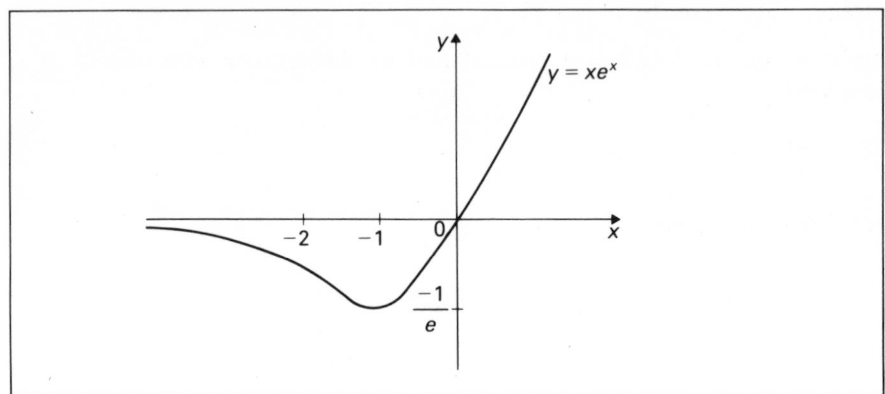

Figure 14.4

Exercise B

1 Find the derivatives of the following functions.

a) $e^{\frac{1}{2}x^2}$ e) $e^{1/x}$

b) e^{-5x} f) $e^{(\sin x + \cos x)}$

c) $e^{\sin 2x}$ g) $e^{2x} + \dfrac{1}{e^{2x}}$

d) $e^{\sqrt{x}}$ h) $\left(e^x + \dfrac{1}{e^x}\right)^3$

2 Determine the expression for $\dfrac{dy}{dx}$, in the simplest possible form, given

a) $y = x^3e^{2x}$ d) $y = \tan(e^x)$ g) $y = \dfrac{e^x}{1 - e^{2x}}$

b) $y = (x + 1)^2e^{-x}$ e) $y = \dfrac{e^x - e^{-x}}{e^x + e^{-x}}$

c) $y = e^{-2x}\cos x$ f) $y = \dfrac{x}{e^x}$

3 Given that $y = e^{4x} \sin 3x$, determine $\dfrac{dy}{dx}$ and express your result in the form $Re^{4x} \cos(3x + \theta)$ where R is a positive constant. Derive the value of R, and $\tan \theta$. Find $\dfrac{d^2y}{dx^2}$ in a similar form.

4 Prove that $\dfrac{d^2}{dx^2}(e^{-x^2}) = 2e^{-x^2}(2x^2 - 1)$.

5 If $y = ae^{nx} + be^{-nx}$ where a, b, n are constants, prove that

$$\frac{d^2y}{dx^2} - n^2y = 0$$

6 Differentiate $y = e^{-2x} \sin 5x$ and express your result in the form $Re^{-2x} \cos(5x + \alpha)$. Write down the value of R and the acute angle α. Hence determine the greatest and least values of y, for $0 \leqslant x \leqslant 2\pi$.

7 An object P moves along a straight line OX so that its displacement, x metres, from O after t seconds, is given by $x = e^{-t}(a \sin t + b \cos t)$ where a, b are constants.

Initially $t = 0$, $x = 1$ and $\dfrac{dx}{dt} = 0$.

Calculate the values of a and b, and the acceleration when $t = \frac{\pi}{4}$ seconds.

8 Given that $y = (x - 1)e^{3x}$ show that

$$\frac{d^2y}{dx^2} - 5\frac{dy}{dx} + 6y = e^{3x}$$

9 If $y = (\sin x)e^x$, prove that

$$\frac{d^2y}{dx^2} - 2\frac{dy}{dx} + 2y = 0$$

10 Show that, if $y = (x^2 + 1)e^{2x}$, then

$$x\frac{d^3y}{dx^3} - 4x\frac{d^2y}{dx^2} + 2(2x - 1)\frac{dy}{dx} + 4y = 0$$

14.4 Integral properties

It is now clear that for $n \in \mathbb{R}$

$$\frac{d}{dx}(e^{nx}) = ne^{nx}$$

and hence $\displaystyle\int e^{nx}\,dx = \frac{1}{n}e^{nx} + c$

Thus $\displaystyle\int e^{2x}\,dx = \tfrac{1}{2}e^{2x} + c$

and $\int e^{-x}\,dx = -e^{-x} + c$

and $\int e^{-x/3}\,dx = \dfrac{e^{-x/3}}{-\frac{1}{3}} + c$

$= -3e^{-x/3} + c$

When the integrals are definite we work as before, for example

$$\int_0^1 e^{3x}\,dx = \left[\frac{e^{3x}}{3}\right]_0^1 = \frac{e^3}{3} - \frac{e^0}{3} = \frac{e^3 - 1}{3}$$

Integrals in which the integrand is a composite function, one part of which is exponential, may be evaluated using the ideas of chapter 12. The following examples show the use of the method of substitution, and the method integration by parts.

Example 1

Evaluate $I = \int xe^{x^2}\,dx$

In the integrand we replace x^2 by u so that $e^{x^2} = e^u$ and

$2x\,\dfrac{dx}{du} = 1 \qquad$ i.e. $x\,dx = \frac{1}{2}\,du$

The integral then can be written as

$$I = \int \tfrac{1}{2}e^u\,du$$

$= \frac{1}{2}e^u + c \quad$ or $\quad \frac{1}{2}e^{x^2} + c$

If we generalise this example we can write down immediately

$$I = \int f'(x)e^{f(x)}\,dx = e^{f(x)} + c$$

Thus $\int \cos xe^{\sin x}\,dx = e^{\sin x} + c$

$$\int (2x - 1)e^{(x2-x)}\,dx = e^{(x2-x)} + c$$

Note that it may be necessary to introduce numerically balancing constants inside the integrand and outside the integral to make the integrand an exact differential.

So if $I = \int x^2e^{-x^3}\,dx$

we would write

$$I = -\frac{1}{3}\int(-3x^2)e^{-x^3}\,dx$$
$$= -\tfrac{1}{3}e^{-x^3} + c$$

Example 2

Evaluate $I = \int_0^1 xe^x\,dx$

In this case, we use the formula for integration by parts:

$$\int u\,dv = uv - \int v\,du$$

with $u = x$ and $dv = e^x\,dx$

then $du = dx$ and $v = \int e^x\,dx = e^x$

So that $I = [xe^x]_0^1 - \int_0^1 e^x\,dx$

$$= [xe^x - e^x]_0^1$$
$$= (e - e) - (0 - e^0)$$
$$= e^0$$
$$= 1$$

Example 3

Evaluate $I = \int x^2e^{2x}\,dx$

The method of approach here follows closely the idea of Example 3 of section 12.4.

We choose $u = x^2$ and $dv = e^{2x}\,dx$ so that
$du = 2x\,dx$ and $v = \tfrac{1}{2}e^{2x}$

Then $I = \tfrac{1}{2}x^2e^{2x} - \int 2x\tfrac{1}{2}e^{2x}\,dx$

$$= \tfrac{1}{2}x^2e^{2x} - \int xe^{2x}\,dx$$

Continuing the process for this second integral

$u = x$ $\quad dv = e^{2x}\,dx$
$du = dx$ $\quad v = \tfrac{1}{2}e^{2x}$

So that

$$\int xe^{2x}\,dx = \tfrac{1}{2}xe^{2x} - \int \tfrac{1}{2}e^{2x}\,dx$$
$$= \tfrac{1}{2}xe^{2x} - \tfrac{1}{4}e^{2x}$$

Thus $I = \frac{1}{2}x^2e^{2x} - (\frac{1}{2}xe^{2x} - \frac{1}{4}e^{2x}) + c$

$= \frac{1}{4}e^{2x}(2x^2 - 2x + 1) + c$

Example 4

Prove that, it $I_n = \int x^n e^x\, dx$, then for $n \geqslant 1$, $I_n = x^n e^x - nI_{n-1}$

Use this reduction formula to evaluate $\int_0^1 x^4 e^x\, dx$.

We again use the ideas outlined in chapter 12 and

let $\quad u = x^n \; dv = e^x\, dx$ in I_n

Then $\quad du = nx^{n-1}\, dx$ and $v = e^x$

Hence $\quad I_n = x^n e^x - \int nx^{n-1}e^x\, dx + c$

$$= x^n e^x - n\int x^{n-1}e^x\, dx + c$$

$$= x^n e^x - nI_{n-1}$$

since $\int x^{n-1}e^x\, dx$ is similar to the original integral with the index n replaced by $n - 1$.

If we now write I_4 etc. for the definite integral

$$
\begin{aligned}
I_4 = \int_0^1 x^4 e^x\, dx \quad &= [x^4e^x]_0^1 - 4I_3 \\
&= e - 4I_3 \\
I_3 = [x^3e^x]_0^1 - 3I_2 \quad &= e - 3I_2 \\
I_2 = [x^2e^x]_0^1 - 2I_1 \quad &= e - 2I_1
\end{aligned}
$$

and

$$
\begin{aligned}
I_1 = [xe^x]_0^1 - I_0 \quad &= e - \int_0^1 e^x\, dx \\
&= e - [e^x]_0^1 \\
&= e - (e - 1) = 1
\end{aligned}
$$

Hence

$$
\begin{aligned}
I_2 = e - 2I_1 \quad &= e - 2 \\
I_3 = e - 3(e - 2) \quad &= -2e + 6
\end{aligned}
$$

And $\quad I_4 = e - 4(-2e + 6) = 9e - 24$

Hence $\quad \int_0^1 x^4 e^x\, dx = 9e - 24$

Example 5

Find $\quad I = \int e^x \sin x\, dx$

In the previous examples, using the methods of integration by parts, it has been clear that by using u for the power of x i.e. x^n, the secondary integral in the result involving du has been a reduced form of the original integral.

In this example, such a procedure is not obvious. So let us choose

$$u = e^x \quad \text{and} \quad dv = \sin x\, dx$$

then $du = e^x\, dx$ and $v = -\cos x$

Thus $I = -e^x \cos x + \int e^x \cos x\, dx$

Repeating the process with this second integral

$$u = e^x \quad \text{and } dv = \cos x\, dx$$
$$du = e^x\, dx \text{ and } v = \sin x$$

Thus $I = -e^x \cos x + \left\{e^x \sin x - \int e^x \sin x\, dx\right\}$

$$= -e^x \cos x + e^x \sin x - \int e^x \sin x\, dx$$

We have still not determined I as a result free from integrals but it is seen that the integral we have now acquired on the R.H.S. is the original one again.

We may therefore write $I = e^x(\sin x - \cos x) - I$
and so $2I = e^x(\sin x - \cos x)$
or $I = \frac{1}{2}e^x(\sin x - \cos x) + c$

Check for yourself that by choosing u and dv differently, the same result would be obtained.

Exercise C

1 Evaluate the indefinite integrals

a) $\int 2e^{-2x}\, dx$

b) $\int 2xe^{x^2+1}\, dx$

c) $\int \sec^2 x e^{\tan x}\, dx$

d) $\int \frac{1}{\sqrt{x}} e^{\sqrt{x}}\, dx$

e) $\int (e^x - 2e^{-x})(e^x + 2e^{-x})\, dx$

f) $\int e^x \sin (e^x)\, dx$

g) $\int e^{-x}(e^{-x} + 1)^3\, dx$

2 Prove, by using Simpson's Rule, with ten strips of equal width 0·1, that

$$\int_1^2 e^{-x^2}\, dx = 0{\cdot}1353$$

when corrected to 4 decimal places.

(**Note** Although this integral is comparatively simple in appearance, it cannot be evaluated by exact methods within the scope of this course.)

3 Evaluate

a) $\int_0^1 e^{-2x}(1 + x^2)\, dx$ c) $\int_0^{\pi/4} e^{3x} \cos x\, dx$

b) $\int_0^1 x^3 e^{2x}\, dx$

4 By using the substitution $y = e^x + 1$ evaluate

$$\int_0^1 \frac{e^x\, dx}{(e^x + 1)^2}$$

5 Given

$$I = \int_{-1}^{1} \frac{x^2\, dx}{e^x + 1} \quad \text{and} \quad J = \int_{-1}^{1} \frac{x^2 e^x\, dx}{e^x + 1}$$

prove, by using the substitution $x = -y$, that $I = J$. Write down $I + J$ as a single integral and hence evaluate I.

6 If $I_n = \int x^n e^{-x}\, dx$, establish a reduction formula connecting I_n and I_{n-1}, and use this formula to derive

a) $\int x^4 e^{-x}\, dx$ b) $\int_0^1 x^5 e^{-x}\, dx$

7 If $I_n = \int e^x \cos^n x\, dx$, prove that

$$(n^2 + 1)I_n = e^x \cos^{n-1} x(\cos x + n \sin x) + n(n-1)I_{n-2}$$

Hence evaluate $\int_0^{\pi/2} e^x \cos^5 x\, dx$

8 Prove that $\int e^{ax} \sin bx\, dx = \dfrac{e^{ax}(a \sin bx - b \cos bx)}{a^2 + b^2}$

where a, b are constants.

Derive a similar result for $\int e^{ax} \cos bx\, dx$

14.5 Napierian or natural logarithms

We return to the problem of evaluating $\int \frac{1}{x}\, dx$,

and substitute $x = e^u$ so that $dx = e^u\, du$. The integral then becomes

$$\int \frac{1}{e^u} \cdot e^u\, du = \int 1\, du$$
$$= u + k$$

Hence, since $x = e^u \Rightarrow u = \log_e x$, our integral is

$$\int \frac{1}{x}\, dx = \log_e x + k$$

Such a logarithm, with base e, is called a **Napierian logarithm**, after John Napier who discovered the function and studied its properties.

We also write $\ln x$ for $\log_e x$ and use the name **natural logarithm**. Tabulated values of $\ln x$ for varying values of x are usually included in sets of mathematical tables. Also many electronic calculators contain an '$\ln x$' key.

We have thus shown that when $y = \ln x$,

$$\frac{dy}{dx} = \frac{1}{x}$$

This standard result is of considerable importance in the study of mathematics at this level.

The basic rules of logarithms when applied to natural logarithms become

$\ln (xy) = \ln x + \ln y$ **Rule 1**

$\ln \left(\frac{x}{y}\right) = \ln x - \ln y$ **Rule 2**

$\ln (x^p) = p \ln x$ **Rule 3**

We go on to apply these ideas to a wider group of functions.

14.6 Differential properties of $\ln x$

Let $\quad y = \ln \{f(x)\}$ and write $u = f(x)$

Then $\quad y = \ln u \quad$ and $\quad \frac{du}{dx} = f'(x)$

Thus $\quad \frac{dy}{du} = \frac{1}{u}$

And $\quad \frac{dy}{dx} = \frac{dy}{du} \cdot \frac{du}{dx} = \frac{1}{u} \cdot f'(x),\quad$ by the chain rule

So, replacing u by $f(x)$

$$\frac{dy}{dx} = \frac{d}{dx}\{f(x)\} = \frac{f'(x)}{f(x)}$$

Again, should we wish to differentiate any exponential function or logarithmic function involving a base a which is not equal to e, we use a process called **logarithmic differentiation**. Here, we first take logarithms and then differentiate. This process is illustrated as follows.

Let $\quad y = a^x$

Take logs to base e

then $\quad \log_e y = \log_e a^x$

i.e. $\quad \ln y = x \ln a \quad$ (by **Rule 3** of logarithms)

Differentiating implicitly,

$$\frac{1}{y}\frac{dy}{dx} = \ln a$$

$$\frac{dy}{dx} = y \ln a = a^x \ln a$$

So, for example,

$$\frac{d}{dx}(10^x) = 10^x \ln 10$$

$$\frac{d}{dx}(3^{x^2}) = 2x \,.\, 3^{x^2} \ln 3 \text{ (By writing } x^2 = u \text{ and applying the chain rule)}$$

Also, given $y = \log_a x$
$a^y = x$

Then taking logs to base e,

$$y \ln a = \ln x$$

Hence, differentiating with respect to x

$$\frac{dy}{dx} \,.\, (\ln a) = \frac{1}{x}$$

So $$\frac{dy}{dx} = \frac{1}{x \ln a}$$

(This result can also be generalised further to give

$$\frac{d}{dx} \log_a f(x) = \frac{f'(x)}{f(x) \ln a}$$

To differentiate other compound functions we shall need to use the basic product and quotient rules.

Example 1

Find the value of $\frac{dy}{dx}$ when y is the given function.

a) $\ln(x^2 - 1)$ d) $\ln(e^x + e^{-x})$
b) $\ln\sqrt{(x+1)}$ e) $(\ln x)^2$
c) $\ln(\sin x)$

a) $y = \ln(x^2 - 1) \Rightarrow \frac{dy}{dx} = \frac{2x}{x^2 - 1}$

b) $y = \ln\sqrt{(x+1)} = \ln(x+1)^{1/2} = \frac{1}{2}\ln(x+1)$

$$\Rightarrow \frac{dy}{dx} = \frac{1}{2} \cdot \frac{1}{x+1} = \frac{1}{2(x+1)}$$

c) $y = \ln(\sin x) \Rightarrow \frac{dy}{dx} = \frac{\cos x}{\sin x} = \cot x$

d) $y = \ln(e^x + e^{-x}) \quad \Rightarrow \quad \dfrac{dy}{dx} = \dfrac{e^x - e^{-x}}{e^x + e^{-x}}$

$$= \frac{e^x - \dfrac{1}{e^x}}{e^x - \dfrac{1}{e^x}}$$

$$= \frac{e^{2x} - 1}{e^{2x} + 1}$$

e) Given $\quad y = (\ln x)^2$

Write $u = \ln x$

Then, $y = u^2, \quad \dfrac{du}{dx} = \dfrac{1}{x} \quad \dfrac{dy}{du} = 2u$

And $\quad \dfrac{dy}{dx} = \dfrac{dy}{du} \cdot \dfrac{du}{dx} = 2u \cdot \dfrac{1}{x} = \dfrac{2 \ln x}{x}$

Example 2
Differentiate

a) $(1 + x) \ln x \qquad$ b) $\dfrac{\ln x}{x} \qquad$ c) $\ln\{x + \sqrt{(x^2 + 1)}\}$

a) Let $y = (1 + x)\ln x = uv$

Then $\quad \dfrac{dy}{dx} = (1 + x) \cdot \dfrac{1}{x} + 1 \cdot \ln x$

$$= 1 + \frac{1}{x} + \ln x$$

b) Let $y = \dfrac{\ln x}{x} = \dfrac{u}{v}$

Then $\quad \dfrac{dy}{dx} = \dfrac{x \cdot \dfrac{1}{x} - 1 \cdot \ln x}{x^2}$

$$= \frac{1 - \ln x}{x^2}$$

c) Let $y = \ln\{x + \sqrt{(x^2 + 1)}\}$

Then $\quad \dfrac{dy}{dx} = \dfrac{1 + \frac{1}{2} \cdot 2x(x^2 + 1)^{-1/2}}{x + \sqrt{(x^2 + 1)}}$

$$= \frac{1 + \dfrac{x}{\sqrt{(x^2 + 1)}}}{x + \sqrt{(x^2 + 1)}}$$

$$= \frac{\{\sqrt{(x^2 + 1)} + x\}}{\sqrt{(x^2 + 1)}\{x + \sqrt{(x^2 + 1)}\}}$$

$$= \frac{1}{\sqrt{(x^2 + 1)}}$$

Example 3
Use logarithmic differentiation to find derivatives of

a) x^x, b) $\sqrt{\left(\frac{x-1}{x+1}\right)}$

a) Let $y = x^x$

Then taking natural logarithms

$\ln y = \ln(x^x) = x \ln x$

Differentiate implicitly, using the product rule

$$\frac{1}{y}\frac{dy}{dx} = x \cdot \frac{1}{x} + 1 \cdot \ln x$$
$$= 1 + \ln x$$

Hence $\dfrac{dy}{dx} = (1 + \ln x)x^x$

b) Let $y = \sqrt{\left(\frac{x-1}{x+1}\right)} = \left(\frac{x-1}{x+1}\right)^{1/2}$

Take natural logarithms,

$$\ln y = \ln\left(\frac{x-1}{x+1}\right)^{1/2} = \tfrac{1}{2}\ln\left(\frac{x-1}{x+1}\right)$$

Hence $2 \ln y = \ln(x-1) - \ln(x+1)$, applying **Rule 2** of logarithms.

Differentiate implicitly

$$\frac{2}{y}\frac{dy}{dx} = \frac{1}{x-1} - \frac{1}{x+1}$$
$$= \frac{(x+1)-(x-1)}{(x-1)(x+1)}$$
$$= \frac{2}{x^2-1}$$

So that

$$\frac{dy}{dx} = \frac{y}{x^2-1}$$
$$= \frac{\sqrt{(x-1)}}{\{\sqrt{(x+1)}\} \times \{(x-1)(x+1)\}}$$
$$= \frac{1}{(x+1)^{3/2}(x-1)^{1/2}}$$

Example 4
Determine the coordinates of the turning point on the graph of

$y = x^3 - \frac{1}{9}\ln x$

and determine the nature of this particular point.

$y = x^3 - \frac{1}{9}\ln x$

So $\dfrac{dy}{dx} = 3x^2 - \dfrac{1}{9x}$

$\dfrac{dy}{dx} = 0$ for turning points, hence $27x^3 - 1 = 0 \Rightarrow x = \frac{1}{3}$

Now $\dfrac{d^2y}{dx^2} = 6x + \dfrac{1}{9x^2}$

When $x = \frac{1}{3}$, $\dfrac{d^2y}{dx^2} > 0$

hence when $x = \frac{1}{3}$ there is a minimum point

And when $x = \frac{1}{3}$

$$\begin{aligned} y &= \tfrac{1}{27} - \tfrac{1}{9} \ln \tfrac{1}{3} \\ &= \tfrac{1}{27} - \tfrac{1}{9} \ln 3^{-1} \\ &= \tfrac{1}{27} + \tfrac{1}{9} \ln 3 \\ &= \tfrac{1}{27} + \tfrac{1}{27} \ln 3^3 \\ &= \frac{1 + \ln 27}{27} \end{aligned}$$

The graph has a minimum point at $\left(\dfrac{1}{3}, \dfrac{1 + \ln 27}{27}\right)$

Exercise D

1 Find the differential coefficients of the following functions.

a) $\ln(x^2 + x + 1)$
b) $\ln \sqrt{(x^2 - x + 1)}$
c) $\ln(\cos x)$
d) $\ln(e^{2x} - 2x)$
e) $\ln \left\{\dfrac{(1 + \sin x)}{\cos x}\right\}$
f) $\ln\{x - \sqrt{(x^2 - 1)}\}$
g) $\log_{10} x^2$
h) $\log_2(2^x + 2^{-x})$
i) $\ln \left\{\tan\left(x + \dfrac{\pi}{4}\right)\right\}$

2 Determine, by logarithmic differentiation, the differential coefficients of

a) 10^{-5x}
b) $\sqrt{\left\{\dfrac{4 - x^2}{4 + x^2}\right\}}$
c) $\dfrac{2x + 1}{\sqrt{(3 - 4x - 4x^2)}}$
d) $(\ln x)^{\ln x}$

3 Find $\frac{dy}{dx}$ when

a) $y = e^x \ln x$

b) $y = x \ln(\cot x)$

c) $y = 4x^4 \ln x - x^4$

d) $y = x^2 \ln(x+3)$

e) $y = \frac{x^2+1}{\ln x}$

f) $y = \frac{\ln x}{x + \ln x}$

4 a) If $y = \cos(\ln x)$, prove that $x^2y'' + xy' + y = 0$
b) If $y = \ln(1 + \sin x)$, show that

i) $\frac{d^3y}{dx^3} + \frac{d^2y}{dx^2} \cdot \frac{dy}{dx} = 0$ ii) $\frac{d^2y}{dx^2} \cdot \frac{d^4y}{dx^4} - \left(\frac{d^3y}{dx^3}\right)^2 + \left(\frac{d^2y}{dx^2}\right)^3 = 0$

5 Prove that $\frac{d^4}{dx^4}(x^3 \ln x) = \frac{3!}{x}$

6 Use the process of logarithmic differentiation to obtain the turning points on the curve $y = \frac{x+4}{(x+2)(1-2x)}$. Sketch the curve.

7 Sketch the graph of $y = x^2 + \ln x$ for $x > 0$. Show that the curve has a point of inflexion at $x = \frac{1}{\sqrt{2}}$, and determine the coordinates of the point where the normal to the curve at $x = \frac{1}{\sqrt{2}}$ crosses the x-axis.

8 Given that $y = x^2 \ln x$, prove that $x = \frac{y'}{-2 + y''}$

14.7 Integral properties

When evaluating definite integrals, it is certain that at some stage we will meet the number $e^{\ln n}$. We consider the simplification of such a number, therefore, before the basic integral problem.

Let $y = e^{\ln n}$

Take logs to base e

Then $\ln y = \ln(e^{\ln n})$
$= (\ln n)(\ln e)$

Now $\ln e = \log_e e = 1$ since we have already proved that for any bases, a, $\log_a a = 1$.

$\ln y = \ln n$
$\Rightarrow y = n$
$\Rightarrow y = e^{\ln n} = n$

So, for example, $e^{\ln 2} = 2$

$$e^{2\ln 2} = e^{\ln 2^2} = 2^2 = 4$$

$$e^{-\frac{1}{2}\ln 2} = e^{\ln 2^{-1/2}} = 2^{-1/2} = \frac{1}{\sqrt{2}}$$

Example 1

Evaluate $I = \int_1^{\ln 2} e^{-x}\, dx$

$$\begin{aligned} I &= [-e^{-x}]_1^{\ln 2} \\ &= -e^{-\ln 2} + e^{-1} \\ &= -e^{\ln 2^{-1}} + e^{-1} \\ &= -(2^{-1}) + e^{-1} \\ &= -\frac{1}{2} + \frac{1}{e} \\ &= \frac{2-e}{2e} \end{aligned}$$

Exercise E

1 Simplify the following exponential forms.

a) $e^{\ln 4}$ e) $\dfrac{e^{\ln a}}{e^{\ln b}}$

b) $e^{\frac{1}{2}\ln 9}$ f) $e^{\int \frac{dx}{x}}$

c) $e^{-\frac{1}{3}\ln 27}$ g) $e^{-\ln(\text{cosec}\, x)}$

d) $e^{\ln x^2}$ h) $e^{x \ln x}$

2 Evaluate the integrals

$$\int_0^{\ln 3} e^x\, dx \qquad \int_0^{\ln 2} e^{-3x}\, dx$$

14.8 The integration of $\frac{1}{x}$ and expressions of the form $\frac{f'(x)}{f(x)}$

1 We have seen that for $x > 0$, the graph of $y = \ln x$ exists and at any point (x, y) on it

$$\frac{dy}{dx} = \frac{1}{x}$$

It is clear, therefore, that for $x > 0$

$$\int \frac{1}{x}\, dx = y + c = \ln x + c$$

However if $x < 0$ we cannot write down the integral in this form, for ln x does not exist for such values.

But $\frac{d}{dx}\{\ln(-x)\} = \frac{-1}{-x}$ where $x < 0$ from the curve reflected in the y-axis.

Hence, $\int \frac{-1}{-x}\,dx = \ln(-x) + c$ when $x < 0$

i.e. $\int \frac{1}{x}\,dx = \ln(-x) + c$ when $x < 0$

So, combining the results, $\int \frac{1}{x}\,dx = \begin{cases} \ln(x) + c, & x > 0 \\ \ln(-x) + c, & x < 0 \end{cases}$

i.e. $$\int \frac{1}{x}\,dx = \ln|x| + c$$

for all real values of $x \neq 0$

Moreover since the constant of integration c is undefined, it is sometimes convenient to write $c = \ln k$ where $k > 0$

Then $\int \frac{1}{x}\,dx = \ln|x| + \ln k = \ln\{k|x|\}$

2 Likewise we can write $\frac{d}{dx}\{\ln(ax + b)\} = \frac{a}{ax + b}$

Thus $\int \frac{dx}{ax + b} = \frac{1}{a}\ln|(ax + b)| + c = \frac{1}{a}\ln\{k|(ax + b)|\}$

And $\frac{d}{dx}\{\ln f(x)\} = \frac{f'(x)}{f(x)}$ gives the general result

$$\int \frac{f'(x)}{f(x)}\,dx = \ln\{k|f(x)|\}$$

So, for example,

$$\int \frac{3}{x}\,dx = 3\int \frac{1}{x}\,dx = 3\ln|x| + c \quad \text{or} \quad \ln\{k|x^3|\}$$

$$\int \frac{1}{2x + 1}\,dx = \frac{1}{2}\int \frac{2}{2x + 1}\,dx$$

$$= \tfrac{1}{2}\ln|2x + 1| + c$$

$$\int \frac{2x}{x^2 + 1}\,dx = \ln|(x^2 + 1)| + c$$

since if $f(x) = x^2 + 1$, $f'(x) = 2x$

Again $\int \tan x \, dx = \int \frac{\sin x}{\cos x} \, dx$

$$= -\int \frac{(-\sin x)}{\cos x} \, dx$$

$$= -\ln|(\cos x)| + c$$

$$= \ln|(\cos x)^{-1}| + c$$

$$= \ln|\sec x| + c$$

$$\int \cot x \, dx = \int \frac{\cos x}{\sin x} \, dx$$

$$= \ln|\sin x| + c$$

$$\int \frac{e^x + e^{-x}}{e^x - e^{-x}} \, dx = \ln|e^x - e^{-x}| + c$$

And $\int \frac{1}{x \ln x} \, dx = \int \frac{1/x}{\ln x} \, dx$

$$= \ln\{|\ln x|\} + c$$

Exercise F

Evaluate the following integrals.

1 $\int \frac{1}{3x} \, dx$

2 $\int \frac{4}{x} \, dx$

3 $\int \frac{1}{1 + 2x} \, dx$

4 $\int \frac{1}{1 - 2x} \, dx$

5 $\int \frac{3x^2}{x^3 + 1} \, dx$

6 $\int \frac{3x + 1}{3x^2 + 2} \, dx$

7 $\int \frac{e^x}{e^x + 1} \, dx$

8 $\int \tan 3x \, dx$

9 $\int \frac{\sec^2 x}{2 + \tan x} \, dx$

10 $\int \frac{\cos x - \sin x}{\cos x + \sin x} \, dx$

11 $\int \frac{\sin x \cos x}{1 + \sin^2 x} \, dx$

12 $\int \frac{x}{2x - 3} \, dx$

13 $\int_1^3 \frac{x - 3}{x^2 - 6x + 5} \, dx$

14 $\int_0^{\pi/2} \frac{\cos x}{5 + 3 \sin x} \, dx$

15 $\int_0^{\ln e} \frac{e^x + 1}{e^x + x} \, dx$

14.9 Integrals involving partial fractions

In chapter 1 we saw that the algebraic expression

$$\frac{x+1}{(x-1)(x+2)}$$

could be expressed in terms of its partial fractions by the relation

$$\frac{x+1}{(x-1)(x+2)} \equiv \frac{2}{3(x-1)} + \frac{1}{3(x+2)}$$

Thus we could write

$$I = \int \frac{x+1}{(x-1)(x+2)}\,dx$$

$$= \int \left\{\frac{2}{3(x-1)} + \frac{1}{3(x+2)}\right\} dx$$

$$= \frac{2}{3}\int \frac{dx}{x-1} + \frac{1}{3}\int \frac{dx}{x+2}$$

$$= \tfrac{2}{3}\ln|(x-1)| + \tfrac{1}{3}\ln|(x+2)| + c \quad \left(\text{using} \int \frac{f'(x)}{f(x)}\,dx\right)$$

$$= \ln|\{(x-1)^{2/3} \,.\, (x+2)^{1/3}\}| + c$$

$$= \ln|\{(x-1)^2(x+2)\}^{1/3}| + c$$

The process of evaluating integrals of the form

$$\int \frac{f(x)}{g(x)}\,dx$$

may always be carried out this way whenever $g(x)$ can be written in terms of real algebraic factors, and

$$\frac{f(x)}{g(x)}$$

is expressible as the sum or difference of the associated partial fractions. Further illustrations are limited to the cases where $g(x)$ possesses single linear factors.

Example 1

Evaluate $I = \displaystyle\int \frac{(2x+1)\,dx}{x^3+2x^2-3x}$

$$\frac{2x+1}{x^3+2x^2-3x} \equiv \frac{2x+1}{x(x^2+2x-3)}$$

$$\equiv \frac{2x+1}{x(x+3)(x-1)}$$

Let $\displaystyle\frac{2x+1}{x^3+2x^2-3x} \equiv \frac{A}{x} + \frac{B}{x+3} + \frac{C}{x-1}$

Then $2x + 1 \equiv A(x+3)(x-1) + Bx(x-1) + Cx(x+3)$
When $x = 0$, $1 = -3A \Rightarrow A = -\frac{1}{3}$
When $x = 1$, $3 = 4C \Rightarrow C = \frac{3}{4}$
When $x = -3$, $-5 = 12B \Rightarrow B = -\frac{5}{12}$

Hence $I \equiv \int \left\{ \frac{-\frac{1}{3}}{x} + \frac{-\frac{5}{12}}{x+3} + \frac{+\frac{3}{4}}{x-1} \right\} dx$

$$= -\tfrac{1}{3} \ln|x| - \tfrac{5}{12} \ln|(x+3)| + \tfrac{3}{4} \ln|(x-1)| + c$$

$$= \ln \left| \frac{(x-1)^{3/4}}{x^{1/3}(x+3)^{5/12}} \right| + c$$

Example 2

Evaluate $I = \int_0^1 \frac{dx}{4 + x - 3x^2}$

$$\frac{1}{4+x-3x^2} \equiv \frac{1}{(4-3x)(1+x)} \equiv \frac{A}{(4-3x)} + \frac{B}{(1+x)}$$

Hence $1 \equiv A(1+x) + B(4-3x)$
When $x = -1$, $1 = 7B \Rightarrow B = \frac{1}{7}$
Coefficients of 1: $1 = A + 4B \Rightarrow A = \frac{3}{7}$

So $I = \int_0^1 \left\{ \frac{\frac{3}{7}}{4-3x} + \frac{\frac{1}{7}}{1+x} \right\} dx$

$$= -\frac{1}{7}\int_0^1 \frac{-3\,dx}{4-3x} + \frac{1}{7}\int_0^1 \frac{dx}{1+x}$$

$$= -\tfrac{1}{7}[\ln|(4-3x)| + \tfrac{1}{7}\ln|(1+x)|]_0^1$$

$$= -\tfrac{1}{7}\{\ln 1 - \ln 4\} + \tfrac{1}{7}\{\ln 2 - \ln 1\}$$

$$= \tfrac{1}{7}\ln 4 + \tfrac{1}{7}\ln 2 \quad (\text{since } \ln 1 = 0)$$

$$= \tfrac{1}{7}\{\ln 4 + \ln 2\} = \tfrac{1}{7}\ln 8$$

$$= \tfrac{3}{7}\ln 2, \quad \text{since } \ln 8 = \ln 2^3 = 3\ln 2$$

14.10 Two special integrals

1 $\int \sec\theta \, d\theta$

For this integral we multiply the integrand by the fraction

$$\frac{\sec\theta + \tan\theta}{\tan\theta + \sec\theta},$$

which is equal to 1 and does not affect the value of the integrand. We obtain

$$\int \sec\theta \left\{ \frac{\sec\theta + \tan\theta}{\tan\theta + \sec\theta} \right\} d\theta = \int \left(\frac{\sec^2\theta + \sec\theta\tan\theta}{\tan\theta + \sec\theta} \right) d\theta$$

It is easily seen that the numerator is the derivative of the denominator, and hence the integral is

$$\ln|\tan\theta + \sec\theta| + k$$

2 $\int \csc\theta \, d\theta$

Here we multiply the integrand by $\dfrac{\csc\theta + \cot\theta}{\cot\theta + \csc\theta}$, and obtain

$$\int\left(\frac{\csc^2\theta + \csc\theta\cot\theta}{\cot\theta + \csc\theta}\right)d\theta = -\int\frac{(-\csc^2\theta - \csc\theta\cot\theta)}{\cot\theta + \csc\theta}\,d\theta$$

$$= -\ln|\csc\theta + \cot\theta| + k$$

These two results and the two for $\int \tan\theta \, d\theta$ and $\int \cot\theta \, d\theta$ obtained earlier in this section complete the list of standard integrals of the six trigonometric forms and should be memorised since they may occur within further integral problems.

Exercise G

1 Evaluate the following indefinite integrals, expressing your answers in their simplest forms.

a) $\displaystyle\int \frac{2}{(x-1)(x+1)}\,dx$ e) $\displaystyle\int \frac{x}{(x+1)(x+2)(x+3)}\,dx$

b) $\displaystyle\int \frac{1}{(x+2)(1-x)}\,dx$ f) $\displaystyle\int \frac{3}{(x-1)(x-2)(2x-1)}\,dx$

c) $\displaystyle\int \frac{2x-3}{(x-1)(x+7)}\,dx$ g) $\displaystyle\int \frac{x}{(x+1)(x+2)^2}\,dx$

d) $\displaystyle\int \frac{x+2}{x^2+x}\,dx$ h) $\displaystyle\int \frac{x+1}{(x-2)(x-1)^2}\,dx$

2 Evaluate

a) $\displaystyle\int_1^3 \frac{5\,dx}{(2x-1)(x+2)}$ c) $\displaystyle\int_1^{10} \frac{3x+1}{3x^2+2x}\,dx$

b) $\displaystyle\int_3^5 \frac{6x^2\,dx}{(x+1)(x-1)(x-2)}$ d) $\displaystyle\int_0^{1/2} \frac{2x^3+1}{(1-x)(1+x)^2}\,dx$

3 Evaluate

$$\int_0^{\pi/4} \sec^5\theta \, d\theta$$

4 It is given that $\displaystyle\int_0^1 \frac{(x-\lambda)}{(x+1)(3x+1)}\,dx = 0.$

Determine the value of λ.

5 Using the substitution $x^2 = u$, evaluate $\int_2^5 \frac{x\,dx}{3x^4 - 7x^2 - 6}$.

6 Given that $\frac{dx}{dt} = k(3 - x)(2 - x)$, and that $x = 0$ when $t = 0$,

determine t in terms of x, and prove that when $x = 1$, $t = \frac{1}{k} \ln \frac{4}{3}$.

14.11 Further examples

Finally we consider some other integrals and applications involving natural logarithms either in the integrand, or in the solution.

Example 1

Evaluate $I = \int_1^e x \ln x \, dx$

This integral may be evaluated using the method of integration by parts.

We choose $u = \ln x$ so that $du = \frac{1}{x}\,dx$

and $dv = x\,dx$ and $v = \frac{x^2}{2}$

Then
$$I = uv - \int v\,du$$
$$= \left[\frac{x^2}{2} \ln x - \int \frac{x^2}{2} \cdot \frac{1}{x}\,dx\right]_1^e$$
$$= \left[\frac{x^2}{2} \ln x - \frac{1}{2}\int x\,dx\right]_1^e$$
$$= \left[\frac{x^2}{2} \ln x - \frac{x^2}{4}\right]_1^e$$
$$= \left(\frac{e^2}{2} - \frac{e^2}{4}\right) - (-\tfrac{1}{4}) \quad \text{since } \ln e = 1 \text{ and } \ln 1 = 0$$
$$= \tfrac{1}{4}(e^2 + 1)$$

Example 2

Further integrals of this type are $\int x^n \ln x \, dx$.

The special case when $n = 0$ is worth noting particularly.

When $I = \int \ln x \, dx$

we choose $u = \ln x, \quad du = \frac{1}{x}\,dx$

and $dv = 1\,.\,dx, \quad v = x$

so that
$$\begin{aligned} I &= \int 1\,.\,\ln x\,dx \\ &= x\ln|x| - \int x\,.\,\frac{1}{x}\,dx \\ &= x\ln|x| - x + c \end{aligned}$$

Example 3

Evaluate $\displaystyle\int_1^e \frac{\ln x}{x}\,dx$

If we write $u = \ln x$, then $du = \frac{1}{x}\,dx$

And when $x = 1, u = 0; \quad x = e, u = 1.$

Hence
$$\begin{aligned} \int_1^e \frac{\ln x}{x}\,dx &= \int_0^1 u\,du = \left[\frac{u^2}{2}\right]_0^1 \\ &= \tfrac{1}{2} \end{aligned}$$

Example 4

Evaluate $\displaystyle\int (1 + e^x)^{1/2}\,dx$

Integrals of this type are frequently evaluated using the substitution $x = \ln u$.

Then $e^x = e^{\ln u} = u$ and $dx = \frac{1}{u}\,du$

Hence $\displaystyle\int (1 + e^x)^{1/2}\,dx = \int (1 + u)^{1/2}\,\frac{1}{u}\,du$

Now write $1 + u = v^2$ so that $(1 + u)^{1/2} = v$

$\dfrac{1}{u} = \dfrac{1}{v^2 - 1}$ and $du = 2v\,dv$

The integral then becomes

$$\begin{aligned} \int \frac{v\,.\,2v}{v^2 - 1}\,dv &= 2\int \frac{v^2}{v^2 - 1}\,dv \\ &= 2\int \left(1 + \frac{1}{v^2 - 1}\right)dv \end{aligned}$$

Writing
$$\begin{aligned} \frac{1}{v^2 - 1} &= \frac{1}{(v - 1)(v + 1)} \\ &= \frac{A}{v - 1} + \frac{B}{v + 1} \\ &= \frac{\frac{1}{2}}{v - 1} - \frac{\frac{1}{2}}{v + 1} \end{aligned}$$

We have $\int (1+e^x)^{1/2}\,dx = 2\int\left(1+\frac{\frac{1}{2}}{v-1}-\frac{\frac{1}{2}}{v+1}\right)dv$

$$= 2\{v + \tfrac{1}{2}\ln|(v-1)| - \tfrac{1}{2}\ln|(v+1)|\}$$

$$= 2v + \ln\left|\frac{v-1}{v+1}\right|$$

$$= 2v + \ln\left|\frac{(v-1)^2}{v^2-1}\right|$$

$$= 2(u+1)^{1/2} + \ln\left|\frac{\{(u+1)^{1/2}-1\}^2}{u}\right|$$

(replacing v by $(u+1)^{1/2}$)

$$= 2(e^x+1)^{1/2} + \ln|\{(e^x+1)^{1/2}-1\}^2| - \ln|(e^x)|$$

$$= 2\{-\tfrac{1}{2}x + (1+e^x)^{1/2} + \ln|\{(1+e^x)^{1/2}-1\}| + c$$

An alternative method which may be used to evaluate this integral is to use the substitution $z^2 = 1 + e^x$. Try this substitution for yourself and show that the results are the same.

Example 5

Evaluate $\displaystyle\int_{\pi/4}^{\pi/2} \frac{\cos^3 x}{\sin x}\,dx$

Writing $s = \sin x$, $ds = \cos x\,dx$

And when $x = \dfrac{\pi}{4}$, $s = \dfrac{1}{\sqrt{2}}$; $x = \dfrac{\pi}{2}$, $s = 1$.

Then $\displaystyle\int_{\pi/4}^{\pi/2} \frac{\cos^2 x}{\sin x}\,dx = \int_{1/\sqrt{2}}^{1} \frac{(1-s^2)}{s}\,ds$

$$= \int_{1/\sqrt{2}}^{1}\left(\frac{1}{s}-s\right)ds$$

$$= \left[\ln|s| - \frac{s^2}{2}\right]_{1/\sqrt{2}}^{1}$$

$$= (\ln 1 - \tfrac{1}{2}) - (\ln 2^{-1/2} - \tfrac{1}{4})$$

$$= \tfrac{1}{2}(\ln 2 - \tfrac{1}{2})$$

Exercise H

1 Use the method of integration by parts to evaluate the following integrals.

a) $\int x^2 \ln x\,dx$ c) $\int x(\ln x)^2\,dx$

b) $\displaystyle\int_1^2 x^{1/2}\ln x\,dx$ d) $\displaystyle\int_1^e \frac{\ln x}{x^2}\,dx$

2 Evaluate

a) $\int_{\pi/4}^{\pi/3} \sec^2 x \ln(\tan x)\, dx$ c) $\int_0^{\ln 2} \frac{dx}{(e^x + 1)^2}$, using $u = e^x$

b) $\int_0^{\pi/4} \frac{\sin^3 x}{\cos x}\, dx$

3 If $I_n = \int (\ln x)^n\, dx$, prove

$$I_n = x(\ln x)^n - nI_{n-1}$$

Hence evaluate $\int_1^e (\ln x)^4\, dx$.

4 Prove that for positive integers p and q,

$$I_{p,q} = \int x^p(\ln x)^q\, dx = \frac{x^{p+1}(\ln x)^q}{p+1} - \frac{q}{p+1} I_{p,q-1}$$

Hence, evaluate $\int_1^e x^3(\ln x)^2\, dx$.

5 Sketch the curve $y = \ln x$ and determine the area of the region between the curve, the x-axis, and the line $x = e^2$.

6 Sketch the curve

$$y^2 = \frac{1}{(x+1)(x-3)}$$

showing that it exists for $x > 3$ and $x < -1$. If the area of the region between the curve and the ordinates $x = 3{\cdot}5$, $x = 5$ is A, use the substitution $z = x - 1$ to prove that

$$A = 2 \ln \left\{1 + \frac{\sqrt{3}}{2}\right\}$$

Mathematical Induction and Further Series

CHAPTER FIFTEEN

15.1 The idea of induction

Let us consider the following process, which aims to achieve a result for the sum of the first n positive integers. We may write

$$\begin{aligned}
&1 &&= 1 = \tfrac{1}{2}(1 \times 2)\\
&1 + 2 &&= 3 = \tfrac{1}{2}(2 \times 3)\\
&1 + 2 + 3 &&= 6 = \tfrac{1}{2}(3 \times 4)\\
&1 + 2 + 3 + 4 &&= 10 = \tfrac{1}{2}(4 \times 5)\\
&1 + 2 + 3 + 4 + 5 &&= 15 = \tfrac{1}{2}(5 \times 6)
\end{aligned}$$

and so on.

From this pattern it would seem to be likely that the result for $1 + 2 + 3 + 4 + \ldots + n$ would be equal to $\frac{1}{2}n(n + 1)$, since on the R.H.S. of the expressions above, the terms in the brackets all involve the corresponding last term of the L.H.S. sum, multiplied by the next consecutive integer. In all cases this product is then halved. However, it is poor mathematical practice to assume the generality of a result, purely on the basis of its truth for some particular numerical value. Nevertheless, if evidence of the possibility of a certain result exists, it is worthwhile heeding this. Using a process called **Mathematical Induction**, it is often possible to confirm the correctness of the assumed general result and to develop an authentic proof. This process is outlined and explained in the following examples.

Example 1

From the illustration we have chosen it is inferred that

$$1 + 2 + 3 + 4 + \cdots + n = \tfrac{1}{2}n(n + 1)$$

Let us assume that the result is true for a particular value of n, say $n = k$, so that we assume

$$1 + 2 + 3 + 4 + \cdots + k = \tfrac{1}{2}k(k + 1)$$

Then proceeding with the L.H.S. sum, to the next term

$$\begin{aligned}
1 + 2 + 3 + 4 + \cdots + k + (k + 1) &= \{1 + 2 + 3 + 4 + \cdots + k\} + (k + 1)\\
&= \tfrac{1}{2}k(k + 1) + (k + 1)\\
&= \tfrac{1}{2}(k + 1)(k + 2)\\
&= \tfrac{1}{2}\overline{(k + 1)}(\overline{k + 1} + 1).
\end{aligned}$$

That is, *if* the result (i.e. $\frac{1}{2}k(k + 1)$) is true for the value $n = k$, then the result for the next consecutive value of n, which is $k + 1$, *is also true*. We now seek to obtain a particular value of n for which the result is true and use the process to develop the proof.

Clearly, when $n = 1$, the L.H.S. = 1 and R.H.S. = $\frac{1}{2}(1)(1 + 1) = 1$
i.e. L.H.S. = R.H.S. and the result is true.

Now if the result is true for $n = k$, it is also true for $n = k + 1$, as has been proved.

So using $k = 1$, the result is true for $n = 1 + 1 = 2$
Now using $k = 2$, the result is true for $n = 2$
Therefore it is true for $n = 2 + 1 = 3$; and so on.

We conclude this special form of proof by stating that by mathematical induction the result is then true for all values of $n \in \mathbb{Z}^+$.

We can now develop alternative methods of proving the standard results established in chapter 8 for
a) the sum of the squares
b) the sum of the cubes
of the first n natural numbers.

Example 2

Prove that $S_2 = \sum_i^N r^2 = \frac{1}{6}N(N + 1)(2N + 1)$

Assume that the result is true for a particular value of N, say k.

Then it is assumed that

$$\sum_1^k r^2 = \tfrac{1}{6}k(k + 1)(2k + 1)$$

So proceeding to the next term in the sum

$$\begin{aligned}\sum_1^{k+1} r^2 &= \sum_1^k r^2 + (k + 1)^2 \\ &= \tfrac{1}{6}k(k + 1)(2k + 1) + (k + 1)^2 \\ &= \tfrac{1}{6}(k + 1)\{k(2k + 1) + 6(k + 1)\} \\ &= \tfrac{1}{6}(k + 1)\{2k^2 + 7k + 6\} \\ &= \tfrac{1}{6}(k + 1)(2k + 3)(k + 2) \\ &= \tfrac{1}{6}(k + 1)(k + 2)(\overline{2k + 1} + 1)\end{aligned}$$

So, since the result on the R.H.S. is $\frac{1}{6}N(N + 1)(2N + 1)$ with N replaced by $k + 1$, we have proved that *if* the result is true for $N = k$, it is also true for $N = k + 1$.

Now when N (or k) = 1

$$\text{L.H.S.} = \sum_1^1 r^2 = 1^2 = 1$$

and the R.H.S. = $\frac{1}{6}(1)(1 + 1)(2 + 1) = \frac{1}{6}(1)(2)(3) = 1$

i.e. L.H.S. = R.H.S.

and the result is true.

But if it is true for $N = 1$ it is also true for $N = 1 + 1 = 2$. Now it is true for $N = 2$; therefore it is also true for $N = 2 + 1 = 3$; and so on.

Therefore by mathematical induction, the result is true for all positive integral values of N.

Example 3

Prove that $S_3 = \sum_1^N r^3 = \{\frac{1}{2}N(N + 1)\}^2$

Assume that the result is true for the particular value of N, say k.

Then it is assumed that

$$\sum_1^k r^3 = \{\tfrac{1}{2}k(k + 1)\}^2$$

So proceeding to the next term in the sum

$$\begin{aligned}\sum_1^{k+1} r^3 &= \sum_1^k r^3 + (k + 1)^3 \\ &= \{\tfrac{1}{2}k(k + 1)\}^2 + (k + 1)^3 \\ &= \tfrac{1}{4}(k + 1)^2\{k^2 + 4(k + 1)\} \\ &= \tfrac{1}{4}(k + 1)^2(k^2 + 4k + 4) \\ &= \tfrac{1}{4}(k + 1)^2(k + 2)^2 \\ &= \tfrac{1}{4}(\overline{k + 1})^2(\overline{k + 1}) + 1)^2 \\ &= \{\tfrac{1}{2}(\overline{k + 1})(\overline{k + 1} + 1)\}^2\end{aligned}$$

So, since the result on the R.H.S. is $\{\frac{1}{2}N(N + 1)\}^2$, with N now replaced by $k + 1$, we have proved that if the result is true for $N = k$, it is also true for $N = k + 1$.

Now when $N = 1$,

$$\text{L.H.S.} = \sum_1^1 r^3 = 1^3 = 1$$

and the R.H.S. $= \{\frac{1}{2}(1)(1 + 1)\}^2 = 1$

Hence L.H.S. = R.H.S. and the result is true for $N = 1$.

Therefore, by virtue of what has been proved, the result is true for $N = 1 + 1 = 2$. Now, the result being true for $N = 2$, by virtue of what has been proved the result is also true for $N = 2 + 1 = 3$, and so on.

Therefore, by mathematical induction, the result is true for all $N \in \mathbb{Z}^+$.

The method of proof by induction may also be used to good effect in other areas of pure mathematics as well as in the summation of series. The next set of examples extend the range of this usage.

Example 4

Prove that the expression $f(n) \equiv 6^n - 5n + 4$ is exactly divisible by 5 for all $n \in \mathbb{Z}^+$.

Suppose that the result is true for the particular value of $n = k$ so that $f(k) = 6^k - 5k + 4 = 5N$, where $N \in \mathbb{Z}^+$.

Then $\quad f(k+1) = 6^{k+1} - 5(k+1) + 4$

And $\quad f(k+1) - f(k) = (6^{k+1} - 6^k) - 5$
$$= 6^k(6-1) - 5$$
$$= 5 \times 6^k - 5$$
$$= 5(6^k - 1)$$

Hence $\quad f(k+1) = f(k) + 5(6^k - 1)$

But by assumption $\quad f(k) = 5N$

So $\quad f(k+1) = 5N + 5(6^k - 1)$
$$= 5\{N + 6^k - 1\}$$

Hence $\quad f(k+1) \div 5 = N + 6^k - 1 =$ another integer

Thus we have proved that if the result is true for $n = k$, it is true also for $n = k + 1$.

But when $\quad n = k = 1$
$$f(1) = 6^1 - 5 + 4 = 5$$

and therefore $f(1)$ is exactly divisible by 5.

So, since the result is true for $n = 1$, by virtue of what has been proved, the result is true for $n = 1 + 1 = 2$.

Now the result is true for $n = 2$, therefore by virtue of what has been proved, the result is true for $n = 2 + 1 = 3$; and so on.

Therefore, by mathematical induction, the result is true for all $n \in \mathbb{Z}^+$.

Example 5

Prove that the number $f(n) \equiv n^3 - n$ contains a factor 6 for all integral values of n such that $n \geqslant 2$.

Clearly, when $n = 2$, $\; f(2) = 2^3 - 2 = 8 - 2 = 6$, and the result is true.

Suppose, therefore, that the result is true for the particular value of $n = k$.

So, $\quad f(k) = k^3 - k = 6p, \quad$ where $p \in \mathbb{Z}^+$.

Then $\quad f(k+1) = (k+1)^3 - (k+1)$

Hence $\quad f(k+1) - f(k) = (k+1)^3 - k^3 - 1$
$$= k^3 + 3k^2 + 3k + 1 - k^3 - 1$$
$$= 3k^2 + 3k$$
$$= 3(k^2 + k)$$

So $\quad f(k+1) = f(k) + 3(k^2 + k)$

Now when k is an odd integer, k^2 is odd and $(k^2 + k)$ is the sum of two odd integers, i.e. $(k^2 + k)$ is even.

Hence $3(k^2 + k) = 6 \times$ (another integer), since we may extract the factor 2 out of the even number $k^2 + k$. Furthermore, if k is an even integer, $k^2 + k$ is the sum of two even numbers and $3(k^2 + k)$ will be equal to $6 \times$ (another integer).

Hence $f(k+1) = f(k) + 6 \times$ (an integer) and since by assumption $f(k) = 6p$, $f(k+1)$ will contain the factor 6.

Thus, if the result is true for $n = k$, it is also true for $n = k + 1$.

But we have already established that the result is true for $n = k = 2$. Therefore the result is also true for $n = 2 + 1 = 3$.

Now the result is true for $n = 3$, therefore by virtue of what we have proved, the result is also true for $n = 3 + 1 = 4$, and so on.

Therefore, by mathematical induction, the result is true for all positive integral values of $n \geqslant 2$.

Example 6

Prove that for all positive integral values of the number n,

$$\frac{d}{dx}(x^n) = nx^{n-1}$$

We know that $\frac{d}{dx}(x^1)$ represents the gradient of the line $y = x$, i.e. is 1.

So, $\frac{d}{dx}(x^1) = 1 = 1 \,.\, x^0$ and the result is true for $n = 1$.

Now suppose the result to be true for the particular value k of n. Then we assume

$$\frac{d}{dx}(x^k) = kx^{k-1}$$

$$\begin{aligned} \text{Now}\quad \frac{d}{dx}(x^{k+1}) &= \frac{d}{dx}(x \times x^k) \\ &= x\frac{d}{dx}(x^k) + x^k\frac{d}{dx}(x^1), \quad \text{(using the product rule)} \\ &= x \,.\, kx^{k-1} + x^k \,.\, 1 \\ &= kx^k + x^k \\ &= (k+1)x^k \end{aligned}$$

So, if the result is true for $n = k$, it is also true for $n = k + 1$.

But we have shown already that the result is true for $n = 1$, so it is also true for $n = 1 + 1 = 2$. Now the result is true for $n = 2$, so by virtue of what we have proved the result is true for $n = 2 + 1 = 3$, and so on. Therefore, by mathematical induction the result is true for all $n \in \mathbb{Z}^+$.

As a final example, to indicate a further extension of the process to series involving the natural numbers, we prove the following result.

Example 7

Prove that $\log(\tfrac{2}{1}) + 2\log(\tfrac{3}{2}) + 3\log(\tfrac{4}{3}) + \cdots + n\log\left(\frac{n+1}{n}\right)$

$$= \log\left\{\frac{(n+1)^n}{n!}\right\} \text{ where } n \in \mathbb{Z}^+.$$

Let $S(n) \equiv \log(\tfrac{2}{1}) + 2\log(\tfrac{3}{2}) + 3\log(\tfrac{4}{3}) + \cdots + n\log\left(\frac{n+1}{n}\right)$

so that for $n = 1$, $S(1) = \log(\tfrac{2}{1}) = \log 2$

But for $n = 1$, R.H.S. $= \log\left\{\frac{(1+1)}{1!}\right\} = \log\frac{2}{1!} = \log 2$

and the result is true.

Now assume that for $n = k$

$$S(k) = \log \left\{\frac{(k+1)^k}{k!}\right\}$$

Then $S(k+1) = S_k + (k+1)\log\left\{\frac{(k+2)}{k+1}\right\}$

$$= \log\left\{\frac{(k+1)^k}{k!}\right\} + \log\left\{\frac{(k+2)^{k+1}}{(k+1)^{k+1}}\right\}$$

$$= \log\left\{\frac{(k+1)^k \times (k+2)^{k+1}}{k! \times (k+1)^{k+1}}\right\}$$

$$= \log\left\{\frac{(k+1)^k \times (k+2)^{k+1}}{k!(k+1) \times (k+1)^k}\right\}$$

$$= \log\left\{\frac{(k+2)^{k+1}}{(k+1)!}\right\} \quad \text{since } (k+1)! = k!(k+1)$$

That is, if the result is true for $n = k$, it is also true for $n = k + 1$.

But we have shown the result to be true for $n = k = 1$, thus it is true also for $n = 1 + 1 = 2$.

Now the result is true for $n = 2$, therefore by virtue of what we have proved the result is true for $n = 3$, and so on. Therefore by mathematical induction, the result is true for all $n \in \mathbb{Z}^+$.

Exercise A

1 Prove the truth of the following results, using the method of mathematical induction

a) $1 + 3 + 5 + 7 + \cdots + (2n - 1) = n^2$

b) $\sum_{1}^{N} r(r+1) = \frac{1}{3}N(N+1)(N+2)$

2 Prove that the sum of the squares of the first n positive odd integers is $\frac{1}{3}n(4n^2 - 1)$.

3 Prove that

a) $\sum_{1}^{N} \frac{1}{r(r+1)} = \frac{N}{N+1}$

b) $\sum_{1}^{N} r^2(r+1) = \frac{1}{12}N(N+1)(N+2)(3N+1)$

4 Prove that the sum of the first n powers of 2,

$1 + 2 + 4 + 8 + \cdots + 2^{n-1}$ is $2^n - 1$

5 Prove that

$1.1! + 2.2! + 3.3! + \ldots n.n! \quad = \quad (n+1)! - 1$

6 Show, by using the standard results for $\sum_{1}^{N} r$ and $\sum_{1}^{N} r^2$, that

$$\sum_{1}^{N} (2r+1)(r-2) = \tfrac{1}{6}n(4n^2 - 3n - 19)$$

Prove the result also, using the method of induction.

7 Prove that, when $n \in \mathbb{Z}^+$,
a) $f(n) \equiv 9^n + 7$ is exactly divisible by 8
b) $f(n) \equiv 2^{3n-1} + 3$ is exactly divisible by 7
c) 64 is a factor of $f(n) \equiv 3^{2n+2} - 8n - 9$

8 Prove that if $\sin x \neq 0$,

$$\cos x + \cos 3x + \cos 5x + \cdots + \cos (2n-1) = \tfrac{1}{2} \sin (2nx) \csc x$$

9 Given that $f(n) \equiv 2^{n+2} + 3^{2n+1}$, simplify the value of the expression $f(n+1) - 2f(n)$. Hence prove by induction that, if $n \in \mathbb{Z}^+$, then $f(n)$ is divisible by 7.

15.2 The binomial expansion

In chapter 2 we saw, using the numerical values given by Pascal's triangle, that it was possible to establish a result for the expansion of the binomial expression $(1+x)^n$ for chosen whole number values of n.

For $n = 1, 2, 3, 4, 5$ in turn, these expansions were

$$(1+x)^1 = 1 + x$$
$$(1+x)^2 = 1 + 2x + x^2$$
$$(1+x)^3 = 1 + 3x + 3x^2 + x^3$$
$$(1+x)^4 = 1 + 4x + 6x^2 + 4x^3 + x^4$$
$$(1+x)^5 = 1 + 5x + 10x^2 + 10x^3 + 5x^4 + x^5$$

Now we can use to suitable effect a symbol which will occur frequently in your further study should you go on to consider the topics of arrangements and selection (i.e. the topic of **permutations** and **combinations**) and **probability**. This is the symbol ^nC_r, which in later work is used for the number of ways of choosing r things from n and which will be justified to stand for the expression

$$\frac{n!}{(n-r)!r!}$$

Here we simply consider the expression

$$\frac{n!}{(n-r)!r!} \equiv {}^nC_r$$

and incorporate the use of the symbol into our study as is now outlined.

Firstly note that when

$$r = 1, \quad \frac{n!}{(n-1)!1!} = n \qquad \Rightarrow {}^nC_1 = n$$

$$r = 2, \quad \frac{n!}{(n-2)!2!} = \frac{n(n-1)}{2!} \qquad \Rightarrow {}^nC_2 = \frac{n(n-1)}{2!}$$

$$r = 3, \quad \frac{n!}{(n-3)!3!} = \frac{n(n-1)(n-2)}{3!} \Rightarrow {}^nC_3 = \frac{n(n-1)(n-2)}{3!}$$

and so on.

Also when

$$r = 0, \quad \frac{n!}{n!0!} = 1 \quad \Rightarrow {}^nC_0 = 1$$

$$r = n, \quad \frac{n!}{(n-n)!n!} = \frac{n!}{0!n!} = 1 \Rightarrow {}^nC_n = 1$$

(since $0! = 1$, as we have already seen in chapter 8).

And generally, $\displaystyle {}^nC_r = \frac{n!}{(n-r)!r!} = \frac{n(n-1)(n-2)\cdots(n-r+1)}{r!}$

We can then build a table determining values of the symbol ${}^nC_r$, where first r is given the values 0, 1, 2, 3, 4, ... and then n the values 1, 2, 3, 4, ... (The components of the value of each ${}^nC_r$, from which its value is calculated, are shown, the term on the L.H.S. of the equals sign in the table.

	${}^nC_0$	${}^nC_1$	${}^nC_2$	${}^nC_3$	${}^nC_4$	${}^nC_5$	${}^nC_6$
$n = 1$	1	1					
$n = 2$	1	2	1				
$n = 3$	1	3	$\frac{(3)(2)}{2!} = 3$	1			
$n = 4$	1	4	$\frac{(4)(3)}{2!} = 6$	$\frac{(4)(3)(2)}{3!} = 4$	1		
$n = 5$	1	5	$\frac{(5)(4)}{2!} = 10$	$\frac{(5)(4)(3)}{3!} = 10$	$\frac{(5)(4)(3)(2)}{4!} = 5$	1	

(As an exercise, carry on and complete the table for the next few values of n, i.e. 6, 7, ...)

Clearly the values in the table correspond to those given in Pascal's triangle, so we could, in fact, write the binomial expansions in the form

$$(1 + x)^1 = 1 + {}^1C_1x$$

$$(1 + x)^2 = 1 + {}^2C_1x + {}^2C_2x^2$$

$$(1 + x)^3 = 1 + {}^3C_1x + {}^3C_2x^2 + {}^3C_3x^3$$

$$(1 + x)^4 = 1 + {}^4C_1x + {}^4C_2x^2 + {}^4C_3x^3 + {}^4C_4x^4$$

$$(1 + x)^5 = 1 + {}^5C_1x + {}^5C_2x^2 + {}^5C_3x^3 + {}^5C_4x^4 + {}^5C_5x^5$$

and so on.

Although the first terms in the expansions are strictly ${}^1C_0$, ${}^2C_0$, ${}^3C_0$, ${}^4C_0$ and ${}^5C_0$ in turn, we have chosen to write 1 for them.

The inference from these expansions, as outlined at the beginning of this chapter, is that we could write in general

$$(1+x)^n = 1 + {}^nC_1x + {}^nC_2x^2 + {}^nC_3x^3 + \cdots + {}^nC_rx^r + \cdots + {}^nC_nx^n$$

or, using the 'sigma' notation, $\sum_0^n {}^nC_rx^r \equiv (1+x)^n$

We now establish the truth of this suggested result using the method of induction. In the proof, as will be seen, we will need to simplify the form of the expression ${}^nC_{r+1} + {}^nC_r$, so we tackle this task first.

$${}^nC_{r+1} + {}^nC_r = \frac{n!}{(n-r-1)!(r+1)!} + \frac{n!}{(n-r)!r!}$$

Now since

$$(r+1)! = (r+1)(r)(r-1)\cdots(2)(1) = (r+1)r!$$

and $(n-r)! = (n-r)(n-r-1)\cdots(2)(1) = (n-r)(n-r-1)!$

the common denominator of the R.H.S. of the expression we are simplifying is

$$(n-r)!(r+1)!$$

Thus

$$\begin{aligned}{}^nC_{r+1} + {}^nC_r &= n!\left\{\frac{(n-r)+(r+1)}{(n-r)!(r+1)!}\right\}\\ &= \frac{n!\{(n+1)\}}{(n-r)!(r+1)!}\\ &= \frac{(n+1)!}{(n+1-\overline{r+1})!(r+1)!}\\ &= {}^{n+1}C_{r+1} \quad \text{by definition}\end{aligned}$$

Now let us assume the truth of the result for the particular value of n, say k, i.e. we assume

$$(1+x)^k = 1 + {}^kC_1x + {}^kC_2x^2 + \cdots + {}^kC_rx^r + \cdots + {}^kC_kx^k$$

Then

$$\begin{aligned}(1+x)^{k+1} &= (1+x)^k(1+x)\\ &= \{1 + {}^kC_1x + {}^kC_2x^2 + \cdots + {}^kC_rx^r + \cdots + {}^kC_kx^k\}(1+x)\\ &= 1 + {}^kC_1x + {}^kC_2x^2 + \cdots + {}^kC_rx^r + \cdots + {}^kC_kx^k\\ &\quad + x + {}^kC_1x^2 + \cdots + {}^kC_{r-1}x^r + \cdots + {}^kC_{k-1}x^k + {}^kC_kx^{k+1}\\ &= 1 + \{{}^kC_1 + 1\}x + \{{}^kC_2 + {}^kC_1\}x^2 + \{{}^kC_3 + {}^kC_2\}x^3 + \ldots\\ &\quad + \{{}^kC_r + {}^kC_{r-1}\}x^r + \cdots + \{{}^kC_k + {}^kC_{k-1}\}x^k + \{{}^kC_kx^{k+1}\}\end{aligned}$$

Now using the result for ${}^nC_{r+1} + {}^nC_r = {}^{n+1}C_{r+1}$

$${}^kC_1 + 1 = {}^kC_1 + {}^kC_0 = {}^{k+1}C_1 \quad (\text{i.e. } r = 0)$$

$${}^kC_2 + {}^kC_1 = {}^{k+1}C_2$$

$${}^kC_3 + {}^kC_2 = {}^{k+1}C_3$$

and

$$^kC_r + {^kC_{r-1}} = {^{k+1}C_r}$$
$$^kC_k + {^kC_{k-1}} = {^{k+1}C_k}$$

Finally, since $^nC_n = 1$, $^kC_k = {^{k+1}C_{k+1}} = 1$ and the result for $(1 + x)^{k+1}$ may now be written as

$$(1 + x)^{k+1} = 1 + {^{k+1}C_1}x + {^{k+1}C_2}x^2 + {^{k+1}C_3}x^3 + \cdots + {^{k+1}C_r}x^r + \cdots + {^{k+1}C_k}x^k + {^{k+1}C_{k+1}}x^{k+1}$$

Thus, if the result is true for $n = k$, it is also true for $n = k + 1$. But when $n = 1$

i.e. $k = 1$ L.H.S. $= (1 + x)^1 = 1 + x$
and R.H.S. $= 1 + {^1C_1}x = 1 + x$

(the R.H.S. series terminates with the term $^nC_n x^n$ for all n).

Thus the L.H.S. = R.H.S. and the result is true for $n = 1$. So, by virtue of what has been proved, the result is true also for $n = 1 + 1 = 2$.

Now, the result being true for $n = 2$, is also true for $n = 2 + 1 = 3$ by virtue of what has been proved, and so on.

Therefore, by mathematical induction, the result is true for all integral values of the number n.

Finally, either by repeating the method for the binomial $(a + b)^n$ or by writing $x = \dfrac{b}{a}$ in the proof we have just carried out and multiplying the result throughout by a^n, we can write

$$(a + b)^n = a^n + {^nC_1}a^{n-1}b + {^nC_2}a^{n-2}b^2 + {^nC_3}a^{n-3}b^3 + \cdots + {^nC_r}a^{n-r}b^r + \cdots + {^nC_n}b^n$$

The following examples amplify the way in which we can use the ^nC_r notation in the binomial expansion, and show how to achieve particular coefficients without referring back to the Pascal triangle.

Example 1

Expand $(2x + 1)^6$

$$(2x + 1)^6 = (2x)^6 + {^6C_1}(2x)^5(1) + {^6C_2}(2x)^4(1)^2 + {^6C_3}(2x)^3(1)^3 + {^6C_4}(2x)^2(1)^4 + {^6C_5}(2x)(1)^5 + {^6C_6}(1)^6$$

$$= 64x^6 + \frac{6!}{5!1!}32x^5 + \frac{6!}{4!2!}16x^4 + \frac{6!}{3!3!}8x^3 + \frac{6!}{2!4!}4x^2 + \frac{6!}{1!5!}2x + 1$$

$$= 64x^6 + 6 \times 32x^5 + 15 \times 16x^4 + 20 \times 8x^3 + 15 \times 4x^2 + 6 \times 2x + 1$$

$$= 64x^6 + 192x^5 + 240x^4 + 160x^3 + 60x^2 + 12x + 1$$

Clearly it is not necessary in practice to write down the various powers of 1. Furthermore, cancellation of the factors in the factorial forms would be carried out as they are being written down. It is especially helpful to note that since

$${}^nC_r = {}^nC_{n-r} \;*$$

${}^6C_1 = {}^6C_5$, ${}^6C_2 = {}^6C_4$, results which also effect quick simplification.

Example 2

Write down the coefficient of the term which does not contain x in the expansion of

$$\left(3x^2 - \frac{1}{2x}\right)^9$$

The general term of $(a+b)^n$ is ${}^nC_r a^{n-r} b^r$

So the general term in the given expression is

$${}^9C_r(3x^2)^{9-r} \times \left(-\frac{1}{2x}\right)^r$$

or $\quad {}^9C_r 3^{9-r}(-\tfrac{1}{2})^r x^{18-3r}$

Thus the term does not involve x, provided that the index $18 - 3r = 0$ (i.e. $x^0 = 1$). So $r = 6$. Then the coefficient of the required term is

$$\begin{aligned}{}^9C_6 3^{9-6}(-\tfrac{1}{2})^6 &= \frac{9!}{3!6!} \times 27 \times \frac{1}{64} = \frac{9 \times 8 \times 7 \times 27}{2 \times 3 \times 64}\\ &= \tfrac{567}{16}\end{aligned}$$

Example 3

In the expansion of $(1 + \lambda x - 3x^2)^6$, the coefficients of the terms involving x^2 and x^3 are 42 and 20 respectively. Determine the value of λ and the coefficient of the term involving x^4.

Since we can write $\quad (1 + \lambda x - 3x^2)^6 = \{1 + x(\lambda - 3x)\}^6$

the general term in the expansion is

$${}^6C_r\{x(\lambda - 3x)\}^r$$

or $\quad {}^6C_r x^r\{\lambda - 3x\}^r$

So those terms involving x^2, x^3 and x^4 are acquired from

$${}^6C_1 x(\lambda - 3x) + {}^6C_2 x^2(\lambda - 3x)^2 + {}^6C_3 x^3(\lambda - 3x)^3 + {}^6C_4 x^4(\lambda - 3x)^4$$

Coefficient of x^2: $\quad 42 = (-3)^6C_1 + (\lambda^2)^6C_2$

$$42 = -18 + 15\lambda^2$$

$$\Rightarrow \lambda^2 = 4, \quad \lambda = \pm 2$$

---

* Proof of ${}^nC_r = {}^nC_{n-r}$

$${}^nC_{n-r} = \frac{n!}{\{n-(n-r)\}!(n-r)!} \quad \text{by definition}$$

$$= \frac{n!}{r!(n-r)!} = \frac{n!}{(n-r)!r!} = {}^nC_r$$

Coefficient of x^3: $\quad 20 = (-6\lambda)^6C_2 + (\lambda^3)^6C_3$

$$20 = -90\lambda + 20\lambda^3$$

or $\quad 2 = -9\lambda + 2\lambda^3$

Now if $\lambda = +2, \quad -9\lambda + 2\lambda^3 = -18 + 16 = -2 \neq$ L.H.S.

But if $\lambda = -2, \quad -9\lambda + 2\lambda^3 = \quad 18 - 16 = \quad 2 =$ L.H.S.

Hence $\lambda = -2$

Furthermore when we expand the brackets $(\lambda - 3x)^2$, $(\lambda - 3x)^3$ and $(\lambda - 3x)^4$, we can pick out the coefficient of x^4 in the subsequent products as follows

$$\begin{aligned}\text{Coefficient of } x^4 &= (9)^6C_2 - (9\lambda^2)^6C_3 + (\lambda^4)^6C_4 \\ &= 135 - 36 \times 20 + 16 \times 15 \\ &= -345\end{aligned}$$

Example 4

Show that when $(1 - 2x)^{18}(1 + 3x)^{17}$ is expanded, the coefficient of the term involving x^2 is zero. Calculate the coefficient of the term involving x^3.

The general term in the product is of the form

$${}^{18}C_m(-2x)^m \times {}^{17}C_n(3x)^n$$

i.e. $\quad {}^{18}C_m \times {}^{17}C_n \times (-2)^m \times 3^n \times x^{m+n}$

Thus, for the term in x^2, $m + n = 2$ where m and $n \in \mathbb{Z}^+$.

Thus $\quad m = 0, \quad n = 2$
$\qquad m = 1, \quad n = 1$
$\qquad m = 2, \quad n = 0$

are the pairs of values satisfying this condition and the coefficient of x^2 is

$$\begin{aligned}&{}^{18}C_0 \times {}^{17}C_2 \times (-2)^0 \times 3^2 + {}^{18}C_1 \times {}^{17}C_1 \times (-2)^1 \times 3^1 \\ &\quad + {}^{18}C_2 \times {}^{17}C_0 \times (-2)^2 \times 3^0 \\ &= 18 \times 17 \times 4 - 18 \times 17 \times 6 + 18 \times 17 \times 2 \\ &= 18 \times 17(4 - 6 + 2) = 0\end{aligned}$$

For the term involving x^3 we must have $m + n = 3$. The pairs of integers satisfying this result are

$m = 0, \quad n = 3$
$m = 1, \quad n = 2$
$m = 2, \quad n = 1$
$m = 3, \quad n = 0.$

The coefficient of the term in x^3 is therefore

$$\begin{aligned}&{}^{18}C_0 \times {}^{17}C_3 \times (-2)^0 \times 3^3 + {}^{18}C_1 \times {}^{17}C_2 \times (-2)^1 \times 3^2 \\ &\quad + {}^{18}C_2 \times {}^{17}C_1 \times (-2)^2 \times 3^1 + {}^{18}C_3 \times {}^{17}C_0 \times (-2)^3 \times 3^0 \\ &= 17 \times 8 \times 15 \times 9 - 18 \times 17 \times 16 \times 9 + 18 \times 17 \times 17 \times 6 \\ &\quad - 6 \times 17 \times 16 \times 4 \\ &= -1020\end{aligned}$$

Example 5

Derive a value for $1 + {}^nC_1 + {}^nC_2 + {}^nC_3 + \cdots + {}^nC_n$

Since $(1 + x)^n = 1 + {}^nC_1 x + {}^nC_2 x^2 + {}^nC_3 x^3 + \cdots + {}^nC_n x^n$

If we replace x by the value 1

$$2^n = 1 + {}^nC_1 + {}^nC_2 + {}^nC_3 + \cdots + {}^nC_n$$

which is the required result.

Moreover, if we write $x = -1$

$$(1 - 1)^n = 0 = 1 - {}^nC_1 + {}^nC_2 - {}^nC_3 + \cdots + (-1)^n\,{}^nC_n$$

Thus $1 + {}^nC_2 + {}^nC_4 + \cdots = {}^nC_1 + {}^nC_3 + {}^nC_5 + \cdots$

But since these two results have sum 2^n, each of them must be equal to $\frac{1}{2} \times 2^n = 2^{n-1}$

i.e. $1 + {}^nC_2 + {}^nC_4 + \cdots = {}^nC_1 + {}^nC_3 + {}^nC_5 + \cdots = 2^{n-1}$

Series of this type are plentiful and are termed as series connecting the binomial coefficients. (Strictly coefficients of the binomial expansion.)

Other results of this nature can be obtained by employing differential and integral calculus.

For example, given

$$(1 + x)^n = 1 + {}^nC_1 x + {}^nC_2 x^2 + {}^nC_3 x^3 + \cdots + {}^nC_n x^n$$

if we differentiate throughout

$$n(1 + x)^{n-1} = {}^nC_1 + (2)^nC_2 x + (3)^nC_3 x^2 + \cdots + (n)^nC_n x^{n-1}$$

So, replacing x by 1,

$$n2^{n-1} = {}^nC_1 + (2)^nC_2 + (3)^nC_3 + \cdots + (n)^nC_n$$

Furthermore, if we integrate both sides of the result between the limits 0 and 1.

$$\int_0^1 (1 + x)^n\,dx = \int_0^1 \{1 + {}^nC_1 x + {}^nC_2 x^2 + {}^nC_3 x^3 + \cdots + {}^nC_n x^n\}\,dx$$

$$\Rightarrow \left[\frac{(1 + x)^{n+1}}{n + 1}\right]_0^1 = \left[x + \frac{{}^nC_1 x^2}{2} + \frac{{}^nC_2 x^3}{3} + \frac{{}^nC_3 x^4}{4} + \cdots + \frac{{}^nC_n x^{n+1}}{n + 1}\right]_0^1$$

Thus $\dfrac{2^{n+1} - 1}{n + 1} = 1 + \dfrac{{}^nC_1}{2} + \dfrac{{}^nC_2}{3} + \dfrac{{}^nC_3}{4} + \cdots + \dfrac{1}{n + 1}\,{}^nC_n$

Exercise B

1 Evaluate
a) ${}^4C_2$ b) ${}^7C_3$ c) ${}^{13}C_6$

2 Show that
a) ${}^8C_5 = {}^7C_5 + {}^7C_4$ b) ${}^8C_4 + {}^8C_3 = {}^9C_4$

3 Find the value of ${}^6C_0 + {}^6C_2 + {}^6C_4 + {}^6C_6$

4 Simplify
a) ${}^nC_r + {}^{n-1}C_r + {}^{n-1}C_{r-1}$ b) ${}^{n-2}C_r + (2)^{n-2}C_{r-1} + {}^{n-2}C_{r-2}$

5 Determine the value of n given
a) ${}^nC_3 = (6)^{n-1}C_2$ b) ${}^nC_4 = (5)^{n-2}C_3$

6 Write down the term in x^3 in the expansion of

a) $(2x+1)^5$ d) $(1-x)^2(2+x)^5$

b) $(x-2)^{10}$ e) $(3-x)(1+2x)^9$

c) $\left(2x^2+\frac{1}{x}\right)^9$ f) $\left(x+\frac{1}{x}\right)^2(1+x)^6$

7 In the expansion of $(3x+2)^{19}$ the coefficients of the terms involving x^r and x^{r+1} are equal. Calculate the value of r.

8 Determine the value of the coefficient of the term which does not contain x in the expansion of

a) $\left(\frac{3x^2}{2}-\frac{1}{3x}\right)^9$ b) $\left(x-\frac{1}{x^3}\right)^{28}$

9 It is given that in the expansion of $(1+\alpha x+\beta x^2)^6$ in ascending powers of x the coefficient of the term in x^2 is zero and the coefficient of the term in x^3 is -440. Determine the values of α and β.

10 In the expansion of

$$\left(2x^2+\frac{a}{x^3}\right)^{10}$$

in ascending powers of x it is given that the coefficients of x^5 and x^{15} are equal. Determine the value of a.

11 Find the coefficients of x^6 and x^7 in the expansion of $(1+2x)(1-x^2)^{10}$ as a series of ascending powers of x.

12 Prove that if T_2 and T_3 are the coefficients of x^2 and x^3 in the expansion of $(2+2x+x^2)^n$ in ascending powers of x then

$$T_3 : T_2 = (n^2-1) : 3n$$

13 Prove that
a) ${}^nC_0 + (2)^nC_1 + (4)^nC_2 + (8)^nC_3 + \cdots + (2^n)^nC_n = 3^n$
b) $1 - (2)^nC_1 + (3)^nC_2 + \cdots + (-1)^n(n+1)^nC_n = 0$

14 Prove that

$$(1+x)^n = 1 + {}^nC_1x + {}^nC_2x^2 + \cdots + {}^nC_nx^n$$
$$= {}^nC_n + {}^nC_{n-1}x + {}^nC_{n-2}x^2 + \cdots + x^n$$

Hence by considering the coefficients of x^n in the product of these two series show that

$$1 + ({}^nC_1)^2 + ({}^nC_2)^2 + ({}^nC_3)^2 + \cdots + ({}^nC_n)^2 = \frac{\{2n\}!}{\{n!\}^2}$$

The questions 7 and 8 of exercise B in chapter 8 were concerned with investigating the sum of a very large number of terms of two given geometric progressions. We now examine this problem in a different way.

Let us start from a practical aspect and suppose that we have a piece of string 2 m in length. Let us cut this in half, put one of the equal pieces on one side, cut the other in half and put one of those pieces on one side, then cut the other in half, and so on.

Clearly, supposing that we could continue to carry out the 'halving' operation indefinitely we would have on one side, ultimately, pieces of string whose length, in metres, would be

$$1, \tfrac{1}{2}, \tfrac{1}{4}, \tfrac{1}{8}, \tfrac{1}{16}, \tfrac{1}{32}, \ldots$$

the rth piece being of length $\dfrac{1}{2^{r-1}}$.

So, we could express the total length of the pieces of string we had cut and collected in the form of a series in which the number of terms is infinite, i.e.

$$1+\frac{1}{2}+\frac{1}{4}+\frac{1}{8}+\frac{1}{16}+\frac{1}{32}+\cdots+\frac{1}{2^{r-1}}+\cdots \text{ to } \infty$$

However we know that this series has a finite sum, i.e. the original length of the string, 2 m. Thus we can write

$$1+\frac{1}{2}+\frac{1}{4}+\frac{1}{8}+\cdots+\frac{1}{2^{r-1}}+\cdots+\text{ to } \infty = 2$$

or $\displaystyle \lim_{n\to\infty} \sum_{r=1}^{n}\left(\frac{1}{2^{r-1}}\right) = 2$

In the case when a series consisting of an infinite number of terms has a finite sum we say that the series **converges** to that sum.

Let us now consider the matter of convergency of a geometric progression more generally.

The series

$$1+x+x^2+x^3+\cdots+x^n$$

is a geometric progression in which the common ratio is x. Thus the sum of the series is

$$\frac{1(1-x^n)}{1-x}$$

that is $1+x+x^2+x^3+\cdots+x^n \equiv \dfrac{1-x^n}{1-x}$

Now if we let $n\to\infty$ so that the L.H.S. becomes an infinite series of powers of x, we can write

$$1+x+x^2+x^3+\cdots+x^r+\cdots = \lim_{n\to\infty}\left\{\frac{1-x^n}{1-x}\right\}$$

Now if $-1<x<1$, i.e. $|x|<1$, when $n\to\infty$, $x^n\to 0$ so that the value of the right-hand limit ultimately becomes

$$\frac{1-0}{1-x}=\frac{1}{1-x}$$

Thus we have established (reversing the order of writing) that

$$\frac{1}{1-x} = 1 + x + x^2 + x^3 + \cdots + x^r \cdots \text{ to } \infty, \text{ if } |x| < 1$$

But $\dfrac{1}{1-x} = (1-x)^{-1}$

which is a particular case of the binomial form $(a+x)^n$ where n is now equal to the value -1.

So we have justified that for a value of n which is no longer such that $n \in \mathbb{Z}^+$ that the expansion of a binomial form exists in the form of a power series provided that $|x| < 1$.

It is worth noting that if we replace x by the value 2 in the result we have been considering, the series $1 + 2 + 2^2 + 2^3 + \cdots$ (which clearly does not have a finite sum) is not equal to

$$\frac{1}{1-x} \text{ with } x = 2, \text{ namely } \frac{1}{1-2} \text{ or } -1.$$

So for this value of x

$1 + x + x^2 + x^3 + \cdots$ **diverges** to ∞

By writing $-x$, instead of x, in our discussion it is clear that

$$\frac{1}{1+x} = (1+x)^{-1}$$
$$= 1 - x + x^2 - x^3 + \cdots + \text{ to } \infty, \text{ provided that } |x| < 1$$

15.4 The binomial series

In section 13.3 we used the method of Maclaurin to establish for the general binomial form $(a+h)^n$ the relation

$$(a+h)^n = \sum_{r=0}^{n} \frac{a^{n-r}h^r n(n-1)\cdots(n-r+1)}{r!}$$

which we see can now also be written as

$$(a+h)^n = \sum_{r=0}^{n} {}^nC_r a^{n-r}h^r, \text{ when } n \in \mathbb{Z}^+$$

since $\dfrac{n(n-1)\cdots(n-r+1)}{r!} = \dfrac{n!}{(n-r)!r!} = {}^nC_r$ if $r \leqslant n$

and is zero if $r > n$.

But if $n \notin \mathbb{Z}^+$, and is more generally chosen so that $n \in \mathbb{R}$, the symbols which we would be caused to introduce, for example, ${}^{-4}C_2$, ${}^{1/2}C_5$, in using the expansion of $(a+h)^{-4}$, $(a+h)^{1/2}$, would have no meaning.

To overcome this obstacle we make use of an alternative form of the symbol, namely $\dbinom{n}{r}$ which we take to stand for the expression

$$\frac{n(n-1)(n-2)\cdots(n-r+1)}{r!}.$$

Thus, for example, the value of

$$\binom{-1}{3} \text{ is } \frac{(-1)(-2)(-3)}{3!} \qquad \text{i.e. } -1$$

$$\binom{\frac{1}{2}}{2} \text{ is } \frac{(\frac{1}{2})(\frac{1}{2}-1)}{2!} \qquad \text{i.e. } -\frac{1}{8}$$

$$\binom{-\frac{3}{4}}{3} \text{ is } \frac{(-\frac{3}{4})(-\frac{7}{4})(-\frac{11}{4})}{3!} \qquad \text{i.e. } -\frac{77}{128}$$

and so on.

We can then write the binomial series for $(a + h)^n$ as

$$(a + h)^n = \sum_{r=0}^{N} \binom{n}{r} a^{n-r} h^r$$

where now $N \to \infty$, provided that h, a satisfy any restrictive conditions for convergency which must apply for such an infinite expansion to exist.

We have seen in the cases where n has been equal to -1, and equal to $\frac{1}{2}$, the special restrictions which must be imposed upon h, a and these conditions being established by the comparison test for convergence.

Approaching the general problem requires rather more knowledge than the needs of this course. Nevertheless, the result itself, i.e. $|h| < a$ has already been confirmed in these particular cases.

We therefore use as a general form of the binomial series

$$(a + x)^n = a^n + \binom{n}{1} a^{n-1} x + \binom{n}{2} a^{n-2} x^2 + \cdots + \binom{n}{r} a^{n-r} x^r + \cdots + \text{ to } \infty,$$

provided that $|x| < a$.

The following series should be well enough known to be quoted from memory whenever they occur.

$$(1 - x)^{-1} = 1 + x + x^2 + x^3 + \cdots + x^r + \cdots$$

$$(1 + x)^{-1} = 1 - x + x^2 - x^3 + \cdots + (-1)^r x^r + \cdots$$

$$\begin{aligned}(1 - x)^{-2} &= 1 + \binom{-2}{1}(-x) + \binom{-2}{2}(-x)^2 + \binom{-2}{3}(-x)^3 \\ &\quad + \cdots + \binom{-2}{r}(-x)^r + \cdots \\ &= 1 + \frac{(-2)(-x)}{1!} + \frac{(-2)(-3)}{2!}(-x)^2 \\ &\quad + \frac{(-2)(-3)(-4)}{3!}(-x)^3 + \cdots \\ &= 1 + 2x + 3x^2 + 4x^3 + 5x^4 + \cdots + (r + 1)x^r + \cdots\end{aligned}$$

And $(1 + x)^{-2} = 1 - 2x + 3x^2 - 4x^3 + 5x^4 + \cdots + (-1)(r + 1)x^r + \cdots$

Example 1

Find the first three terms and the general terms in the expansion of the following expressions in ascending powers of x. State the range of values of x for which the expansions are valid.

a) $f(x) \equiv (1+3x)^{-1}$

$$f(x) = 1 - (3x) + (3x)^2 - (3x)^3 + \cdots + (-1)^r(3x)^r + \cdots$$

(using $(1+x)^{-1}$)

$$= 1 - 3x + 9x^2 - 27x^3 + \cdots + (-1)^r \,.\, 3^r x^r + \cdots$$

Expansion is valid if $|3x| < 1$, i.e. $|x| < \frac{1}{3}$

b) $f(x) \equiv \dfrac{1}{(3-x)^2}$

Write in the form $\dfrac{1}{3^2\left(1-\dfrac{x}{3}\right)^2} = \dfrac{1}{9}\left(1-\dfrac{x}{3}\right)^{-2}$

Then $f(x) = \dfrac{1}{(3-x)^2}$

$$= \frac{1}{9}\left\{1 + 2\left(\frac{x}{3}\right) + 3\left(\frac{x}{3}\right)^2 + \cdots + (r+1)\left(\frac{x}{3}\right)^r + \cdots\right\}$$

(using $(1-x)^{-2}$)

$$= \frac{1}{9}\left\{1 + \frac{2}{3}x + \frac{x^2}{3} + \cdots + \frac{(r+1)}{3^r}x^r + \cdots\right\}$$

Expansion is valid if $|x| < 3$

c) $f(x) = \sqrt{(1+x)}$

Expression is

$$f(x) \equiv (1+x)^{1/2} = 1 + \binom{\frac{1}{2}}{1}x + \binom{\frac{1}{2}}{2}x^2 + \cdots + \binom{\frac{1}{2}}{r}x^r + \cdots$$

$$= 1 + \frac{(\frac{1}{2})}{1!}x + \frac{(\frac{1}{2})(-\frac{1}{2})}{2!}x^2 + \cdots + \binom{\frac{1}{2}}{r}x^r + \cdots$$

$$= 1 + \frac{x}{2} - \frac{x^2}{8} + \cdots + \binom{\frac{1}{2}}{r}x^r + \cdots$$

Expansion is valid if $|x| < 1$.
Note that as it is clear that the general term does not simplify, there is little to be gained by writing

$$\binom{\frac{1}{2}}{r} = \frac{(\frac{1}{2})(-\frac{3}{2})(-\frac{5}{2})\cdots\left(-\dfrac{(2r-1)}{2}\right)}{r!}.$$

d) $f(x) \equiv \dfrac{1-x}{1+x}$

Expression is $(1-x)(1+x)^{-1}$ which is valid for $|x| < 1$ since $(1+x)^{-1}$ is to be expanded.

Now $f(x) \equiv (1-x)(1+x)^{-1} = (1-x)(1 - x + x^2 \cdots (-1)^r x^r + \cdots)$

$$\begin{aligned} &= 1 - x + x^2 + \cdots + (-1)^r x^r + \cdots \\ &\quad\;\; - x + x^2 + \cdots + (-1)^{r-1}x^r + \cdots \\ &= 1 - 2x + 2x^2 + \cdots + 2(-1)^r x^r + \cdots \end{aligned}$$

Example 2

Write down the first four terms in the expansion of

$$f(x) \equiv \frac{3}{(1-x)(1+2x)}$$

State the range of values of x for which the expansion is valid.

We could write $\dfrac{3}{(1-x)(1+2x)} = 3(1-x)^{-1}(1+2x)^{-1}$

and multiply together the two series expansions of $(1-x)^{-1}$ and $(1+2x)^{-1}$.

However, it is more usual to use the ideas of partial fractions in this type of example since ultimately the series corresponding to the factors in the denominator of the expression can be isolated by this method.

Let $\quad f(x) \equiv \dfrac{A}{1-x} + \dfrac{B}{1+2x}$

Then $\quad 3 \equiv A(1+2x) + B(1-x)$

When $\quad x = 1,\ 3 = 3A \Rightarrow A = 1$

$\quad x = -\frac{1}{2},\ 3 = \frac{3}{2}B \Rightarrow B = 2$

Hence $\quad f(x) = \dfrac{1}{1-x} + \dfrac{2}{1+2x}$

$$= (1-x)^{-1} + 2(1+2x)^{-1}$$

$$= 1 + x + x^2 + x^3 + \cdots$$

$$+2\{1 - 2x + 4x^2 - 8x^3 + \cdots\}$$

$$= 3 - 3x + 9x^2 - 15x^3 + \cdots$$

Since $(1-x)^{-1}$ is valid for $|x| < 1$ and $(1+2x)^{-1}$ is valid for $|2x| < 1$ i.e. $|x| < \frac{1}{2}$ the *combined* expansion will be true for the more limited range $|x| < \frac{1}{2}$ as shown in figure 15.1.

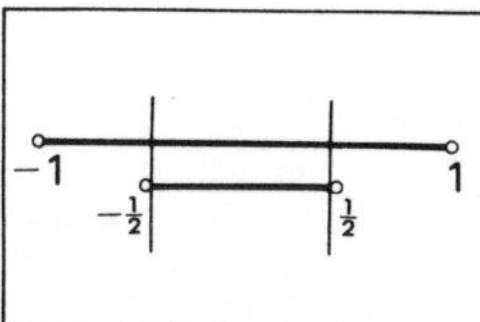

Figure 15.1

Example 3

Expand $f(x) \equiv \dfrac{2x+4}{(x-1)(x+3)}$

up to the term including x^4. Give the range of values of x for which the expansion is valid.

Let $f(x) \equiv \dfrac{A}{x-1} + \dfrac{B}{x+3}$

so that $2x + 4 \equiv A(x+3) + B(x-1)$

Then when $\quad x = 1, \quad 6 = 4A \Rightarrow A = \frac{3}{2}$

$\quad x = -3, \quad -2 = -4B \Rightarrow B = \frac{1}{2}$

Thus $f(x) = \dfrac{\frac{3}{2}}{(x-1)} + \dfrac{\frac{1}{2}}{(x+3)}$

$$= -\tfrac{3}{2}(1-x)^{-1} + \tfrac{1}{6}\left(1+\frac{x}{3}\right)^{-1}$$

(**Note** the manipulation to achieve the form $(1+x)^{-1}$ in each case.)

Hence $f(x) = -\tfrac{3}{2}\{1 + x + x^2 + x^3 + x^4 + \cdots\}$

$$+\frac{1}{6}\left\{1 - \frac{x}{3} + \frac{x^2}{9} - \frac{x^3}{27} + \frac{x^4}{81} + \cdots\right\}$$

$$= (-\tfrac{3}{2} + \tfrac{1}{6}) + x(-\tfrac{3}{2} - \tfrac{1}{18}) + x^2(-\tfrac{3}{2} + \tfrac{1}{54})$$

$$+x^3(-\tfrac{3}{2} - \tfrac{1}{162}) + x^4(-\tfrac{3}{2} + \tfrac{1}{486}) + \cdots$$

$$= -\tfrac{4}{3} - \tfrac{14}{9}x - \tfrac{40}{27}x^2 - \tfrac{122}{81}x^3 - \tfrac{364}{243}x^4 - \cdots$$

Since $(1-x)^{-1}$ is valid for $|x| < 1$ and $\left(1+\dfrac{x}{3}\right)^{-1}$ is valid for $\left|\dfrac{x}{3}\right| < 1$, the limited range which includes both is $|x| < 1$.

Example 4

This example illustrates the fact that when x is very small in the use of the binomial series, the first few terms in the expansion generally lead to approximations which are sufficiently adequate for certain numerical conclusions to be reached.

Write down the first four terms of the expansion of $(1+x)^{1/4}$ in ascending powers of x. Hence derive the fourth root of 1·08 to five places of decimals.

We have $(1+x)^{1/4} = 1 + \dbinom{\frac{1}{4}}{1}x + \dbinom{\frac{1}{4}}{2}x^2 + \dbinom{\frac{1}{4}}{3}x^3 + \cdots$

$$= 1 + \tfrac{1}{4}x + \frac{(\frac{1}{4})(-\frac{3}{4})}{2!}x^2 + \frac{(\frac{1}{4})(-\frac{3}{4})(-\frac{7}{4})}{3!}x^3 + \cdots$$

$$= 1 + \tfrac{1}{4}x - \tfrac{3}{32}x^2 + \tfrac{7}{128}x^3 + \cdots$$

Hence $\sqrt[4]{1{\cdot}08} = \sqrt[4]{(1+0{\cdot}08)} = (1+0{\cdot}08)^{1/4}$

$$= 1 + \tfrac{1}{4}(0{\cdot}08) - \tfrac{3}{32}(0{\cdot}08)^2 + \tfrac{7}{128}(0{\cdot}08)^3 + \cdots$$

$$= 1 + 0{\cdot}02 - 0{\cdot}0006 + 0{\cdot}000\,028$$

$$= 1{\cdot}019\,42(8) \text{ or } 1{\cdot}01943$$

Example 5

If x is so small that x^4 and higher powers of x are negligible, show that

$$f(x) \equiv (1-4x)^{1/2} - (1+3x)^{-2/3}$$

is approximately equal to $Ax^2 + Bx^3$, where A and B are constants.

Hence evaluate $\sqrt{(0{\cdot}96)} - \dfrac{1}{\sqrt[3]{\{(1{\cdot}03)^2\}}}$ to five places of decimals.

$$f(x) = 1 + \binom{\frac{1}{2}}{1}(-4x) + \binom{\frac{1}{2}}{2}(-4x)^2 + \binom{\frac{1}{2}}{3}(-4x)^3 + \cdots$$

$$-1 - \binom{-\frac{2}{3}}{1}(3x) - \binom{-\frac{2}{3}}{2}(3x)^2 - \binom{-\frac{2}{3}}{3}(3x)^3 - \cdots$$

$$= 1 - 2x + \frac{(\frac{1}{2})(-\frac{1}{2})}{2!}(-4x)^2 + \frac{(\frac{1}{2})(-\frac{1}{2})(-\frac{3}{2})}{3!}(-4x)^3 + \cdots$$

$$-1 + 2x - \frac{(-\frac{2}{3})(-\frac{5}{3})}{2!}(3x)^2 - \frac{(-\frac{2}{3})(-\frac{5}{3})(-\frac{7}{3})}{3!}(3x)^3 - \cdots$$

$$= -2x^2 - 4x^3 + \cdots - 5x^2 + \tfrac{35}{3}x^3$$

$$= -7x^2 + \tfrac{23}{3}x^3$$

$\Rightarrow A = -7$ and $B = \frac{23}{3}$

The number $\sqrt{0{\cdot}96} - \dfrac{1}{\sqrt[3]{\{(1{\cdot}03)^2\}}}$ is $f(x)$ with $x = 0{\cdot}01$

And $f(0{\cdot}01) \approx -7(0{\cdot}0001) + \frac{23}{3}(0{\cdot}000\,001)$
$\approx -0{\cdot}0007 \quad + 0{\cdot}000\,008$
$\approx -0{\cdot}000\,69$

Example 6

A quantity z is expressed in terms of the variables x and y by means of the expression $z = 2x^{1/2} . y^{-3}$. When the value of z is calculated, mistakes of $+1$ per cent and $+0{\cdot}5$ per cent are made in the values of x and y. Calculate the corresponding percentage error in z.

The values of x and y substituted in the expression in error are

$x(1 + \frac{1}{100})$ i.e. $x(1 + 0{\cdot}01)$ and
$y(1 + \frac{1}{200})$ i.e. $y(1 + 0{\cdot}005)$

So the value of z which will be calculated is

$$z = 2x^{1/2}y^{-3}(1 + 0{\cdot}01)^{1/2}(1 + 0{\cdot}005)^{-3}$$

$$= 2x^{1/2}y^{-3}\left\{1 + 0{\cdot}005 + \frac{(\frac{1}{2})(-\frac{1}{2})}{2!}(0{\cdot}01)^2 + \cdots\right\}$$

$$\times\left\{1 - 0{\cdot}015 + \frac{(-3)(-4)}{2!}(0{\cdot}005)^2 + \cdots\right\}$$

$$= 2x^{1/2}y^{-3}\{1 + 0{\cdot}005 - 0{\cdot}015 - \tfrac{1}{8} \times 0{\cdot}0001 + 6 \times (0{\cdot}005)^2$$

$$-0{\cdot}000\,075 + \cdots\}$$

$$= 2x^{1/2}y^{-3}\{1 - 0{\cdot}01 + \cdots\}$$

i.e. Error in $z = -0{\cdot}01 \times 2x^{1/2}y^{-3}$
or $\quad -0{\cdot}01 \times 100 = -1\%$

Exercise C

1 Evaluate

a) $\binom{-1}{2}$ b) $\binom{\frac{1}{2}}{4}$ c) $\binom{-\frac{3}{4}}{2}$ d) $\binom{-p}{p}$ where $p > 0$

2 Write down the first four terms, the general term, and the range of the variable x for which the expansions of the following are valid.

a) $(1+2x)^{-1}$ c) $(3+2x)^{1/3}$

b) $\left(1-\dfrac{x}{2}\right)^{-2}$ d) $(1-x^2)^{-2/3}$

3 Write down the $(r+1)$th term in the following expansions, assuming that the value of x chosen is one which fulfils the condition of convergency.

a) $(1-x)^{-3}$ b) $(1+2x)^{-3/2}$ c) $(1-x)^{-n}$

4 Expand the following expressions up to and including the term in x^4. Write down the coefficient of the term involving x^r and also state the range of values of x for which the expansion is valid.

a) $(1+x)(1-x)^{-2}$ c) $\left(\dfrac{1+x}{1-x}\right)^2$

b) $(1+x)\sqrt{(1-x)}$

5 In the expansion of

$$\frac{5-6x}{(1-x)^2}$$

as a series of ascending powers of x determine the coefficient of the term involving x^r in the expansion. Prove that when $r=5$ this coefficient is zero.

6 If x is very large, prove that

$$\sqrt{(x+1)}-\sqrt{(x-1)}\approx\frac{1}{\sqrt{x}}\left(1+\frac{1}{8x^2}\right)$$

(Suggestion: write $x+1=x\left(1+\dfrac{1}{x}\right)$ so that $\dfrac{1}{x}$ is now very small.)

7 By expressing

$$\frac{5}{1-x-6x^2}$$

in terms of its partial fractions, expand the function as a series of ascending powers of x as far as the term containing x^4. Write down the coefficient of x^r and the range of values of x for which the expansion is valid.

8 Express $\dfrac{11x+12}{(2x-3)(x+2)(x-3)}$

in terms of its partial fractions and determine the coefficient of x^3 in the expansion of the function as a series of ascending powers of x. Is your result valid if $x=-\frac{5}{4}$?

9 Evaluate to five places of decimals

a) $\sqrt{1{\cdot}01}$ c) $\sqrt[5]{0{\cdot}99}$

b) $\sqrt[3]{1006}$ d) $\dfrac{1}{(0{\cdot}998)^2}$

10 The formula

$$T = 2\pi\sqrt{\frac{l}{g}}$$

gives the time of oscillation T sec of a pendulum of length l metres. (π and g are constants.) This pendulum oscillates completely in 2 seconds. If the length of the pendulum is decreased by 4 per cent, determine the percentage change in the time of an oscillation. Determine how many seconds per hour this new pendulum gains or loses.

11 a) Calculate the range, or ranges, of values of x for which $|3x - 5| > 7$.
b) If x has a value satisfying this condition, find the sum to infinity of

$$1 + 2\left(\frac{7}{3x-5}\right) + 3\left(\frac{7}{3x-5}\right)^2 + 4\left(\frac{7}{3x-5}\right)^3 + \cdots$$

12 Show that the roots of the quadratic equation

$$ax^2 + 2bx + c = 0$$

may be written in the form

$$-\frac{b}{a}\left\{1 \pm \left(1 - \frac{ac}{b^2}\right)^{1/2}\right\}$$

If ac is extremely small compared with the value of b^2, prove that the roots of the equation are x_1, x_2 where

$$x_1 \approx -\frac{2b}{a} + \frac{c}{2b} \text{ and } x_2 \approx \frac{-c}{2b}$$

13 Neglecting all powers of x above the first, obtain an approximation to the value of the function

$$f(x) \equiv \frac{(1-2x)^{1/2}(8-x)^{1/3}}{(1-x)^2}$$

14 Express the quotient $\dfrac{(2{\cdot}004)^4}{\sqrt{4{\cdot}02}}$

in a general form in which the numerator and denominator involve a binomial form using the same numerical value of the variable x. Expand this form as a series of ascending powers of x and hence derive a numerical value of the given quotient correct to three places of decimals.

CHAPTER SIXTEEN

Differential Equations

16.1 Introduction

A differential equation is the name given to any equation which contains at least one derivative. Such equations arise from a variety of situations in all kinds of subjects. Problems in physics often give rise to differential equations, and these usually occur in problems arising in dynamics, hydrodynamics and electrical or magnetic theory. In chemistry, we often need to measure the speed of chemical reactions in relation to the chemicals involved. In biology, much interest is centred round ideas of rates of growth and rates of decay of bacteria, animals and plants. Clearly there are many other examples where a 'rate of change' is involved, and these always lead to a differential equation.

We shall consider here some of the simpler types of equation, where only the first derivatives occur, and where these are raised to no power above the first.

The 'solution' of a differential equation involves finding an equation connecting the variables involved, without containing a derivative. There are no specific values for the variables, as would normally be expected in a 'solution'.

The 'first order' equations (i.e. those involving just first derivatives) considered here fall into two categories

a) those in which the variables separate, and

b) those of the form $\dfrac{dy}{dx} + Py = Q$, which are called **linear.**

16.2 Separation of the variables

Example 1

If we are told that $\dfrac{dy}{dx} = x$, then it is clear that

$$y = \tfrac{1}{2}x^2 + k, \text{ where } k \text{ is some constant.}$$

The integration of x is straightforward. We say that the solution of the differential equation $\dfrac{dy}{dx} = x$ is $y = \tfrac{1}{2}x^2 + k$

We could represent this solution graphically, by taking different values of the constant k as shown in figure 16.1.

We see, then, that the differential equation represents a family of curves, and that the solution of the differential equation is a parametric representation of the family, where the parameter is k.

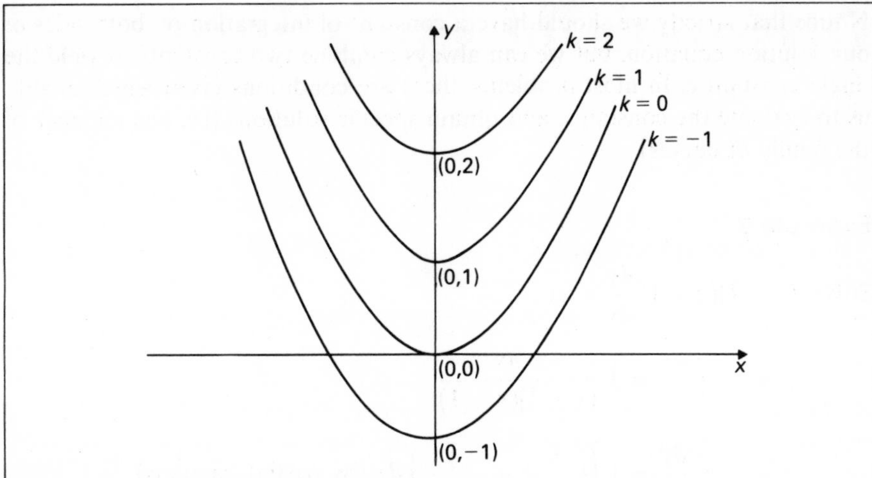

Figure 16.1

Let us return to the original equation $\dfrac{dy}{dx} = x$

we could arrange our working more precisely by noting that this is equivalent to

$$dy = x\,dx \qquad \textbf{i}$$

and hence

$$\int dy = \int x\,dx \qquad \textbf{ii}$$

$$\Rightarrow \qquad y = \tfrac{1}{2}x^2 + c$$

We have treated $\dfrac{dy}{dx}$ as if it were a fraction, and have separated the variables x and y so that all the terms in x are on one side of the equation, and all the terms in y are on the other side. Notice too that equation **i** is really meaningless, since we have not defined anywhere what is meant by the symbol 'dx' or 'dy' standing by itself. The equation only has a meaning in our work when we insert the integral signs, as in equation **ii**.

Example 2

Solve $\dfrac{dy}{dx} = xy^2$

separating the variables, we have

$$\frac{dy}{y^2} = x\,dx$$

hence $\displaystyle\int \frac{dy}{y^2} = \int x\,dx$

$$\Rightarrow \qquad -\frac{1}{y} = \frac{x^2}{2} + c$$

or $\qquad y = \dfrac{-2}{x^2 + 2c}$

Notice that strictly we should have a constant of integration on both sides of our solution equation, but we can always combine two constants to yield the single constant c. In most problems, there are conditions given which enable us to evaluate the constants, and obtain specific solutions (i.e. one member of the family of curves).

Example 3

Solve $(x-1)(x+1)\dfrac{dy}{dx} = xy$

$$\Rightarrow \quad \int \frac{dy}{y} = \int \frac{x\,dx}{(x-1)(x+1)}$$

$$\Rightarrow \quad \int \frac{dy}{y} = \int \left(\frac{\frac{1}{2}}{x-1} + \frac{\frac{1}{2}}{x+1}\right) dx \text{ (by partial fractions)}$$

$$\Rightarrow \quad \ln|y| = \tfrac{1}{2}\ln|x-1| + \tfrac{1}{2}\ln|x+1| + c$$

In this case, we can obtain a more simplified form of the answer if we write $c = \ln C$.

Hence the solution becomes

$$\ln|y| = \tfrac{1}{2}\ln|x-1| + \tfrac{1}{2}\ln|x+1| + \ln C$$

$$\Rightarrow \quad y = C\sqrt{\{(x-1)(x+1)\}}$$

Exercise A

Find the general solutions of the following differential equations

1 $\dfrac{dy}{dx} = x^2 + x$

2 $y\dfrac{dy}{dx} = \sec^2 x$

3 $(1+x^2)\dfrac{dy}{dx} + y^2 = 0$

4 $\sin y\dfrac{dy}{dx} + \cos x = 0$

5 $\dfrac{dy}{dx} = \dfrac{y}{x}$

6 $(1-x^2)\dfrac{dy}{dx} = 1 - y^2$

7 $(x+1)\dfrac{dy}{dx} = y$

8 $x(x+1)\dfrac{dy}{dx} = y(y+1)$

9 $\dfrac{dy}{dx} - 1 = \cot^2 x$

10 $\dfrac{dy}{dx} - 1 = \dfrac{1}{y+1}$

In all of these questions, you have included a constant as a result of the integration, and you have obtained the general solution in each case.

In most practical situations, however, some initial values are known, and we can use these to determine the value of the constant in each situation. These initial values are often called 'boundary conditions'. When we have a specific value for the constant, the solution is no longer general, but becomes the particular solution for the given set of boundary conditions.

Example 4

Solve the differential equation

$$(1 + x^2)\frac{dy}{dx} = 2xy, \text{ given that } y = 2 \text{ when } x = 0$$

Separating the variables, we have

$$\int \frac{dy}{y} = \int \frac{2x\,dx}{1 + x^2}$$

$$\Rightarrow \ln|y| = \ln|1 + x^2| + c$$

Now $\quad y = 2$ when $x = 0$ gives

$$\ln 2 = \ln 1 + c$$

$$\Rightarrow \quad c = \ln 2 \text{ (since } \ln 1 = 0)$$

Hence the particular solution is

$$\ln|y| = \ln(1 + x^2) + \ln 2$$

Note that we can drop the modulus signs on $1 + x^2$, since $1 + x^2$ is positive for all $x \in \mathbb{R}$.

Since every term is logarithmic, we may simplify this answer as follows

$$\ln|y| = \ln(1 + x^2) + \ln 2$$

$$\Rightarrow \ln|y| = \ln 2(1 + x^2)$$

$$\Rightarrow \quad y = 2(1 + x^2)$$

Example 5

Solve $\cot x\frac{dy}{dx} = \tan y, \quad$ given that $y = \frac{\pi}{2}$ when $x = \frac{\pi}{4}$

Separating the variables gives

$$\int \frac{dy}{\tan y} = \int \frac{dx}{\cot x}$$

$$\Rightarrow \int \cot y\,dy = \int \tan x\,dx$$

$$\Rightarrow \ln|\sin y| = \ln|\sec x| + c$$

The boundary conditions give

$$\ln\left(\sin\frac{\pi}{2}\right) = \ln\left(\sec\frac{\pi}{4}\right) + c$$

$$\Rightarrow \quad \ln 1 = \ln\sqrt{2} + c$$

$$\Rightarrow \quad c = -\ln\sqrt{2}$$

Hence the particular solution is

$$\ln|\sin y| = \ln|\sec x| - \ln\sqrt{2}$$

Which simplifies to give

$$\sin y = \frac{\sec x}{\sqrt{2}}$$

What justification is there for not including modulus signs at this stage?

Exercise B

Find the particular solutions of the following differential equations, using the given boundary conditions.

1 $\sec x \dfrac{dy}{dx} = \csc\left(y + \dfrac{\pi}{4}\right)$, $y = \dfrac{\pi}{4}$ when $x = \dfrac{\pi}{2}$

2 $(1 + x^2)\dfrac{dy}{dx} = 2x(y^2 + y)$, $y = 2$ when $x = 0$

3 $xy\dfrac{dy}{dx} = 2x + 3$, $y = 0$ when $x = 1$

4 $\dfrac{dy}{dx} = -\dfrac{x}{y^2}$, $y = 1$ when $x = 1$

5 $\cot xy \dfrac{dy}{dx} = \sin x \sin y$, $y = \dfrac{\pi}{2}$ when $x = 0$

6 $x(x - 1)\dfrac{dy}{dx} = y(y - 1)$, $y = 4$ when $x = 3$

7 $x\dfrac{dy}{dx} = y + xy$, $y = 2$ when $x = 2$

8 $x(1 + x)\dfrac{dy}{dx} = 1 - y$, $y = 0$ when $x = 3$

9 $\sin y \dfrac{dy}{dx} = \cos x$, $y = 0$ when $x = 0$

10 $\dfrac{dy}{dx} = \dfrac{1 + \sin x}{2y}$, $y = \sqrt{2}$ when $x = 0$

Although solution of problems by means of differential equations is strictly beyond the scope of this course, a brief study of some examples is included here for completeness.

We look at some problems arising from various situations, and whose solutions are found by means of differential equations. We need to make a 'mathematical model', which represents the physical situation, and solve the resulting mathematical problem. Some examples will illustrate the approach.

Example 6

Certain bacteria in a culture are thought to increase at a rate proportional to the square root of the number present at any time. Initially, there are 25, and after one hour there are 81. Find

a) how many there will be after two hours, and
b) how long it takes for the number to reach 625.

Let x = number of bacteria at time t

then $$\frac{dx}{dt} \propto \sqrt{x}$$

i.e. $\dfrac{dx}{dt} = k\sqrt{x}$ and $\begin{cases} x = 25 \text{ when } t = 0 \\ x = 81 \text{ when } t = 1 \end{cases}$

$$\Rightarrow \quad \int \frac{dx}{\sqrt{x}} = k \int dt$$

$$2\sqrt{x} = kt + c$$

The boundary conditions give

$$2\sqrt{25} = 0 + c \Rightarrow c = 10$$
$$2\sqrt{81} = k + c \Rightarrow k = 8$$

Hence the particular solution is

$$2\sqrt{x} = 8t + 10, \text{ i.e. } \sqrt{x} = 4t + 5.$$

We are asked to calculate
a) x when $t = 2$,

hence $\sqrt{x} = 8 + 5$
$\Rightarrow \sqrt{x} = 13$
$\Rightarrow x = 169$

For b), we need to calculate the value of t when $x = 625$

Hence $\sqrt{625} = 4t + 5$
$\Rightarrow 4t = \sqrt{625} - 5$
$\Rightarrow 4t = 25 - 5$

i.e. $4t = 20$
i.e. $t = 5$

So we see that after two hours the number of bacteria has risen to 169, and that it will take 5 hours before the 'population' reaches 625.

Example 7

Newton's law of cooling states that the temperature of a body in a room (at constant temperature) reduces at a rate proportional to the difference between the body's temperature and the temperature of the room. A body with temperature 90° is placed in a room with temperature 18°, and is then cooling at a rate of 3° per minute. Find how long it takes for the body's temperature to drop to 60°. What will be the temperature 24 minutes from the time the body was put in the room?

Let the body's temperature be $\theta°$ at time t minutes from the start. Then we know that when

$$t = 0, \theta = 90° \text{ and } \frac{d\theta}{dt} = -3,$$

Our differential equation can be established from

$$\frac{d\theta}{dt} \propto (\theta - 18)$$

i.e. $$\frac{d\theta}{dt} = k(\theta - 18)$$

From the initial 'cooling rate', we have

$$-3 = k(90 - 18)$$
$$\Rightarrow \quad k = -\tfrac{1}{24}$$

Hence the differential equation is

$$\frac{d\theta}{dt} = -\tfrac{1}{24}(\theta - 18)$$

$$\Rightarrow \quad \int \frac{d\theta}{(\theta - 18)} = -\tfrac{1}{24}\int dt$$

$$\ln|\theta - 18| = -\tfrac{1}{24}t + c$$

The boundary condition that $\theta = 90°$ when $t = 0$ enables us to find c:

$$\ln(90 - 18) = 0 + c \quad \Rightarrow \quad c = \ln 72$$

Hence the particular solution is

$$\ln|\theta - 18| = -\tfrac{1}{24}t + \ln 72$$

$$\Rightarrow 24 \ln\left|\frac{72}{\theta - 18}\right| = t^{*}$$

We require to find a) t, when $\theta = 60$
and b) θ, when $t = 24$

For a), we have

$$24 \ln\left(\frac{72}{60 - 18}\right) = t$$

$$\text{i.e.} \quad 24 \ln\left(\frac{72}{42}\right) = t$$

$\Rightarrow \quad t = 12{\cdot}94$ or 13 minutes, approximately

For b), we have

$$24 \ln\left(\frac{72}{\theta - 18}\right) = 24$$

$$\Rightarrow \quad \ln\left(\frac{72}{\theta - 18}\right) = 1$$

$$\Rightarrow \quad \frac{72}{\theta - 18} = e = 2{\cdot}718$$

$$\Rightarrow \quad \theta = \frac{72}{2{\cdot}718} + 18$$

$\Rightarrow \quad \theta = 44{\cdot}5°$ approximately

---

* **Note** There are times when it is more convenient to express the solution without the use of ln. In this case the solution would be $\theta = 18 + 72e^{-t/24}$.

1 A bullet is fired from a gun at a speed of 210 ms^{-1}, and its retardation is known to be kV ms^{-2}, where V ms^{-1} is its speed at time t secs. When $t = 0{\cdot}5$, $V = 105$. Find the value of k and find when the speed is reduced to 10 ms^{-1}. Find also how far the bullet travels from time $t = 1$ to $t = 2$.

2 A cylindrical water tank which is initially full of water, has a small hole in its horizontal base. The water leaks out at a rate proportional to the square of the depth of water at any time. If it is half full after $1\frac{1}{2}$ hours, find how long it takes for it to become a quarter full.

3 A conical beaker (with vertex downwards) has its height equal to the diameter of its circular opening. It is being filled at a rate of $7t$ cm^3/minute where t is the time in minutes since the start. Find in terms of the height, h cm, of the cone, the time taken to

a) reach a depth of $\frac{1}{2}h$ cm, and

b) fill the beaker completely.

4 In a certain chemical reaction, a compound is being made from two other compounds. At time t hours, there is an amount x of the required substance present.

The speed of the reaction $\frac{dx}{dt}$ is defined by

$$\frac{dx}{dt} = 2(x - 3)(x - 6)$$

Initially, $x = 0$.

Find approximately how many *seconds* elapse before $x = 2$. Explain why x can never reach the value of 3.

5 A certain radioactive element is thought to decay at a rate proportional to the quantity remaining. Initially there was 20 g of the element. Show that after time t hours, there will be $20e^{-kt}$ g.

If half the original quantity has vanished after time $t = \frac{1}{2}$, find the value of t for which the amount remaining is a quarter of the initial amount. Explain why the element will never completely vanish.

16.3 Linear equations of the first order

We consider equations which can be written in the form

$$\frac{dy}{dx} + Py = Q$$

where P and Q are functions of x.

In general such an equation cannot be solved by separation of the variables, and we seek a method which can always be applied to such equations. We recall that the product rule for differentiation gave the sum of two terms,

i.e. $$\frac{d}{dx}(uv) = u\frac{dv}{dx} + v\frac{du}{dx}$$

We use this as our starting point.

Consider $\frac{d}{dx}(Ry)$, where R is a function of x.

Then $\frac{d}{dx}(Ry) = R\frac{dy}{dx} + y\frac{dR}{dx} \dots$ **i**

Our differential equation, if multiplied by R is

$R\frac{dy}{dx} + RPy = RQ$ **ii**

Comparing **i** with **ii**, we see that

$$R\frac{dy}{dx} + RPy = \frac{d}{dx}(Ry) \text{ if } RP = \frac{dR}{dx}$$

Hence if we multiply the original differential equation by a function R, such that $RP = \frac{dR}{dx}$, then the differential equation becomes easily integrable.

We need to solve the subsidiary equation $RP = \frac{dR}{dx}$ to find the function R in terms of P. Separating the variables we have

$$\int P\,dx = \int \frac{dR}{R}$$

$$\Rightarrow \quad \int P\,dx = \ln R$$

$$\Rightarrow \quad R = e^{\int P\,dx}$$

Since multiplication by this function R makes the differential equation integrable, it is often called an **integrating factor**. We consider two examples to illustrate the technique.

Example 1

Solve $\frac{dy}{dx} + \frac{2x}{1+x^2}y = \frac{1}{(1+x^2)^2}$

The integrating factor $R = e^{\int \frac{2xdx}{1+x^2}}$

$= e^{\ln(1+x^2)}$

$= (1+x^2)$

Hence the differential equation becomes

$$(1+x^2)\frac{dy}{dx} + 2xy = \frac{1}{1+x^2}$$

$$\Rightarrow \quad \frac{d}{dx}\{y(1+x^2)\} = \frac{1}{1+x^2}$$

$$\Rightarrow \quad y(1+x^2) = \int \frac{1}{1+x^2}\,dx$$

$$\Rightarrow \quad y(1+x^2) = \tan^{-1} x + c$$

Example 2

Solve $\dfrac{dy}{dx} - y \sec x = 1$

The integrating factor, $R = e^{\int (-\sec x)\,dx}$

$$\Rightarrow \quad R = e^{\{-\ln(\sec x + \tan x)\}}$$

$$\Rightarrow \quad R = \frac{1}{\sec x + \tan x}$$

Hence multiplying by R, the differential equation becomes

$$\left(\frac{1}{\sec x + \tan x}\right)\frac{dy}{dx} - \frac{\sec x}{\sec x + \tan x}\,y = \frac{1}{\sec x + \tan x}$$

$$\Rightarrow \quad \frac{d}{dx}\left(y \cdot \frac{1}{\sec x + \tan x}\right) = \frac{1}{\sec x + \tan x}$$

$$\Rightarrow \quad \frac{y}{\sec x + \tan x} = \int \frac{dx}{\sec x + \tan x}$$

$$\Rightarrow \quad \frac{y}{\sec x + \tan x} = \int \frac{\cos x}{1 + \sin x}\,dx$$

$$\Rightarrow \quad \frac{y}{\sec x + \tan x} = \ln|1 + \sin x| + c$$

Boundary conditions to enable us to evaluate c are used in the same way as before.

Notice that it is always a relatively simple task to check that your integrating factor is correct, by differentiating your product and comparing the result with the original equation.

Exercise D

1 Solve $\dfrac{dy}{dx} + 2xy = e^{-x^2}$

2 Find the particular solution of the differential equation

$$\frac{dy}{dx} - y = x$$

for which $y = 1$ when $x = 0$.

3 Find the general solution of the differential equation

$$\cos^2 x \frac{dy}{dx} + y = e^{\cot x}$$

4 Solve

$$\sin x \frac{dy}{dx} - y = \sin^2 x$$

5 Find the particular solution of

$$(1+x^2)^4\frac{dy}{dx}+8(1+x^2)^3xy=\tan^2 x$$

given that $y=2$ when $x=0$.

6 Find the particular solution of

$$\frac{dy}{dx}+\frac{2y}{100+2x}=12$$

given that $y=50$ when $x=0$

7 Find the general solution of

$$x\frac{dy}{dx}+3y=\frac{1}{x^2}\cos x$$

8 Find the equation of the particular member of the family of curves defined by

$$\frac{dy}{dx}-\frac{3y}{x}+3x=0$$

which passes through the point (1, 2). Sketch the curve.

9 Rewrite the differential equation

$$(y-x-y^3)\frac{dy}{dx}=4y$$

so that it becomes a differential equation in $\frac{dx}{dy}$, and hence find the general solution. (**Note** Since the y's and x's become interchanged in the standard equation, care must be used to find the integrating factor.)

10 Find the particular solution of

$$e^x\frac{dy}{dx}+e^xy=x$$

given that $y=2{\cdot}5$ when $x=0$.

11 Find the particular solution of

$$\frac{dy}{dx}+3y=2e^{-3x}\sin 2x$$

given that $y=4$ when $x=0$.

12 Solve $\frac{dy}{dx}=y+1$ a) by separating the variables, and
b) by using an integrating factor.

Prove that your two answers are equivalent.

13 Find the general solution of

$$\frac{dy}{dx}+y=e^x\cos 2x$$

(Hint: to find $\int e^{2x}\cos 2x\,dx$, let I represent this integral, integrate by parts twice, and obtain an algebraic equation for I, as in Chapter 14 example 5, page 296.)

14 Given that $v\dfrac{dv}{dx} = -4x$, find an expression for v in terms of x, given that $v = 0$ when $x = 0$.

In your solution, write $v = \dfrac{dx}{dt}$, and solve the resulting differential equation to obtain x in terms of t. Hence write down a formula for x in terms of t given that $x = 1$ when $t = \dfrac{\pi}{4}$.

Miscellaneous Revision Examples

The following examples, selected from examination papers set in recent years, have been grouped together to give further practice. Although the questions have not been linked together to form any special pattern, content has been chosen to offer wide variety of knowledge and application. Those questions in the 'A' papers are of the shorter variety and are commonly found in the Section A of those examination papers which employ the Section A/Section B division.

The questions in the 'B' papers are longer in their general nature and frequently contain a number of requests for answers at part way stages.

Paper A1

A.E.B.

1 If the roots of the quadratic equation $x^2 - 3px + p^2 = 0$ are α and β where $\alpha > \beta$ find the values of $\alpha^2 + \beta^2$ and $\alpha - \beta$ when p is positive. Find, in terms of p, a quadratic equation whose roots are $\dfrac{\alpha^3}{\beta}$ and $\dfrac{-\beta^3}{\alpha}$.

A.E.B.

2 When the expression $x^3 + ax^2 + bx + c$ is divided by $x^2 - 4$ the remainder is $18 - x$ and when it is divided by $x + 3$ the remainder is 21. Find the remainder when the expression is divided by $x + 1$.

L.U.

3 Leaving all answers as multiples of π, find the values of x for which $0 < x < \pi$ and $\cos x + \cos 3x = 0$.

L.U.

4 Find the equation of the circle which has its centre on the line $x - 2y + 2 = 0$ and touches the y-axis at the point (0, 3).

L.U.

5 Given that $\lg y = 1 - 0{\cdot}5 \lg x$, express y explicitly in terms of x and sketch the graph of y against x. Only the general shape of the graph and its position relative to the axes are required.

[lg p means $\log_{10} p$]

O. and C.

6 Prove, by induction, that the sum of the first n terms of the series $1^2 + 4^2 + 7^2 + 10^2 + \cdots$ is $\frac{1}{2}n(6n^2 - 3n - 1)$.

O. and C.

7 Given that $f(x) = \tan x$, write expressions for $f'(x)$, $f''(x)$ and $f'''(x)$. Deduce a Taylor approximation for $\tan \alpha$ when α is small, in the form

$$\tan \alpha \approx a + b\alpha + c\alpha^2 + d\alpha^3,$$

where a, b, c, d are constants.

J.M.B.

8 Use Simpson's rule with 5 ordinates (4 strips) to find an approximate value of $\int_0^4 xe^{-x}\,dx$. Give your answer correct to two decimal places.

L.U.

9 The function f is periodic with period 2 and $f(x) = \dfrac{4}{x}$ for $1 < x \leqslant 2$, $f(x) = 2(x - 1)$ for $2 < x \leqslant 3$. Sketch the graph of f when $-3 \leqslant x \leqslant 3$.

L.U. 10 The first two non-zero terms of the expansion of

$$(1 + ax)(1 + bx)^9$$

as a series of ascending powers of x are 1 and $-5x^2$. Given that $a > 0$ and $b < 0$, find the values of a and b.

Paper A2

L.U. 1 Solve the differential equation $\frac{dy}{dx} = 4y^2$ given that $y = \frac{1}{2}$ when $x = -2$.

L.U. 2 Find the set of values of x for which $\frac{4x}{(x+2)} > 1$

L.U. 3 The remainder when $f(x)$, a polynomial in x, is divided by $(x - a)(x - b)$ is $Px + Q$, where P and Q are constants. Show that, when $a \neq b$,

$$P = [f(a) - f(b)]/(a - b)$$

O. and C. 4 Given that $z_1 = 1 + 2i$, $z_2 = 2 + i$, obtain, in the form $a + bi$ (a, b real), the complex numbers

i) $z_1 z_2$ ii) $\frac{z_1}{z_2}$ iii) $z_1^2 - z_2^2$

and find their moduli, and arguments (in degrees, to the nearest degree).

A.E.B. 5 Given that $\log_2 (x - 5y + 4) = 0$ and $\log_2 (x + 1) - 1 = 2 \log_2 y$, find the values of x and y.

A.E.B. 6 a) Evaluate $\int_0^{\pi/4} (\sin x + \cos x)^2 \, dx$. b) Find $\int \frac{dx}{e^x + e^{-x}}$.

J.M.B. 7 Determine the ranges of the following functions whose domains are as given:

i) $f: x \to x^2 + 3$, x real, $0 \leqslant x \leqslant 3$;
ii) $y: x \to x^2 - 2x + 3$, x real, $0 \leqslant x \leqslant 3$;
iii) $h: x \to 2 \sin x + \cos x$, x real.

State whether f^{-1}, g^{-1}, h^{-1} exist, giving in each case brief reasons for your answers.

O. and C. 8 Use standard formulae to prove that, for any angles A, B, $\cos (A + B) - \sin (A - B) = (\cos A - \sin A - \sin A)(\cos B + \sin B)$.

If $\cos (A + B)$ is equal to $\sin (A - B)$, what can you deduce about the values, in degrees, of A and/or B?

L.U. 9 The points A, B, C have position vectors **a**, **b**, **c** respectively referred to an origin O. State the position vector of P, the mid-point of AB. Find the position vector of the point which divides the line segment PC internally in the ratio 3 : 1.

L.U. 10 Sketch on the same diagram the circle with polar equation $r = 4 \cos \theta$ and the line with polar equation $r = 2 \sec \theta$. State the polar coordinates of their points of intersection.

Paper A3

O. and C. 1 Given that $f(x) = x^4 + 3x^3 - 5x^2 - 9x - 2$ find by trial two integer solutions of the equation $f(x) = 0$. Hence factorise $f(x)$ and solve the equation completely.

O. and C. 2 A region of the plane is bounded by parts of the x-axis and of the line $x = k$, and by that part of the curve with equation $y = 1 - \dfrac{1}{\sqrt{x}}$ for which $1 < x < k$. Find in terms of k, the volume formed when this region is rotated through a whole turn about the x-axis.

L.U. 3 Express $\dfrac{(3+x)}{(1-x)(1+3x)}$ in partial fractions

and find $\displaystyle\int_0^{1/2} \frac{(3+x)}{(1-x)(1+3x)}\,dx.$

J.M.B. 4 If $\log_y x + 2\log_x y = 5$, show that $\log_y x$ is either $\frac{1}{2}$ or 2. Hence find all pairs of values of x and y which satisfy simultaneously the equation given above and the equation $xy = 27$.

A.E.B. 5 Expand $\sqrt{(4-x)}$ as a series of ascending powers of x up to and including x^2.
If terms x^n, $n \geqslant 3$, can be neglected find the quadratic approximation to $\sqrt{\left\{\dfrac{4-x}{1-2x}\right\}}$. State the range of values of x for which this approximation is valid.

J.M.B. 6 Show that $(i+1)^4 = -4$.
By raising each side of the above result to the power n, where n is a positive integer and then using the binomial theorem to expand the left-hand side, prove that

$$1 - \binom{4n}{2} + \binom{4n}{4} - \cdots - \binom{4n}{4n-2} + \binom{4n}{4n} = (-1)^n 2^{2n}$$

L.U. 7 Shade on a sketch the domain for which

$$y^2 - x < 0 \text{ and } x^2 - y < 0.$$

L.U. 8 Given that $x = at^2$ and $y = 2at$, where a is a constant, find $\dfrac{d^2y}{dx^2}$ in terms of a and t.

J.M.B. 9 It is given that

$$\frac{1}{b+c}, \frac{1}{c+a}, \frac{1}{a+b}$$

are three consecutive terms of an arithmetic series. Show that a^2, b^2 and c^2 are also three consecutive terms of an arithmetic series.

A.E.B. 10 A plane is inclined at an angle α to the horizontal and a line PQ on the plane makes an acute angle β with PR which is a line of greatest slope on the plane. Show that the inclination θ of PQ to the horizontal is given by $\sin\theta = \sin\alpha\cos\beta$. Show that the angle ϕ between the vertical plane through PQ and the vertical plane through PR is given by

$$\cos\phi\cos\theta = \cos\alpha\cos\beta.$$

Paper B1

J.M.B. 1 In the ΔABC, AB is of unit length and $BC = CA = P$. The point p lies on AB at a distance x from A and is such that $\angle ACP = \theta$ and $\angle BCP = 2\theta$. By using the sine rule, or otherwise, show that

$$\cos\theta = \frac{1-x}{2x}$$

State the ranges of possible values of θ as p varies, and deduce that $\frac{1}{3} < x < \frac{1}{2}$. Express $\cos 3\theta$ in terms of p. Hence, or otherwise, find the value of x correct to two decimal places when $p = \dfrac{1}{\sqrt{2}}$.

J.M.B. 2 Find the x-coordinate of the turning point of the curve whose equation is $y = \dfrac{a}{x} + \ln x$ where $x > 0$ and $a > 0$, and determine whether this turning point is a maximum or a minimum. Deduce the range of values of the constant a for which $y > 0$ for all $a > 0$.

In the case when $a = 1$, find the area and the x-coordinate of the centroid of the region bounded by the curve, the x-axis and the ordinates $x = 1$ and $x = 2$. Express both answers in terms of $\ln 2$.

L.U. 3 Shade the region C for which the polar coordinates r, θ satisfy $r \leqslant 4\cos 2\theta$ for $-\pi/4 \leqslant \theta \leqslant \pi/4$.

Find the area of C.

State the equations of the tangents to the curve $r = 4\cos 2\theta$ at the pole.

O. and C. 4 i) Differentiate with respect to x

a) $\dfrac{e^{2x}}{e^{2x}+1}$ b) $x^{1/2}\sin^2(x^{1/2})$

ii) Find the values of the first four derivatives of $\tan x$ with respect to x at $x = 0$ and hence prove that $\tan x = x + \frac{1}{3}x^3$, if x is so small that powers of x greater than the fourth may be neglected. Hence, assuming x is sufficiently small, expand $\{(1 + \tan x)e^{-x}\}$ in ascending powers of x as far as the term in x^3.

L.U. 5 Show that if $(x + t)$ is a common factor of $x^3 + px^2 + q$ and $ax^3 + bx + c$, then it is a factor of $apx^2 - bx + aq - c$. Show that $x^3 + \sqrt{7}x^2 - 14\sqrt{7}$ and $2x^3 - 13x - \sqrt{7}$ have a common factor and hence find all the roots of the equation $2x^3 - 13x - \sqrt{7} = 0$.

O. and C. 6 Given that $y = (2x+1)/(x^2+2)$, where x is real, show algebraically that y can only take values in the range $-\frac{1}{2} \leqslant y \leqslant 1$. Hence find the turning points on the curve with equation $y = (2x+1)/(x^2+2)$ and sketch the curve. Explain how the intersection of this curve with a suitable straight line can show that the cubic equation $2x^3 + 2x - 1 = 0$ has only one real root. Using any method, find this root correct to two decimal places.

L.U. 7 i) Evaluate a) $\displaystyle\int_1^3 \frac{x+2}{x+1}\,dx$ b) $\displaystyle\int_0^{\pi/2} \sin^4 x \cos x\,dx$.

ii) Sketch the curve $y = x + \dfrac{1}{x}$. Show that the area of the finite region enclosed by the curve and the line $y = 2\frac{1}{2}$ is $1\frac{7}{8} - \ln 4$.

L.U.

8 i) If $z = r(\cos\theta + i\sin\theta)$, find the modulus and argument of each of the two values of w such that $w^2 = z$. Find the two square roots of $(2 - i2\sqrt{3})$ in the form $r(\cos\theta + i\sin\theta)$ and represent this number and its square roots by points in an Argand diagram.

ii) If w is one of the non-real cube roots of unity, prove that the other is w^2. Prove also that

a) $1 + w + w^2 = 0$ b) $1 + w$ is a cube root of -1.

A.E.B.

9 Express $\cot(A + B)$ in terms of $\cot A$ and $\cot B$ and hence, without the use of tables, express $\cot 22\frac{1}{2}°$ in surd form.

Deduce $\cot 67\frac{1}{2}°$ in surd form.

Show that $\cot^{-1}(1 + x) + \cot^{-1}(1 - x) = \cot^{-1}\left(-\frac{x^2}{2}\right)$.

O. and C.

10 The position vectors of two points A and B relative to an origin O are **a** and **b**. You are given that **a** and **b** have unit length, and that the angle between these vectors is 60°. Write down the values of the products $\mathbf{a} \cdot \mathbf{a}$, $\mathbf{b} \cdot \mathbf{b}$ and $\mathbf{a} \cdot \mathbf{b}$. The point C on the line-segment AB is such that $AC = 2CB$. If **c** is the position vector of C, express **c** in terms of **a** and **b**. Hence calculate

i) the length of **c**;

ii) the cosine of the angle between **a** and **c**.

Paper B2

A.E.B.

1 a) Solve the simultaneous equations

$$a \log_4 128 - b \log_8 2 = 6,$$
$$\log_2 a + \tfrac{1}{3}\log_2 (b^3) = 2 \log_4 6.$$

b) The first term of an arithmetic series is $\log_e x$ and the rth term is $\log_e(xc^{r-1})$. Show that the sum s_n of the first n terms of the series is $\frac{n}{2}\log_e(x^2c^{n-1})$.

If $c = 1$, find the sum of the series

$$e^{s_1} + e^{s_2} + e^{s_3} + \cdots + e^{s_r} + \cdots + e^{s_n}.$$

A.E.B.

2 The tangent to the parabola $y^2 = 4ax$ at the point $P(at^2, 2at)$ meets the x-axis at T. The straight line through P parallel to the axis of the parabola meets the directrix at Q. If S is the focus of the parabola, show that $PQTS$ is a rhombus.

If M is the mid-point of PT and N is the mid-point of PM, find the equation of the locus of i) M, ii) N.

O. and C.

3 Find by trial the two consecutive integers between which the solution of the equation $x + 2e^x = 0$ lies. Of these two integers, take as a first approximation the one which makes the absolute value of the left-hand side smaller. Then find a second approximation to the solution by a single application of the Newton-Raphson method. Give two places of decimals in your answer.

J.M.B.

4 State the derivative of $\sin x$ and $\cos x$, and use these results to show that the derivative of $\tan x$ is $\sec^2 x$.

Show further that $\frac{d}{dx}(\arc\tan x) = \frac{1}{1 + x^2}$.

A vertical rod AB of length 3 units is held with its lower end B at a distance of 1 unit vertically above a point O. The angle subtended by AB at a variable point P on the horizontal plane through O is θ. Show that

$$\theta = \arc \tan x - \arc \tan \frac{x}{4}, \text{ where } x = OP.$$

Prove that, as x varies, θ is a maximum when $x = 2$, and that the maximum value of θ can be expressed as arc tan $\frac{3}{4}$.

.M.B.

5 Prove that the equation of the tangent at the point $(x_1 y_1)$ on the ellipse $\frac{x^2}{a^2} + \frac{y^2}{b^2} = 1$ is $\frac{xx_1}{a^2} + \frac{yy_1}{b^2} = 1$.

The tangent at the point $(2 \cos \theta, \sqrt{3} \sin \theta)$ on the ellipse $\frac{x^2}{4} + \frac{y^2}{3} = 1$ passes through the point (2, 1). Show that $\sqrt{3} \cos \theta + \sin \theta = \sqrt{3}$.

Without using tables, calculator or a slide rule, find all the solutions of this equation which are in the range $0° \leqslant \theta < 360°$. Hence obtain the coordinates of the points of contact Q and R of the tangent to the ellipse from R. Verify that the line through the origin and the point P passes through the mid-point of the line QR.

..U.

6 i) Find $\int \frac{dx}{\sqrt{(x^2 - 2x + 10)}}$.

ii) Find $\int \frac{dx}{x^2 - 2x + 10}$.

iii) By using the substitution $x = \sin \theta$, show that

$$\int_0^{1/2} \frac{x^4 \, dx}{\sqrt{(1 - x^2)}} = (4\pi - 7\sqrt{3})/64.$$

.E.B.

7 Given that $y = \frac{x^2 + \lambda}{x + 2}$ and x is real, find

i) the set of possible values of y when $\lambda = 5$;
ii) the set of values of λ for which y can take all real values.

Find the values of m for which the line $y = mx$ touches the curve

$$y = \frac{2x^2 + 1}{2(x + 2)}$$

and state the equations of the tangents from the origin to this curve.

). and C.

8 a) Show that $(1 + 3i)^3 = -(26 + 18i)$.
b) Find the three roots z_1, z_2, z_3 of the equation $z^3 = -1$.
c) Find in the form $a + ib$, the three roots z'_1, z'_2, z'_3 of the equation $z^3 = 26 + 18i$.
d) Indicate in the same Argand diagram the points represented by z_r and z'_r for $r = 1, 2, 3$ and prove that the roots of the equation may be paired so that

$$|z_1 - z'_1| = |z_2 - z'_2| = |z_3 - z'_3| = 3$$

.E.B.

9 a) Solve the differential equation

$$y\sqrt{(1 - x^2)}\frac{dy}{dx} = \frac{1}{y} - y^2,$$

given that $y = \frac{1}{2}$ when $x = 0$.

b) Use Simpson's rule with five ordinates to evaluate approximately

$$\int_0^{1/2} (1 - x^2)^{1/2}\, dx.$$

By using a suitable substitution, show that

$$\int_0^{1/2} (1 - x^2)^{1/2}\, dx = \frac{\pi}{12} + \frac{\sqrt{3}}{8}.$$

O. and C.

10 Obtain $\int x^2 \cos 2x\, dx$ and hence prove that $\int_0^{\pi} x^2 \cos 2x\, dx = \frac{1}{2}\pi$.

Find the area of the region enclosed by the curve $y = x \sin x$, for $0 \leqslant x \leqslant \pi$, and the x-axis. The region is rotated through 2π radians about the x-axis. Find the volume of the solid of revolution thus generated. From your results, or otherwise, find the volume of the solid of revolution generated by rotation of the same region through 2π radians about the line $y = -\frac{1}{8}$.

[Leave your answers in terms of π.]

Paper B3

A.E.B.

1 a) Given that $y = \log_a (x^3)$ and $z = \log_x a$, show that $yz = 3$. Hence find the numerical values of y and z when
$\log_a (3 \log_a x) - \log_a (\log_x a) = \log_a 27$.

b) Simplify the expression $\dfrac{\sin 2x - \sin 2y}{\cos 2x + \cos 2y}$

and hence, or otherwise, evaluate it without using tables when $x = \text{arc tan } \frac{1}{3}$ and $y = \text{arc tan } \frac{1}{4}$. Hence, or otherwise, find the value of cosec $(x - y)$, without using tables.

A.E.B.

2 Sketch the curve $y = \ln (x - 2)$.

The inner surface of a bowl is of the shape formed by rotating completely about the y-axis that part of the x-axis between $x = 0$ and $x =$ and that part of the curve $y = \ln (x - 2)$ between $y = 0$ and $y = 2$. The bowl is placed with its axis vertical and water is poured in. Calculate the volume of water in the bowl when the bowl is filled to a depth h (<2).

If water is poured into the bowl at the rate of 50 cubic units per second find the rate at which the water level is rising when the graph of the water is 1.5 units.

L.U.

3 For the curve with equation $y = \dfrac{x}{x - 2}$, find

i) the equation of each of the asymptotes,
ii) the equation of the tangent at the origin.

Sketch the curve, paying particular attention to its behaviour at the origin and as it approaches its asymptotes.

On a separate diagram, sketch the curve with equation

$$y = \left| \frac{x}{x - 2} \right|.$$

L.U. 4 i) Using the substitution $t = \tan \frac{x}{2}$, or otherwise, find without using tables, the two possible values of $\tan x$ when $\sin x - 7 \cos x + 5 = 0$.

ii) Express $(\cos 3\theta + \sin 3\theta)$ in the form $R \cos (3\theta - \alpha)$, where $R > 0$ and $0 < \alpha < \pi/2$. Hence, or otherwise, find the general solution in radians of the equation $\cos 3\theta + \sin 3\theta = 1$.

L.U. 5 i) Show that $(a + b)$ is a root of the equation

$$x^3 - 3abx - (a^3 + b^3) = 0.$$

Express the equation $x^3 - 6x - 6 = 0$ in the above form, giving your values for a^3 and b^3. Hence find a real root of the equation $x^3 - 6x - 6 = 0$, expressing your answer in the form $\sqrt[3]{m} + \sqrt[3]{n}$, where m and n are positive integers.

ii) Given that the equation $x^3 + px^2 + qx + r = 0$

has three roots α, β, γ where $\alpha + \beta = \gamma$, show that $p^3 + 8r = 4pq$.

A.E.B. 6 The parametric equations of a curve are $x = 2t + \sin 2t$, $y = \cos 2t$. Show that

i) $\dfrac{dy}{dx} = -\tan t$

ii) $\dfrac{d^2y}{dx^2} = -\frac{1}{4} \sec^4 t$

iii) $\displaystyle\int_{t=\pi/4}^{t=3\pi/4} y \, dx = \frac{\pi}{2} - 2$

J.M.B. 7 The ellipse

$$\frac{x^2}{a^2} + \frac{y^2}{b^2} = 1$$

intersects the positive x-axis at A and the positive y-axis at B. Determine the equation of the perpendicular bisector of AB.

i) Given that this line intersects the x-axis at P and that M is the mid-point of AB, prove that the area of $\triangle PMA$ is

$$\frac{b(a^2 + b^2)}{8a}.$$

ii) If $a^2 = 3b^2$, find in terms of b, the coordinates of the points where the perpendicular bisector of AB intersects the ellipse.

A.E.B. 8 a) Solve the differential equation

$$\cos y \frac{dy}{dx} = (\cot x)(1 + \sin y),$$

given that $y = \pi/2$ when $x = \pi/2$.

b) For $0 \leqslant x \leqslant \pi$ the curves $y = \sin x$ and $y = \sin \frac{1}{2}x$ intersect at the origin O and at the point A. Find the coordinates of A. The region enclosed by the arcs of the two curves between O and A is rotated completely about the x-axis. Find the volume of the solid formed.

O. and C.

9 The following steps give a method for finding a numerical approximation to π. You are asked to carry them out.

i) The diagram represents a circle of unit radius. Show that the shaded area is $\frac{1}{12}\pi$ square units.

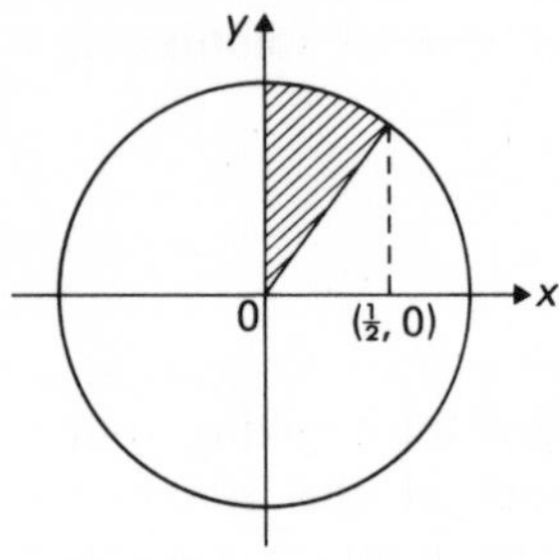

ii) Deduce that $\pi = 12\int_0^{1/2} \sqrt{(1-x^2)}\,dx - \frac{1}{2}\sqrt{27}$.

iii) Give a Taylor approximation for $\sqrt{(1+z)}$, neglecting powers of z above the second. Deduce an approximation for $\sqrt{(1-x^2)}$, neglecting powers of x above the fourth.

iv) Use your approximations to estimate both

$$\int_0^{1/2} \sqrt{(1-x^2)}\,dx \text{ and } \sqrt{(25+2)}.$$

v) Hence estimate π, giving your answer to 3 significant figures.

J.M.B.

10 The vertices of a quadrilateral lie on a circle of radius r. The angles subtended at the centre of the circle by the sides of the quadrilateral taken in order, are in arithmetical progression with the first term α and common difference β. Show that $2\alpha + 3\beta = \pi$ and interpret this geometrically.

Prove that the area of the quadrilateral is $2r^2 \cos\beta \cos\frac{1}{2}\beta$. In the case when $\alpha = \dfrac{\pi}{6}$, show that the shortest side of the quadrilateral divides the area of the circle in the ratio $(\pi - 3):(11\pi + 3)$.

Paper B4

L.U.

1 If α, β, γ are the roots of the equation $x^3 - px^2 - q = 0$, prove that $\alpha^2 + \beta^2 + \gamma^2 = p^2$ and express $\beta^2\gamma^2 + \gamma^2\alpha^2 + \alpha^2\beta^2$ in terms of p and q Use the Remainder Theorem to find the remainder when $x^3 - 7x^2 + 36$ is divided by $x + 2$. Hence, solve the equation $x^3 - 7x^2 + 36 = 0$ and verify that the expression for $\beta^2\gamma^2 + \gamma^2\alpha^2 + \alpha^2\beta^2$ is correct in this case.

J.M.B.

2 The points A $(-8, 9)$ and C $(1, 2)$ are opposite vertices of a parallelogram $ABCD$. The sides BC, CD of the parallelogram lie along the lines $x + 7y - 15 = 0$, $x - y + 1 = 0$, respectively. Calculate

i) the coordinates of D,

ii) the tangent of the acute angle between the diagonal of the parallelogram,

iii) the length of the perpendicular from A to the side CD,

iv) the area of the parallelogram.

J.M.B.

3 Sketch the curve whose equation is $y = \dfrac{x+2}{x}$ and state the equation of its asymptotes. By considering this sketch, or otherwise, show that the curve whose equation is $y = \ln\left(\dfrac{x+2}{x}\right)$ has no point whose x-coordinates lie in the interval $-2 < x < 0$. Sketch the curve on a separate diagram. Prove that the area bounded by the *second* curve, the x-axis and the lines $x = 1$ and $x = 2$ is $3 \ln \frac{4}{3}$.

A.E.B.

4 a) Find the sum of the first n terms of the series whose rth term is $2^r + 2r - 1$.

b) If x is so small that terms in x^n, $n \geqslant 3$, can be neglected and $\frac{3+ax}{3+bx} = (1-x)^{1/3}$, find the values of a and b.

Hence, without the use of tables, find an approximation in the form $\frac{p}{q}$, where p and q are integers, for $\sqrt[3]{0{\cdot}96}$.

A.E.B.

5 a) Solve the differential equation

$$\frac{dy}{dx} + y \cot x = \cos 2x,$$

given that $y = 1$ when $x = \pi/4$.

b) Given that $I_n = \int x(\ln x)^n \, dx$, show that

$$2\, I_n + nI_{n-1} = x^2(\ln x)^n.$$

Hence, or otherwise, evaluate $\int_1^2 x(\ln x)^2 \, dx$.

L.U.

6 Find the equation of the tangent to the parabola $y^2 = 4ax$ at the point $T(at^2, 2at)$. If S is the focus, find the equation of the chord QSR which is parallel to the tangent at T.

Prove that $QR = 4TS$.

L.U.

7 i) Show that, in general,

$$\frac{\sin 2A - \cos 2A + 1}{\sin 2A + \cos 2A + 1} = \tan A, \text{ and } \frac{\sin 3B + \sin B}{\cos 3B + \cos B} = \tan 2B.$$

ii) The point P lies in the first quadrant of the x-y plane, with origin O, and the points A $(a, 0)$ and B $(0, b)$ are such that $a > 0$, $b > 0$. If $\angle APO = \alpha$ $(\alpha \neq \pi/4)$ and $\angle BPO = \beta$ $(\beta \neq \pi/4)$ use the sine rule to show that the gradient of OP is

$$\frac{b \cot \beta - a}{a \cot \alpha - b}.$$

O. and C.

8 a) If all the angles are acute, show that

$$\arctan\left(\tfrac{1}{5}\right) + \arctan\left(\tfrac{1}{8}\right) = \arctan\left(\tfrac{1}{3}\right).$$

and hence that

$$\arctan\left(\tfrac{1}{2}\right) + \arctan\left(\tfrac{1}{5}\right) + \arctan\left(\tfrac{1}{8}\right) = \frac{\pi}{4}.$$

b) Solve the following equations for θ, giving all the values in the range $0° \leqslant \theta \leqslant 360°$;

i) $\cos 3\theta + \cos 2\theta + \sin 3\theta + \sin 2\theta = 0$,

ii) $\tan^2 \theta = \sec \theta + 5$.

A.E.B.

9 a) Find $\frac{dy}{dx}$ when i) $y = \ln(\sec 2x + \tan 2x)$,

ii) $y = \frac{(1+2x^2)}{(1+x^2)}$

and simplify your answers.

b) If $y = \cos(e^x + \frac{1}{4}\pi)$ show that $\frac{d^2y}{dx^2} + e^{2x}y = \frac{dy}{dx}$.

Find the least positive value of x for which y is a minimum.

J.M.B.

10 Given that ω denotes either one of the non-real roots of the equation $z^3 = 1$, show that
i) $1 + \omega + \omega^2 = 0$, and
ii) the other non-real root is ω^2.
Show that the non-real root of the equation $\left(\dfrac{1-u}{u}\right)^3 = 1$ can be expressed in the form $A\omega$ and $B\omega^2$, where A and B are real numbers, and find A and B.

Paper B5

J.M.B.

1 Write down the expansion of $(1 + y)^n$ in ascending powers of y, giving the first four terms. Given that x is so small that its cube and higher powers are negligible compared with unity, find the constants a, b and c in the approximate formula

$$\left(\frac{1+3x}{8-3x}\right)^{1/3} \approx a + bx + cx^2.$$

L.U.

2 i) If a, b, c are constants and $a > 0$, derive the condition that $ax^2 + bx + c$ should be positive for all real values of x. Prove that the equation

$$x - 1 = K(x-2)(x+1)$$

has real roots for all non-zero values of K.
ii) If $\log_a 45 + 4 \log_a 2 - \frac{1}{2} \log_a 81 - \log_a 10 = \frac{3}{2}$, find the base a.

L.U.

3 Sketch the curve $y = 1 + 2e^{-x}$, showing clearly the behaviour of the curve as $x \to +\infty$. Find the area of the finite region enclosed by the curve and the lines $x = 0$, $x = 1$, $y = 1$. Find also the volume formed when this region is rotated completely about the line $y = 1$.

A.E.B.

4 a) Use Simpson's rule with five ordinates to obtain an approximate value of $\int_1^2 x \lg x \, dx$. Work as accurately as your tables will allow.

b) Evaluate i) $\displaystyle\int_0^{\pi/2} \sin 2\theta \sqrt{(\sin \theta)} \, d\theta$.

ii) $\displaystyle\int_0^1 x(2 + e^{-x^2}) \, dx$.

A.E.B.

5 a) Express $1 + i\sqrt{3}$ in the $r(\cos\theta + i \sin\theta)$ form. Hence simplify $(1 + i\sqrt{3})^4$ and express the cube roots of $1 + i\sqrt{3}$ in the $r(\cos\theta + i\sin\theta)$ form. Illustrate these cube roots on an Argand diagram.
b) Use the method of induction to show that, if n is a positive integer,

$$\left(\frac{\cos\theta - i\sin\theta}{\cos\theta + i\sin\theta}\right)^n = \cos 2n\theta - i \sin 2n\theta.$$

O. and C.

6 In a certain chemical reaction two reagents A and B combine to form a new substance C. Within a given unit volume the quantities (in moles) of A, B and C are initially a, $2a$ and 0 and t seconds later are $a - x$, $2a - x$ and x respectively. At time t, the rate of increase of x is 0·5 times the

equation for x and by writing the equation in the form product of the quantities A and B then present. Obtain a differential

$$\frac{P}{a-x}+\frac{Q}{2a-x}=\frac{dt}{dx}$$

(where P and Q are constant), integrate to obtain t as a function of x. Find the time T taken for the quantity of A to be reduced to $\frac{1}{2}a$. Why will the quantity A never be reduced to zero? How much of B is present after time $2T$?

J.M.B. 7 A straight line of gradient m passes through the point (1, 1) and cuts the x- and y-axes at A and B respectively. The point P lies on AB and is such that $AP:PB = 1:2$. Show that, as m varies, P moves on the curve whose equation is $3xy - x - 2y = 0$. Find the perpendicular distance of the point (1, 1) from the tangent to this curve at the origin. A new origin is chosen at the point O' with coordinates $x = \frac{2}{3}$, $y = \frac{1}{3}$, and axes $O'X$ and $O'Y$ are drawn parallel to the x- and y-axes respectively. Express the equation of the curve on which P moves, in terms of the coordinates of P relative to the axes $O'X$, $O'Y$.

J.M.B. 8 Show that, in the triangle ABC, the length of the perpendicular from C to AB is given by $\frac{ab}{c}\sin C$. The triangle ABC, in which $\angle ACB = 150°$, lies in a horizontal plane and D is the point at a distance 2 units vertically above C. Given that $\angle DBC = 30°$ and $\angle DAC = 45°$, find the lengths of the sides of the triangle ABC and the length of the perpendicular from C to AB.

A.E.B. 9 a) Draw the curves $y = \sin x$ and $y = 1/x$ for $0 < x \leqslant \pi$ and hence show that the smallest positive root of the equation $x \sin x = 1$ is approximately 1·1. Use Newton's method once to find a closer approximation.

b) Given that $\tan\theta + \tan\phi = \tan\alpha$,
$\cot\theta + \cot\phi = \cot\beta$,
show that $\cot\alpha - \tan\beta = \cot(\theta + \phi)$.

O. and C. 10 The position vectors with respect to a point O of four non-collinear points A, B, C, D are $\mathbf{a}$, $\mathbf{b}$, $\frac{2}{3}\mathbf{a}$, $\frac{1}{3}\mathbf{b}$, respectively. Show that any point on AD has position vector given parametrically by $(1-t)\mathbf{a} + \frac{1}{3}t\mathbf{b}$ and write down the position of any point on BC. Hence find the position vector of the point of intersection E of BC and AD.

If $\triangle OAB$ is equilateral and has sides of unit length, prove that $\mathbf{a}\,.\,\mathbf{b} = \frac{1}{2}$ and, by considering the scalar product of **ED** and **EC** calculate $\angle CED$.

Paper B6

L.U. 1 i) Solve the simultaneous equations

$$\log_2 x - \log_4 y = 4,$$
$$\log_2(x-2y) = 5.$$

ii) Prove by induction or otherwise, that

$$\frac{1}{1.2}+\frac{1}{2.3}+\frac{1}{3.4}+\cdots+\frac{1}{n(n+1)}=\frac{n}{n+1}.$$

J.M.B.

2 Given that $\sec\theta + \tan\theta = x$ and $\operatorname{cosec}\theta + \cot\theta = y$, show that $x + \dfrac{1}{x} = 2\sec\theta$ and $y + \dfrac{1}{y} = 2\operatorname{cosec}\theta$.

Find $\dfrac{dx}{d\theta}$ and $\dfrac{dy}{d\theta}$ in terms of θ, and hence show that $\dfrac{dy}{dx} = -\dfrac{1+y^2}{1+x^2}$

J.M.B.

3 Prove that the equation of the tangent to the parabola $y^2 = 4ax$ at the point P $(at^2, 2at)$ is $ty - x = at^2$. Write down, or obtain, the equation of the normal to the parabola at P. The tangent and the normal meet the x-axis at T and N respectively. Express PT^2 and PN^2 in terms of a and t.

O. and C.

4 Find the solution of the differential equation

$$\frac{dy}{dx} = xy \ln x$$

which satisfies the initial conditions $x = 1$, $y = 1$, giving $\ln y$ in terms of x.

O. and C.

5 The position vectors of the vertices of a triangle ABC referred to a given origin are **a**, **b**, **c**. P is a point on AB such that $\dfrac{AP}{PB} = \frac{1}{2}$, Q is a point on AC such that $\dfrac{AQ}{QC} = 2$, and R is a point on PQ such that $\dfrac{PR}{RQ} = 2$.

Prove that the position vector of R is $\frac{4}{9}\mathbf{a} + \frac{1}{9}\mathbf{b} + \frac{4}{9}\mathbf{c}$.

Prove that R lies on the median BM of the triangle ABC, and state the value of $\dfrac{BR}{RM}$.

A.E.B.

6 a) Show that over the range $0 \leqslant x \leqslant 6$ the mean value of the function $9x(6 - x)$ is two-thirds of the maximum value of the function.
b) A piece of wire 80 cm in length is cut into three parts, two of which are bent into equal circles and the third into a square. Find the radius of the circles if the sum of the enclosed area is a minimum.

A.E.B.

7 a) Express $7\sin x - 24\cos x$ in the form $R\sin(x - \alpha)$, where R is positive and α is an acute angle. Hence, or otherwise, solve the equation $7\sin x - 24\cos x = 15$, for $0° < x < 360°$.
b) Solve the simultaneous equation $\cos x + \cos y = 1$, $\sec x + \sec y = 4$, for $0° < x < 180°$, $0° < y < 180°$.

O. and C.

8 Sketch on the same diagram the graphs of
i) $y = \ln x$,
ii) $y = \alpha x$ where α is a small positive number.
Explain why the equation $\ln x - \alpha x = 0$ has a solution close to $x = 1$.

Using the Newton-Raphson process once, or otherwise, find a closer approximation to the solution in terms of α.

L.U.

9 Prove that in *any* triangle ABC, with the usual notation,

$$a^2 = b^2 + c^2 - 2bc\cos A.$$

A convex quadrilateral $ABCD$ has $AB = 5$ cm, $BC = 8$ cm, $CD = 3$ cm, $DA = 3$ cm and $BD = 7$ cm. Find the angles DAB, BCD and show that the quadrilateral is cyclic. Show also that $AC = 5\frac{4}{7}$ cm.

J.M.B.

10 The complex numbers z_1 and z_2 satisfy the equation

$$z_2^2 - z_1 z_2 + z_1^2 = 0.$$

Find the ratio z_2/z_1, given that its imaginary part is positive. If $z_1 = a + ib$, where a and b are real, show that

$$z_2 = \tfrac{1}{2}(a - b\sqrt{3}) + \tfrac{1}{2}(b + a\sqrt{3})i.$$

In an Argand diagram, the points P and Q represent z_1 and z_2 respectively and O is the origin. Show that the triangle OPQ is equilateral.

Paper B7

J.M.B.

1 a) Differentiate with respect to x
   i) $\ln(3x - 1)^3$,
   ii) xe^{x^2}
b) Given that $y = \arcsin\sqrt{x}$, prove that

$$\frac{dy}{dx} = \frac{1}{\sin 2y}$$

O. and C.

2 i) Expand $(1 + x)^{1/2}$ in ascending powers of x as far as the term in x^3. Show that the exact difference between the first two terms and $(1 + x)^{1/2}$ is $\frac{1}{4}x^2/\{1 + \frac{1}{2}x + (1 + x)^{1/2}\}$.
   By substituting $x = 0{\cdot}024$ in the first two terms of the expansion of $(1 + x)^{1/2}$ show that $10^{1/2}$ is $3{\cdot}162$ approximately and that this is in fact correct to 3 decimal places.
ii) Expand $\ln(1 + \cos x)$ in ascending powers of x as far as the term in x^4.

A.E.B.

3 a) Prove by induction, or otherwise, that

$$\sum_{r=1}^{n} r^3 = \tfrac{1}{4}n^2(n + 1)^2.$$

Find $\displaystyle\sum_{r=1}^{n} r(r - 2)(r + 2)$.

b) Express $\dfrac{1}{r(r + 1)}$ in partial fractions and hence, or otherwise, find

$$\sum_{r=1}^{\infty} \frac{(-1)^{r+1}}{r(r + 1)}$$

L.U.

4 Given that α and β are the roots of the equation $x^2 - px + q = 0$, prove that $\alpha + \beta = p$ and $\alpha\beta = q$. Prove also that
a) $\alpha^{2n} + \beta^{2n} = (\alpha^n + \beta^n)^2 - 2q^n$,
b) $\alpha^4 + \beta^4 = p^4 - 4p^2q + 2q^n$.
Hence, or otherwise, form the quadratic equation whose roots are the fourth powers of those of the equation $x^2 - 3x + 1 = 0$.

L.U.

5 i) If $I = \int_0^1 x^n e^{-x}\,dx$, where $n \geqslant 0$, find the relation between I_n and I_{n-1} where $n \geqslant 1$. Express $\int_0^1 x^4 e^{-x}\,dx$ in terms of e.
ii) Find the general solution of the differential equation

$$\frac{d^2y}{dx^2} + \frac{dy}{dx} + y = 0$$

O. and C.

6 Sketch the curve given parametrically by $x = 4\cos\theta$, $y = 3\sin\theta$, for $0 \leqslant \theta \leqslant 2\pi$. A rectangle is inscribed in the curve. It has one corner at the

point $(4 \cos \phi, 3 \sin \phi)$, where $0 \leqslant \phi \leqslant \frac{\pi}{2}$, its sides parallel to the axes and its centre at the origin. Find as ϕ varies, the maximum area of the rectangle.

Write $4 \cos \phi + 3 \sin \phi$ in the form $C \sin (\phi + \alpha)$, with $C > 0$, giving the values of C and $\tan \alpha$. Use your result to find the maximum value of the perimeter of the rectangle, and to show that no such rectangle has a perimeter of less than 12 units.

J.M.B.

7 Sketch the curve whose equation is $y + 3 = \dfrac{6}{x - 1}$.

Find the coordinates of the points where the line $y + 3x = 9$ intersects the curve and show that the area of the region enclosed between the curve and the line is $\frac{3}{2}(3 - 4 \ln 2)$. Determine the equation of the two tangents to the curve which are parallel to the line.

A.E.B.

8 Find the centre and the radius of the circle C which passes through the points (4, 2), (2, 4) and (2, 6). If the line $y = mx$ is a tangent to C, obtain the quadratic equation satisfied by m. Hence, or otherwise, find the equations of the tangents to C which pass through the origin O.

Find also

i) the angle between the two tangents.

ii) the equations of the circle which is the reflection of C in the line $y = 3x$.

L.U.

9 i) Simplify $(1 + w)(1 + w^2)$ where

$$w = \cos (2\pi/3) + i \sin (2\pi/3).$$

Express in the form $r (\cos \theta + i \sin \theta)$ each of the three roots of the equation $(z - w)^3 = 1$.

ii) Express in terms of θ the roots of α and β of the equation $z + z^{-1} = 2 \cos \theta$.

In the Argand diagram the points P and Q represent the numbers $(\alpha^n + \beta^n)$ and $(\alpha^n - \beta^n)$ respectively. Show that the length of PQ does not depend on the integer n.

L.U.

10 With the usual notation for any triangle ABC prove that

$$\frac{a}{\sin A} = \frac{b}{\sin B} = \frac{c}{\sin C} = 2R.$$

AD, BE, CF are altitudes of any triangle ABC. Show that $EF = a \cos A$. Show also that the circumradius of the triangle EFC is $\frac{1}{2}a$. Find the circumradius of the triangle DEF in terms of R.

Answers

Chapter 1

Exercise A

1 a) $5x^2 - 3x + 3$
b) $3x^2 - 2x - 1$
c) $5a - b - c$

2 a) $-4x^2 + x + 6$
b) $-12x^3 + 11x^2 - 5x$
c) $-5x^2 - 2xy + 4y^2 + 3$

3 a) $2x^2 + x + 2$
b) 4
c) $13x^2 + 6x$

4 a) $x^5 - 3x^4 + 2x^3 + 3x^2 - 9x + 10$
b) $x^5 + 5x^4 + 2x^3 - 14x^2 - 4x + 8$
c) $x^3 + y^3 + z^3 - 3xyz$

5 a) $x^2 - x + 1$
b) $3x^2 - 4x + 6$, Rem -3
c) $x^2 + 6x + 9$, Rem 19
d) $3x - 2$, Rem -1
e) $x^2 - 3x + 2$

6 a) $3x^2 + 2x - 4$
b) $2x^2 - x + 3$
c) $x^2 + 3x + 1$
d) $x^3 + x^2 + 1$

Exercise B

1 $2x^2 + 7x + 7$, Rem 24

2 $3x^2 + 4$

3 $\frac{x^2}{3} - \frac{4x}{9} - \frac{8}{27}$, Rem $-\frac{151}{27}$

4 $3x^3 - 4x^2 - 6x + 2$, Rem -3

5 $x^5 + x^4 + x^3 - x^2 - x - 1$

Exercise C

1 a) 0
b) 20
c) 98
d) 0

3 a) $(x - 2)(3x + 2)(x + 1)$
b) $(x - 3)(2x + 1)(x + 3)$
c) $(x - 3)(4x - 3)(x + 2)$
d) $(x - 2)(x + 4)(x^2 + 2x + 4)$

4 a) $(x + 3)(x^2 - 3x + 9)$
b) $(x - 1)(x + 1)(x^2 - x + 1)(x^2 + x + 1)$
c) $x(x - y)(x^2 + xy + y^2)$
d) $(5x + 2y)(25x^2 - 10xy + 4y^2)$
e) $(x + y)(x^2 - xy + y^2 + 1)$
f) $(x + y)^3$

5 a) $\frac{x^2 + xy + y^2}{y}$
b) 1
c) $\frac{y(x^2 + y^2)}{x^3 - y^3}$
d) $\frac{3xy}{x - y}$

Exercise D

1 a) $\pm 1{\cdot}5$
b) $0, -\frac{4}{3}$
c) $\pm 2{\cdot}5$
d) ± 2

2 a) $5, -3$
b) $-5, 1{\cdot}5$
c) $-2, \frac{3}{4}$
d) $-2, -\frac{2}{3}$
e) $5, \frac{1}{3}$

3 a) $9{\cdot}83, -1{\cdot}83$
b) $2{\cdot}387, 0{\cdot}279$

4 a) $-2{\cdot}87, 0{\cdot}87$
b) $0{\cdot}5, -1{\cdot}5$
c) $2{\cdot}54, 0{\cdot}13$
d) $3, 0{\cdot}8$

5 a) Real, distinct
b) Real, distinct
c) Real, equal
d) Imaginary
e) Imaginary

Exercise E

1 $-0{\cdot}5, 2$

2 $3{\cdot}56, -0{\cdot}56$

3 $1{\cdot}703, -1{\cdot}37$

Exercise F

1 a) 2
b) -16
c) $-\frac{1}{6}$
d) $\dfrac{44\sqrt{10}}{27}$
e) $-\dfrac{8\sqrt{10}}{9}$

2 a) $x^2 + 7x - 4 = 0$
b) $4x^2 + 57x + 4 = 0$
c) $x^2 = 0$
d) $2x^2 + 7x - 67 = 0$

3 a) $x^2 + 8x + 9 = 0$
b) $2x^2 - 6x + 5 = 0$, 4

4 a) $x^2 + (a - 2b)x + b(b - a + 1) = 0$
b) $bx^2 + a(b + 1)x + a^2 + (b - 1)^2 = 0$
c) $b^2x^2 + (2b - a^2)x + 1 = 0$
d) $x^2 + a(a^2 - 3b)x + b^3 = 0$

5 ± 10

6 $b^2x^2 + (2b - a^2)(b^2 + 1)x + (b^2 + 1)^2 = 0$

Exercise G

2 a) 11
b) 13
c) 4
$x^3 - 11x^2 + 13x - 4 = 0$

3 a) $x^3 - x + 4 = 0$
b) $x^3 + 9x^2 + 26x + 28 = 0$

Exercise H

1 $\dfrac{1}{3(x-1)} + \dfrac{2}{3(x+2)}$

2 $\dfrac{2}{x+3} + \dfrac{3}{2x-1}$

3 $\dfrac{1}{2(x+1)} - \dfrac{7}{10(x+3)} + \dfrac{1}{5(x-2)}$

4 $\dfrac{1}{4(x-2)} + \dfrac{3}{4(x+2)}$

5 $-\dfrac{4}{3x} + \dfrac{7}{12(x-3)} + \dfrac{7}{4(x+1)}$

6 $-\dfrac{2}{x} - \dfrac{1}{x^2} + \dfrac{2}{x-1}$

7 $\dfrac{1}{x-1} - \dfrac{1}{x+2} - \dfrac{3}{(x+2)^2}$

8 $-\dfrac{1}{3(2x-1)} + \dfrac{5}{3(x-2)} + \dfrac{4}{(x-2)^2}$

9 $\dfrac{2}{x} - \dfrac{1}{x^2} - \dfrac{3}{2x+1}$

10 $\dfrac{1}{x^2} - \dfrac{3}{4(x-1)} + \dfrac{1}{4(x-1)^2} + \dfrac{3}{4(x+1)} + \dfrac{1}{4(x+1)^2}$

Chapter 2

Exercise A

1 a) 8
b) $\pm\frac{1}{3}$
c) 243
d) $\frac{1}{625}$
e) 4
f) $\frac{2}{3}$
g) $\frac{1}{64}$
h) 1
i) ± 1
j) $\pm\frac{1}{2}$

2 a) x^{16}
b) x^2
c) $9x^6$
d) $8x^{-1}$
e) $5x^{-1}$
f) $\dfrac{5x}{2}$
g) $\dfrac{3(2 - x^2)}{x^4}$
h) $\dfrac{(x+3)(x-1)}{x^3}$
i) $\dfrac{2y^2}{3x}$

Exercise B

1 a) $2\sqrt{7}$
b) $5\sqrt{3}$
c) $4\sqrt{3}$
d) $4\sqrt{3}$
e) $6\sqrt{2}$
f) 2
g) $\sqrt{5}$
h) $9\sqrt{3}$
i) 0
j) $2\sqrt{2}$
k) 3
l) $3 - 2\sqrt{2}$
m) $11 - \sqrt{2}$

2 a) $\dfrac{3\sqrt{2}}{2}$
b) $\dfrac{4\sqrt{3}}{3}$
c) $3\sqrt{3}$
d) $\sqrt{2}$
e) $3(\sqrt{2} + 1)$
f) $2 - \sqrt{3}$
g) $3 + \sqrt{7}$
h) $7 + 4\sqrt{3}$
i) $2 - \sqrt{3}$
j) $4\sqrt{3}$

3 a) $\dfrac{x-\sqrt{x}}{x-1}$ c) $\sqrt{(x^2+1)}-x$ e) $\dfrac{x^2-(x+1)\sqrt{x-1}}{x}$

b) $\dfrac{x\sqrt{x}-2\sqrt{x}+x}{x}$ d) $\sqrt{x}-\sqrt{y}$ f) $\dfrac{3}{1-x}$

Exercise C

1) 10 2) 2 3) 7 4) 1, 9 5) 7 6) 4

Exercise D

1 a) $\log_3 27 = 3$ b) $\log_2 32 = 5$ c) $\log_6 216 = 3$ d) $\log_8 4 = \frac{2}{3}$ e) $\log_{1/3} \frac{1}{9} = 2$ f) $\log_5 0{\cdot}2 = -1$ g) $\log_2 \frac{1}{4} = -2$ h) $\log_{2/3} 2{\cdot}25 = -2$ i) $\log_4 1 = 0$ j) $\log_4 0{\cdot}5 = -\frac{1}{2}$

2 a) 3 b) 1 c) -3 d) $\frac{1}{2}$ e) $\frac{1}{2}$ f) -2 g) -2 h) 6 i) -3 j) $-\frac{4}{3}$

Exercise E

1 a) -1 b) $-\log_x 2$ c) 1 d) 1 2 a) 12 b) 3 c) 1

Exercise F

1 a) 10·19 b) 12·28 c) 15·51 d) 183 e) 1·293 f) 7·159 g) 0·306 h) 0·4305 i) 2·303 j) 3·136

2 a) 2·807 b) 1·431 c) 2·044 d) $-1{\cdot}727$ e) $-1{\cdot}850$

Exercise G

1 1·292
2 $-0{\cdot}1587$
3 1·984
4 0, 1·585
5 0, 1·465
6 $\pm 0{\cdot}3010$
7 0, 1, $\log_3 2$

Exercise H

1 a) 2, $-\frac{1}{16}$ b) 4 c) 512 d) 4 and 1 2 a) 1·6, 2·5 b) 25, 4 c) 2, 8

Exercise I

1 a) $1+10x+40x^2+80x^3+80x^4+32x^5$
b) $16x^4-96x^3+216x^2-216x+81$
c) $\dfrac{1}{x^6}+\dfrac{6}{x^4}+\dfrac{15}{x^2}+20+15x^2+6x^4+x^6$
d) $16-16x+6x^2-x^3+\dfrac{x^4}{16}$
e) $17x^4-24x^2+48+\dfrac{24}{x^2}+\dfrac{17}{x^4}$

2 -20 3 a) 25 b) 544 4 362 209

5 $1+4x+10x^2+16x^3+19x^4+16x^5+10x^6+4x^7+x^8$ 6 $-1+46\sqrt{3}$

Chapter 3

Exercise A

1) 2 2) 3 3) $\frac{1}{2}$ 4) 0 5) 0 6) -2 7) -1 8) 1 9) 3 10) 3

Exercise B

1 1 2 $-2x$ 3 2 4 $1+2x$ 5 $4x^3$
6 $6x$ 7 $6x+2$ 8 0 9 $3x^2-2x$ 10 $9x^2-6x$

Exercise C

1 a) $2x-2$
b) $3x^2-27$
c) $2x-9$
d) $1-\frac{1}{x^2}$
e) 1
f) $\frac{1}{2}x^{-1/2}-\frac{1}{2}x^{-3/2}$
g) $32x-\frac{32}{x^3}$
h) $-9x^{-4}+4x^{-3}$
i) $4x-5$
j) $2x$

2 a) $f(2)=0;\ f'(2)=0$
b) $f(2)=-8;\ f'(2)=0$
c) $f(2)=4\frac{1}{2};\ f'(2)=3\frac{3}{4}$
d) $f(2)=2;\ f'(2)=1$
e) $f(2)=1;\ f'(2)=0$
f) $f(2)=0;\ f'(2)=7$
g) $f(2)=21;\ f'(2)=14$
h) $f(2)=0;\ f'(2)=16$
i) $f(2)=0;\ f'(2)=2$
j) $f(2)=0;\ f'(2)=4$

Exercise D

1 Maximum at (3, 11)

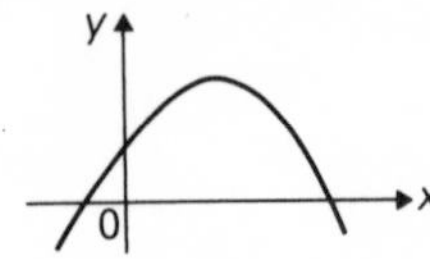

2 Minimum at (3, 6·5), maximum at (2, 7)

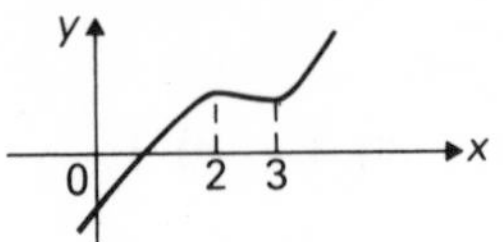

3 Minimum at $(\frac{1}{2}, 4)$, maximum at $(-\frac{1}{2}, -4)$

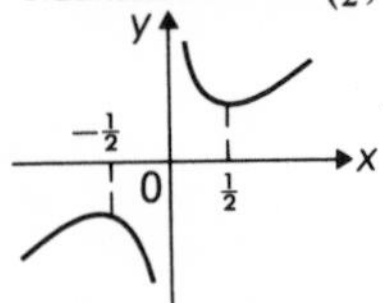

4 Maximum at (0, 0), minimum at $(-1, -1)$, minimum at $(1, -1)$

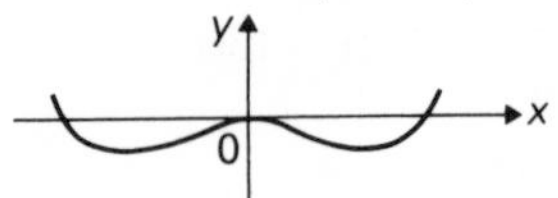

5 No turning points

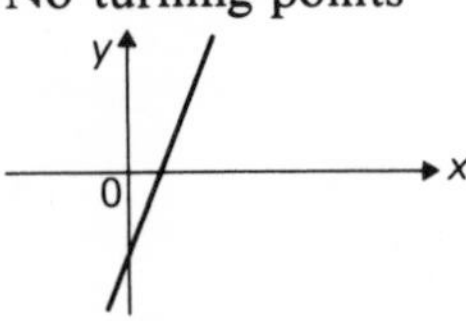

6 Minimum at $(0, -3)$

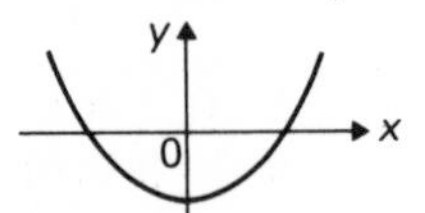

7 Maximum at (0, 9), minimum at $(\frac{2}{3}, 8\frac{23}{27})$

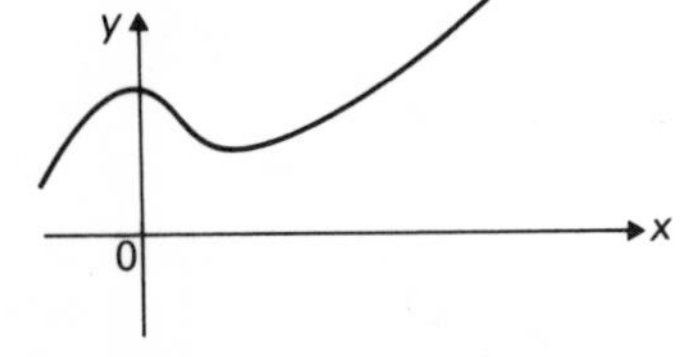

8 Maximum at (0, 0), minimum at $\left(-\frac{1}{\sqrt{2}}, -\frac{1}{4}\right)$
minimum at $\left(\frac{1}{\sqrt{2}}, -\frac{1}{4}\right)$

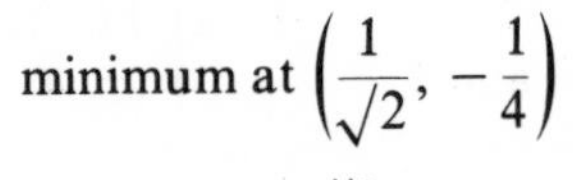

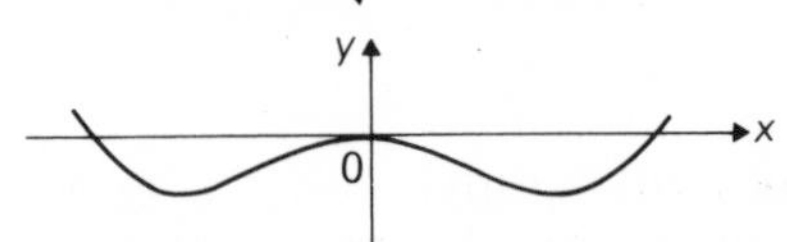

Exercise E

1 $A=x(20-x)$

2 $h=250/(\pi x^2)$; $S=2\pi x^2+\frac{500}{x}$; $S_{min}\approx 220\text{ cm}^2$

3 3 m × 3 m × 1·5 m; cost = £40·50

Exercise F

1 a) $2x+3$
b) $1\frac{1}{2}x^{-1/2}-1$
c) $7x$
d) $-\frac{18}{x^2}-1$
e) $-\frac{6}{x^3}-2$
f) $-\frac{2}{x^2}$

2 a) $2t$
b) $8t^3$
c) $4x$; $\frac{dy}{dx}=\frac{dy}{dt}\Big/\frac{dx}{dt}$

3 a) speed $=\frac{1}{24}(-3t^2+24t+36)$
acceleration $=\frac{1}{24}(-6t+24)$
b) $v=\frac{19}{8}$ m/s; $a=\frac{1}{2}$ m/s^2
c) $3\frac{1}{2}$ m/s

Exercise G

1 48

2 a) $24t^2$
b) 2
c) 4; no

3 a) $3x^2$
b) $6x$
c) 6
values when $x = 0$: 0; 0; 6
inflexion at (0, 0)

4 a) $4x^3$
b) $12x^2$
c) $24x$
values when $x = 0$: 0; 0; 0
minimum at (0, 0), yet usual 'rules' do not apply

5 6th derivative zero for all x
$x = 2$

Chapter 4

Exercise A

1 a) $\frac{\pi}{6}$
b) $\frac{3\pi}{2}$
c) $\frac{\pi}{4}$
d) $\frac{49\pi}{36}$

2 a) 57·3°
b) 128·9°
c) 90°
d) 60°

3 a) 3 cm
b) 3π cm
c) π cm
d) 9 cm

4 $\frac{30}{\pi} = 9{\cdot}55$ cm

5 a) $1\frac{1}{2}$
b) 85·9°

Exercise B

1 a) 25 cm^2
b) 27 cm^2
c) 1·77 cm^2

2 a) 1·75
b) 0·8
c) $\frac{1}{3}$

3 a) 3·74
b) 4·12
c) 2

Exercise C

1 a) −0·5
b) −0·9455
c) 0·9511
d) 0·7071
e) 0·5
f) −0·8660
g) 0·0698
h) 0·9976
i) −0·7071
j) −0·8660

3 a) $\cos\alpha$
b) $\sin\alpha$
c) $-\cos\alpha$
d) $-\sin\alpha$

Exercise D

	sin	cos	tan	cot	sec	csc
0	0	1	0	∞	1	∞
30	$\frac{1}{2}$	$\frac{\sqrt{3}}{2}$	$\frac{1}{\sqrt{3}}$	$\sqrt{3}$	$\frac{2}{\sqrt{3}}$	2
45	$\frac{1}{\sqrt{2}}$	$\frac{1}{\sqrt{2}}$	1	1	$\sqrt{2}$	$\sqrt{2}$
60	$\frac{\sqrt{3}}{2}$	$\frac{1}{2}$	$\sqrt{3}$	$\frac{1}{\sqrt{3}}$	2	$\frac{2}{\sqrt{3}}$
90	1	0	∞	0	∞	1

	sin	cos	tan	cot	sec	csc
120	$\frac{\sqrt{3}}{2}$	$-\frac{1}{2}$	$-\sqrt{3}$	$-\frac{1}{\sqrt{3}}$	-2	$\frac{2}{\sqrt{3}}$
135	$\frac{1}{\sqrt{2}}$	$-\frac{1}{\sqrt{2}}$	-1	-1	$-\sqrt{2}$	$\sqrt{2}$
150	$\frac{1}{2}$	$-\frac{\sqrt{3}}{2}$	$-\frac{1}{\sqrt{3}}$	$-\sqrt{3}$	$-\frac{2}{\sqrt{3}}$	2
180	0	-1	0	∞	-1	∞

Exercise F

1 $\sin 3\theta = 3\sin\theta - 4\sin^3\theta$
2 $\cos 3\theta = 4\cos^3\theta - 3\cos\theta$
6 $\sin 4\theta = \frac{336}{625}$, $\cos 4\theta = -\frac{527}{625}$
7 a) $\frac{117}{125}$ b) $-\frac{44}{125}$
c) $-\frac{237}{3125}$ d) $-\frac{3116}{3125}$
8 60°

Exercise G

5 a) $\tan\left(\dfrac{A+B}{2}\right)$
b) $-\cot\left(\dfrac{A+B}{2}\right)$
6 0
7 0°, 60°, 120°, 180°, 240°, 300°, 360°

Exercise H

1 maximum value = 1 when $x = \dfrac{\pi}{2}$, minimum value $= -1$, when $x = \dfrac{3\pi}{2}$
2 maximum value = 1 when $x = 2\pi$; minimum value $= -1$, when $x = \pi$ or 3π
3 a) $3 + \cos x$ b) $-4\sin x$; $x = 0, \pi, 2\pi$, etc.
4 When $x = \dfrac{\pi}{4}$, maximum value $= \sqrt{2}$; next when $x = \dfrac{5\pi}{4}$, minimum value $-\sqrt{2}$

Chapter 5

Exercise A

1 a) $4\sqrt{2}$
b) $3\sqrt{10}$
c) $\sqrt{74}$
3 $\left(\dfrac{8}{5}, \dfrac{22}{5}\right)$
5 $\dfrac{a(p^2+1)^2}{p^2}$
6 a) $\dfrac{3}{5}$ b) $\dfrac{3}{5}$ c) $\dfrac{1}{2}$ d) $\dfrac{2}{p+q}$
9 $5\sqrt{2}$
10 square
11 a) 135°

Exercise B

1 $x - y + 1 = 0$
2 $x^2 + y^2 - 8x - 6y = 0$
3 $y^2 = 4(x-1)$
4 $x^2 + y^2 - 6x - 6y + 5 = 0$
5 $x - 2y = 0$

Exercise C

1 a) $3x - y - 2 = 0$
b) $x - 2y + 6 = 0$
c) $3x + 4y = 0$
2 a) $x + y + 1 = 0$
b) $4x - 3y - 7 = 0$
c) $x + y + 1 = 0$
3 a) $5x - 3y - 15 = 0$
b) $2x + 3y + 1 = 0$
c) $tx - y + c = 0$
4 a) $x - y + 1 = 0$
b) $x + y - 4 = 0$
5 a) $x - 2y + 8 = 0$
b) $3x + y - 10 = 0$
7 $x + y - 5 = 0$

Exercise D

1 $3x + 4y - 15 = 0$, $4x - 3y - 1 = 0$
2 $x + 4y - 9 = 0$, $x_1 = -3$
3 a) $B(3, 1)$, $D(3, 6)$ b) $x + 2y - 5 = 0$, $2x - y - 5 = 0$ c) $AB = \sqrt{5}$. $BC = 2\sqrt{5}$

4 S is at $(7, 8)$

5 PR is $14x - 4y - 45 = 0$, QS is $2x - 8y + 27 = 0$; $(4\frac{1}{2}, 4\frac{1}{2})$.
$\left(\frac{47}{13}, \frac{54}{13}\right)$; 0·9499

6 a) $7x - 2y - 7 = 0$, $3x + y - 2 = 0$ b) $\left(\frac{11}{13}, -\frac{7}{13}\right)$ c) 13

7 $x - 4y + 10 = 0$, $(1, 1\frac{1}{2})$, $y + x = 0$, $(-2, 2)$

9 $4x - 3y - 3 = 0$, 10, $3x + 4y + 4 = 0$, $(-4, 2)$, $(4, -4)$

10 $x + 2y - 5 = 0$, $\frac{\sqrt{5}}{5}$, $\frac{1}{2}$

11 $3x + 4y - 6 = 0$, $(10, -6)$, $(-6, 6)$

12 $2x + y - 1 = 0$, $\left(\frac{6}{5}, -\frac{7}{5}\right)$, $x + 3y + 3 = 0$

13 $x + y - 1 = 0$

14 $\left(-\frac{4}{3}, \frac{14}{3}\right)$, $(-3, 6)$, $\left(-\frac{7}{3}, \frac{11}{3}\right)$, $\left(-\frac{2}{3}, \frac{7}{3}\right)$

15 $(0, 0)$, $(5, 2)$, $(3, 7)$. $3x + 7y - 29 = 0$, $9x - 8y = 0$, $12x - y - 29 = 0$,
$\left(\frac{8}{3}, 3\right)$

Exercise E

1 a) $x^2 + y^2 = 25$
b) $x^2 + y^2 - 4x + 6y + 4 = 0$
c) $x^2 + y^2 = 25$
d) $x^2 + y^2 + 2x - 4y - 20 = 0$
e) $x^2 + y^2 - 6x - 2y + 5 = 0$

2 a) $(-1, 2)$, $\sqrt{5}$
b) $(1, 3)$, $\sqrt{5}$
c) $(-2, 3)$, 4

3 a) $x^2 + y^2 - x - 5y + 4 = 0$
b) $x^2 + y^2 - 2x - 6y + 5 = 0$
c) $x^2 + y^2 + 6x - 2y - 15 = 0$

Exercise F

1 $(3, 1)$, $(-2, 6)$ 2 $\sqrt{10}$ 4 $4\frac{1}{2}$ 5 $\frac{3}{4}$, $6\frac{1}{4}$ 6 4, $(2, 0)$

Exercise G

1 $5x - 2y + 11 = 0$ 2 $4x - 3y - 10 = 0$ 3 $xy = 2$
4 $y^2 = 2x$ 5 $x - 2y + 1 = 0$

Exercise H

1 a) $y^2 = 8x$ b) $4y + x^2 = 0$ c) $x^2 - 2x - 6y + 4 = 0$

2 a) $y^2 = 25x$ b) $x^2 = 9y$ c) $(y + 1)^2 = 4x$

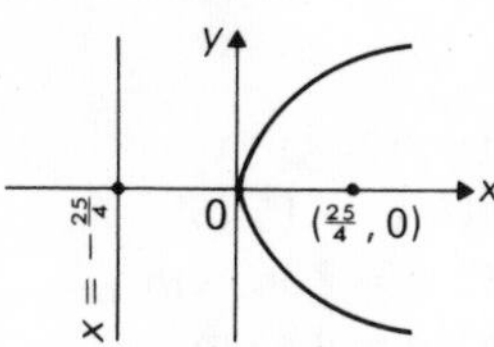

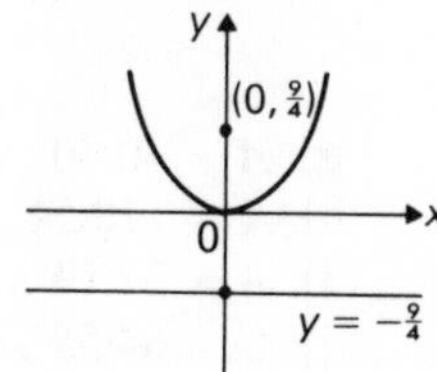

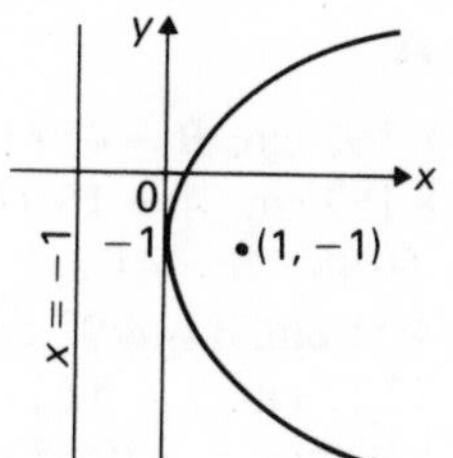

3 $(y + 2)^2 = 8(x - 1)$, $(1, -2)$, $(3, -2)$

4 a) 16 b) 8 c) 2

5 20, 80

6 $x^2 - 24x + 4y^2 - 12y - 4xy + 24 = 0$

7 a) $y^2 = x$ b) $5x + 2y^2 = 0$ c) $x + 1 = y^2$

8 $x = 3$

10 $x = \pm 2t$, $y = -t^2$

Exercise I

1 a) $\dfrac{x^2}{25}+\dfrac{y^2}{9}=1$; 10, 6 b) $\dfrac{x^2}{16}+\dfrac{y^2}{4}=1$; 8, 4 c) $4x^2+9y^2=36$; 6, 4

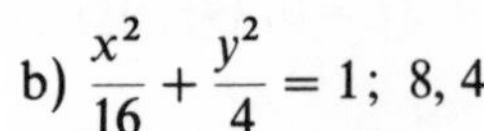

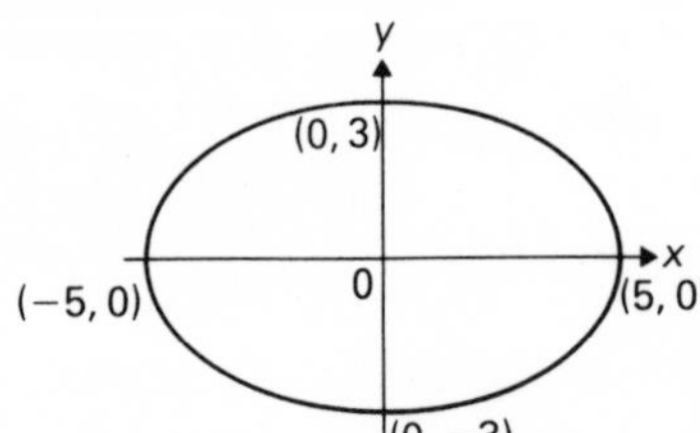

(b) y; (0, 2); x; (−4, 0); 0; (4, 0); (0, −2)

(c) y; (0, 2); x; (−3, 0); 0; (3, 0); (0, −2)

2 a) $(\pm 4, 0)$, $x=\pm\dfrac{25}{4}$
b) $(\pm 2\sqrt{3}, 0)$, $x=\pm\dfrac{8\sqrt{3}}{3}$
c) $(\pm\sqrt{5}, 0)$, $x=\pm\dfrac{9\sqrt{5}}{5}$

3 a) $\dfrac{x^2}{36}+\dfrac{y^2}{25}=1$
b) $\dfrac{x^2}{25}+\dfrac{y^2}{16}=1$
c) $\dfrac{4x^2}{25}+\dfrac{4y^2}{9}=1$

4 a) $(2, 1)$, $(2\pm 2\sqrt{2}, 1)$, $x=2\pm\dfrac{9\sqrt{2}}{4}$
b) $(-1, 2)$, $(-1\pm\sqrt{5}, 2)$, $x=-1\pm\dfrac{9\sqrt{5}}{5}$

5 a) $x=2\cos\theta$, $y=\sin\theta$
b) $x=\frac{3}{2}\cos\theta$, $y=\sin\theta$
c) $x=1+3\cos\theta$,
$y=2+2\sin\theta$

6 a) $\dfrac{x^2}{16}+\dfrac{y^2}{9}=1$
b) $\dfrac{x^2}{144}+\dfrac{y^2}{9}=1$
c) $\dfrac{x^2}{9}+\dfrac{(y-1)^2}{4}=1$
d) $(x-1)^2+4(y-1)^2=4$

Exercise J

1 a) $x+y+a=0$
b) $3x-4y+12=0$
c) $x-2y+7=0$

2 a) $x-y-3a=0$
b) $4x+3y-34=0$
c) $2x+y-11=0$

3 a) $(1, 2)$
b) $\left(\dfrac{2}{9}, \dfrac{4}{3}\right)$
c) $(9, 3)$
d) $(5, 2)$

4 a) $2x+y-3=0$
b) $2x-y-12=0$
c) $2x+y-14=0$

5 a) $x-2y+16=0$, $2x+y-48=0$
b) $4x-2y+5=0$, $4x+8y-45=0$
c) $2x+2y+1=0$, $2x-2y-3=0$

6 a) $x-y-5=0$, $x+y-1=0$
b) $x+2y+4=0$, $2x-y+3=0$
c) $x+2y-8=0$, $2x-y-1=0$

7 a) $3x+5\sqrt{3}y-30=0$, $5x-\sqrt{3}y-8=0$
b) $x+y-5=0$, $5x-5y-7=0$
c) $x+2y-2\sqrt{2}=0$, $2\sqrt{2}x-\sqrt{2}y-3=0$

Chapter 6

Exercise A

1 a) $a=6{\cdot}362$ cm, $B=47{\cdot}61°$, $C=62{\cdot}39°$
b) $c=8{\cdot}159$ cm, $A=15{\cdot}07°$, $B=119{\cdot}93°$
c) $A=50{\cdot}98°$, $B=41{\cdot}75°$, $C=87{\cdot}27°$
d) $b=4{\cdot}51$ cm, $c=4{\cdot}97$ cm, $A=43{\cdot}1°$
e) $a=12{\cdot}17$ cm, $b=5{\cdot}61$ cm, $C=38{\cdot}5°$
f) $a=8{\cdot}99$ cm, $c=19{\cdot}65$ cm, $B=86{\cdot}48°$
g) $A=40{\cdot}40°$, $B=84{\cdot}87°$, $C=54{\cdot}73°$
h) $A=110{\cdot}54°$, $B=54{\cdot}01°$, $C=15{\cdot}45°$
i) $A=27{\cdot}19°$, $C=86{\cdot}33°$, $c=2{\cdot}363$ cm
j) $C=64{\cdot}55°$, $A=77{\cdot}45°$, $a=4{\cdot}8$ cm;
$C=115{\cdot}45°$, $A=26{\cdot}55°$, $a=2{\cdot}2$ cm;
Difference 2·6 cm.

3 a) BD = 13·01 cm
b) 113·55°, 66·45°, 113·55°, 66·45°
c) 81°

4 1·415 cm

Exercise B

1 90° 2 135°, 315° 3 90° 4 199·47°, 340·53° 5 0°, 30°, −30° 6 30°
7 60°, 90°, 270°, 300° 8 180° 9 51·33°, 128·67° 10 14·48°, 90°, 165·52°, 270°
11 58·27°, 121·73° 12 195°, 255° 13 0°, 30°, 150°, 180°, 360°
14 0° 15 33·69°, 45°, 213·69°, 225°

Exercise C

1 0°, 45°, 135° 2 −90°, −45°, −30°, 30°, 45°, 90°
3 20°, 45°, 100°, 135°, 140°, 220°, 225°, 260°, 315°, 340° 4 0°, 36°, 90°, 108°, 180°
5 0°, 20°, 30°, 60°, 90° 6 −180°, −90°, −60°, 0°, 60°, 90°, 180°
7 0°, 20°, 45°, 90°, 100°, 135°, 140°, 180° 8 −180°, −120°, −60°, 0°, 60°, 120°, 180°
9 $-67\frac{1}{2}°$, −60°, $-22\frac{1}{2}°$, $22\frac{1}{2}°$, 60°, $67\frac{1}{2}°$ 10 0°, 60°, 105°, 120°, 165°, 180°

Exercise D

1 90°, 323·13° 2 −27·66° 3 257·59°, 349·79° 4 72·41°, 220·21°
5 −13·3°, 119·56° 6 $\theta = (360n + 110{\cdot}6)°$, or $(360n + 216{\cdot}9)°$ 7 48·88°, 171·72°
8 $\sqrt{10}\cos(\theta - 71{\cdot}56°)$, maximum $\sqrt{10}$, minimum $-\sqrt{10}$
9 No,

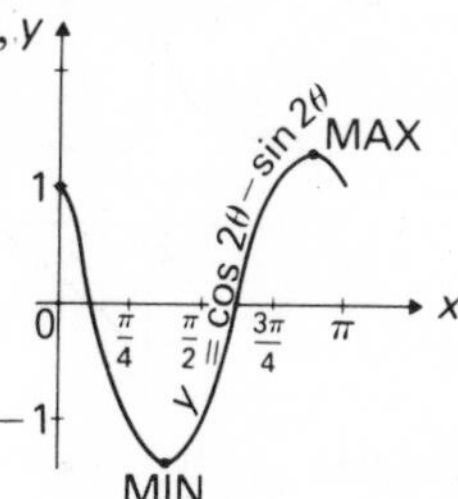

Chapter 7

Exercise A

1
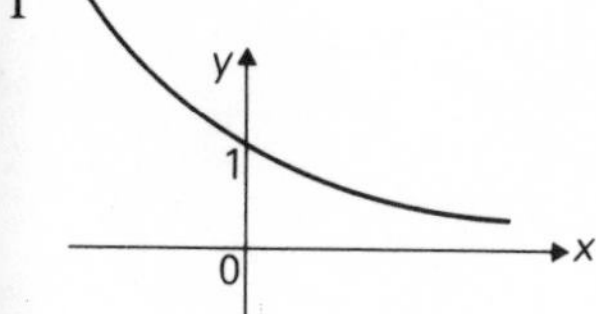

2
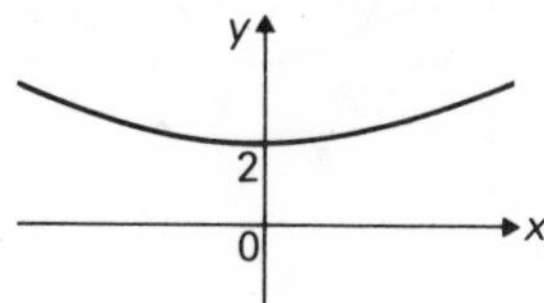

3
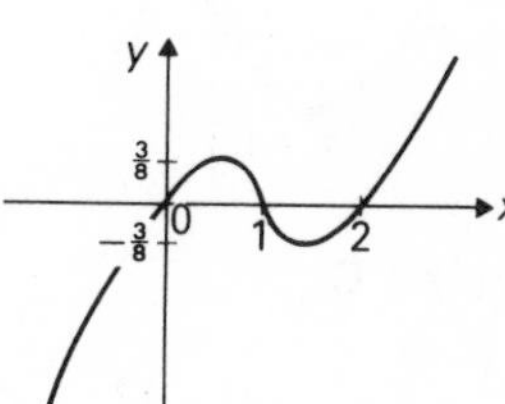

4
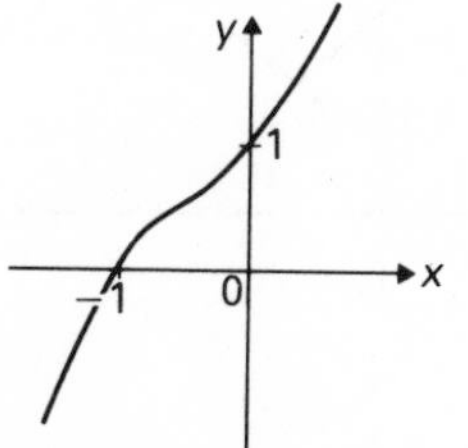

5
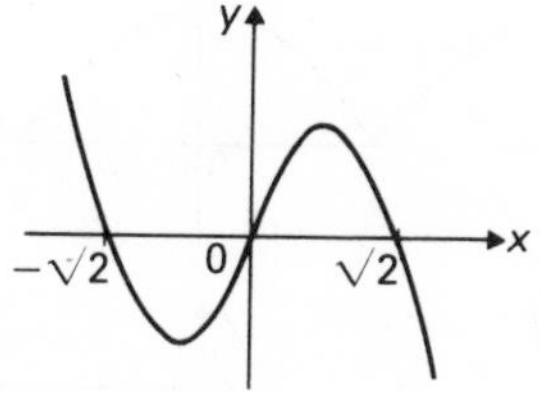

6
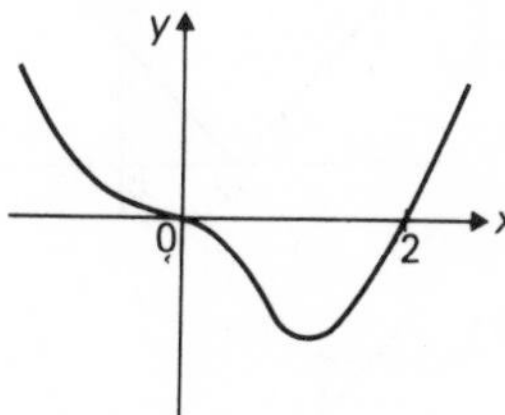

7
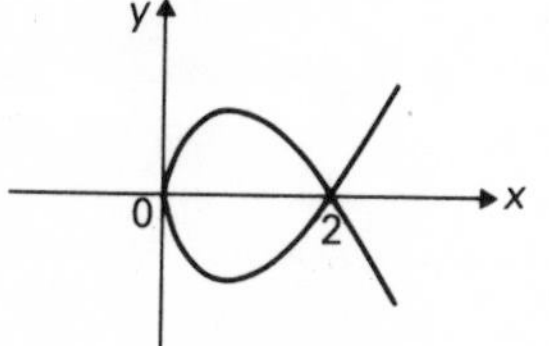

8
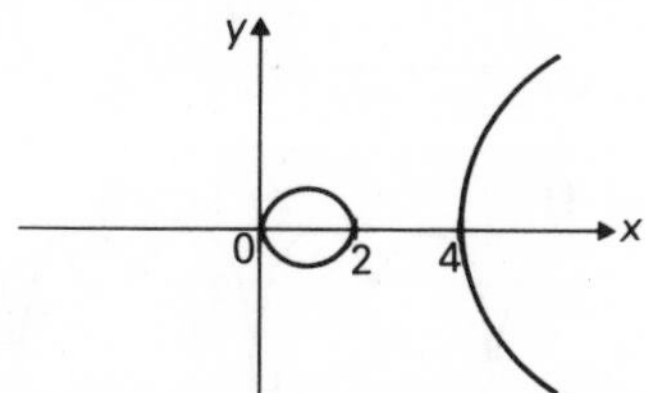

Exercise B

1 a)

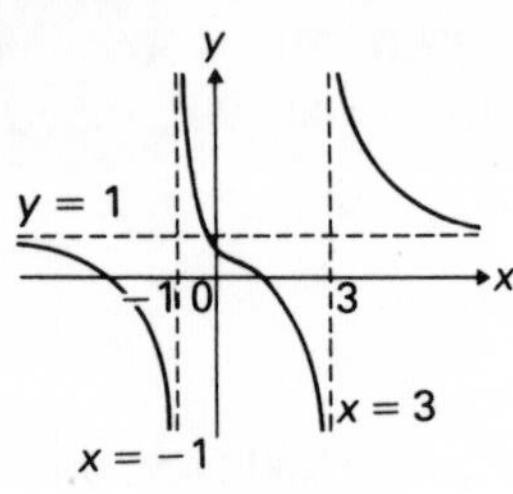

e)

y
y = 1
x
0
2
x = 2

i)

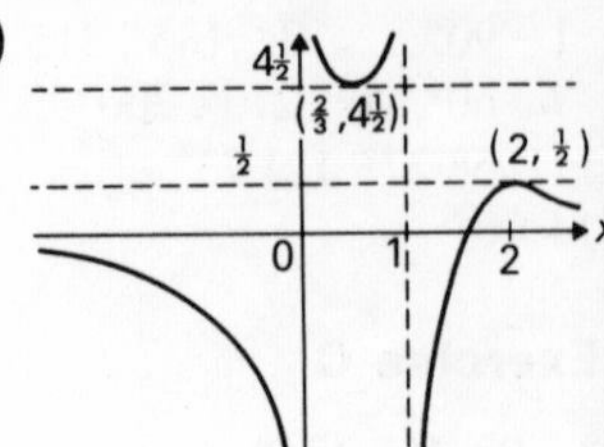

b)

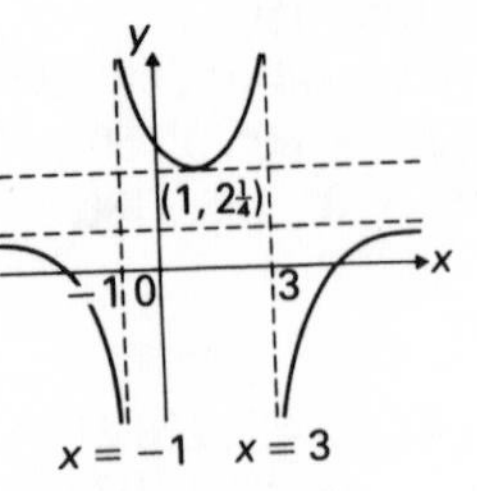

f)

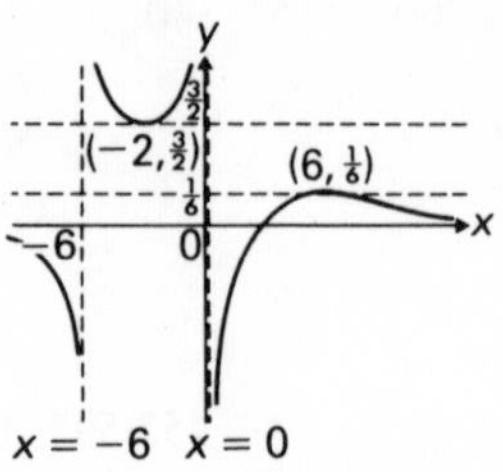

j)

y
$y = \dfrac{5\sqrt{2} - 1}{\sqrt{2}}$
(0, 2)
$\sqrt{2} - 1$
x
$-\sqrt{2} - 1$
0
(3, 0)
$y = \dfrac{5\sqrt{2} - 1}{2}$
(0, −3)

c)

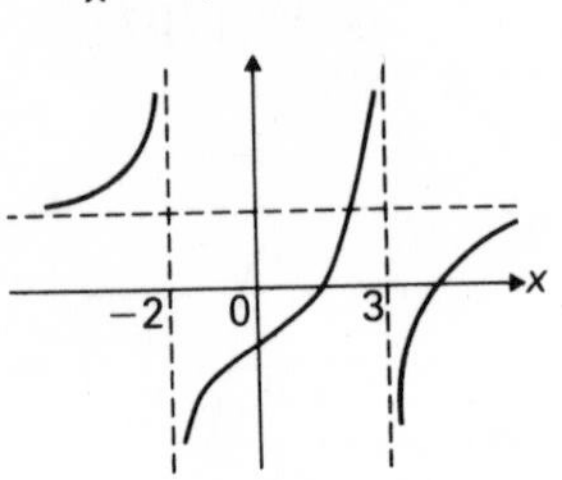

g)

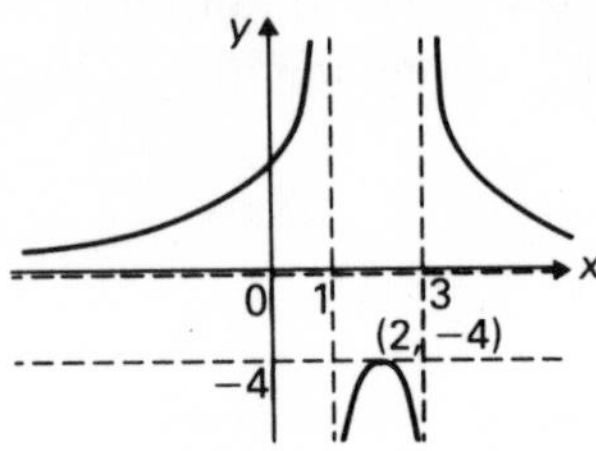

2 Maximum point at $(1, \frac{2}{3})$, minimum at $(-1, -2)$

3

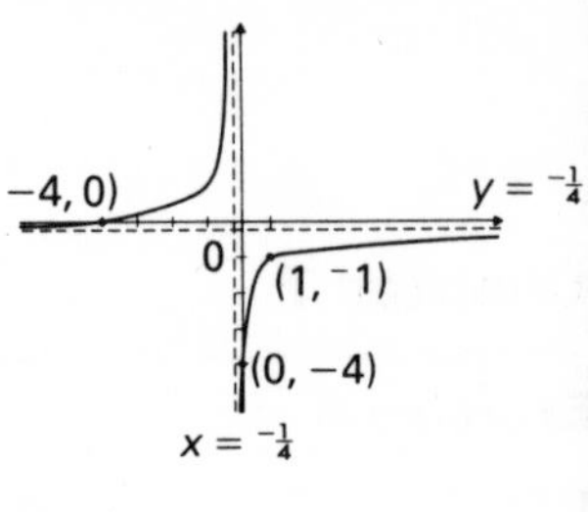

d) 

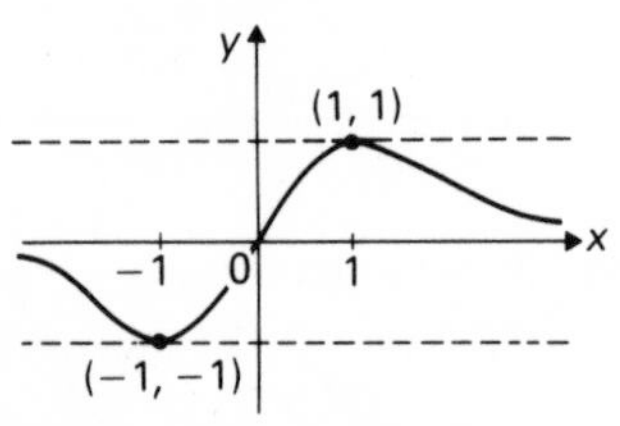

h)

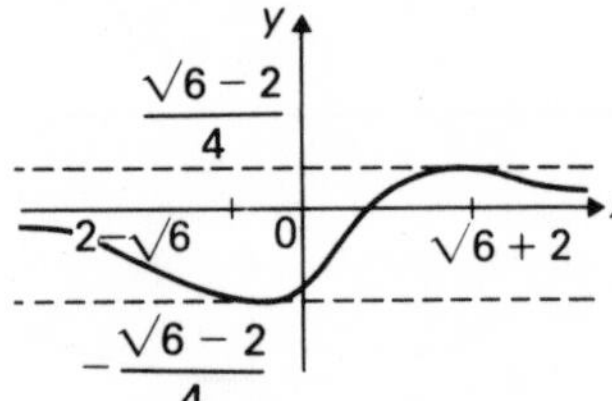

Exercise C

1 a)

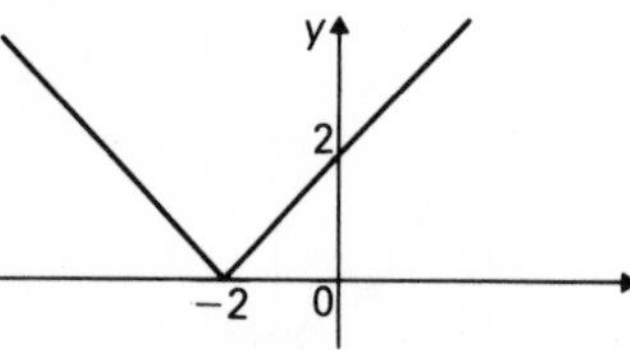

d)

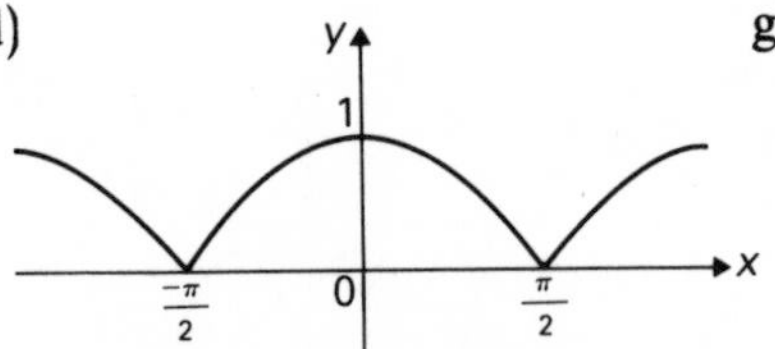

g)

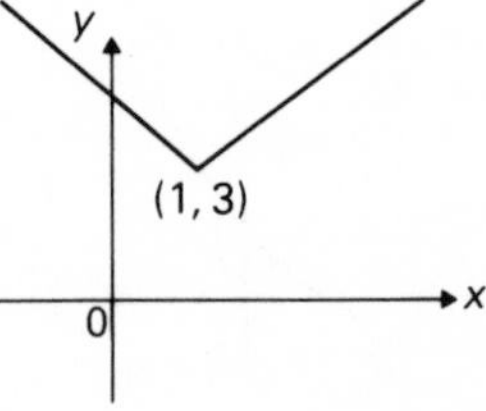

b)

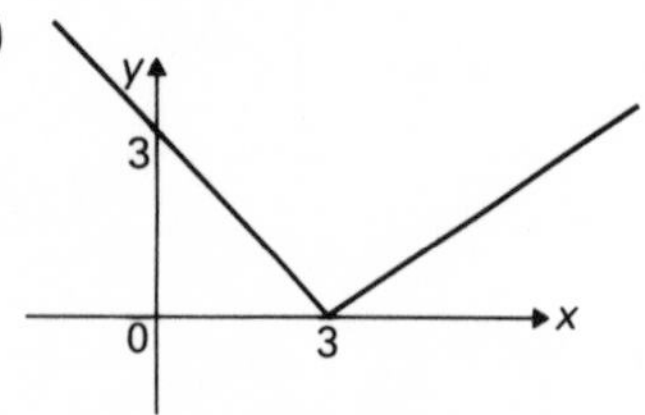

e)

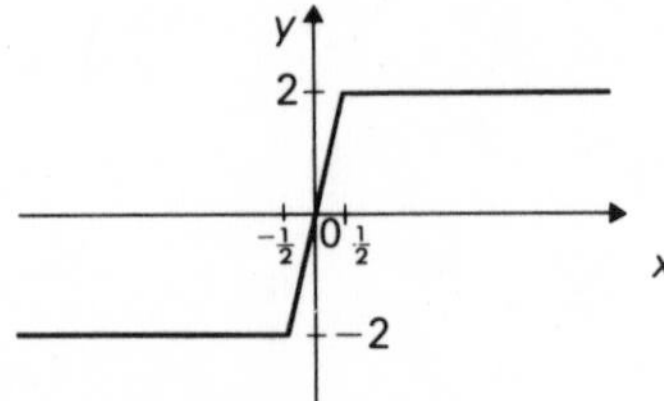

c)

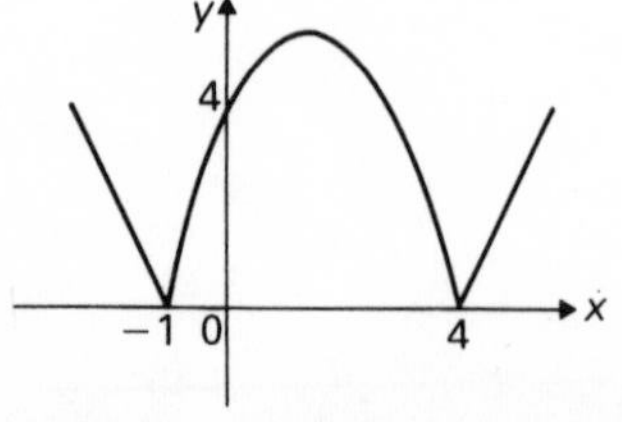

f)

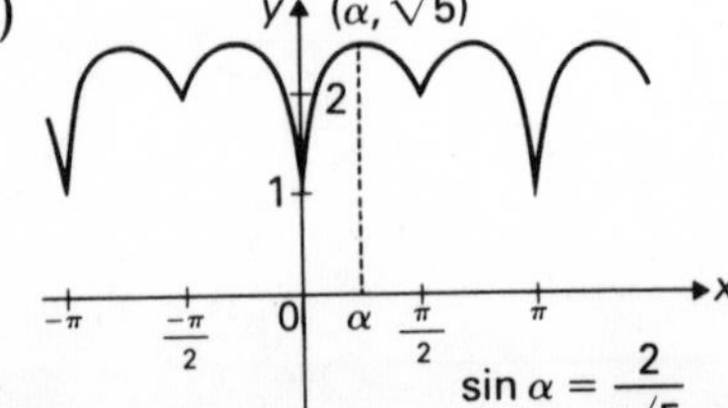

$\sin \alpha = \dfrac{2}{\sqrt{5}}$

2 a)

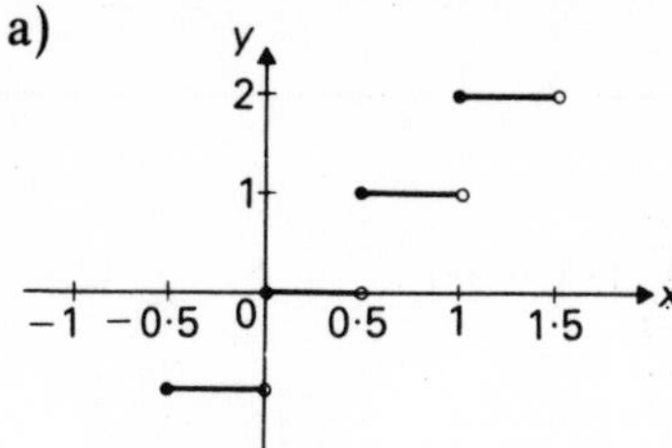

b)

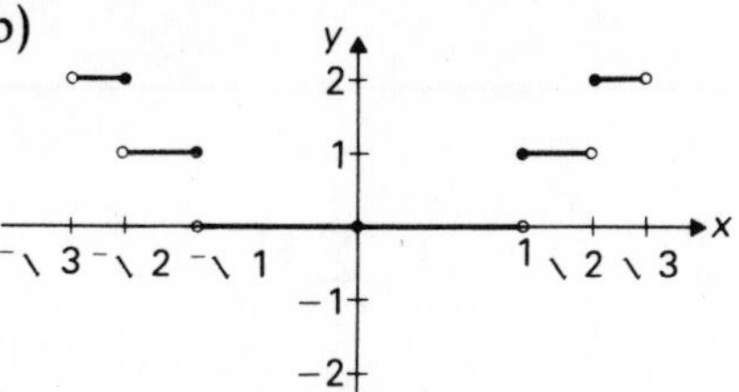

c)

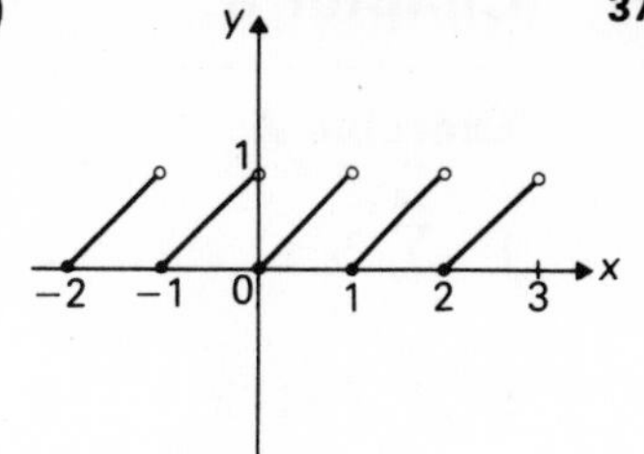

3 a) for all x
b) for all x excepting $x = 0$
c) for all x
d) for all x excepting $x \in \mathbb{Z}$

Exercise D

2 They all lie on a circle radius 2 units.
3 They all lie on a line passing through the pole, inclined at 30° to the initial line.

Exercise E

1

(r, θ)

0

(a, θ)

2

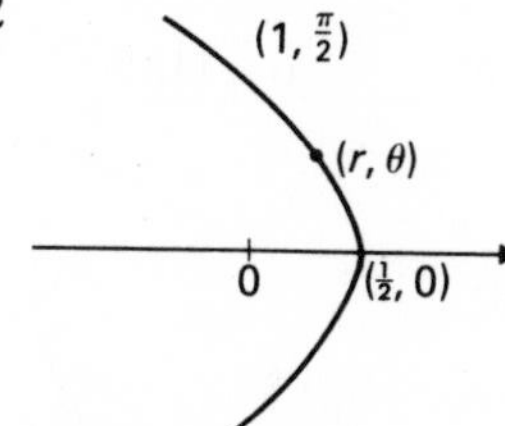

3

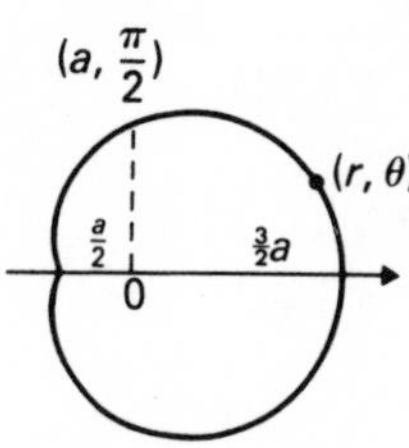

4

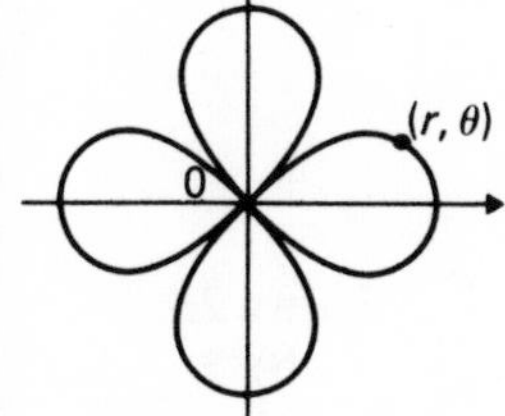

5

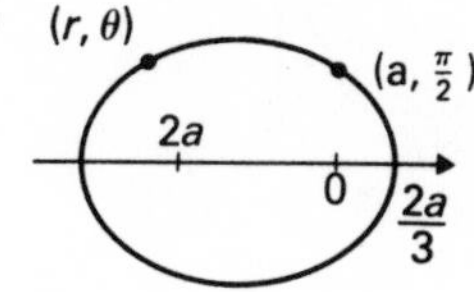

6

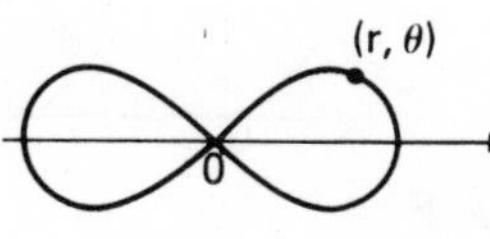

7

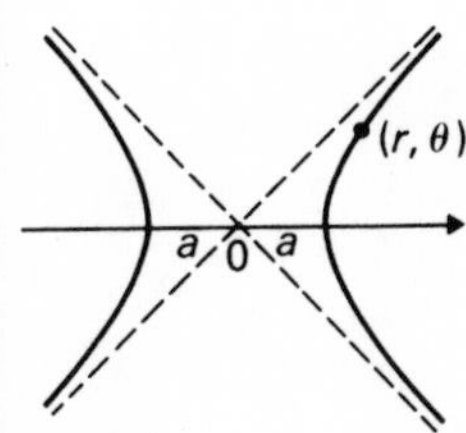

8

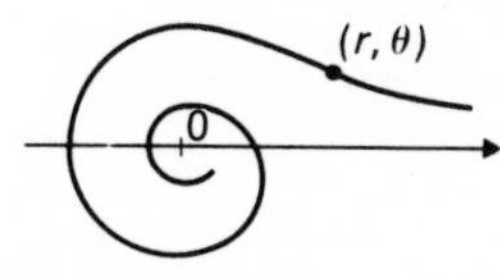

9

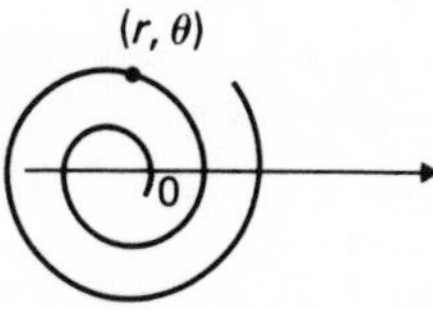

10

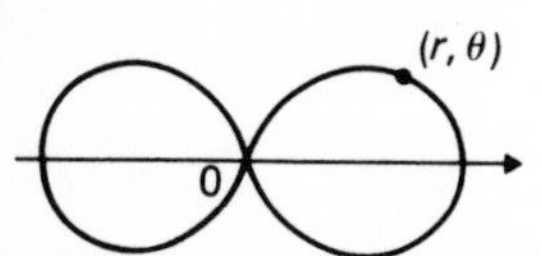

Exercise A

1 $\sum_{r=1}^{14} 3r$ 2 $\sum_{r=0}^{5} \frac{1}{2^r}$ 3 $\sum_{r=1}^{7} r^3$ 4 $\sum_{r=1}^{7} (14-3r)$ 5 $\sum_{r=1}^{N} (2+3r)$

6 $\sum_{r=0}^{6} (-1)^r 2^r$ 7 $\sum_{r=1}^{N} \frac{1}{r(r+1)}$ 8 $\sum_{r=1}^{N} x^r$ 9 $\sum_{r=0}^{7} 12^r$ 10 $\sum_{r=1}^{10} \frac{1}{r}$

11 $0+2+6+12+20+30+42+56+72+90$ 12 $5+8+11+14+17$

13 $\frac{1}{1}+\frac{1}{4}+\frac{1}{9}+\frac{1}{16}+\frac{1}{25}+\frac{1}{36}+\frac{1}{49}$ 14 $-2-4-6-8-10-12-14-16-18$

15 $10+9+8+7+6+5+4+3+2+1$

16 $1+2+3+4+5+6+7+8+9+10$ (the same!) 17 $1+8+27+64+125$

18 $125+64+27+8+1$ (the same!) 19 $-6-3+4+15$

20 $-5-2+3+10+19+30$

Exercise B

1 5050 2 2500 3 325 4 1365 5 1·9995 7 2 8 $\frac{1}{1-x}$ 10 396

Exercise C

1 385 2 784 3 5505 4 15 6 1185 8 1

Exercise D

1 $\frac{3}{4}$ 2 $\frac{26}{3}$ 3 40 4 12

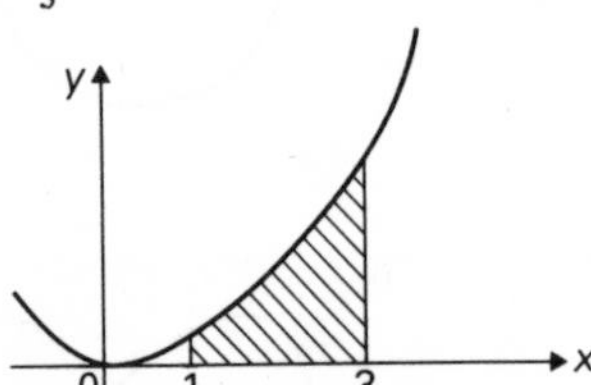

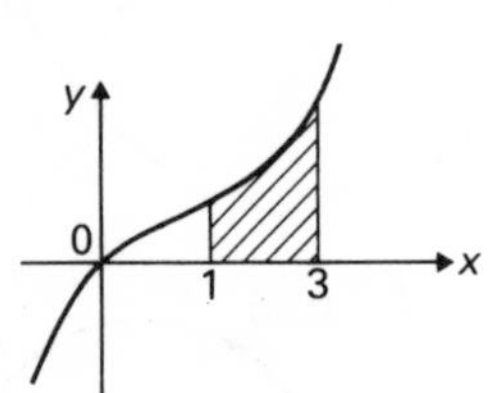

Exercise E

1 a) $\frac{x^2}{2}+\frac{x^3}{3}+c$
b) x^3+x^4+c
c) $4x+2x^2+\frac{1}{3}x^3+c$
d) $\frac{1}{3}x^3-x^2+c$
e) $\frac{1}{2}x^2-\cos x+c$
f) $\sin x+\cos x+c$
g) $x+\frac{1}{2}x^2+\frac{1}{3}x^3+c$
h) x^3+x^2+x+c
i) $\frac{2}{3}x^{3/2}+\frac{2}{5}x^{5/2}+c$
j) $\frac{2}{3}x^{3/2}+2x^{1/2}+c$

2 a) $\frac{14}{3}$
b) $\frac{26}{3}$
c) 0
d) 0

3 a) 2
b) 2
c) symmetry

4

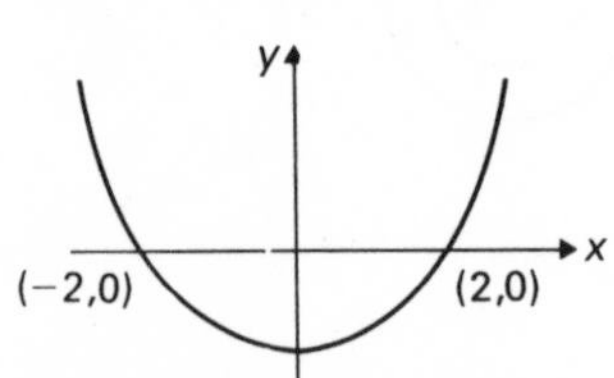

negative;
$\int_{-2}^{2} (x^2-4)\,dx = -\frac{32}{3}$

5 2
6 10

Exercise F

2 $\frac{16\pi}{15}$ 3 $\frac{4}{3}\pi ab^2$ 4 $\frac{192\pi}{5}$ 5 $\frac{\pi}{3}(2a^3-3a^2b+b^3)$

Exercise G

1 $\frac{3}{4}h$ from vertex 2 $(\frac{8}{3}, 0)$ 3 $(\frac{27}{20}, 0)$ 4 $\dfrac{3(2r-h)^2}{4(3r-h)}$ from the centre of the sphere

Chapter 9

Exercise A

1 $PQRS$ is a rhombus 2 $(-1, 3, 3)$; 3; $\sqrt{38}$

3 a) i) $4\hat{\mathbf{i}} - 6\hat{\mathbf{j}}$
ii) $-8\hat{\mathbf{k}}$
iii) $12\hat{\mathbf{i}} - 25\hat{\mathbf{j}} - 6\hat{\mathbf{k}}$

b) i) $\sqrt{14}$ ii) $\sqrt{29}$ iii) $\sqrt{27}$ iv) $\sqrt{11}$ v) $\sqrt{82}$ vi) 8

4 $-\hat{\mathbf{i}} + 2\hat{\mathbf{j}} + 2\hat{\mathbf{k}}$; 3 units 5 $6\hat{\mathbf{i}} + 4\hat{\mathbf{j}}$

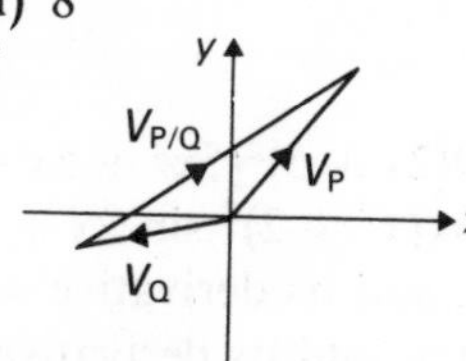

6 Not collinear

8 $\sqrt{41}$; $3\sqrt{41}$. CD parallel to AB and $CD = 3AB$

Exercise B

2 $(\frac{8}{3}, \frac{4}{3}, -\frac{5}{3})$; $\mathbf{r} = \begin{pmatrix}7\\2\\3\end{pmatrix} + \lambda\begin{pmatrix}13\\2\\14\end{pmatrix}$ 3 Collinear; 3 : 2 4 Collinear

5 $\mathbf{r} = \begin{pmatrix}0\\-1\\1\end{pmatrix} + t\begin{pmatrix}2\\5\\7\end{pmatrix}$; $\mathbf{r} = s\begin{pmatrix}1\\1\\2\end{pmatrix}$

Exercise C

$\begin{pmatrix}1\\-2\\-1\end{pmatrix} \cdot \begin{pmatrix}1\\-3\\7\end{pmatrix} = 0$ hence result 2 $\mathbf{r} = \begin{pmatrix}a\\b\\c\end{pmatrix} + \lambda\begin{pmatrix}m\\n\\p\end{pmatrix}$ 4 $3x + y + 8z + 7 = 0$; $\begin{pmatrix}3\\1\\8\end{pmatrix}$

79·7° 7 Yes.
Equation of plane is $7x - 3y + 2z = 0$ 8 a) $\lambda = 1$, $\mu = -1$
b) $x - y + z = 2$

i) plane
ii) line
iii) line
line iii) lies in the plane and is perpendicular to line ii) which is also normal to plane 10 One unit; $8x + 9y - 12z = 34$

1 $5\frac{5}{9}$ 12 $A = 46{\cdot}2°$; $B = 109{\cdot}2°$; $C = 24{\cdot}6°$ 13 $\cos^{-1}\left(\frac{1}{3}\right)$

Chapter 10

Exercise A

$12x(3x^2 + 1)$ 2 $\dfrac{-(3 + 2x)}{(3x + x^2)^2}$ 3 $2x \cos (x^2 + 2)$

$-3 \sin (3x - 1)$ 5 $3(4x^3 + 15x^2)(x^4 + 5x^3)^2$

$6(x + 1)(x^2 + 2x + 1)^2 = 6(x + 1)^5$ 7 $4x(x^2 - 4)$; rewrite first as $(x^2 - 4)^2$

$-2(9x^2 + 2)(3x^3 + 2x)^{-3}$ 9 $-2 \sin x \cos x$

0 $-\dfrac{\cos x}{\sin^2 x}$

Exercise B

1 $\dfrac{28\,000}{\pi r^2}$ cm/min 2 $\dfrac{K}{\pi\sqrt{2}}$ 3 $\dfrac{1}{x\sqrt{2}}$ cm/min 4 $\dfrac{1}{30x}$ cm/sec 5 a) $1{\cdot}6\,\pi$ cm^2/sec b) $16\,\pi$ cm^3/sec

Exercise C

1 $9x^2 + 8x + 6$ 2 $(x+1)^2(x-2)(5x-4)$ 3 $2x \sin x + x^2 \cos x$
4 $2 \cos x - (2x+1) \sin x$ 5 $\cos^2 x - \sin^2 x$ (or $\cos 2x$) 6 $-2 \sin x \cos x$ (or $-\sin 2x$)
7 $4(x+5)^2(7x+11)$ 8 $2(x+8) \sin x + (x+8)^2 \cos x$ 9 $4x^3$
10 $(x+1)^4(8x^3 + 10x^2 + 8x + 1)$

Exercise D

1 $2(x^2+1)(5x^2+6x+1)$ 2 $2(2x+5)^2(5x^2+5x+3)$ 3 $2 \sin^2 x + 2(2x+1) \sin x \cos x$
4 $\sin 2x \cos 2x$ 5 $6x(x^2+2)^2 \sin 2x + 2(x^2+2)^3 \cos 2x$
6 $\sec^2 x$ (**Note** $\sin x \sec x = \tan x$, and its derivative $= \sec^2 x$)
7 $-\csc^2 x$ (**Note** $\cos x \csc x = \cot x$, and its derivative $= -\csc^2 x$)
8 $4(11x+13)(13x-11)(143x+24)$ 9 $2352x^3(196x^4-1)^2$
10 $2x \sin^2 2x(\sin 2x + 3x \cos 2x)$ 11 $2 \sin x \cos x$; $2 \cos 2x$. $-2 \cos x \sin x$; $-2 \sin 2x$

Exercise E

1 a) $\dfrac{1-x^2}{(x^2+1)^2}$ b) $\dfrac{1}{(x+1)^2}$ c) $\dfrac{(x+1)^2(2x-1)}{x^2}$ d) $-\dfrac{(x+2)}{x^3}$ e) $\dfrac{4x^3+x^2-4x-1}{(2x-1)^2}$ f) $-\sin x$

2 $-\csc^2 x$

3 $\dfrac{-2x}{(x^2+1)^2}$ in both cases

4 a) $\dfrac{2-6x^2}{(x^2+1)^3}$
b) $\sin x(1 + \sec^2 x)$
c) $2 \sec^2 2x$
d) $\dfrac{(x^3+x^2+1)(3x^3+7x^2+4x-3)}{(x+1)^4}$
e) $\dfrac{(1+\sin x)(3-\sin x)\cos x}{(1-\sin x)^2}$
f) $4 \tan x \sec^2 x$
g) $\dfrac{6x^3-12x^2+24x-16}{(x^2-4)^3}$
h) $\dfrac{-13}{(2x-3)^2}$
i) $\dfrac{x \cos x - \sin x}{x^2}$
j) $\dfrac{\cos x + x \sin x}{\cos^2 x}$

5 Maximum point when $x = -\sqrt{6}$
Minimum point when $x = \sqrt{6}$

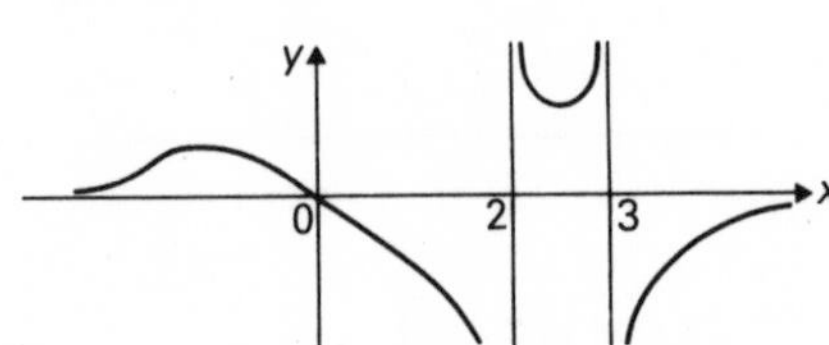

Exercise F

1 $\frac{3}{4}$ 2 $\dfrac{2x \sin y - y}{x - x^2 \cos^2 y}$ 3 $\frac{3}{16}$ 4 $-\dfrac{y+2x}{x+2y}$

Chapter 11

Exercise A

1 a) $\pm 3i$
b) $\pm 2{\cdot}5i$
c) $\pm \frac{i}{7}$
d) $\pm i\sqrt{11}$
e) $\pm i\sqrt{\left(\frac{17}{5}\right)}$
f) $\pm i\sqrt{10}$

2 a) $-1 \pm i$
b) $\frac{1}{2} \pm \frac{3i}{2}$
c) $-7 \pm i$
d) $-\frac{1}{2} \pm i\frac{\sqrt{3}}{2}$
e) $\frac{3}{2} \pm i\frac{\sqrt{5}}{2}$
f) $\frac{4}{5} \pm i\frac{\sqrt{14}}{5}$

Exercise B

1 a) $5 - i$
b) $8 - 3i$
c) $6 + i$

2 a) $25i$
b) 7
c) $17 - i$
d) $1 + 2i$
e) -4
f) $2 - 2i$

3 a) $\frac{3}{25} - \frac{4}{25}i$
b) $1 - 2i$
c) $\frac{-1}{58} - \frac{17}{58}i$
d) 5
e) $-\frac{31}{50} - \frac{17}{50}i$
f) $-\frac{6}{5}i$
g) $\frac{2}{5} - \frac{4}{5}i$
h) 0

4 a) $\frac{8}{13} - \frac{12}{13}i$
b) $\frac{3}{11} + \frac{\sqrt{2}}{11}i$
c) $\frac{21}{170} - \frac{33}{170}i$
d) $\pm(1 + i)$
e) $\pm(2 + 3i)$
f) $\pm(p + i)$

Exercise C

1 a) $2; 0$
b) $3; \frac{\pi}{2}(= 1{\cdot}57)$
c) $0{\cdot}5; -\frac{\pi}{2}(= -1{\cdot}57)$
d) $13; 0{\cdot}395$
e) $\sqrt{13}; 0{\cdot}588$
f) $\sqrt{2}; -\frac{\pi}{4}(= -0{\cdot}785)$
g) $\sqrt{29}; 1{\cdot}951$
h) $2; \frac{\pi}{3}(= 1{\cdot}047)$
i) $\sqrt{5}; -0{\cdot}464$

2 $|z_1 z_2| = 3$, $\arg z_1 z_2 = -\frac{\pi}{6}$

$\left|\frac{z_1}{z_2}\right| = \frac{4}{3}$, $\arg \frac{z_1}{z_2} = -\frac{\pi}{2}$

3 2

5 a) 5, 1 b) $\frac{\pi}{2}, 0{\cdot}644$

6 13, 0

7 a) $\left(r + \frac{1}{r}\right)\cos\theta + i\left(r - \frac{1}{r}\right)\sin\theta$, b) $\left(r^2 + \frac{1}{r^2}\right)\cos 2\theta + i\left(r^2 - \frac{1}{r^2}\right)\sin 2\theta$, $r + \frac{1}{r}, \theta$

Exercise D

1 a) -1
b) 3
c) 0

3 $A = B = C = \frac{1}{3}$

4 $p^2 + q^2 + r^2 - pq - pr - qr$

Exercise E

1 $z_1 = \frac{1}{\sqrt{2}}(1 + i)$, $z_2 = \frac{1}{\sqrt{2}}(-1 + i)$,

$z_3 = \frac{1}{\sqrt{2}}(-1 - i)$, $z_4 = \frac{1}{\sqrt{2}}(1 - i)$

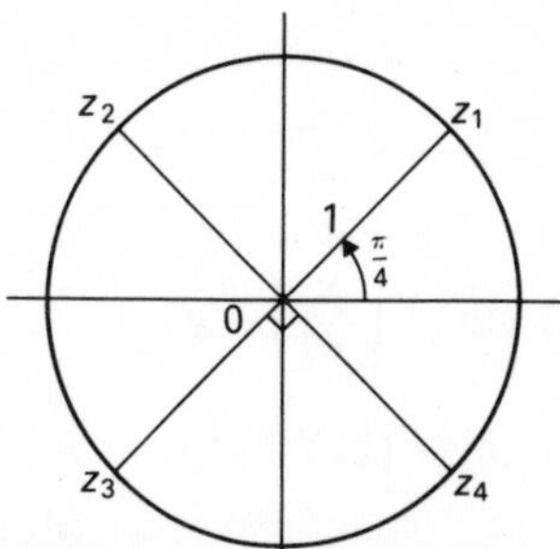

2 $1, -1$ (twice), $\frac{1}{2} \pm i\frac{\sqrt{3}}{2}$, $\pm i$

3 $(x^2+1)(x^2+\sqrt{3}x+1)(x^2-\sqrt{3}x+1)$

4 a) $2, -1 \pm 2i$
b) $2{\cdot}44$
$2{\cdot}29\omega + 0{\cdot}15\omega^2$
$0{\cdot}15\omega + 2{\cdot}29\omega^2$
c) $4{\cdot}07$
$3{\cdot}11\omega + 0{\cdot}96\omega^2$
$0{\cdot}96\omega + 3{\cdot}11\omega^2$

5 a) $1, 3, \pm i$
b) $-\frac{1}{2} \pm i\frac{\sqrt{3}}{2}$, $\frac{1}{2} \pm i\frac{\sqrt{3}}{2}$
c) $-1 \pm i\sqrt{2}$, $1 \pm 2i$

Chapter 12

Exercise A

1 a) $\frac{1}{3}(x^2+1)^3 + c$
b) $\frac{1}{3}\cos^3 x + c$
c) $\frac{1}{3}(x^3+7x)^3 + c$
d) $-\frac{1}{3(x^2+1)^3} + c$
e) $\frac{1}{11}(x^6+x^7+x^8)^{11} + c$

2 a) $\frac{1}{2}(x^3+x^2)^2 + c$
b) $\frac{1}{3}\sin^3 x + c$
c) $-\frac{1}{(x^3+1)} + c$
d) $\frac{1}{3}(x^2+2x+3)^3 + c$
e) $\frac{1}{4}(x^3+5)^4 + c$
f) $\frac{1}{4}(\cos x + \sin x)^4 + c$
g) $\frac{1}{3}\left(x+\frac{1}{x}\right)^3 + c$

Exercise B

1 a) $\frac{1}{64}(4x+3)^4 + \frac{1}{48}(4x+3)^3 + c$
b) $\frac{7}{45}(3x+5)^5 - \frac{41}{36}(3x+5)^4 + c$
c) $\frac{1}{12}(3x-2)^4 + \frac{4}{9}(3x-2)^3 + c$
d) $-\frac{1}{2}\cos(x^2) + c$
e) $\frac{1}{3}\sin^3(2x+5) + c$
f) $\frac{7}{2(x^2+1)^2} - \frac{11}{2(x^2+1)} + c$

2 a) $\frac{3}{16}(2x+1)^4 + \frac{7}{12}(2x+1)^3 + c$
b) $\frac{2}{45}(3x-5)^5 + \frac{13}{36}(3x-5)^4 + c$
c) $\frac{1}{12}\cos^4(3x+4) + c$
d) $\frac{1}{2}\tan^2(2x-1) + c$
e) $c - \frac{7}{6(2x^2-1)^3}$

Exercise C

1	18	2	$0{\cdot}104$	3	$0{\cdot}05$	4	$\frac{1}{18}$	5	$\frac{1}{2}$	6	$273{\cdot}6$	7	$0{\cdot}707$
8	$0{\cdot}919$	9	$20\frac{1}{3}$	10	1	11	1	12	$8{\cdot}0529$	13	$\frac{\pi}{4}$	14	$\frac{\pi}{4}$
15	$32\frac{2}{3}$	16	$\frac{43}{30}$	17	0, by symmetry					18	a) $-\frac{1}{4}$ b) 0	19	250

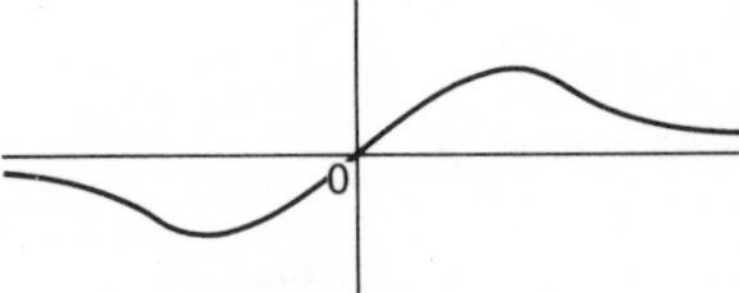

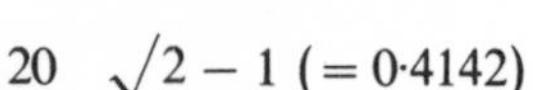
20 $\sqrt{2} - 1$ $(= 0{\cdot}4142)$

Exercise D

1 a) $\sin^{-1}\left(\frac{x}{3}\right) + c$
b) $\sin^{-1}\left(\frac{x}{2}\right) + c$
c) $\sin^{-1}\left(\frac{x}{\sqrt{2}}\right) + c$
d) $\sin^{-1}(2x) + c$
e) $\frac{1}{2}\sin^{-1}\left(\frac{2x}{3}\right) + c$
f) $2\sin^{-1}\left(\frac{3x}{2}\right) + c$

2 a) $-\cos^{-1}\left(\frac{x}{4}\right)+c$ c) $-2\cos^{-1}\left(\frac{x}{2}\right)+c$ e) $-\frac{1}{\sqrt{3}}\cos^{-1}\left(\frac{x}{2}\right)+c$

b) $-\frac{1}{2}\cos^{-1}(2x)+c$ d) $-3\cos^{-1}\left(\frac{x}{3}\right)+c$ f) $-\frac{1}{2\sqrt{2}}\cos^{-1}(2x)+c$

3 a) $\frac{1}{2}\tan^{-1}(2x)+c$ c) $2\tan^{-1}(2x)+c$ e) $\frac{1}{2}\tan^{-1}\left(\frac{x}{2}\right)+c$

b) $\frac{1}{6}\tan^{-1}\left(\frac{3x}{2}\right)+c$ d) $\frac{1}{6}\tan^{-1}\left(\frac{x}{2}\right)+c$ f) $\frac{3}{4\sqrt{2}}\tan^{-1}\left(\frac{x}{\sqrt{2}}\right)+c$

4 a) $\tan^{-1}(x-1)+c$ c) $-\sin^{-1}(2-x)+c$ e) $\sin^{-1}\left(\frac{2+x}{3}\right)+c$

b) $\sin^{-1}\left(\frac{x+2}{2}\right)+c$ d) $\tan^{-1}(x+2)+c$ f) $\frac{1}{2}\tan^{-1}(2x+3)+c$

5 a) 0·2318 c) $\frac{\pi}{2}$ e) $\frac{2\pi}{9}$

b) $\frac{\pi}{6}$ d) $\frac{\pi}{8}$ f) 0·208

Exercise E

1 $\frac{1}{8}x(x+10)^8-\frac{1}{72}(x+10)^9+c$
2 $\frac{1}{18}(2x+3)(3x+2)^6-\frac{1}{189}(3x+2)^7+c$
3 $\frac{1}{2}(x^2+1)\tan^{-1}x-\frac{1}{2}x+c$
4 $\frac{1}{3}\sec^3 x+c$
5 $\left(\frac{x^2}{2}-\frac{1}{4}\right)\sin^{-1}x+\frac{1}{4}x\sqrt{(1-x^2)}+c$
6 $(x^2+6x+7)\sin x+2(x+3)\cos x+c$
7 $\frac{1}{8}(x-1)(2x+7)^4-\frac{1}{80}(2x+7)^5+c$
8 $x^3(x+4)^4-\frac{3}{5}x^2(x+4)^5+\frac{1}{5}x(x+4)^6-\frac{1}{35}(x+4)^7+c$
9 $\sin x+c$
10 $(6x-x^3)\cos x+(3x^2-6)\sin x+c$

Exercise F

1 a) $\frac{1}{2}x-\frac{1}{4}\sin 2x+c$
b) $\frac{1}{2}x+\frac{1}{4}\sin 2x+c$

2 a) $\frac{1}{3}\cos^3 x-\cos x+c$
b) $\sin x-\frac{1}{3}\sin^3 x+c$

3 a) $\frac{1}{3}(x^2+1)^{3/2}$
b) $\sqrt{(x^2+1)}+c$

4 a) 0
b) 0·6023

5 π, (a) π,
(b) π

6 $\sqrt{(1-x^2)}-\cos^{-1}x+c$

7 a) -2π
b) $-\frac{1}{2}$

8 One graph is reflected in the line $x=\frac{\pi}{4}$ to give the other graph.

9 a) $\frac{1}{3}(x^2-3)^3+c$
b) $\frac{1}{2}\tan^2 x+c$

10 a) $\frac{2}{5}(x+3)^{5/2}-2(x+3)^{3/2}+c$
b) $\frac{1}{6}$

12 1

Chapter 13

Exercise A

1 2·943
2 1·896
4 0·618 and −1·618
5 0·464
6 1·452

Exercise B

1 5·0133; 5·0266; 4·9867; 4·9734
2 1·0349; 0·9651
3 $y=13x-16$;
approximately within the range $1{\cdot}8<x<2{\cdot}2$

Exercise C

1 $1-\frac{x^2}{2!}+\frac{x^4}{4!}-\frac{x^6}{6!}$

3 $(a+h)^n \approx a^n + na^{n-1}h + n(n-1)a^{n-2}\frac{h^2}{2!} + n(n-1)(n-2)a^{n-3}\frac{h^3}{3!} +$

4 $\tan^{-1} x = x - \frac{1}{3}x^3 + \frac{1}{5}x^5 \cdots$

5 $y = \sqrt{2} - \frac{\left(x-\frac{\pi}{4}\right)^2}{\sqrt{2}}$ (or $y = 0{\cdot}978 + 1{\cdot}11x - 0{\cdot}71x^2$)
Approximately between $x = 0{\cdot}5$ and $x = 1$

Exercise D

1 $1 - 2h + 3h^2 - 4h^3 + 5h^4 - \cdots$
convergent for $|h| < 1$

2 $1 - \frac{1}{2}h + \frac{3}{8}h^2 - \frac{5}{16}h^3 + \cdots$, 0·9853

3 $1 - 3h + 6h^2 - 10h^3 + 15h^4 - \cdots$
convergent for $|h| < 1$

4 $n = \frac{1}{4}$ or $\frac{3}{4}$; $n = \frac{1}{4}$

5 $\int_0^x (1 - h^2 + h^4 - h^6 + h^8 - \cdots)\,dh = x - \frac{x^3}{3} + \frac{x^5}{5} - \frac{x^7}{7} + \frac{x^9}{9} - \cdots$
and $\int_0^x \frac{1}{1+h^2}\,dh = \tan^{-1}(x)$

6 $1 - \frac{1}{2}x^2 - \frac{1}{8}x^4 - \frac{1}{16}x^6 - \cdots$
when integrated, becomes $x - \frac{1}{6}x^3 - \frac{1}{40}x^5 - \frac{1}{112}x^7 - \cdots$
a series for $\sin^{-1} x$.

Exercise E

1 18 (straight line graph yields exact result) 2 0·1745 3 −0·697

4 a) 1·0865
b) 1·0892

5 a) 0·349 76
b) 0·347 99
c) 0·347 37

Exercise F

1 a) 0·346 67
b) 0·346 58

2 trapezium rule gives 1·106 312
Simpson's rule gives 1·107 14
$[\tan^{-1} 2 - \tan^{-1} 0] = 1{\cdot}1071$

3 Quadratic approximation will always be exact for quadratic curve.
a) 6
b) $\frac{5}{6}$

4 a) 0·2028
b) 0·2027

5 2 6 1·105 13 7 Exact value = 1

8 $4\int_0^a \sqrt{(a^2 - x^2)}\,dx$

10 $\int_{-a}^{a} x^3\,dx$ is zero because of the rotational symmetry of the function x^3.
Whereas $\int_{-a}^{a} x^2\,dx = \frac{2a^3}{3}$ by Simpson and exact integration.
Trapezium rule approximates to this value, depending on number of strips used.

Chapter 14

Exercise A

1 a) 0·368
b) 1·105
c) 0·607

2 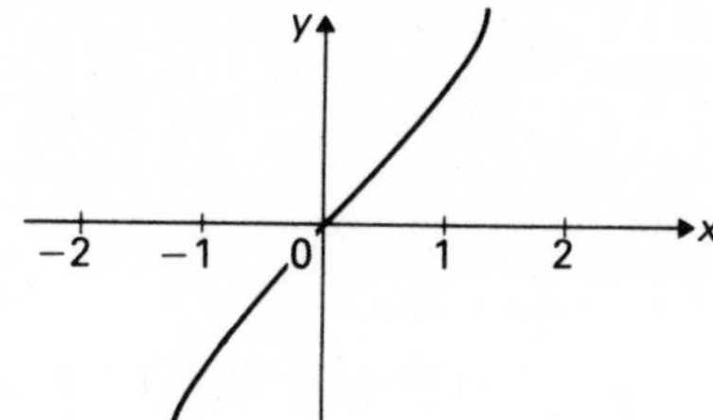

3 a) $1+\frac{x}{2}+\frac{x^2}{8}+\frac{x^3}{48}+\frac{x^4}{384}$
b) $1+2x+2x^2+\frac{4}{3}x^3+\frac{2}{3}x^4$
c) $1-2x+2x^2-\frac{4}{3}x^3+\frac{2}{3}x^4$
d) $1+\frac{x^2}{2}+\frac{x^4}{24}$

4 $\frac{1}{2}\left(e-\frac{1}{e}\right)$

5 a) $2-3x+\frac{5}{2}x^2-\frac{3}{2}x^3$
b) $4x+\frac{8}{3!}x^3+\frac{12}{5!}x^5+\frac{16}{7!}x^7$
c) $x+x^2+\frac{x^3}{3}-\frac{x^5}{30}$

7 TN = 5

8 $\frac{2}{e^x+e^{-x}}$

9 e^2-e^{-2}

10 a) $x^2-y^2=4$
b) $xy=y+2$

Exercise B

1 a) $\frac{x}{2}e^{(x^2/4)}$
b) $-5e^{-5x}$
c) $2\cos 2xe^{\sin 2x}$
d) $\frac{1}{2\sqrt{x}}e^{\sqrt{x}}$
e) $-\frac{1}{x^2}e^{1/x}$
f) $(\cos x-\sin x)\,e^{(\sin x+\cos x)}$
g) $2e^{2x}-\frac{2}{e^{2x}}$
h) $3\left(e^x+\frac{1}{e^x}\right)^2\left(e^x-\frac{1}{e^x}\right)$

2 a) $x^2(3+2x)e^{2x}$
b) $(1+x^2)e^{-x}$
c) $-(2\cos x+\sin x)e^{-2x}$
d) $e^x\sec^2(e^x)$
e) $\frac{4}{(e^x+e^{-x})^2}$
f) $\frac{1-x}{e^x}$
g) $\frac{e^x(1+e^{2x})}{(1-e^{2x})^2}$

3 $R=5,\ \theta=\tan^{-1}\left(\frac{4}{3}\right)$, $25e^{4x}\cos(3x+\theta+\phi)$ where $\tan\phi=\frac{3}{4}$

6 $R=\sqrt{29},\ \alpha=\tan^{-1}\frac{2}{5}$ max, min $=\frac{5}{\sqrt{29}}e^{-[(\pi/5)-(2\alpha/5)]},\ \frac{-5}{\sqrt{29}}e^{-[(3\pi/5)-(2\alpha/5)]}$

7 $a=b=1$; acceleration $=0$

Exercise C

1 a) $-e^{-2x}+c$
b) $e^{x^2+1}+c$
c) $e^{\tan x}+c$
d) $2e^{\sqrt{x}}+c$
e) $\frac{1}{2}(e^{2x}+4e^{-2x})+c$
f) $-\cos(e^x)+c$
g) $\frac{-(e^{-x}+1)^4}{4}+c$

3 a) $\frac{3-7e^{-2}}{4}$
b) $\frac{e^2+3}{8}$
c) $\frac{1}{10}(2\sqrt{2}\,e^{3\pi/4}-3)$

4 $\frac{e-1}{2(1+e)}$

5 $\frac{1}{3}$

6 a) $-e^{-x}(x^4+4x^3+12x^2+24x+24)$ b) $120-\frac{326}{e}$

7 $\frac{3}{26}(2e^{\pi/2}-3)$

8 $\frac{e^{ax}(a\cos bx+b\sin bx)}{(a^2+b^2)}$

Exercise D

1 a) $\frac{2x+1}{x^2+x+1}$
b) $\frac{2x-1}{2(x^2-x+1)}$
c) $-\tan x$
d) $\frac{2(e^{2x}-1)}{e^{2x}-2x}$
e) $\sec x$
f) $-\frac{1}{\sqrt{(x^2-1)}}$
g) $\frac{2}{x\ln 10}$
h) $\frac{2^{2x}-1}{2^{2x}+1}$
i) $2\sec 2x$

2 a) $-10^{-5x}(5\ln 10)$ c) $8(3-4x-4x^2)^{-3/2}$

b) $-\dfrac{8x}{(4+x^2)\sqrt{(16-x^4)}}$ d) $\dfrac{(\ln x)^{\ln x}}{x}\{\ln(\ln x)+1\}$

3 a) $e^x\left(\ln x+\dfrac{1}{x}\right)$ d) $2x\ln(x+3)+\dfrac{x^2}{x+3}$ f) $\dfrac{1-\ln x}{(x+\ln x)^2}$

b) $\ln(\cot x)-\dfrac{2x}{\sin 2x}$ e) $\dfrac{2x^2\ln x-x^2-1}{x(\ln x)^2}$

c) $16x^3\ln x$

6 $(-1, 1), (-4, 0), (-7, \frac{1}{25})$ 7 $y = x^2 + \ln x$

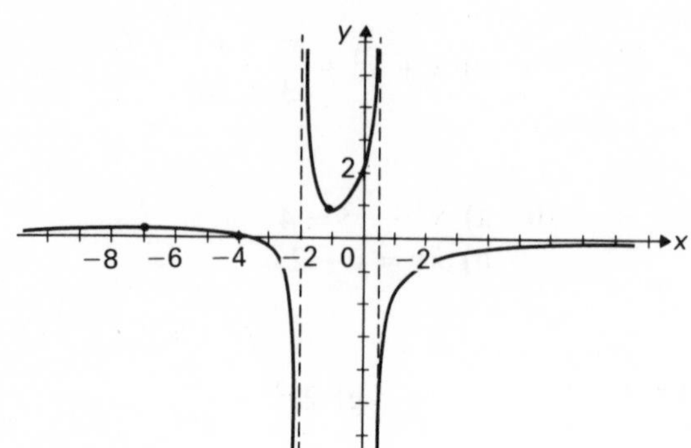

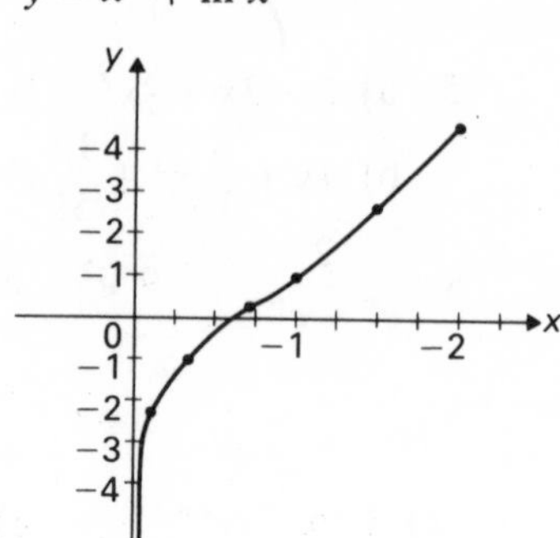

Exercise E

1 a) 4 c) $\frac{1}{3}$ e) $\dfrac{a}{b}$ g) $\sin x$ 2 a) 2

b) 3 d) x^2 f) x h) x^x b) $\frac{7}{24}$

Exercise F

1 $\frac{1}{3}\ln|x|+c$ 2 $4\ln|x|+c$ 3 $\frac{1}{2}\ln|(1+2x)|+c$

4 $-\frac{1}{2}\ln|(1-2x)|+c$ 5 $\ln|(x^3+1)|+c$ 6 $\frac{1}{2}\ln(3x^2+2)+\dfrac{\sqrt{6}}{6}\tan^{-1}\dfrac{\sqrt{6}x}{6}+c$

7 $\ln(e^x+1)+c$ 8 $-\frac{1}{3}\ln|(\cos 3x)|+c$ 9 $\ln|(2+\tan x)|+c$

10 $\ln|(\cos x+\sin x)|+c$ 11 $\frac{1}{2}\ln(1+\sin^2 x)+c$ 12 $\frac{3}{4}\ln|(2x-3)|+\frac{1}{2}x+c$

13 $\frac{1}{2}\ln\frac{4}{3}$ 14 $\frac{1}{3}\ln\frac{8}{5}$ 15 $\ln(e+1)$

Exercise G

1 a) $\ln\left|\dfrac{x-1}{x+1}\right|+c$ d) $\ln\left|\dfrac{x^2}{x+1}\right|+c$ g) $\ln\left|\dfrac{(x+2)}{x+1}\right|-\dfrac{2}{x+2}+c$

b) $\ln\left|\left(\dfrac{x+2}{1-x}\right)^{1/3}\right|+c$ e) $\ln\left|\dfrac{(x+2)^2}{\sqrt{(x+1)}(\sqrt{(x+3)^3})}\right|+c$ h) $\ln\left|\left(\dfrac{x-2}{x-1}\right)^3\right|+\dfrac{2}{x-1}+c$

c) $\frac{1}{8}\ln\left|\dfrac{(x+7)^{17}}{x-1}\right|+c$ f) $\ln\left|\dfrac{(x-2)(2x-1)^2}{(x-1)^3}\right|+c$

2 a) $\ln 3$ c) $\ln 8$

b) $\ln\dfrac{3^9}{2^4}$ d) $\frac{11}{4}\ln 3-2\ln 2-\frac{7}{6}$

3 $\frac{1}{8}\{7\sqrt{2}+3\ln(\sqrt{2}+1)\}$ 4 $\frac{1}{3}$ 5 $\frac{1}{11}\ln 2$

6 $\dfrac{1}{k}\ln\dfrac{2(3-x)}{3(2-x)}=t$

Exercise H

1 a) $\frac{1}{9}x^3(3\ln|x| - 1) + c$
b) $\frac{2^{5/2}}{3}\left(\ln 2 - \frac{2}{3}\right) + \frac{4}{9}$

c) $\frac{x^2}{2}\left\{(\ln|x|)^2 - (\ln|x|) + \frac{1}{2}\right\} + c$
d) $1 - \frac{2}{e}$

2 a) $\sqrt{3}(\ln\sqrt{3} - 1) + 1$
b) $\frac{1}{4}(2\ln 2 - 1)$
c) $\ln\frac{4}{3} - \frac{1}{6}$

3 $9e - 24$

4 $\frac{1}{32}(5e^4 - 1)$

5 $e^2 + 1$

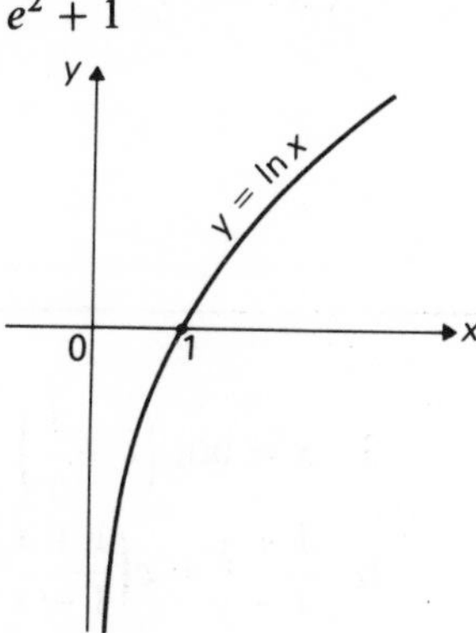

6 $y^2 = \frac{1}{(x+1)(x-3)}$

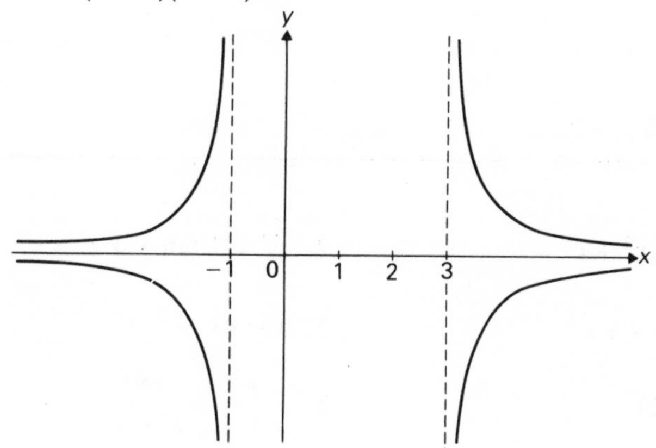

Chapter 15

Exercise B

1 a) 6
b) 35
c) 1716

3 32

4 a) $2\,{}^nC_r$
b) ${}^nC_r$

5 a) 18
b) 5, 16

6 a) $80x^3$
b) $-15\,360x^3$
c) $2016x^3$
d) $-40x^3$
e) $1872x^3$
f) $52x^3$

7 11

8 a) $\frac{7}{18}$
b) $-1\,184\,040$

9 $\alpha = 2, \beta = -10$

10 $\pm\frac{\sqrt{3}}{3}$

11 $-120, -240$

Exercise C

1 a) 1
b) $-\frac{5}{128}$
c) $\frac{21}{32}$
d) $\frac{(-1)^p(2p-1)!}{p!(p-1)!}$

2 a) $1 - 2x + 4x^2 - 8x^3,\ (-1)^r2^rx^r,\ |x| < \frac{1}{2}$
b) $1 + x + \frac{3x^2}{4} + \frac{x^3}{2},\ \frac{(r+1)x^r}{2^r},\ |x| < 2$
c) $3^{1/3}\left(1 + \frac{2x}{9} - \frac{4x^2}{81} + \frac{40}{2187}x^3\right),$
$\frac{(-1)^{r+1}3^{1/3-2r}\,.\,(2)(5)\ldots(3r-4)\,.\,2^rx^r}{r!},\ |x| < \frac{3}{2}$
d) $1 + \frac{2x^2}{3} + \frac{5x^4}{9} + \frac{40x^6}{81},\ \frac{(2)(5)\ldots(3r-1)x^{2r}}{3^rr!},\ |x| < 1$

3 a) $\frac{(r+2)(r+1)x^r}{2}$
b) $\frac{(-1)^r\,.\,(3)(5)\ldots(2r+1)x^r}{r!}$
c) $\frac{(n+r-1)!x^r}{(n-1)!r!}$

4 a) $1 + 3x + 5x^2 + 7x^3 + 9x^4,\ (2r+1),\ |x| < 1$
b) $1 + \frac{1}{2}x - \frac{5}{8}x^2 - \frac{3}{16}x^3 - \frac{13}{128}x^4,\ -\frac{(2r-4)!(4r-3)}{2^{2r-2}r!(r-2)!},\ |x| < 1$
c) $1 + 4x + 8x^2 + 12x^3 + 16x^4,\ 4r,\ |x| < 1$

5 $5-r$

7 $\frac{3}{1-3x}+\frac{2}{1+2x}$, $5+5x+35x^2+65x^3+275x^4$, $3^{r+1}+(-1)^r 2^{r+1}$, $|x|<\frac{1}{3}$

8 $\frac{2}{2x+3}-\frac{2}{x+2}+\frac{1}{x-3}$, $-\frac{55}{648}$, yes

9 a) 1·004 99 c) 0·997 99
b) 10·019 96 d) 1·004 01

10 2·0 204%, gains 73 s

11 a) $x>4$ or $x<-\frac{2}{3}$
b) $\left(\frac{3x-12}{3x-5}\right)^{-2}$

13 $2(1+\frac{23}{24}x)$

14 8·044

Chapter 16

Exercise A

1 $y=\frac{1}{3}x^3+\frac{1}{2}x^2+c$

2 $y^2=2\tan x+c$

3 $x=\tan\left(c+\frac{1}{y}\right)$

4 $\cos y=\sin x+c$

5 $y=cx$

6 $\frac{1+y}{1-y}=c\left(\frac{1+x}{1-x}\right)$

7 $y=c(x+1)$

8 $\frac{y}{y+1}=\frac{cx}{x+1}$

9 $y=-\cot x+c$

10 $y-\ln|y+2|=x+c$

Exercise B

1 $1-\cos\left(y+\frac{\pi}{4}\right)=\sin x$

2 $y=\frac{2(x^2+1)}{1-2x^2}$

3 $y^2=4x+6\ln|x|-4$

4 $2y^3=5-3x^2$

5 $\tan\left(\frac{y}{2}\right)=(\sec x+\tan x)\,e^{-\sin x}$

6 $y=\frac{8x}{9-x}$

7 $y=xe^{x-2}$

8 $y=\frac{x-3}{4x}$

9 $\cos y=1-\sin x$

10 $y^2=x-\cos x+3$

Exercise C

1 $k=\ln 4$ (= 1·3863); 2·19 sec; 28·4 m

2 $4\frac{1}{2}$ hours

3 a) $\sqrt{\left(\frac{\pi h^3}{336}\right)}$
b) $\sqrt{\left(\frac{\pi h^3}{42}\right)}$

4 416 seconds; $x\to 3$ as $t\to\infty$

5 $t=1$; $x\to 0\Rightarrow t\to\infty$

Exercise D

1 $y=e^{-x^2}(x+c)$

2 $y=2e^x-1-x$

3 $y=e^{-\tan x}(\tan x+c)$

4 $y(\csc x+\cot x)=x+\sin x+c$

5 $y(1+x^2)^4=\tan x-x+2$

6 $y(50+x)=600x+6x^2+2500$

7 $yx^3=\sin x+c$

8 $y=3x^2-x^3$

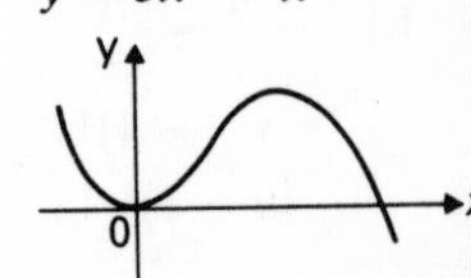

9 $x=\frac{1}{5}y-\frac{1}{13}y^3+cy^{-1/4}$

10 $y=\frac{1}{2}e^{-x}(x^2+5)$

11 $y=e^{-3x}(5-\cos 2x)$

12 $y=ce^x-1$

13 $y=\frac{1}{4}e^x(\cos 2x+\sin 2x)+ce^{-x}$

14 $x=\sin 2t\pm\sqrt{3}\cos 2t$

Index